States of Immersion Across Media

States of Immersion Across Media

Bodies, Techniques, Practices

Edited by
Philippe Bédard,
Alanna Thain,
Carl Therrien

Routledge
Taylor & Francis Group
LONDON AND NEW YORK

The publication of this book is made possible by a grant from the Social Sciences and Humanities Research Council of Canada

First published in 2025 by Amsterdam University Press Ltd.

Published 2025 by Routledge
4 Park Square, Milton Park, Abingdon, Oxon OX14 4RN
605 Third Avenue, New York, NY 10158

Routledge is an imprint of the Taylor & Francis Group, an informa business

ISBN: 9789048558766 (hbk)
ISBN: 9781041186601 (pbk)
ISBN: 9781003704324 (ebk)
NUR 670

Cover illustration: Olivia McGilchrist

Cover design: Coördesign, Leiden

DOI: 10.5117/9789048558766

Table of Contents

Section 2 Finding New Frames

Section 3 The Edge of Sense(s)

Acknowledgements

States of Immersion Across Media: Bodies, Techniques, Practices emerged from the 2020 conference *Immersivité et Innovations Technologiques / Immersivity and Technological Innovations,* at a moment where a global lockdown shunted most of our academic labour to online spaces. Delayed by several months as we adjusted to a new reality, the event allowed us to experiment with certain propositions of immersivity at a time when we were more immersed in our screens than ever, and more alienated from these forms of capture by the absent tonic of other scenes of labour, leisure and life. At and around the edges of the online encounter we were eventually able to mount, the richly historical and embodied forms of attention that our participants shared through their research, curiosity and engagement were real-time negotiation of our deceptions and desires around immersivity itself. Our first thanks go to all the presenters of that conference, including many of the authors now collected in this book: Anaïs Bernard, Martin Bonnard, Felipe Carelli, Natalja Chestopalova, Laureline Chiapello, Jason Coley, Daniel Cross, Maxime Deslongchamps-Gagnon, Kester Dyer, Diego Gachadouat Ranz, Radhanatha Gagnon, Maria Gladitsch, Louis-Philippe Hamel, Stephanie Janes, Caroline Klimek, Marco Luna, Irina Lyubchenko, Silvia Mascia, Liz Miller, Claudia Mitchell, Michaela Pňačeková, Chélanie Beaudin-Quintin, Yosr Ben Romdhane, Ariel Rogers, Irene Rozsa, Hani Sadati, Julia Salles, Kata Szita, Philippe Theophanidis and Aude Weber-Houde. Thanks also to the four invited artists who adapted their own practices to Zoom, providing us each day with a sensory and somatic warmup, allowing us to stay online: Nadège Grebmeier Forget, Ivetta Sunyoung Kang, Emilie Morin and Olivia McGilchrist (whose fabulous VR installation *JONKONNU – GENS INCONNUS* provides our cover image for this book). We also thank the artists who took part in two roundtables. The first addressed the research and creation of immersivity, with Louise Boisclair, Jean-Pierre Flayeux, and Allison Moore. The second—held in a virtual venue designed by the PHI Centre—with Sandra Rodriguez, Marie-Pier Gauthier, and Maude Thibodeau, confirmed the intuition, developed in this volume's interview section, that the voices of practitioners are central to any worthwhile conversation on immersive media. We also thank our keynote speakers Alexa Weik von Mossner, Deborah Levitt, Charles Acland and our guest of honour, the brilliant artist and technological visionary Douglas Trumbull, who passed away just a few months afterwards.

The conference straddled the online territory of platform capitalism online, and the pragmatic, lived site of organizational labour in Montreal/

Tio'tia:ke/ Mooniyang, located on the unceded territories of the Haudenosaunee and Anishinaabe peoples. We benefited from the profound expertise of AbTeC (Aboriginal Territories in Cyberspace) who have long navigated embodiment, situatedness and the speculative potential of immersion in their artistic and pedagogic practices. They welcomed us to AbTeC island for a wild tour through the world of avatar and world design in Second Life, facilitated by AbTeC co-founder Skawennati (whose interview with Carl Therrien closes this book), Maize Longboat and Swarm, creator of a custom world tour. It is that spirit of collaboration and a generous margin of experimentation that we seek to extend with the writings gathered here.

Our particular thanks go to our collaborators in organizing the conference who have continued to support the work of this volume: Thomas Carrier-Lafleur, Joël Lehmann and André Gaudreault from GRAFIM. We gratefully acknowledge the material support of the Social Sciences and Humanities Research Council of Canada, the Fonds de recherche du Québec – Société et Culture (FRQSC), GRAFIM (Groupe de recherche sur l'avènement et la formation des identités médiatiques), the Labo CinéMédias, and the Moving Image Research Lab (MIRL). Many thanks to the team from Amsterdam University Press who helped bring this volume to fruition: Maryse Elliott, Chantal Nicolaes, Santiago Hidalgo, Katharina Loew, Mike Sanders and Marnie Mariscalchi. Thank you also to our assiduous copy-editors Émilie Bauduin and Lou Andrysiak, to Jean-Sébastien Houle and to Adam Hill for their meticulous work. Our sincere thanks to the small legion of peer reviewers for this book, and also for the press, whose precious feedback during an overcharged period for many academics is gratefully received.

Alanna wishes to thank collaborators who have shared thoughts and experiences around immersion that have shaped her thinking, in particular concerning the moving, dancing body and immersive media: Noémie Solomon, Catherine Lavoie-Marcus, Reşat Fuac Çam, Dayna McLeod, Julia Bee, Toni Pape, Christoph Brunner, Sissel Marie Tonn and Jonathan Reus, and collaborators and guests of the Moving Image Research Lab. She is deeply grateful to have Carl and Philippe as co-editors and dear friends, for putting the solid in solidarity, for their rich knowledge, kindness and professionalism, and their serious playfulness.

Philippe is thankful for getting the opportunity to work within—and in the sidelines of—Québec's rich immersive industry during the process of editing this volume, notably while working for Xn Québec, the NFB's interactive studios, Festival du Nouveau Cinéma and SODEC. He also wishes to thank those who either through their work or their fruitful conversations have influenced his thinking about virtual reality and immersive media writ large: Caroline Klimek, Aubrey Anable, Da Ye Kim, Paul Roquet, Olivia McGilchrist, Hilary Bergen, and Kent Bye. Finally, he thanks Carl and Alanna for giving him the opportunity to take on this project with them.

Carl is deeply grateful for the opportunity to collaborate with Alanna and Philippe on this project over the last five years. He wishes to thank the virtual conference participants and authors of this anthology, whose contributions will fuel so many ongoing conversations with colleagues, collaborators and friends who have inspired him over the years: Dominic Arsenault, Sylvain Beausoleil, Anouck Belthoise, Alexis Blanchet, Alenda Chang, Santiago Hidalgo, Mat Lamarche, François Martel Lacoursière, Pierre-Yves Montpetit, Pierre-Louis Patoine, Bernard Perron, Samuel Poirier-Poulin, Cindy Poremba, Jean-Charles Ray, Laurie Mei Ross-Dionne, Fabienne Sacy, Michaël Turcotte-Talbot, and Alex Wade. He wishes to express his deepest gratitude to Noëlline Létourneau, who taught him to stand up for himself, walk, and swim in murky waters. She remains forever by his side.

Introduction

Carl Therrien, Philippe Bédard, and Alanna Thain

In our contemporary media culture, it seems difficult to disconnect from immersion. The concept features prominently in much of the marketing for new audiovisual tech, from earbuds to video game controllers and 4K projectors. Exhibition spaces, design studios, and event production firms frontload it in their branding strategies. Immersion sells, in theory. 2016 was widely presented as "The Year of Virtual Reality" in the trade press, fuelled by the techno-optimism of major industrial players finally jumping headfirst into the VR adventure, hoping to make their new headsets the most desirable option for moviegoers, video game players, and office procrastinators. But these audiences didn't embrace innovation as expected. Then 2017 became the year of VR… and then 2018. In 2019, Leighton Evans published "The Re-Emergence of Virtual Reality," highlighting that the medium is still breaking through, facing barriers to mass consumer adoption.[1] Writing about the latest augmented reality headsets released by Apple in 2024, Wagner James Au amplifies Avi Bar-Zeev's challenge to tech companies, inviting them to discuss publicly how many people are nauseated by these new technologies, a percentage estimated to be rather high.[2]

To this day, VR's coming of age is still not exactly what the venture capitalists hoped for. *The Void*, a novel mixed-reality attraction using VR headsets and haptic feedback to augment its generic walls, fans, and heat lamps with the wonders of major blockbuster franchises from Disney, was severely impacted by the COVID-19 pandemic; all venues were closed by 2020. On the domestic front, the pandemic created an incentive for everyone to upgrade their access point to the metaverse(s). Yet again, the big breakthrough didn't occur as expected. In an article published in *The Guardian* on May 13, 2023, John Naughton comically underscores these shattered ambitions with a solemn tone: "A moment's silence, please, for the

1 Evans, "The Re-Emergence of Virtual Reality."
2 Au, "With Vision Pro Launched."

Bédard, Philippe, Alanna Thain, and Carl Therrien, eds. *States of Immersion Across Media: Bodies, Techniques, Practices*. Taylor & Francis Group, 2025
DOI 10.5117/9789048558766_INTRO

death of Mark Zuckerberg's Metaverse."[3] Yet with so much money invested, permadeath is often unacceptable for these players, who frequently hint that a comeback might be lurking around the corner.

Can a medium be defined through its inability to fully emerge, or rather its ability to constantly be born again? The new VR or XR—Extended Reality—gear has been linked to venerable genealogies. Evans, building on the work of Stephan Oettermann and Jaron Lanier (to name only a few), connects it directly to the development of the Panorama and other mass amusements of the eighteenth and nineteenth centuries. Oliver Grau goes back much further in *Virtual Art* (2003).[4] Beyond tech and across media, many recent contributions emphasize the essential role of storytelling to induce states of immersion, expanding classic narratological models to account for simulation, interactivity, and platform economies.[5] In doing so, these contributions re-emphasize the so-called "book problem": this bare-bones media format remains one of the most immersive and popular to this day. Since all our media forms have revealed their incredibly immersive nature, and since these extensions pervade every moment of our lives, can we even achieve a state outside of immersion at this point? As Frank Rose underscores with his book title, building on the original definition of the term: this world of ours, hypermediated by stories, is the "sea we swim in."

States of Immersion Across Media: Bodies, Techniques, Practices examines the continual reinvention of immersive interactions between users and technologies, as well as the diverse forms that emerge in the process. In particular, this book attends to the affordances of immersion's rough edges—the limits of graphically elaborated game space, the thin air where a fully animated object in VR encounters a hand reaching out in an actual environment, the mismatch between memory and mediation, or just the bad fit between bodies and worlds—to carefully address the question: what do we want from immersion? The pursuit of immersive experience has a long and rich history, one accelerating today in the wake of multiple media interfaces and formats that promise ever more complete and compelling encounters, as illustrated by VR artist and producer Félix Lajeunesse's description of virtual reality as aiming "to literally emulate reality in its full sensorial range."[6] The deep and pervasive force of media in our lives,

3 Naughton, "A Moment's Silence."

4 Grau, *Virtual Art*.

5 See for instance Koenitz, *Understanding Interactive Digital Narrative*; and Rose, *The Sea We Swim In*. Readers who want to explore these connections should revisit Marie-Laure Ryan's rigorous contributions *Narrative as Virtual Reality* (2001) and *Avatars of Story* (2006).

6 See the interview with the filmmaker in this volume.

re-attuning our bodies to be perpetually on call to a dynamic of attention and interruption, drawing us into expanded ecologies of work, play, and consumption through the portals of smartphones, earbuds, and chatbots, has generated an ambiguous ecology of immersion that exceeds our capacities to neatly delimit off and on, in or out. At the same time, the accelerated production of devices and forms that claim the label of "immersive" is continually interrupted by false starts, failures, and dead ends. Immersion is always haunted by the spectre of half-life, whether as a slow but steady disintegration of excitement, or as an "inadequate" copy of the so-called real. As we will see below, its metaphoric and actual conditions are seen to feed off the life force of its users—their attention, labour, affective range, or time.

In the first section of this introduction, we propose a brief necrology of immersion, not to put a stake in it, but to explore the tension between the discursive categories and claims for immersion's ontologies against the lived encounter with its forms. The goal is simply to acknowledge recent cultural manifestations and academic contributions that question classic definitions and theories of immersion, namely: immersion as illusion of non-mediation, as flow, as embodied presence, as a consequence of perceptual realism, as escapism, as a gateway to addiction, and even as identificatory emancipation. In the second section, we present the content of this anthology in order to highlight how our contributors rejuvenate the concept of immersion through their ongoing reflections, analyses and creative practices. At a moment where the fantasy of a fully "liveable" immersive world has re-emerged once again, in the cacophony of a confusing remix of audio-visual forms, what can the many re-animations of immersive practices tell us about dead ends, renewals, and the bodies, media, and technologies we desire?

1. A Brief Necrology of Immersion

According to Mat Lamarche, former lead level designer for Behaviour Interactive's tremendously popular slasher movie simulator *Dead by Daylight* (2016), "immersion is dead."[7] In this online multiplayer gore fest, gameplay is asymmetrical: people can play from the perspective of the killer or in a team of survivors. Some members of the online community created a trope to deride poor victims who don't play the game according to their rules: the "immersed Claudette," referring to players who are so taken by the illusion that they forget about team objectives, and hide as if their throat

7 Conversation with Therrien, ca. 2018.

was about to be sliced in real life. For this group, the "immersed Claudette" is also a drag, her stealth and stillness keeping other players from being fully immersed in the game. Beyond the immersive qualities of ludic worlds, the real endgame appears to be elsewhere, in the "meta," in outsmarting the system, its illusions, and its immersed players. Or perhaps engaging in this social media metagame can also be understood as a form of immersion? Over twenty years ago, in *Narrative as Virtual Reality* (2001), Marie-Laure Ryan hinted at the possibility of "medium-aware" forms of engagement.[8] This hypothesis contradicts one of the most common conceptions of immersion: as perceptual illusion of non-mediation or even mediated presence, closely related to strategies of immediacy or transparency.[9] What counts as "play" at the edges of an immersive world? And since developers quickly become unable to inform and control play in an online setting, who has the privilege of imposing the proper terms of engagement?

In our contemporary media ecosystem, are players escaping one immersed state only to reach another one on social media platforms? Much like a dramatic ending in a popular slasher simulator, immersion seems inescapable. As Braxton Soderman notes in revision of Mihály Csíkszentmihályi's concept of flow, or optimal engagement, and its ideological underpinnings, "the use of flow in games can be read as symptomatic of a broader culture that seeks to capture consumers in streams of profitable consumption."[10] Rose concurs: in a data-driven world, stories are being used tactically, strategically.[11] They can be weaponized as the ultimate device to capture, extract, manipulate, and control, and some books purport to teach you how to get better at this "game." Play, once perceived as a liberating force from alienating working conditions, has morphed into an imperative in many gamified Western cultures. In our contemporary media ecosystem, so-called freedom simply routes choice into subscription-based streams of lost time, from binge-watching Netflix, to playing *Grand Theft Auto: Online* (Rockstar North, 2013), to commenting on play sessions on Twitch.

In these contemporary immersive contexts, we are far removed from the optimal experience that Csíkszentmihályi associated with flow, in which user skills and challenges are reasonably attuned in order to prevent dysphoric emotions like boredom and anxiety. We are bored out of our minds scrolling down on social media, or continuously anxious and frustrated

8 Ryan, *Narrative as Virtual Reality*, p. 351.

9 See Lombard and Ditton, "At the Heart of It All"; and Bolter and Grusin, *Remediation*.

10 Soderman, *Against Flow*, p. 17.

11 Rose, *The Sea We Swim In*, pp. xv-xvi.

while getting blown up in massively multiplayer online matches. In the last decade, two popular and engaging video game genres elevated both boredom and anxiety as desirable immersive experiences, respectively the hyper-contemplative walking simulator genre and the hyper-agonistic *Souls* series of games (FromSoftware, 2009–2018). In the former, progression is a breeze; in the latter, failure and humiliation dominate. What are we engaging in when we immerse ourselves in these flattened affects? Where are we going, when flow simply pools into stagnation?

In an era of pervasive media engagement and its shadow, an always-on, yet always-shifting, manically inconsistent meta-immersive subjectivity, traditional definitions of immersion show their limits. Media innovators are flooding us with new technological devices, candidates for the next big immersive apparatus of the twenty-first century. But this exciting new medium, fantasized for millennia through countless devices of polysensorial encapsulation, never seems to get out of the womb long enough to grow. As the guest editors from a recent issue of the *Journal of Visual Culture* note: "Many of us are 'over' virtual reality (VR), even if it has yet to happen in the way we imagine it."[12] Apparently, increased bodily presence and interactivity are not attractive enough to seduce audiences en masse. VR is already offering a new body for moving images, yet some of the most successful VR outings, for instance documentaries produced by Felix & Paul Studios, leave most of the body out of the equation; users turn their heads around to gaze at spherical movie shots. Not far removed from watching a movie in IMAX, but with the added sense of rubber on your face. In time we may become as inured to this feeling as we are to the dodgy hygiene of movie seats or to the intergenerational stickiness of cinema floors: TBD. It remains to be seen if VR can attain the cinematic superpower of constant re-animation via technological innovation, or if it is in danger of becoming the greatest "forever unborn" media.[13] What kind of desires and what kind of bodies meet in the awkward adaptations of immersion's new enclosures?

Video games have long fantasized their own VR future. Several VR arcade cabinets and headsets were already produced in the 1990s. *Wired* magazine once described Atari's *Battlezone* (1980) as "Virtual Reality for a quarter,"

12 Belisle and Roquet, "Guest Editors' Introduction."

13 In *The End of Cinema?*, André Gaudreault and Philippe Marion highlight how the medium has survived numerous "deaths" throughout its history, underscoring its uncanny power as a medium of re-animation. Perhaps VR could find a place in our media ecosystem by renewing the moviegoing experience? Multiple VR apps have moved backwards to an expanded form of cinema: in CINEVR, Plex VR, and Bigscreen VR, one can watch a real flick and reconnect with friends in a virtual theatre space.

though we have long since moved beyond the green wireframe 3D shapes into photorealistic CGI territory.[14] Even so, while a new generation of consoles generate visual illusions indistinguishable from the latest Pixar movie in real time, the technostalgia of pixelated, retro-inspired game worlds has become as desirable as the glorified photorealistic set pieces seen in contemporary video games. We now use our gigantic 4K HD screens to get immersed in boxy low-poly worlds and chromatically restricted 2D vintage shoot 'em ups. As Jesper Juul's recent book on the indie video game sector (2019) makes perfectly clear, old-school, "authentic," and "handmade" artificially overblown pixels have made a significant comeback in the twenty-first century.[15] We are caught in convoluted media intermediaries, oscillating between progression and regression in any given image, and the only certitude about our innovation-obsessed culture is that the latest gear will always sell us new ways to simulate and relive our previous media experiences. Sensual memories and nostalgic affects are more powerful triggers of immersion than audio-visual verisimilitude. The risk of such repackaging is that we increasingly pay not just for immersive encounters, but for access to our feelings and lived experience itself.

In *Playing Nature* (2019), Alenda Y. Chang discusses the refinement of environmental representation in video games, including the mathematical simulation of natural patterns which improve the surface appearance of virtual landscape and fauna (a process she refers to as digital morphogenesis).[16] But Chang also highlights the environmental storytelling qualities instilled by William Crowther and later Don Woods in their spelunking textual simulator *Colossal Cave Adventure* (1975–1977).[17] "Text games remind us that game worlds are not just substitutive or compensatory simulations," Chang argues, "but also evocative spaces in their own right. Sophisticated computer graphics do not necessarily immersive games make."[18] In a key moment where more ecocritical voices point to the environmental impacts of our media ecosystem, perhaps it is time to question this constant production of "next-generation" media, and the "descriptive technology" of this biological metaphor that naturalizes it.

Predictably, the industry doesn't have much time to focus on these critiques. It is accelerating aggressively towards the next embodiment of our

14 Wen, "Retrocadia."

15 Juul, *Handmade Pixels*.

16 Chang, *Playing Nature*, p. 118.

17 Interestingly, *Colossal Cave* has recently been faithfully recreated in 3D as well as in VR, an effort led by adventure game pioneer Roberta Williams.

18 Chang, *Playing Nature*, p. 31.

encapsulating and inescapable media ecosystem, be it XR gear or AI-driven personal assistants. Now that users are in on the meta level, corporations will inevitably try to capture this realm above all other media realms. Facebook moved swiftly towards this common cyberpunk dystopia in October 2021 by rebranding as, simply, "Meta." The transition to this new brand image was a convenient way for the corporation to distance itself from the ongoing chaos it is accused of accelerating, but even more clearly a PR opportunity for its next big product: the metaverse, a 36 billion dollar investment that was supposed to be the killer app to sell Oculus headsets. Many of us tried, again, to dream about docking into a massively multiplayer, ray-traced version of *Second Life* (Linden Lab, 2003), manipulating media and virtual desktop icons in mid-air with our fingers, as in *Minority Report*, or wearing a full headset and running on a treadmill on our docking station in the back of our van, just like the protagonist of *Ready Player One* (Spielberg, 2002 and 2018 respectively). Could our bodies finally be at ease in our shared collective hallucinations? The metaverse promises to duplicate all aspects of life, yet remain always connected to the territory below, in fact monitoring and exploiting it ever more extensively. To date, of course, almost no one is there. Unlike our current physical bodies, the requirements for having a body in the Meta-verse are prohibitive, and the promise of total capture belied at every turn by experience, less a fully realized second life and more of an all-inclusive resort. Forty years ago, David Cronenberg imagined that alongside the immersive headsets of virtual reality and the new flesh of *Videodrome* (1983) would be Cathode Ray Missions, providing charitable doses of immersive media for the less fortunate.

Even as we might drag our feet when faced with the awkward interfaces of bodily immersion, we cannot help dreaming about perfectly natural, fully embodied escapism in virtual simulations of a better world. Other speculative fictions bet on some form of disembodiment, further removing us from another source of physical trauma—our very own bodies—, much like the human beings one discovers at the end of *Enslaved: Odyssey to the West* (Ninja Theory, 2010, co-written by Alex Garland and Tameem Antoniades), thousands upon thousands of immersed zombies floating idly in large walled sections. In this immersive pyramidal shrine/enclosure, our protagonists Monkey and Tripitaka meet the antagonist, who feeds illusions to all these stranded bodies. The illusions are built from the memories of a man who lived "before the war," which appears close to our time. "I am more than an archive. I am an ark," says the keeper. Compared to the decayed wasteland outside, "Pyramid is a world in which you can live." In the end, this weirdly stranded, disembodied state is presented as salutary and the titular enslavement is revealed to be desirable in the face of a war-torn wasteland. Addressing the player's avatar Monkey more directly, the illusion-maker continues: "See for yourself. See what they have before you take it away. Go on… go on, put on the mask." We only see a glimpse of this colorful and comforting simulated world reflected on Monkey's gleeful face before Tripitaka decides to sever the cords of the illusion-making machine. It appears our hero would have stayed in. Can immersive enclosures really manage to offworld us from the collapse of our current material world, from planetary precarity and the brutality and banality of much contemporary life? Beyond the utopian/dystopian binary, terms often artificially determined by the agendas of immersive media's delimited appeal to "participation," where might we find a protopian impulse that could start from a different place?[19]

Enslaved's twist ending reveals the comforting, desirable nature of disconnecting through illusions. Similarly, at the end of Ari Folman's *The Congress* (2013), we discover stranded human beings, using hallucinogenic drugs to immerse themselves in a colorful animated world. As it plays out, all this could be related to the "Lotus-Eater Machine trope," whose name comes from Homer's *Odyssey*, in which "Odysseus meets a society living off narcotic plants, to the extent that anyone who eats of the lotus no longer cares about anything else, including going home."[20] In 2001, Marie-Laure Ryan noted that addiction can be conceptualized as the ultimate stage of immersion. The language of addiction is often deployed (metaphorically or not) to underscore the felt reality of immersion's pull towards a constant feeding of the "next" that sustains an endless now of engagement, whether from constantly upgrading gear and software or the mapping and capture in media of every inch of lived experience. It takes a lot on the part of consumers and producers alike to maintain addiction's unsustainability. Against the promise of an expansive field of being, immersion can also name a state at the narrow end of survival.

As Tripitaka destroys the illusion-making device at the end of *Enslaved*, waking up all the floating zombies, one can almost hear Morpheus and his infamous Jean Baudrillard paraphrase: "Welcome to the desert of the real," except there are no more player mechanics to save the day as the credits roll by. Evil-doing is always predictable in story worlds like these: this immersive redemption is destroying our very nature, or designed for the elites, or corrupting our souls. Fictional escapism almost always comes at a cost for our environment, our social body, or our moral integrity. Living in the matrix

19 On the notion of "protopia," see Bielskyte, "Protopia Futures [Framework]."

20 "Lotus-Eater Machine." This is apparently one of the oldest tropes in the book.

is a tragedy, and many people can't even afford to live there. Thankfully, the actual *Matrix* movie is epic, and even hopeful. Twenty years and two transitions later, both Lilly Wachowski's comments and Lana Wachowski's latest offering in the franchise (*The Matrix Resurrections*, 2021) affirm a reading of the first film as a metaphor for trans lives and liberation. After acting as the most universal sci-fi emancipation story for Gen-Xers. And after being appropriated by the "alt-right" since 2012.[21] Against authorial intent, some still see "choosing the red pill" as a call to action for disenfranchised men to see the truth behind the illusion: that feminist, trans, non-white, queer, non-Christian, and secular radicals are truly in control of the world or rising to power. This interpretive malaise is unfathomable: incels are enamoured with the most visible trans-liberation fiction ever produced.

We constantly immerse ourselves in fictions that tell us to "stick it to the man," stand up for equality, for the environment, for our dignity. Apparently, the message just doesn't come through, and sometimes it can even be completely and toxically subverted. While psychologists can provide hard evidence that being immersed in stories is the perfect mediation to increase empathy in our world,[22] faith in our media ecosystem's ability to foster real empathy and awareness is beginning to wane in marginalized circles, if it was ever active. Bo Ruberg highlights this critique in their recent book of interviews with queer video game avant-garde creators.[23] Lisa Nakamura has consistently documented the "racialized crisis" of labor in technology, concluding that "[r]ather than trying to automate compassion for those who suffer as a result of racialized violence, indifference, and hyper-automation, we need forms of critique that show us how emotions like empathy and compassion have alibi-ed untenable material conditions of labor for racialized and gendered people long before VR claimed them."[24] In *Rising* (2018), famous performance artist Marina Abramović uses VR technology to connect her audience with the climate crisis, placing them in front of a collapsing virtual glacier. The effect beggars belief, most notably that we could mass produce such encounters without accelerating the collapse. If immersive technology fails as a medium of transparency both materially and as an identificatory access to "otherness," to what extent can immersion's rough edges serve to shift empathy from an identificatory

21 See Saunders, "Lilly Wachowski's Latest Project Is Mentoring Up-and-Coming Trans Filmmakers of Color"; and Dignam and Rohlinger, "Misogynistic Men Online."

22 See Rose, *The Sea We Swim In*, part I: "The Narrative Turn."

23 Ruberg, *The Queer Games Avant-Garde.*

24 Nakamura, "Feeling Good About Feeling Bad."

and emotional "stand-in" towards a resistance to an extractive and all-encompassing experience economy?

Is the metaverse less a "next-gen" media and more a temporary escape pod for the privileged few who can access it? Who can afford immersion and, perhaps more crucially in this age of constant connectivity, who can afford not to be immersed? Whose desires are accommodated by our contemporary media ecosystem and whose survival remains obfuscated behind the screens?[25] Beyond reincorporation through new technology, the re-animation of immersion theories needs to be grounded in the lived encounters between actual bodies and media art.

2. Structure and Content of the Anthology

As we have seen, traditional means for guiding us into the shifting terrain of immersivity—as illusion, as flow, as mediated natural embodiment, as compensatory disembodiment, as audiovisual verisimilitude, or as empathy—are proving insufficient at this moment of media history, and in the face of our own desires. This anthology represents an opportunity to explore our immersive assumptions and fantasies in a new light. To return to the immersed Claudette: what if she is in it for the long game? While disruptive to some players, she still plays by the rule of her real fictional self, making a tactic of stillness and stealth, blending into the surroundings in part as an affordance, on the one hand, due to the adjustable luminosity of her dark skin, and on the other, her botanist kinship with the nonhuman environment. The contempt and vitriol for this character restages the too often presumed answer to the question: who is immersion for? How is immersion reimagined from the point of view of people and practices that do not take world-making for granted, but who attend to the work itself? New analytical frameworks are necessary to understand the complicated mental states of immersion, the forms it takes in our cultures, the conflictual interactions with the tech industry, and its potential to foster empathy or just to represent alien realities. Each section from *States of Immersion* will focus on some of these elements more specifically.

The chapters of the first section, "Altered States," introduce and expand on key concepts for the study of immersion, as theorized by Marie-Laure

25 On this crucial point, see Nakamura, "Feeling Good About Feeling Bad," for an expansive and critical interrogation of labor—intellectual, physical, and emotional—in VR that makes claim to empathy, or Steyerl, "Mean Images," which underscores the ongoing lack of attention to those who produce and monitor digital and immersive worlds.

Ryan, Janet H. Murray, Gordon Calleja and many others. At the same time, it covers a broad range of media and technologies, thereby highlighting the intermedial nature of immersion. Following the editor's introduction, Tom Poljanšek and Laura Katharina Mücke wade right in, challenging both the "where" and "when" of immersion. In their chapter "Immersive Situativity: Perspectives on Being in/out of Sync," the authors take various approaches to discussing immersion in terms of feeling oneself in sync with "the wave of a situation." Notably, the wave appears as a recurring motif throughout the chapter to recognize the ongoing process necessary for immersing oneself in any given context, be it a social or mediated experience. Next, Charles Acland dedicates a chapter to the recently departed Douglas Trumbull and his epic *Brainstorm* (1983). In "Douglas Trumbull and the Technological Aesthetic of *Brainstorm*," the author shows how immersivity can be seen as both the topic of the film, as well as the guiding principle in its mode of production. While *Brainstorm* may not enjoy the same level of recognition as *The Matrix*, the movie remains a crucial example in discussions and analyses of how virtual reality and immersive media have been imagined in our culture.

In his chapter "Dual Cognitive Approach," Eugene Kukshinov offers a rigorous overview of some of the most crucial concepts in the field of immersive media. Notably, he highlights a point of contention at the heart of this collection: "Immersion is not a feature of technology; it is a psychological state of blurring/blending actual and mediated experiences."[26] Maxime Deslongchamps-Gagnon offers a logical continuation to Kukshinov's argument by taking emotions as a point of departure in "Emotional and Immersive Experiences in Films and Games: Looking into the Perceptual Model and Striving for a Motivational Theory." Bridging the gap between video games and cinema, Deslongchamps-Gagnon aims to complexify our understanding of immersion by highlighting the intricate nature of emotional responses to fictional media. By extension, this allows him to challenge perceived notions about the way immersion functions in fiction. The theme of water, which runs throughout the volume and represents the essential element in etymological definitions of immersion, continues in the next chapter on "The Logics of Interaction and Immersion in *Encounters*." In this contribution, Claire Chatelet collaborates with artist Mathieu Pradat to address his project *Encounters*, a series of VR experiences where users are asked to walk barefoot in water, echoing the trek through sand in Alejandro González Iñárritu's *Carne y Arena* (2017). This chapter is also the first to

26 See Kukshinov, "Dual Cognitive Approach," in this volume.

include the contribution of a practitioner of immersive arts, an approach which allows the authors to demonstrate the value of the rigorous theoretical overview they propose in the first half of their chapter.

In "Facing Oneself: Out-of-Body Experience, Immersion, and the Remediation of Cinema Through Virtual Reality," Pietro Conte further expands this first section's theoretical overview by considering how films and virtual reality pieces have made use of the trope of out-of-body experiences. By this he demonstrates how previous media influence the budding medium of VR, an argument which will also be explored in the following section. Finally, contrasting the analogy of the ever-moving wave that opened this section, Gabriel Tremblay-Gaudette returns to stillness in his chapter, "The Time It Takes to Tend Virtual Succulents: Immersing Oneself in a Playthrough of *Viridi*." As a uniquely slow and contemplative video game, *Viridi* allows the author to focus on the affective components of his immersive experience. Tremblay-Gaudette thus deftly demonstrates a repeated claim in this section, namely that immersion need not be limited to the pursuit of ever more complex technological mediation.

In the following section, "Finding New Frames," we reframe immersive media as they are typically known, as well as their canonical history. Several chapters take a larger historical perspective on the present moment, either by focusing on technical developments that prefigure current happenings in virtual reality and immersive media writ large or by arguing for a more diffuse understanding of immersion. First, co-editor Philippe Bédard revisits the "The Many Births of Virtual Reality" to contextualize a crucial moment in the emergence of a still nascent medium. Echoing André Gaudreault and Philippe Marion's study of the period of early cinema before its evolution into an institutionally recognized medium with its own artistic conventions, Bédard studies the phase of "early VR" in which we still find ourselves at the time of writing.[27] More importantly, the chapter offers a focused look at some of the ways VR has borrowed the codes and conventions of cinema, while still arguing that VR's identity as a medium has yet to come into focus.

Michael Stock reframes one of early film history's better-known curiosities, the phantom ride, in his chapter "Seeing by Automobility: Protoautomobility, Immersion, the Traveller-Spectator, and the Phantom Ride, 1897–1912." Stock unearths documents relating to practices not typically thought of in studies on immersion, thereby encouraging readers to broaden their perspectives as to what immersion might *look* like. Here, contemporary reviews of early phantom ride films allow the author to study a uniquely

27 Gaudreault and Marion, *The End of Cinema?*

cinematic form of immersion, but also to explore precursors to some of the gimmicks that are prevalent in immersive amusement park rides to this day. Such gimmicks are also the focus of Aude Weber-Houde's chapter, "From Participation to Immersion: Promotional Tactics around Tricked and Dynamic Seats in the Film Theatre." Comparing the promotional materials for William Castle's *The Tingler* (1959) and for more recent examples of "rigged seats," Weber-Houde highlights the rhetorical underpinnings of immersion, framed here in terms of an active and willing participation. In "Surface and Depth: The Picture Plane as a Site of Immersivity in the Digital Age," C. E. Harris continues this trend of looking at the way the notion of "emersion"—as opposed to immersion—has been made operational in the context of the cinematographic apparatus. Specifically, the author reclaims the notion of picture plane—typically used in Renaissance painting—to discuss how digital cinema, new media art, and augmented reality play with depth and (negative) parallax.

Ágnes Bakk draws theatre into the discussion in "'Let's Cross This Portal Together': Immersion Mechanisms in VR Performances and Their Current Limitations," adding it to the list of "precursors" to contemporary immersive media.[28] Bakk compares the VR theatrical experiences of the 1980s and 1990s with recent advances in the medium, highlighting how changes in technological affordances and the exceptional context of the COVID-19 pandemic have affected the practice. Interestingly, while the performances Bakk discusses are often described in terms of "puppeteering," this actual practice is the subject of Dan Leberg and Sabrina Sauer's chapter "Immersive Materiality and Television Production in *The Dark Crystal: Age of Resistance* (2019)." The authors draw from interviews conducted with some of the show's puppeteers "to help contextualize immersion during production from the perspective of the media-makers themselves."[29] In so doing, Leberg and Sauer make a convincing case for understanding immersion beyond reception, by attending to the lived experience during production and its creative and artisanal practices. Concluding this section, Daniele Iannucci likewise proposes to see immersion as an active process, rather than as the passive envelopment the term's etymology might initially suggest. Coming back to the topic of video games in his chapter "'More Mbappé Than Me?':

28 Evans points out that one of the earliest mentions of the expression "virtual reality" came from Antonin Artaud's description of the illusionistic abilities of theatre as "la réalité virtuelle." Evans, *The Re-Emergence of Virtual Reality*, p. 8.

29 Leberg and Sauer, "Immersive Materiality and Television Production in The Dark Crystal: Age of Resistance (2019)," in this volume.

Soccer Video Games, Athletic Bodies, and the Performative Displacement," Iannucci shines a light on franchised sports games in which players are asked to embody real athletes. The author notably borrows Vivian Sobchack's concept of the four bodies to argue for a more nuanced understanding of identity and personification, as both video game players and the football players they embody become intertwined.

The third section of this collection, "The Edge of Sense(s)," refocuses the discussion around immersive media by taking stock of the way they have been using to "make sense" of the world around us. Here, we refer to "sense" in a polysemic manner. First, we consider how immersive forms have always implied novel experiences of sensation and perception, which both solicit our habitual ways of knowing the world and also challenge them. Second, we consider how immersive media draw on and remix established media forms in their appeal to players, spectators, and users. At the same time, for an "always new" format, there are nonetheless new norms that can be traced across contemporary and historical examples. For instance, this section's opening chapter, Sonny Walbrou's "Walter Benjamin on the Threshold of Immersion," offers a unique perspective on Benjamin's thoughts on contemporary "immersive" experiences. Looking at stereoscopes or the *Kaiserpanorama*, Walbrou unearths new ways of thinking immersion as a productive relation between an individual and a form of representation.

Meanwhile, Michał Mochocki and Ignacy Nowicki develop our understanding of key notions through a framework articulating Sarah Lynne Bowman's subtypes of immersion with Ning Wang's categories of authenticity, in their chapter "Correlations of Immersion and Authenticity in the Experience of Fencing in *Hellish Quart* and *For Honor.*" Quite distinct from the other videogame genres discussed so far in the anthology, *Hellish Quart* and *For Honor* allow the authors to offer a multifaceted overview of the components playing an important role in our immersive encounters, including spatial, narrative, affective, and kinaesthetic involvement. Interestingly, both games under review in this chapter also bring to light the larger question of immersive media's relation to history, and this line of inquiry is picked up in Elina Lex's "New Heritage Ecologies: Embracing Entanglement and Complexity in Virtual Heritage Immersion." Lex explores the epistemological challenges of using VR to convey historically significant information. Using a variety of non-fiction augmented and virtual reality projects as case studies, she highlights how immersive media are transforming the landscape of heritage preservation, display, and education. Shifting gears from depictions of real history to the (re)construction of a fictional narrative, Gabrielle Trépanier-Jobin and Mathilde Savoie return to video games as an object of study in their chapter on "The Contribution of Narrative Puzzles to Player Challenge-Based Immersion: A Case Study of *Abzû*." Drawing from a rich body of scholarship on immersion in video games, Trépanier-Jobin and Savoie also conducted qualitative research with actual players of the underwater exploration game to shed light on narrative puzzle-solving.

In "The Opposite of Elsewhere: Home Video as Virtual Reality," Madison Brown invites us on a thought experiment by considering home video as an avatar for virtual reality. In so doing, Brown gives credence to the idea that immersion need not be thought in terms of an opposition between "here" and "there," a proposition that echoes arguments made throughout this volume. Finally, Andrea Andiloro's "Video Game Immersion as Bodily Atmospheric Attunement" adds yet another layer to the discussion of immersion in video games by approaching it from the perspective of attunement. This alternative to Csíkszentmihályi's oft-quoted concept of flow allows Andiloro to highlight once more the active process by which players attune to a given context. Immersion here is therefore understood as the ability to feel in sync with one's (virtual) environment. Andiloro's chapter suggestively offers a logical continuation of, and a conclusion to, Poljanšek and Mücke's "Perspectives on Being in/out of Sync," which opened this volume.

States of Immersion concludes with a series of interviews and texts written by—and with—practitioners working in the field of virtual reality and immersive media arts more broadly. Our first interview is with Félix Lajeunesse, co-founder and chief creative officer of Felix & Paul Studios, a Montreal-based production company recognized internationally for producing some of the most influential virtual reality experiences of the past decade. Lajeunesse not only recounts some of the company's lesser-known immersive experiments, but he also shares his perspective on immersive media at large, both current and yet-to-be. In "Producing *Intravene*: An Experiment in Immersive Audio," Brenda Longfellow describes her recent collaboration with Darkfield, a UK theatre company specializing in immersive binaural theatre. Proposing a glimpse into the creative process behind an urgent project—both in terms of the timeline of production and its subject, the current opioid crises—, the text reminds readers to discover the oft-forgotten role of audio in mediated immersion. Next, Caroline Klimek's aptly titled "A Much-Needed Conversation with Women Working in VR" brings together five women working in different capacities in the immersive sector. These include artists (Nadine Valcin and Cat Bluemke), gamers (Paloma Dawkins), community researchers (Linda Zhang), and an industry insider (Michelle Cortese). As well as being both honest and open—describing the trials and tribulations of working in VR—, the conversation proves inspiring for marginalized individuals wishing to carve their place in this industry. In "Espace & Présence," Guillaume Perreault-Roy shares his theory on the "three pillars of immersion," a philosophy he developed through years of working both in the industry and as an independent artist-researcher. Perreault-Roy's various projects, most notably the titular *Espace & Présence*, serve as fascinating case studies, showcasing aspects of immersion related to technical developments, design strategies, and user experience.

Next, the section offers an in-depth discussion with Kent Bye, host of the *Voices of VR* podcast and oral historian for the past decade of immersive media. With over 1500 interviews published at the time of writing, Bye has worked tirelessly to document the transformations of this constantly evolving medium, with inputs from some of the field's biggest technology manufacturers, artists, curators, policy makers, and theorists. It is only appropriate that our own interrogation of immersive media should address Bye's question of the "ultimate potential of VR," a question which gets to the heart of the goals and hopes for these technologies. Finally, an interview with the Kanien'kehá:ka (Mohawk) artist Skawennati serves as the concluding piece of our volume. For over twenty years, Skawennati has explored the transformative potential of cyberspace, bridging our virtual and actual worlds through the many reincarnations of her avatar and artworks. In this exchange, she brings a crucial perspective to the mix, both in terms of her pioneering work in Second Life and as an artist putting Indigenous narratives in the spotlight. Full of humor and insight, this interview sheds light on practices that are often omitted in canonical histories of immersive media.

Bibliography

Au, Wagner James. "With Vision Pro Launched, Companies Must Talk about XR, Nausea and Gender", *VentureBeat*, March 12, 2024, https://venturebeat.com/games/with-vision-pro-launched-companies-must-talk-about-xr-nausea-and-gender/.

Belisle, Brooke, and Paul Roquet. "Guest Editors' Introduction: Virtual Reality, Immersion, and Empathy." *Journal of Visual Culture* 19, no. 1 (2020): 3–10.

Bielskyte, Monica. "Protopia Futures [Framework]." *Medium*, May 18, 2021. https://medium.com/protopia-futures/protopia-futures-framework-f3c2a5d09a1e.

Bolter, Jay David, and Richard Grusin. *Remediation: Understanding New Media*. Cambridge: The MIT Press, 2000.

Chang, Alenda Y. *Playing Nature: Ecology in Video Games*. Minneapolis, MN: University of Minnesota Press, 2019.

Dignam, Pierce Alexander, and Deana A Rohlinger. "Misogynistic Men Online: How the Red Pill Helped Elect Trump." *Signs: Journal of Women in Culture and Society* 44, no. 3 (2019): 589–612.

Evans, Leighton. *The Re-Emergence of Virtual Reality.* London: Routledge, 2019.

Gaudreault, André, and Philippe Marion. *The End of Cinema? A Medium in Crisis in the Digital Age.* New York, NY: Columbia University Press, 2015.

Grau, Oliver. *Virtual Art: From Illusion to Immersion.* Cambridge, MA: MIT Press, 2003.

Juul, Jesper. *Handmade Pixels: Independent Video Games and the Quest for Authenticity.* Cambridge, MA: MIT Press, 2019.

Koenitz, Hartmut. *Understanding Interactive Digital Narrative: Immersive Expressions for a Complex Time.* London: Routledge, 2023.

Lombard, Matthew, and Theresa Ditton. "At the Heart of It All: The Concept of Presence." *Journal of Computer-Mediated Communication* 3, no. 2 (September, 1997). https://doi.org/10.1111/j.1083-6101.1997.tb00072.x.

Nakamura, Lisa. "Feeling Good about Feeling Bad: Virtuous Virtual Reality and the Automation of Racial Empathy." *Journal of Visual Culture* 9, no. 1 (2020): 47–64.

Naughton, John. "A Moment's Silence, Please, for the Death of Mark Zuckerberg's Metaverse." *The Guardian*, May 13, 2023. https://www.theguardian.com/technology/commentisfree/2023/may/13/death-of-mark-zuckerberg-metaverse-meta-facebook-virtual-reality-ai.

Rose, Frank. *The Sea We Swin In: How Stories Work in a Data-Driven World.* Harrisonburg: W. W. Norton & Company, 2021.

Ruberg, Bonnie. *The Queer Games Avant-Garde: How LGBTQ Game Makers Are Reimagining the Medium of Video Games.* Durham: Duke University Press, 2020.

Ryan, Marie-Laure. *Avatars of Story.* Minneapolis: University of Minnesota Press, 2006.

Ryan, Marie-Laure. *Narrative as Virtual Reality: Immersion and Interactivity in Literature and Electronic Media.* Baltimore, MD: Johns Hopkins University Press, 2001.

Saunders, Wren. "Lilly Wachowski's Latest Project Is Mentoring Up-and-Coming Trans Filmmakers of Color." *Them*, April 27, 2023. https://www.them.us/story/lilly-wachowski-mentoring-the-matrix-interview.

Soderman, Braxton. *Against Flow: Video Games and the Flowing Subject.* Cambridge, MA: MIT Press, 2021.

Steyerl, Hito. "Mean Images." *New Left Review* 140–141 (March–June 2023). https://newleftreview.org/issues/ii140/articles/hito-steyerl-mean-images.

Wen, Howard. "Retrocadia." *Wired*, September 1, 1994. https://www.wired.com/1994/09/retrocadia/.

About the Authors

Carl Therrien is Professor of Cinema and Game Studies at Université de Montréal. He has explored the notion of immersion and affective ecologies in his PhD dissertation and in multiple papers and book chapters over the years. He has published extensively on the current challenges of video game historiography through case studies of specific genres such as visual novels, horror games, and first-person shooters. In *The Media Snatcher* (Platform Studies, MIT Press, 2019), Therrien builds upon previous findings through a comparative study of the TurboGrafx-16 / PC Engine, confronting American and Japanese perspectives of this expansive technological platform. He has co-founded the History of Games international conference series and has contributed to the annual Game History symposium in Montreal since 2014.

Philippe Bédard is an independent researcher with a fascination for moving image media and their technologies. As part of his postdoctoral research, he focused on virtual reality, with an emphasis on the notion of empathy. Finally, he has worked extensively in and around the immersive media industry in Montreal, including work with Xn Québec, Festival du Nouveau Cinéma, the National Film Board of Canada's interactive studio, and SODEC.

Alanna Thain is Associate Professor of Cultural Studies, World Cinemas and Gender, Sexuality and Feminist Studies at McGill University. She directs the Moving Image Research Lab, devoted to the study of the body in moving image media and research-creation practices. She is author of *Bodies in Suspense: Time and Affect in Cinema* (2017) and co-editor of *Lo Tech Pop Cult: Screendance Remixed* (2024). She co-directs the research project The Sociability of Sleep, and is director of the research team CORÉRISC, on epistemologies of embodied risk. Her work on immersive practices concerns the intersections of mobile media, dance movement, soft bodies, and ambient ecologies, in projects such as Cinema Out of the Box, a bike-powered mobile cinema, and in essays such as "Anarchival Cinemas." She is also a member of GRAFIM.

Section 1

Altered States

1. Immersive Situativity: Perspectives on Being In/Out of Sync

Tom Poljanšek and Laura Katharina Mücke

Abstract: Immersive experiences are typically associated with entering alternative worlds, often represented as distinct spatial environments. This paper challenges such conventional notions of immersion by arguing that the contexts of immersion aren't spatially articulated "worlds" but temporally structured "situations" of limited scope. We propose a novel concept of immersion, asserting that being immersed in a situation involves being experientially and actively *in sync* with its unfolding temporal structure, rhythm, vibe, and pace. Drawing on the phenomenology of (film) experience and following the wave metaphor, we examine the underlying dynamics of time experience in immersive encounters. This expanded understanding of immersion has profound implications, bridging the gap between film experience and everyday life and revealing the political dimensions of immersive experience.

Keywords: situation, wave, temporality, immersion, phenomenology, frame theory

> "Qu'est-ce qu'on fait? On ne cesse pas de s'insinuer dans les plis de la nature. Pour nous, la nature c'est un ensemble de plis mobiles. On s'insinue dans le pli de la vague. Habiter le pli de la vague, c'est ça notre tâche à nous."
>
> —Gilles Deleuze[1]

1 Deleuze, "C comme culture." Deleuze paraphrases a letter he received from surfers regarding his book on Leibniz and *The Fold.*

Bédard, Philippe, Alanna Thain, and Carl Therrien, eds. *States of Immersion Across Media: Bodies, Techniques, Practices*. Taylor & Francis Group, 2025
DOI 10.5117/9789048558766_CH01

Inhabiting the Waves: The Fluid Qualities of Immersion

Janet H. Murray's famous definition of immersion as the state of being plunged into a swimming pool is still the metaphorical key reference in speaking about immersion as a felt, intense experience of "being surrounded by a completely other reality."[2] Yet, as often as our underwater experiences are compared to immersive experiences, little has been written about what it means for conceptions of immersion that water very rarely is a static and homogeneous medium. Whether a flowing river, the sea with its currents, a waterfall, or rain, water is usually in motion. Can consequences for the theoretical understanding of immersion be drawn from this metaphorical implication?

In this chapter, we take the fluid nature of water as a starting point for a different approach to the conception of immersion. Usually, theories of immersion focus on the experienced change of the spatial environment. Based on the image of the wave, we propose that a much deeper understanding of immersion is gained by focusing instead on the way we dynamically relate to the rhythm, vibe, pace, and temporal unfolding structure of the different *situations* we find ourselves confronted with. Following Deleuze, our suggestion is that *to feel immersed* means to dynamically inhabit the unfolding "waves" of a situation, to be "in sync" (instead of "out of sync") with them. To articulate this *primarily temporal and rhythmical notion of immersion*,[3] we choose the metaphor of the *wave*, a fitting image for a form that dynamically retains the unity of its shape for a particular time while also being in flux, constantly changing during its emergence and passing.

As Lithuanian phenomenologist Aron Gurwitsch (1932) describes it, when being in the world, we constantly "find ourselves in a situation and are interwoven with it, encompassed by it, indeed just 'absorbed' into it."[4] However, we are not always fully immersed in the situations we find ourselves in; instead, our experience is characterized by the fact that we almost always maintain a certain phenomenological distance towards each situation. In such cases, we are not fully "in sync" with the situation we are facing; a part of us or our attention is elsewhere, distributed on different waves, with each of which we are more or less in sync. As will

2 Murray, *Hamlet on the Holodeck*, p. 98.

3 Henri Maldiney's conception of "rhythm" as the ability to dynamically insert oneself into a situation in combination with his conceptions of "transpassibility" and "transpossibility" would be well compatible with the thoughts developed here: see, e.g., Maldiney, *Drei Beiträge zum Wahnsinn*, p. 44, p. 124ff.

4 Gurwitsch, *Human Encounters in the Social World*, p. 67.

become apparent in the course of our considerations, the balance between being in sync and out of sync with a given situation is itself dynamic. Just as we are rarely entirely in sync with the totality of situations we are faced with at any given moment, we are also rarely completely out of sync with it. Usually, we move on a dynamic spectrum between these two extreme poles (and we do not simply "oscillate" between both states): normally, we are neither completely here nor completely elsewhere.[5]

One of the fundamental claims of this chapter is that immersion as a phenomenon is not limited to experiences mediated by aesthetic artifacts or technical media. Instead, we understand immersion as a fundamental anthropological phenomenon that refers to how people dynamically relate to and inhabit the situations they are confronted with, including media experiences. Therefore, we also challenge the basic ontological delineation of media immersion as an experience of being away from "here" (the "ordinary world" or "paramount reality"[6]) and instead being "there" or "somewhere else" (in a somehow "mediated world"). Thus, we are interested in the dynamic ways in which people immerse themselves in (and are taken in by) *situations*. By doing so, we also take immersion seriously as a social and political phenomenon, insofar as the intersubjective dynamics of getting in and out of sync appear as a site of continuous negotiation of the specific forms and rhythms of the situations that structure our shared worlds of meaning. The wave metaphor will recur throughout this chapter to tie together the different threads of discourse that we are trying to intertwine.

Goffman's Frame Analysis

Erving Goffman's sociological concept of "frame analysis" offers a helpful description for determining what experiences of immersion relate to.[7] Goffman is concerned with the conditions under which we—as socially situated subjects and as subjects who continuously socially situate themselves—make sense of our social reality. Like Gurwitsch, Goffman conceives social reality not as a homogeneous space-time continuum but as a complex ensemble of "frames" or situations whose boundaries and determinations seem predetermined and defined by society itself. Whether shopping, going to school, to a concert, to the cinema, talking or walking with a friend, or

5 See Poljanšek, "Nie ganz bei den Sachen"; Liptay, *Neither Here Nor There*.
6 Schütz, *The Problem of Social Reality*, pp. 341–342.
7 Goffman, *Frame Analysis*.

simply passing anonymous people on the street, each time, the situation one finds oneself confronted with can be described as a specific *frame*. Frames are "conditions" familiar to the members of a community, which divide the social world into different spatiotemporal units, for each of which certain norms, behaviours, and interpretations are considered correct and appropriate.[8] Each frame gives a specific (social) meaning to the objects and events it contains and delineates a range of what is to be expected and what is and is not to be done in normative terms. As Gurwitsch writes, "[w]hat is imposed on us to do is [...] prescribed by the situation and its own structure; and we do more and greater justice to it the more we let ourselves be guided by it, i.e., the less reserved we are in immersing ourselves in it and subordinating ourselves to it."[9]

Goffman acknowledges that the "normality" of our everyday life is commonly based on people's tacit conviction that under "normal conditions," they grasp the frames they are faced with in their supposed "right" (and thus intersubjectively binding) shape. He adds that, although from time to time, "we personally negotiate aspects of all the arrangements under which we live, once these are negotiated, we continue on mechanically as though the matter had always been settled."[10] We propose to reinterpret Goffman's analysis of social worlds as more or less static and intersubjectively fixed and binding ensembles of frames as a helpful—albeit very rigid and schematized—idealization of immersive social dynamics, which are, at their core, always fluid and dynamic and whose shape depends in each case on how (specific) people interact and get in or out of sync with each other.

In terms of our understanding of immersion, the social dynamics of the conscious (dis)placement in and out of different situations can be deepened by Goffman's analysis of frames: *to be immersed* in a social situation means to be *in sync* with the (normative) orientation, style, and rhythm of a frame. It means to feel bodily and cognitively in resonance with a frame, to feel a kind of familiarity with it, and thus to experience oneself as a contributing part of the unfolding situation, in which one's orientation strategies and practical abilities match the situational demands.[11] Accordingly, we understand *non-immersion* as being *out of sync* with a frame, as the dephasing from a situational frame, which is as common as being in sync. Being out of sync

8 Goffman, *Frame Analysis*, p. 3.

9 Gurwitsch, *Human Encounters in the Social World*, p. 67. See also Goffman, *Frame Analysis*, pp. 1–2.

10 Goffman, *Frame Analysis*, p. 2.

11 Our concept of frames as immersive settings contains some structural analogies to what Hartmut Rosa has called "resonance." See Rosa, *Resonanz*.

is a potential trigger for the renegotiation of the normative structure and rhythms of the frames themselves. The changes and the transformations of our social reality are, to a large extent at least, motivated by such micropolitical renegotiations of frames initiated by experiences of being out of sync. In this regard, frame analysis can be understood as a contribution to a theory of immersion *and* of non-immersion in everyday (and media) experiences, as well as a principle of social orientation. Thus, we understand immersion not only as a descriptive term for an intensified experience of a clearly delineated space that includes cognitive as well as bodily aspects, but as a central dimension of communication and of interaction in our social reality.

Now, if we combine Goffman's frame analysis with the image of situations as ensembles of flowing waves, we get an image of the immersive dynamics of social realities as a kind of dynamic and polyphonic (and often also cacophonous) *score*. If the different voices and instruments performing this score are sufficiently in sync (i.e., if the different subjects have sufficiently similar grasps of the normative rhythms and dynamics of a situation), they realize social "music" together. However, if they are not sufficiently in sync, "discordant tones" will result, requiring a retuning, an aborting, or a changing of the score.

The Phenomenology of Immersion

To deepen our account of immersion as an experience of being in sync with the wave of a situation, we now turn to the phenomenology of ordinary experience. One of the most basic phenomenological observations is that consciousness is characterized by "intentionality." Specifically, intentionality names the fact that our conscious states are always directed toward "something." This "something" is usually interpreted as either a concrete thing or a state of affairs. In the following, we focus on the temporal structure and dynamics of intentionality. Further, we will propose that the primary objects of intentionality are not concrete, material things but *situations*. *Intentional reference to concrete things* presupposes *situationally localizing intentionality*. To put it somewhat crudely, before individual objects are conceived *as* this or that, consciousness has to orient itself in terms of "What's going on here?": "What's the context?", "What's the situation?"[12] If this claim proves valid, immersion can be understood in terms of a situationally localizing intentionality that logically precedes object-directed intentionality. This

12 Goffman, *Frame Analysis*, p. 8.

would fit with the fact that immersive experiences are usually described by spatial metaphors, for example, by explaining the process of immersion as a "recentering" of attention through which "consciousness relocates itself to *another world*."[13] When people have immersive experiences, they usually experience a kind of "leap"[14] or "trans-situation" from one "world" or "situation" to another. The core meaning of immersive experience thus lies in the idea that the situation toward which one is intentionally directed is fundamentally substituted by another (e.g., fictional) situation, which is different from one's previous situation.[15] We do not want to abandon this understanding of immersion entirely, but two aspects seem problematic. On the one hand, as we will see below, the situations between which we switch our attention in immersive experiences are not primarily to be understood in *spatial* but in *temporal terms*. On the other hand, such trans-situations occur much more routinely, not only when we switch our "world" of attentional reference from ordinary lifeworld to, for example, the fictional "world" of a film, but whenever attention moves from one (social) situation or frame to another.[16] To speak of whole "worlds" as the intentional objects that change in an immersive experience is to overlook the many small shifts between situations or frames that structure our everyday experience.[17]

Describing everyday experience phenomenologically, every experientially given situation or object consists, on the one hand, of impressionally given parts, of a "core of presentation," as Edmund Husserl describes it.[18] On the other hand, it consists of spatiotemporal "surroundings" or "horizons" that are not impressionally given in perception (or "apperception") but rather passively supplemented—or, as Husserl says, "appresented"—through the faculty of the subject's imagination (*Einbildungskraft*).[19] However, these appresented temporal and spatial parts belong constitutively to each situation or object as they are given in experience. In every perception of something

13 Ryan, *Narrative as Virtual Reality*, p. 103, our emphasis. An exception to this rule is the concept of "flow," understood as a special ludic form of immersion. See Csíkszentmihályi, *Beyond Boredom and Anxiety*.

14 Schütz, *The Problem of Social Reality*, p. 232.

15 Or, as Janet Murray also put it, "as different as water is from air." Murray, *Hamlet on the Holodeck*, p. 98.

16 Psychological evidence and theoretical support for this claim can also be found in the approach of "event segmentation theory," see Radvansky and Zacks, *Event Cognition*, p. 52. One could also compare the concept of "situation" we put forward with the notion of "situation models" as proposed, e.g., by Jan-Noël Thon. See Thon, "Immersion Revisited."

17 See also Schütz, *The Problem of Social Reality*, p. 342.

18 Husserl, *Cartesian Meditations*, p. 122.

19 Husserl, *Cartesian Meditations*, p. 122.

we can thus distinguish a *perceptual core that is presented as actual* (e.g., the facing side of an object) from an internal *horizon or surrounding of co-perceived or appresented spatial and temporal parts that are presented as virtual or "empty"* (e.g., hidden spatial parts of the object or horizons of likely or probable possibilities). Now, both of these, the core that is presented and the horizon that is virtually appresented, are "so fused that they stand in the *functional community of one perception.*"[20] Therefore, in every perception of an object or situation, the "genuinely perceived" can phenomenally be distinguished from the "rest [*Überschuss*], which is not strictly perceived and yet is indeed there too."[21] If we perceive a massive object facing us from a particular perspective, we are not merely presented with a surface. Instead, further possible views or sides of the object are, phenomenologically speaking, "appresented" or "anticipated" within perception as (virtual) parts of this very object.

There is no given object or situation in perception without such a virtual supplement, and it is these *threads of possible further successions of experiences* as experiences of the same thing or situation that let the objects or situations appear as the ones we experience them *as*. Thus, as Deleuze writes, "the virtual must be defined as strictly a part of the real object—as though the object had one part of itself in the virtual into which it plunged as though into an objective dimension."[22] A phenomenological description of what is really given in experience is thus never complete if it does not also include the virtual horizons and "surroundings of which there is also consciousness" without them being directly or impressionally perceived.[23] This horizon, supplemented in perception, has spatial and temporal aspects: We do not simply appresent the occluded sides of a solid object all simultaneously as in a cubist painting. Rather, we schematically anticipate temporally *ordered sequences* of imminent impressional experiences in which the different perspectives of the solid object might appear in the further course of perception *one after the other.* Our ordinary perception thus exhibits something like a syntactic unfolding structure: when we make sense of situations, when we tacitly conceive of them *as situations of this or of that type*, we appresent and anticipate specifically ordered successions of possible concordant courses of events.

20 Husserl, *Cartesian Meditations*, p. 122, original emphasis.

21 Husserl, *Cartesian Meditations*, p. 122.

22 Deleuze, *Difference and Repetition*, p. 209.

23 Husserl, *Ideas Pertaining to a Pure Phenomenology and to a Phenomenological Philosophy*, p. 52.

Now, what is appresented in perception depends on what the situation or object is *(ap)perceived as*, the latter depending on the individual subject's social and cultural experiential tacit knowledge. As Husserl describes it, if we are confronted with a new situation or object in perception and apperceive it as a situation or object of a certain familiar *type*, the "institution" of "a line of harmony [*Einstimmigkeit*] and disagreement [*Unstimmigkeit*]" (with regard to the imminent continuation of perception) takes place.[24] The things and situations of the "intuited surrounding world" in its "flowing present," "have, so to speak, their 'habits'—they behave similarly under typically similar circumstances."[25] Every object or situation has its own typical "cognitive style" (Schütz) or several typical styles in which it fluidly unfolds.[26] As in the case of a wave, then, the unity of units of experience consists of typical patterns, rhythms, or forms that persist through (and are realized throughout) the just past segments of experience of the unit in question, through the moment of impressional present to the anticipated next segments of the same unit. The various objects and situations given to us in experience are thus each characterized by the particular way in which they usually flow and unfold in our experience and through our bodily actions, their structure, style, and rhythm. Now, according to Husserl, just past segments of what is experienced are present in perception through what he calls "retentions,"[27] while anticipated segments of what is experienced are present in perception through what he calls "protentions."[28] To say it metaphorically, each unit of experience can be seen *as a wave*, which we can follow along with our eyes, although it is composed of different water particles at every moment. During its emergence and passing, the wave itself even changes its shape, while each of its phases is seen by us as a phase of the same wave passing and unfolding. Thus, being in sync with the wave of a situation means tacitly *moving along* with one's protentions and retentions, as well as with one's actions, in accordance with its specific rhythm and style of unfolding.

From such a perspective, the various differentiable units of our experience, the range of possible objects that human subjects are able to experience, appear as spatiotemporal threads or waves that form "unities of sense" (*Einheiten des Sinns*) with which one can be more or less in sync.[29] An essential consequence

24 Husserl, *Phantasy, Image Consciousness, and Memory (1898–1925)*, p. 681.

25 Husserl, *The Crisis of European Sciences and Transcendental Phenomenology*, p. 31.

26 Schütz, *The Problem of Social Reality*, p. 232. See also Gurwitsch, *Collected Works III*, p. 390.

27 Husserl, *On the Phenomenology of the Consciousness of Internal Time (1893–1917)*, p. 25.

28 Husserl, *On the Phenomenology of the Consciousness of Internal Time (1893–1917)*, pp. 40–41.

29 Husserl, *Ideas Pertaining to a Pure Phenomenology and to a Phenomenological Philosophy*, p. 128.

of this phenomenological perspective is that it opens the ontology of the units of experience to a plurality of very different kinds of entities, of which concrete objects represent only a special form. Everything that can be apperceived as such a wave or a structured thread unravelling concordantly in experience is a possible "unity of sense" and thus a possible frame of reference for immersive experience, whether it is a concrete object, an interaction or an event, a piece of music, a mere repetitive rhythm, or a scene in a film.

Toward a Temporal Conception of Immersion

This makes it clear that immersion is to be understood not only *spatially*, but above all *temporally*. To be immersed in a particular situation, according to the previous phenomenological explanations, means that one's retentions and protentions, as well as one's own actions, are in sync with the concordant unraveling of the thread or wave of a situation. However, the temporal as well as the spatial scopes of the present, with their appresented horizons of retentions and protentions, *are always limited*. As claimed earlier, the change of frame of reference in immersion usually does not refer to complete "worlds" but to "incomplete" situations of limited scope, which Schütz thus describes as "finite provinces of meaning."[30] Situations are each composed of present, past, and anticipated elements that form a context or unit that is experienced by the subject as somewhat cohesive or accordant.[31] A unit grasped in experience in which a subject can be immersed may be spatially articulated, but it may also be a unit composed of successive sounds, a unit composed of conversational elements, or a unit composed of a series of feelings, affects, thoughts, or fantasies. In any case, this unit will have the temporal structure of a horizon of possible continuations anticipated against the background of the past experiences described in the previous section. This temporal unfolding structure of successive elements—based on *what came before* and oriented toward *what might come next*—is primary to immersion.

In fact, people are much more willing to tolerate gaps and incompleteness in their experience of lived space than in their experience of lived time.

30 Schütz, *The Problem of Social Reality*, p. 342.

31 "Situations," here, can also be described with Michael Thompson as "limited wholes." See Thompson, *Life and Action*, p. 158. Thompson explains this concept according to the relationship in which different partial acts, such as *buying ingredients*, *mixing dough*, and *baking*, stand in relation to the inclusive action of *baking a cake*.

Normally, a perceiving subject does not homogeneously and completely appresent in his experience the not directly perceived, "real" space that physically surrounds them, but only those areas that are somehow in their interest. Likewise, the space of cinematic action we imagine and appresent when we watch a film is often a somewhat scattered space split up in multiple shots, more like a maze of tubes or pathways connecting different places of action than a continuous Cartesian coordinate system. When we see a scene in a film that takes place in a hotel room and in the hallway in front of it, our attention and our imagination are usually not focused on what may be happening in the adjacent hotel rooms. These rooms instead form empty voids in the field of experience, which, however, do not present themselves as deficiencies. Likewise, while watching a film, we do not completely lose sight of where we "really" or "physically" are. We are still partially aware of the situation of the screening as well as of our immediate surroundings (where we can reach for our drink, chat with our companion, and so on). We can even be driven further away from the film's action by being mentally caught up in a scene that reminds us of something else.[32] French film theorist Roger Odin, for example, recounts how he was reminded of his vacation somewhere else in France when he saw a screening of the *Tour de France* in a Paris café.[33] Space thus often plays a prominent role in immersion, but the temporal unfolding structure of successive elements is essential.

Therefore, we can also immersively lose ourselves in temporally structured frames that do not diegetically represent any kind of spatial "world"—as long as we can grasp the respective temporal unfolding structure of the frame through our cultural and imaginary dispositions and flow concordantly with it. The various voices and successive sounds and tones that together make up a piece of music, for example, do not exhibit any spatial order in the sense of external perceptual space. For immersion to take place, followable succession in time is structure enough.[34] Our proposed conception of immersion as being in sync with the unfolding wave of a situation thus allows us to explain immersion even in those situations that do not have a specific spatial structure, such as immersion in music, feelings, or conversation, as complex temporally structured trajectories of elements given successively in experience.

32 See Hanich, "When Viewers Drift Off."

33 Odin, *De la fiction*, pp. 124–128.

34 On this, also see Alfred Schütz's account of the temporal nature of music. Schütz, *Studies in Social Theory*, p. 170.

Time, Space, Rhythm, and the Ontologies of Film Experience

As mentioned earlier, immersion is often described as an experience in which the ordinary environment is phenomenally replaced by another, mediated environment—"another world." This distinctiveness of experienced worlds (or realities) is also often assumed to be a central ontological feature of film: we see the mediated stories "in the films" as something other than "our reality." Not only do we *know* about their often fictional (or at least partly artificial) characters, but we also do not seem to see them as "real" because they "only" reach us through (moving) images.[35] Since images are said to have a different materiality and ontology, we supposedly understand them as mere representations—only secondary images of the real. In many discussions on film ontologies or on the "effect" of films on audiences, however, these dualistically structured charges against the non-reality of film have since been dropped. In documentary film theory and in postcinematic theories, for example, it is widely accepted that films are part of human realities, even if they are mediated forms of communication. At the same time, it is precisely the discussion of immersion as a key concept of media experience in which binary thinking about a hierarchy of differently framed realities still persists as *the* main defining principle. The image of an immersive experience presented at the beginning, as if one were thrown into water, seems to bear major blame for this persistence. Describing film spatially as if one were jumping "into a film" or "into a film experience" that is "as different as water is from air," without taking into account its temporal quality, adheres to the firm fixation of an imagined ontological demarcation between "inside" and "outside" and between analogue and mediated forms of realities.[36] In our view, the interaction between us and other realities and situations, including filmic ones, is much more complex. Therefore, we challenge the distinction between an immersive "inside" (the film as "world") and an "outside" (our everyday experience) by basing our approach instead on a definition of immersion that takes films seriously as being real (in their impact) *and* mediated (in their form).[37] While in film studies, immersion

35 It is essential to mention here the notion of "willing suspension of disbelief," a concept coined in fiction theory that implies that during film reception, we willingly abandon our distrust and start to experience a film (and its story) "as real." The concept's origin is unknown, even if it reappeared first in Samuel Taylor Coleridge's writings in 1817 and has since been highly contested. We can, however, not further elaborate on this here.

36 Murray, *Hamlet on the Holodeck*, p. 98.

37 The ontological state of a film had always been under deep discussion, especially since film has left the classical dispositive of the cinema to become intertwined with other media formats

is usually ascribed to either a) an entirely fictional world; or b) a promised effect of a technical device, we want to encourage a third perspective: a thinking about immersive film experiences that rather focuses on their media- and story-based *offer of social and situational orientation.*

Thus, to approach immersion as an orienting and structuring principle of film experience, we believe one must dig deeper by relying on the sheer fact that films, even fictional films, are real in multiple ways, and not just in the dimensions of an indirect "as if." Therefore, we assume with Goffman that filmic immersion is based on the same anthropological capacities that underlie immersive experience in everyday life. Having this in mind, we will further apply our concept of immersion as being in sync with frames to the idea of being in sync with mediated situations. Working with film theoretical approaches in this second half of our chapter not only has the advantage of allowing us to focus on a particular kind of experienced situation. It also gives us the opportunity to demonstrate how discourses about immersion can be revised when focussing on films as (compilations of) small immersive (and non-immersive) situations instead of uniform "worlds." Hence, we conceive of a single film as a mediated compilation of experience-enabling conditions and structures. A film serves not only as a single situation, but as a comprehensive bundle of different situations embedded in a (moving) image, which is (also) a technical frame, the screen. The fact that our use of film and media has become an inevitable and inseparable part of our lives today seems to us one more reason to *not* conceptualize it *in contrast to* them.

It is helpful here to take up the metaphor of the wave once again, as film itself is not still water but a "moving image," an indexical art that unfolds in space *and* time. Our claim here is that if we focus more on the temporal unfolding of a film, it becomes possible to put down the ontological charges and to conceive of filmic experience as one that unfolds, stops, and changes—in many respects just like everyday situations, even if the filmic experience has another underlying materiality. As only few studies of immersion have mentioned in the past, immersion had already been considered through both a temporal and a spatial lens: In addition to her aforementioned and more familiar definition of immersion as a complete recentreing of attention in another "world," literature scholar Marie-Laure Ryan has also addressed a—less often cited—temporal dimension of immersion. Speaking from a literary studies perspective about films as narratives and of narratives as "steady flow[s] of pictures," Ryan compares temporal immersive

and digitally mediated spaces of living. This is where our approach is located. See, for example, Denson and Leyda, *Post-Cinema.*

experiences to a "quasi-musical experience," "in which the present is not simply a moving point but a moving window that *encompasses memories of the past and premonitions of the future* [...]. [I]t is because past events cast a shadow on the future and restrict the range of what can happen next that we perceive narrative lines and experience suspense."[38] Her division of each moment of a filmic experience into arrangements of time frames fits with the above-described assumption that each present moment derives its experienced meaning from a context of anticipations and recollections, protentions and retentions. Especially in experiencing narrative situations of a film, past and future information are integrated into the respective understanding of "what is going on." Ryan's image of each filmic moment as a moving window is very close to our image of the wave-like structure of immersive situations, where temporal and spatial dimensions of our past experiences come together to be channelled and matched with the things we see on the screen.[39] Both our past experiences and the film's "past" together provide for the supplementation of protentions and retentions. Immersive experiences are, therefore, always based on temporal unfolding structures, each of which instantiates typical "cognitive styles" (Schütz) that the subject can tacitly follow through appropriate anticipations, retentions, and mental or bodily actions.

From this perspective, Étienne Souriau's concept of "diegesis" can also be more accurately understood as a description of the *temporal* way immersive situations are narratively unfolding. Souriau calls the synthesis of temporal and spatial audiovisual knowledge fragments into a coherent filmic narration a process of "diegetization," even if attention has less often been paid to the temporal dimension of filmic diegesis.[40] His use of the term "diegesis" points to the fact that all objects visible within the filmic frame are directly related to other objects and spaces, which lie beyond what is visible on the screen but do, in the same way, belong to the story. They also always need to be appresented while constructing a whole narrative out of the visible frame. His description of the film experience as an imaginative synthesis is consistent with the phenomenological perspective offered in the first part of this chapter, even if neither Souriau nor Ryan connect their approaches to bodily or social experiences. On the contrary, we claim that when we watch a movie, our experiential apparatus operates under

38 Ryan, *Narrative as Virtual Reality*, p. 120 and p. 141, our emphasis.

39 Julian Hanich has equally written about dimensions of the imaginary or protentions and retentions while watching a film in the cinema. See Hanich, "When Viewers Drift Off."

40 Souriau, "La structure de l'univers filmique et le vocabulaire de la filmologie," p. 233.

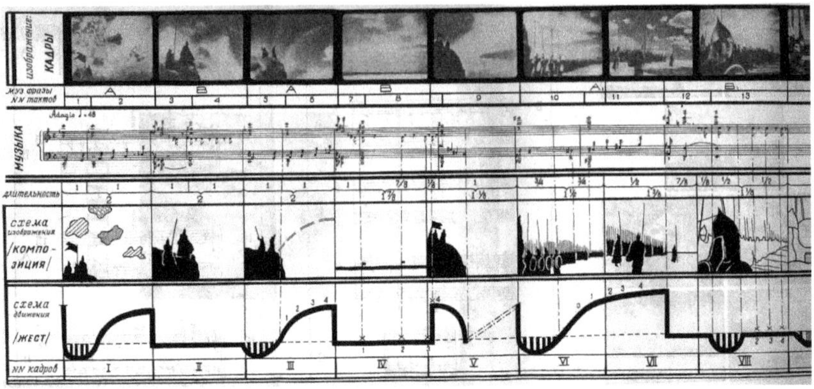

Figure 1.1: A score by Sergej Eisenstein. Eisenstein, "Die Vertikalmontage (1940-1941)," pp. 267-269.

the same perceptual conditions as it does in everyday experience: based on protentions and retentions, which may rely on familiarity with certain already-known genres or types of frames, *all* incoming sensual information is bundled (or channelled) into "unities of sense" or situations of limited scope that we follow with our minds and bodies as they unfold. Especially because of its vivid, affective, and touching quality, film can evoke lasting memories. The immersive experience of "accordance" (what Husserl calls *Einstimmigkeit*) is thereby not so much (and not necessarily) based on the creation of complete spatial contexts, but rather on getting in sync with the *unfolding style, mood, and logic of a (filmic) situation*. Films—like our everyday lives—confront us with a multitude of perceptions framed visually, auditorily, rhythmically, aesthetically, socially, and narratologically. As each observer tries to grasp the perceptible that flows toward them, using the situations and styles of unfolding with which they are familiar, they try to find an orientation that will enable them to come into sync with what they are experiencing—to locate themselves in the situations. The moments in which these synesthetic synchronizations succeed are what we experience as (fleeting) moments of immersion.

The complexity of simultaneously and successively perceived information has often been attributed to film because no other media configuration before it offered so many different technical possibilities for combining experienceable "somethings" such as textual, visual, and auditory bits. In 1940/1941, Russian filmmaker Sergej Eisenstein famously captured those filmic spatiotemporal rhythms within "audiovisual scores." (fig. 1.1)

The graphic depicts filmic experience as a complex and heterogeneous process, much like everyday life. It seems not by chance that the graphic depiction resembles a musical score with different lines, voices, and their intermingling into a concerting assemblage of impulses in the shape of waves. Pictures, sounds, colors, camera movements, cuts, architectures, settings, character movement, etc., are necessarily experienced by viewers through at least some of the categories and tacit knowledge structures which also underlie their everyday experience: as we experience film, so much information (including that this is a filmic experience) simultaneously pelts us that we have to develop strategies to channel it, sort it, and make sense of it by approaching it through patterns and frames we are already familiar with. In this respect, films' audiovisual and, thereby, technical constellations of impressions are analogous to the manifold of impressions we are confronted with in everyday life—and so are its immersivities.[41]

The interweaving of the seen and unseen objects, the known and the unknown stories within a diegesis, the protentions and retentions, and the working of our imagination in understanding a filmic situation is also what Roger Odin seeks to do. His approach can help to explore the *rhythmical* structures of immersive filmic and situational experiences further. Odin is interested in the composite relations of which a filmic situation is constructed. His semio-pragmatic approach specifically aims to capture the *reciprocal functioning* of the various parameters involved in the filmic-communicative sense-making process—including the viewer. Using the same wording as Goffman, calling communication situations "faisceau de contraintes," Odin tries to understand through which "conditions" films are *understood*.[42] Just as we as human beings might apply different readings to different stimuli, or just as we can read a feature film as a documentary and vice versa, so the semantics of the other stimuli we encounter are polysemous.

Politicizing *Déphasage*

In the context of defining immersion as being in sync with the wave of a frame, it becomes crucial to consider Roger Odin's terms for the *success* and *failure* of filmic communication—*mise en phase* and *déphasage*, both of which he understands as common effects of experiencing situations. According to Odin, the concept of the *mise en phase* refers to "le processus

41 It was Michaela Ott, in the introduction of her book on "dividuations," who had held the multiplicity of cinematic impulses responsible for forming individuals *as social subjects* themselves. See Ott, *Dividuationen*, pp. 7–64.

42 Odin, *Les espaces de communication*, p. 20.

qui me conduit à vibrer au rythme de ce que le film me donne à voir et à entendre."[43] He defines it as a (rhythmic) "vibration" *within* the mood, tempo, and construction of the (filmic) situation—a term he incidentally derives from Jean-Marie Schaeffer's earlier definition of the term "immersion."[44] Meanwhile, his second and opposite concept, *déphasage*, is key in referring to the experience of falling out of sync with a film, "[J]e ne vibre plus aux événements racontés,"[45]—of not being able or willing to follow the film's flow. Close to the experience of being out of sync with a frame, a felt *déphasage* from a (film) experience, for Odin, leads to processes of *renegotiating* frames. It leads us to apply other (new or old) established frames to a (filmic) situation. For example, if we do not understand why an actor is playing badly in a film, we may consider that we are not witnessing a fictional, but a documentary film.[46] Each (media) frame[47] thereby becomes intelligible only against a background of presupposed social conditions of understanding that it does not contain itself. In films, social and aesthetic, as well as situational norms, become intertwined and medially reconfigured. Every situation or (filmic) frame must then be seen as socially pre-structured and re-negotiable in its "meaning(s)."

We think Odin's urge to focus on the individual's position within societal frames, on the one hand, and on the failure of communication as a constructive part of immersive orientation, on the other, needs to be taken more seriously concerning a theory of immersion in films, but also in general. Insofar as the frame in which every (filmic) experience is located is, in turn, the frame of a social reality, the question of the conditions of coming in or out of sync with (filmic) situations becomes a political one itself—even if for Odin *mise-en-phase* and *déphasage* remain purely operative and apolitical categories. One can say that his conceptions allow for a much more flexible and not too ontologically restrictive description of the dynamics of immersive experiences in films and media than the one often suggested by the abovementioned rather rigid opposition of "inside" and "outside," or of "here" and "there." Since moving images have become essential to our experiential realities and are impossible to erase, we finally urgently need to highlight the common experiential and anthropological foundations of

43 Odin, *De la fiction*, p. 39.
44 Odin, *De la fiction*, p. 39.
45 Odin, *De la fiction*, p. 42.
46 Odin, *De la fiction*, pp. 43–44.
47 *Media frames* encompass various characteristics, including the cinema as a viewing situation, streaming platforms, the film's aspect ratio, image resolution, the screen, and more. Each of these frames entails its own sense-making regimes and potential (non-)immersions.

filmic experience and everyday experience to understand how immersive experiences generally shape the way we perceive the world socially and politically. Hence, our vision encompasses a broader understanding of immersion, which refers to cognitive and bodily, social and political *dynamics of (re-)orientation* within a multitude of spatial-temporal and social (as well as normative) structured situations, where various ensembles of co-existing waves converge.

The above-depicted score's referential system of rows and columns as points of orientation, opening up experiential spaces and waves for the subject to temporarily inhabit, thus also invites further investigation of the *inter- (or intra-)subjective dynamics* of the propagation and transformation of such waves. Films are always directed toward more than one person, which lets the "meaningfulness" of their frames become a question of shared waves and situations, a question about potentially shared situational immersion. Insofar as the concept of immersion, in the sense we have described, encompasses both filmic and everyday processes of sense-making, it also always has political implications.[48]

Reflecting on this ineradicable social mutability of frames and situations might also offer a continuous confrontation with one's own sense-making processes and the impossibility of establishing a fixed, once-and-for-all meaning or totalizing frame. As we have already hinted at the beginning of this chapter, such reflections are often triggered by the experience of being out of sync with existing social frames. At the same time, complete immersion remains reserved for the fleeting moments of successful synchronicity with the plurality of waves we find ourselves confronted with at any given moment. Both, moments of complete (or, as is often said, "total") immersion and those of total non-immersion, seem to be rare liminal cases within a large field of possible frame negotiations. Through various immersive "drives," we test our ability to inhabit the seemingly fixed waves of meaning unwinding in time—testing how best to swim, how to navigate in the ever-moving environments in which our encounters take place. When we get out of sync with the waves of socially established situations, however, two things might become visible: a) our skills and expectations may not match social reality; b) our social reality does not have to be as it usually

48 We would suggest further addressing the questions of the intersubjective shareability of frames in socially binding or political contexts with the film theoretical concept of "suture." Since its first appearance in the context of the apparatus theory in 1977, the term has been used not only in film but also in political theories to relate to the (im)possibility of intersubjective fixations of meanings. See Lummerding, *Agency@?*, pp. 159–165.

appears to us—and this is definitely a social and often a political issue. So, in studying the social dynamics of immersion, we must further consider that the sense-making we engage in when we relate to situations or frames already has a political dimension on an individual level. An individual subject's ability to get in sync with the particular wave of a situation always says something about the internalized normative expectations and experiences with which it tacitly interprets social, political, and filmic realities. This also means that when I say that I find a film particularly immersive, I am saying something about my own conditions for developing immersion, about the way I (as a socially cultured subject) make sense of certain situations. When I say that I am not immersed in a film, that I cannot (or refuse to) engage with it, I am also referring to the film's inability to relate to my own situational frameworks of understanding. I feel a sense of *déphasage*, which nevertheless opens up possibilities for me to invent or learn new framings.

Conclusion. Political Outlook: Inhabiting the Waves of Our Shared Present

So far, our considerations have primarily focused on the possible synchronization of a single subject with the typical unravelling of a potentially immersive situation. The social world, however, turns out to be composed not only of a multiplicity of different situations or frames through which the individual subject can navigate in solitary wanderings. Instead, many typical situations we experience in everyday life profoundly *require a reciprocal dynamic* of turn-taking and a mutually shared symbolic order and backgrounds of understanding to work out. As Alfred Schütz argues, *joint action* can, in many cases, be illustrated by the paradigm of "making music together."[49] In such "mutual tuning-in relationships," subjects commonly create arrangements of "successive elements," which they can experience as "meaningful," as far as their respective backgrounds of understanding allow them to tacitly "interrelate" these elements through "an interplay of recollections, retentions, protentions, and anticipations."[50] They can thus collectively create waves that they dynamically "inhabit" together.

Our considerations about being in or out of sync are thus also fruitful for a theory of immersive sociality. When subjects realize the waves of situations through joint actions, they can not only experience themselves

49 Schütz, *Studies in Social Theory*, p. 159.
50 Schütz, *Studies in Social Theory*, p. 177 and p. 170.

as being immersively in sync *on their own*, but also as being in sync *with each other*. To experience oneself as being jointly in sync, then, is not so much to know about the inner states of others: it is to be dynamically and interactively in sync with them by tacitly knowing one's way around in the wave of a shared situation. From here, a perspective opens up on the political dimensions of immersion as an intersubjective mutual synchronization and desynchronization, of inclusion and exclusion, as further elaborated, for example, in considerations of "frame alignment" and "frame resonance," following the perspectives of Erving Goffman, Daniel Stern, Henri Lefebvre, or Hartmut Rosa, or as implied in Judith Butler's "performative theory of assembly."[51] In particular, it would be necessary to further clarify through which intersubjective processes frames are stabilized, negotiated, changed, or even newly created, and how not being in sync can even help shape these processes. Working this out, however, would be the task of another paper.[52]

As we have shown, a much more nuanced notion of immersion can be obtained by looking at its fluid and dynamic, temporal and phenomenological aspects, rather than understanding immersion primarily spatially as a reorientation of attention from one "world" to another. In doing so, we recognize that trying to conceptualize these phenomenological and affective qualities presents both conceptual and theoretical challenges. The attempts we have made in the foregoing are, therefore, also attempts *to conceptually fix what is fluid by its very nature*—the many different and complex ways in which we experience ourselves as being (drawn) "into" something or someone, in which we manage to inhabit the flowing waves of our shared present.

Bibliography

Butler, Judith. *Notes toward a Performative Theory of Assembly*. Cambridge, MA: Harvard University Press, 2015.

Csíkszentmihályi, Mihály. *Beyond Boredom and Anxiety: Experiencing Flow in Work and Play*. San Francisco, CA: Jossey-Bass, 1975.

51 See Snow and Benford, "Ideology, Frame Resonance, and Participant Mobilization"; Lefebvre, *Éléments de rythmanalyse*; and Butler, *Notes toward a Performative Theory of Assembly*.

52 We have written about the political dimensions of being in and out of sync and about their consequences for immersion theory in a related article. See Poljanšek and Mücke, "Co-Immersions."

Deleuze, Gilles. "C comme culture." In *L'abécédaire de Gilles Deleuze, avec Claire Parnet*, directed by Pierre-André Boutang, 00:46:21. France: SODAPERAGA, Guy Seligmann, 1996.

Deleuze, Gilles. *Difference and Repetition*. New York, NY: Columbia University Press, 1994.

Denson, Shane, and Julia Leyda. *Post-Cinema: Theorizing 21st-Century Film*. Falmer: REFRAME Books, 2016.

Eisenstein, Sergej M. "Die Vertikalmontage (1940–1941)." In *Jenseits der Einstellung. Schriften zur Filmtheorie*, edited by Felix Lenz, 238–300. Frankfurt am Main: Suhrkamp, 2006.

Goffman, Erving. *Frame Analysis: An Essay on the Organization of Experience*. Boston, MA: Northeastern University Press, 1986.

Gurwitch, Aron. *Human Encounters in the Social World*. Pittsburgh, PA: Duquesne University Press, 1979.

Gurwitsch, Aron. *The Collected Works of Aron Gurwitsch (1901–1973). Volume III. The Field of Consciousness: Theme, Thematic Field, and Margin*. Dordrecht: Springer, 2010.

Hanich, Julian. "When Viewers Drift Off: A Brief Phenomenology of Cinematic Daydreaming." In *The Structure of Film Experience by Jean-Pierre Meunier: Historical Assessments and Phenomenological Expansions*, edited by Julian Hanich and Daniel Fairfax, 336–352. Amsterdam: Amsterdam University Press, 2019.

Husserl, Edmund. *Cartesian Meditations: An Introduction to Phenomenology*. Dordrecht: Springer-Science, 1960.

Husserl, Edmund. *The Crisis of European Sciences and Transcendental Phenomenology: An Introduction to Phenomenological Philosophy*. Evanston, IL: Northwestern University Press, 1970.

Husserl, Edmund. *Ideas Pertaining to a Pure Phenomenology and to a Phenomenological Philosophy. First Book. General Introduction to a Pure Phenomenology*. The Hague: Martinus Nijhoff, 1983.

Husserl, Edmund. *Phantasy, Image Consciousness, and Memory (1898–1925)*. Dordrecht: Springer, 2005.

Lefebvre, Henri. *Éléments de rythmanalyse. Introduction à la connaissance des rythmes*. Paris: Syllepse, 1992.

Liptay, Fabienne. "Neither Here Nor There: The Paradoxes of Immersion." In *Immersion in the Visual Arts and Media*, edited by Fabienne Liptay and Burcu Dogramaci, 87–110. Leiden: Brill Rodopi, 2016.

Lummerding, Susanne. *Agency@? Cyber-Diskurse. Subjektkonstituierung und Handlungsfähigkeit im Feld des Politischen*. Vienna: Böhlau, 2005.

Maldiney, Henri. *Drei Beiträge zum Wahnsinn*. Edited by Till Grohmann. Vienna: Turia + Kant, 2018.

Murray, Janet H. *Hamlet on the Holodeck: The Future of Narrative in Cyberspace.* New York, NY: Free Press, 1997.

Odin, Roger. *De la fiction.* Louvain-la-Neuve: De Boeck Supérieur, 2000.

Odin, Roger. *Les espaces de communication. Introduction á la sémio-pragmatique.* Grenoble: Presses universitaires de Grenoble, 2011.

Ott, Michaela. *Dividuationen. Theorien der Teilhabe.* Berlin: B-Books, 2015.

Poljanšek, Tom, and Laura Katharina Mücke. "Co-Immersions: Toward a Political Conception of Immersivity." *Azimuth: Philosophical Coordinates in Modern and Contemporary Age* 20, no. 2 (2022): 181–196.

Poljanšek, Tom. "Nie ganz bei den Sachen. Zur Phänomenologie der Immersion." In *Technik-Ästhetik*, edited by Christian Grabbe and Oliver Ruf, 183–202. Bielefeld: Transcript, 2022.

Radvansky, Gabriel A., and Jeffrey M. Zacks. *Event Cognition.* Oxford: Oxford University Press, 2014.

Rosa, Hartmut. *Resonanz. Eine Soziologie der Weltbeziehung.* Berlin: Suhrkamp, 2019.

Ryan, Marie-Laure. *Narrative as Virtual Reality. Immersion and Interactivity in Literature and Electronic Media.* Baltimore, MD: Johns Hopkins University Press, 2001.

Schütz, Alfred. *Collected Papers I. The Problem of Social Reality.* The Hague: Martinus Nijhoff, 1962.

Schütz, Alfred. *Collected Papers II. Studies in Social Theory.* The Hague: Martinus Nijhoff, 1976.

Snow, David A., and Robert D. Benford. "Ideology, Frame Resonance, and Participant Mobilization." *International Social Movement Research* 1 (1988): 197–217.

Souriau, Étienne. "La structure de l'univers filmique et le vocabulaire de la filmologie." *Revue internationale de filmologie* 2, no. 7–8 (1951): 231–240.

Thompson, Michael. *Life and Action. Elementary Structures of Practices and Practical Thought.* Cambridge, MA, London: Harvard University Press, 2008.

Thon, Jan-Noël. "Immersion Revisited: On the Value of a Contested Concept." In *Extending Experiences: Structure, Analysis and Design of Computer Game Player Experience*, edited by Amyris Fernandez, Olli Leino and Hanna Wirman, 29–43. Rovaniemi: Lapland University Press 2008.

About the Authors

Tom Poljanšek is a postdoctoral researcher in Philosophy at Georg-August-University Göttingen. His research areas include phenomenology, social ontology, philosophy of technology, and aesthetics. His dissertation, "Realität und Wirklichkeit. Zur Ontologie geteilter Welten" (2022), explores the plurality of the shared worlds of meaning we inhabit and their relation to mind-independent reality.

Laura Katharina Mücke has been an assistant to the working unit "Everyday Media and Digital Cultures" at the University of Mainz, Germany. Her PhD project, based in Vienna and Groningen, aims to formulate a critical and feminist perspective on the discursivity of seemingly omnipotent concepts of media experience, e.g., "immersion."

2. Douglas Trumbull and the Technological Aesthetic of *Brainstorm*

Charles R. Acland

Abstract: This research situates Douglas Trumbull's movie *Brainstorm* (1983) in the context of his career, with consideration of the critical, promotional, and trade paratextual material that circulated at the time of the film's initial release. It is argued that *Brainstorm* presents three layers of technological reflexivity: first, its primary diegetic focus involves the technology of sensory transport and intersubjectivity; second, the film was itself initially designed to introduce a new immersive cinema technology, Trumbull's Showscan format; and third, the actual media format of the released version incorporated alternations between 35mm and 70mm. In the end, *Brainstorm* delivers a miniature manifesto for technological aesthetics that reveals the material and discursive conditions in which we experience and understand ideas about immersivity and the future of entertainment technology.

Keywords: *Brainstorm*, blockbuster, technological reflexivity, media format, Hollywood

For all the conceptual challenges the term "immersivity" poses—and the chapters in this book are a fine indication of how generative the idea has been for media scholarship—there is a general popular agreement about what it references in media technology and culture. According to this broad conception, immersivity in media refers to a manifestation—a text, an event, an apparatus—that surrounds and occupies the senses. It describes an interaction between technology and the human sensory system in the production of an environment, which is thought to be better realized when it more completely engulfs individuals and audiences. The ultimate horizon of the most immersive experience possible, then, is the total artwork, one that erases sensory possibilities outside of the artistic construction. Residing in this dominant understanding is a measure of relative success, where the

Bédard, Philippe, Alanna Thain, and Carl Therrien, eds. *States of Immersion Across Media: Bodies, Techniques, Practices*. Taylor & Francis Group, 2025
DOI 10.5117/9789048558766_CH02

most immersive media situation is that which seals off our senses from the world outside of the artwork effectively and completely.

Yet, a troubling feature resides in this conceptualization of immersivity: it is a particularly functionalist way to think about media and artistic experiences, one that proposes a metric or a scale for such erasure of the difference between artwork and experience. This notion treats the immersive experience rather like a light dimmer one can turn up—allowing less sensory interference from the outside world for maximum immersiveness—or down—allowing more interference, thus reducing sensory melding with the work. It is a critical and philosophical dead-end to get caught in evaluating greater and lesser sensory capture. Instead, our critical stance might avoid this impasse by approaching "immersivity" with concerted attention to the environment or, in other words, to the dimmer switch and the appended apparatus rather than to some problematic metric of media experience. Bringing the dulled, unobserved infrastructural materiality of media experience to the front of our critical consciousness allows us to examine it, to share it, and to investigate the mechanisms that direct us to see something as a unified and complete rendition. In approaching immersivity, however compelled we are to embrace the sublimity of totality, I propose that we be attentive to the technological, discursive, and material conditions that structure what we describe as immersive, which include primary attention to the limits, repetitions, borders, conventions, and exclusions elided by that supposed experiential totality.

Especially instructive in the analysis of the material aspects of the technologies of immersion is Ariel Rogers's essay "'You Don't So Much Watch It as Download It': Conceptualizations of Digital Spectatorship" (2012).[1] In this essay, Rogers shows that what we understand of the changing cinematic experience is very much an encounter between technological conditions and how they are described for and by critics, commentators, and audience members alike. As she explains, there are always attendant discourses about experience in play with media engagement. We never confront a media experience without having that encounter already mediated by any number of paratextual and discursive formations. Directly relevant to the present task of studying "immersivity," Rogers charts the repetition of the dreams of implanting images and experiences as a special brand of immersivity, one bound up with ideas of intersubjectivity, transmission, and embodiment. She elaborates by discussing the figurations of digital cinema in the 1990s around the releases of *Conceiving Ada* (Lynn Hershman Leeson, 1997), *Star*

1 Rogers, "'You Don't So Much Watch It as Download It'."

Wars: Episode 1. The Phantom Menace (George Lucas, 1999), *Final Fantasy: The Spirits Within* (Hironobu Sakaguchi, 2001), and *The Celebration* (Thomas Vinterberg, 1998). But Rogers importantly points to two precursors of this emergent formation of the digital—*Brainstorm* (Douglas Trumbull, 1983) and *Strange Days* (Kathryn Bigelow, 1995)—which are both dramas about new media technology and intersubjective possibility. She makes apparent that new media forms, for all the supposed advancements they announce, tend to reinstate, rather than upturn, dominant sensibilities and priorities about technologies, communities, embodiment, and experience. Rogers notes internal contradictions in the discourses of the digital in the 1990s, where "the relationship between perceptual experience and bodily materiality had become unmoored, with the sensual experience of the digital image associated, alternately, with a detachment from the material world and a return to it."[2] In this chapter, I similarly focus on the material and discursive settings of the abstract category of "immersive media" in order to examine the singularly influential work of special effects innovator Douglas Trumbull.

Trumbull's unparalleled career as a motion picture producer, director, and entrepreneur presents us with a body of work that has essentially established how the cinematic future can be imagined. He has had an astounding impact on the modern blockbuster, helping shape what an expensive effects-driven movie looks like today. Such movies are a dominant and stable fixture of global popular imagination, occupying an expansive part of the economies of theatrical exhibition, production financing, and corporate strategy. Trumbull helped guide a larger historical turn in motion picture aesthetics, one that moved technology, special effects, and ideas about immersivity to the centre of the highly capitalized mainstream American movie experience. As I have argued in *American Blockbuster: Movies, Technology, Wonder* (2020), the contemporary big-budget movie is as much a showcase for technological possibilities—which invariably means experiential promises—as it is a narrative, franchise, or star vehicle.[3] In appreciating and examining Trumbull's works, attention typically focuses on his effects work in the justifiably celebrated films *2001: A Space Odyssey* (Stanley Kubrick, 1968), *Close Encounters of the Third Kind* (Steven Spielberg, 1977), and *Blade Runner* (Ridley Scott, 1982), as well as on his debut feature film as a director and producer, *Silent Running* (1972). I want to take a closer look at one of his more underappreciated efforts, *Brainstorm*, which he also directed and produced. As this chapter elaborates, though the

2 Rogers, "'You Don't So Much Watch It as Download It'," p. 231.
3 See Acland, *American Blockbuster*.

movie itself is not a blockbuster, and though it was a troubled production for many reasons, *Brainstorm*—the diegetic story, the critical discussion about it, and the technological conditions according to which it circulated in the early 1980s—is a miniature manifesto for a technological aesthetic that privileges attention not just to immersivity but to *the mechanics of the immersive experience*. In this way, it is an unusually rich text and case study through which we can outline the core discursive elements of what we consider immersive in popular moving image performance.

Trumbull's artistic priorities were not just to represent but to construct audience experiences, and he was quite explicit about this goal. Sensations have been key to the design of his effects and he would often reference, in interviews, writings, and various capacities as a public expert on cinematic technology, scientific literature on the human sensory system to support his projects. The title of a career retrospective interview he gave to Simon Howson in 2011 captured this theme perfectly: "Douglas Trumbull: The Quest for Immersion."[4] As the well-chosen word "quest" indicates, metaphors of transportation are frequently used to describe the result of Trumbull's efforts. In this respect, with both his aesthetic agenda and the reputation he built, he helped define and delimit what was in fact a century-old filmic and para-filmic idea in popular films: the ride film or sequence. Moreover, when Trumbull stepped away from making feature films in the 1980s, he worked on moving image productions for site-specific theme park attractions and, later, at IMAX, he developed ride simulation experiences. In these productions, the visualization of spectator movement was done in such a way that viewers would feel the movement as they were taken to some other place or time through the experience. This focus on travel and cinematic immersion is evident in the visual effects he did on his very first project, *To the Moon and Beyond* (1964), produced for the 1964 New York World's Fair, and has since resulted in some memorable cinematic moments of future or extra-dimensional transportation. Perhaps his most remembered contribution is the "Star Gate" sequence in Stanley Kubrick's *2001: A Space Odyssey*. The impact of that abstract psychedelic interlude has been such that it has glued Trumbull's reputation to that film, to the point where the Italian title for his directorial-feature debut, the science fiction parable *Silent Running*, was *2002: La Seconda Odessia*. But, extra-dimensional transportation can be found in many of his works, even in his perhaps least remembered contribution, which was released the same year as *2001*; the sequences that bookend the camp classic *Candy* (1968)

4 Howson, "Douglas Trumbull."

Figure 2.1: *2001: A Space Odyssey* (Stanley Kubrick, 1968), "Star Gate" sequence.

present a similar trippy journey through space and mind as *2001*.[5] As Julie A. Turnock has authoritatively documented, contributions of this kind, merging an experimental focus on non-representation and visual dynamism in popular films, rose to dominance in the 1970s and have been fixed ever since as what she calls the "blockbuster aesthetic."[6]

The reputational weight of Trumbull was solidified at an early stage in his career, initially as a result of the unusual visibility that Kubrick had provided for his film experiments. For example, Trumbull spoke at the Independent Film Producers of America conference in October 1968, in Palm Springs, discussing the effects work he had done for *2001: A Space Odyssey*. This conference was held six months after its release and Trumbull appeared with Linwood Dunn, one of the most accomplished and well-respected Hollywood effects artists. Dunn had been involved in an ongoing roadshow to celebrate and educate the public about special effects, and the work he did in this capacity helped inspire a generation of budding filmmakers and film fans.[7] Interestingly, on this particular occasion, Edmund Carpenter,

5 Other feature films of the period experimented with abstract, bent, and reflected light to represent transportation to other times and dimensions. We see this imagery, for example, in the opening credit sequence of *Planet of the Apes* (Franklin Schaffner, 1968). This aesthetic was also common in other contemporary art practices, like event-based multi-media "happenings." Trumbull's aesthetic, then, is a notable continuation of the visual and sensory experiments across media from the 1960s.

6 See Turnock, *Plastic Reality*. Other key works on special effects, sensation, and media history include Bukatman, *Matters of Gravity*; Ndalianis, *Neo-Baroque Aesthetics and Contemporary Entertainment*; Pierson, *Special Effects*; and Prince, *Digital Visual Effects in Cinema*.

7 Gregory, "Creativity, Involvement Keynote New Look, Name at IFPA Conference." For a commentary on the unparalleled importance of Linwood Dunn's contributions, see Turnock, "Before Industrial Light and Magic."

famed anthropologist and longtime collaborator of Marshall McLuhan, gave a keynote address on contemporary media, education, and sensation. In other words, very early on, Trumbull was already being situated not only as an innovative visual effects creator but as an interpreter, a pedagogue, and an interlocutor on the future of the moving image experience.

This association of Trumbull with the future of media continued throughout his career. Shortly after the release of *Brainstorm*, in 1984, he appeared in a print advertisement for a Hewlett Packard programmable calculator. There again, referencing his groundbreaking work for Kubrick, the advertisement's tagline was: "In 2001, we reached the stars. Now we can practically touch them."[8] Ironically playing on the year 2001 as something of the past in 1984, the copy presented Trumbull's accomplishment as one of transportation, rearticulating one of the dominant themes of his approach to film effects: "The year was 1968. But for the audience, the year was 2001. And they were not in a movie theatre, they were in deep space—propelled by the stunning special effects of Douglas Trumbull."[9] Apart from plugging the new film experiences of his ride film *New Magic* (1984), the only link between Trumbull and the product in the advertisement is an idea of technological innovation; he is a visionary and this product is the most advanced calculator of its kind. Trumbull was a sign, in and of himself, of the technological future.

His status as a public pedagogue and as an emblematic representative of new technological possibilities shaped the construction and circulation of his ambitious film *Brainstorm*. Most explicitly, the film makes immersivity and technology central thematic and narrative concerns. Echoing Trumbull's highly visible association with ideas on new media, transportation, and film experience, it is a feature entirely devoted to the representation and the experience of extra-dimensional travel. It also addresses the implications of sharing experiences so intense that your own ontological state dissolves temporarily as you experience that of another. The story involves a team of technology innovators developing a helmet that, when worn, can record brain activity with absolute fidelity and intensity. This recorded experience can then be played back to the users of the device. This second or later

8 "In 2001, We Reached the Stars," p. 1.

9 "In 2001, We Reached the Stars," p. 1. This advertisement played a role in a conflict between Kubrick and Trumbull. Kubrick took issue with the impression it gave that Trumbull was solely responsible for the visual effects in *2001*; he published an open letter to that effect and threatened legal action. Trumbull only ever maintained that he was part of a team, and reasonably expected appropriate credit. Hewlett Packard pulled the ad. Bartlett, "Douglas Trumbull, Visual Effects Maestro."

Figure 2.2: Head gear for recording brain activity in *Brainstorm* (Douglas Trumbull, 1983).

viewing effectively transmits the original experience—the sensations, thoughts, moods—to whomever dons the helmet. In this way, *Brainstorm* is an exploration of the dream of a total immersive artwork, perfect and complete, or rather it is an example of the aforementioned dominant discursive formation about immersive media. The plot elements encompass corporate interference in innovation, military appropriation of new technology for the purposes of human manipulation, and romantic complications both eased and amplified by technological means. As Scott Bukatman pointedly described the tale: "*Brainstorm*'s sensory-recording apparatus is co-opted by the military (surprise!), which want to use it for the usual evil purposes. A fantasy of the ultimate effects technology—and its heroic inventors—turns into a nightmare of bottom-line psychic exploitation."[10] While the focus of the film is on a new medium for recording and sharing experiences, it also presents a representation of computer hacking through telephone lines, which would have appeared in 1983 as an advanced technological concern, one shared at the time with another film, *WarGames* (John Badham, 1983).

The film displays a technological obsession with visceral thrills; the characters use the brain device to experience rollercoasters, sex, and death.[11] The plot engine for the second half of the movie involves the head

10 Bukatman, *Matters of Gravity*, p. 239.

11 Bruce Joel Rubin received story credit for *Brainstorm*. A few years later, he would win an Oscar for the screenplay of another tale of love, corporate crime, and the afterlife, *Ghost* (Jerry Zucker, 1990).

Figure 2.3, 2.4, 2.5: Lillian's (Fletcher) death recording.

scientist, Lillian (Louise Fletcher), who records her own death. A more junior member of her team, the movie's lead Mike (Christopher Walken), then fights to access, replay, and experience Lillian's death. When he does, this moment comes as a climactic effects sequence and, of course, becomes an opportunity for the audience to see an audiovisual rendering of death and of the mysteries beyond life, too. The film depicts this death experience as floating through bubbles that contain memories and, ultimately, as being greeted by angel-like figures. Critics mostly agreed that this segment was one of the least satisfying of the film and many saw its imagery as cliché. In response, I ask: were critics so taken by the film's concept of perfect recordable and transportable individual experience that they really expected a report on life after death… only to be let down by what turns out to be a notably Earth-bound special effects sequence? Perhaps a more unsettling conclusion resides here, one that no critic dared to approach: maybe the afterlife is disappointing and a little dull.

Brainstorm's general concept is inspiring and one's imagination is easily sparked by the wild possibilities it presents. Nowadays, it is also a fairly conventional science fiction trope, having been explored in later films, most familiarly in *Strange Days*. Unsurprisingly, the film sticks with a recognizable narrative structure, focusing on a white man's pursuit of universal understanding, his career obsession, and his complicated relationship with his beautiful wife, Karen (Natalie Wood). A radical version might have explored some sort of cross-identity exchange of subjectivity between genders, ethnicities, abilities, or even species. There is a tantalizing moment of such unconventional intersubjectivity when the brain helmet is surreptitiously connected to a chimpanzee by Mike's mischievous lab colleagues while he is wearing a linked helmet. His experience is immediately and painfully supplanted by that of the monkey, represented as a point of view (POV) shot from inside the chimp's cage. This cross-species intersubjectivity is abhorrent and dangerous, and Mike tears his helmet off in an instant, effectively shutting down the experiential transgression. As if to bracket such imaginative uses, this moment of radical intersubjectivity happens within the first minutes of the film, effectively bumping us against the limits of the conceit from the start. Strikingly, the primary narrative focus—Mike's pursuit of the recording of Lillian's death—involves a transmission of experience across genders and ages, but the logic of the film does not allow this to be addressed directly as a cross-gender and inter-age experience, but only as a universal human event.

Trumbull designed the first part of *Brainstorm* to be like an "instructional teaching film" in which the helmet technology and its immersive, transmissive, and recording capabilities would be laid out and explained.[12] To capture the contrasting media forms that constitute *Brainstorm*, the film establishes two distinct representational modes in its opening moments—that of the story and that of the brain recording device. But several instances break this down, making the brain record seep into the representation of the film world. This instability is most pronounced in one jolting shot that

12 Studlar, "Trumbull on Technology," p. 7.

Figure 2.6: POV, chimpanzee experience.

suggests a lasting change in the helmet-wearer's brain structure resulting from their use of the device; following an experience of a taped psychotic episode, Mike is shown shifting to the brain-tape mode even though he is no longer wearing the headgear. Throughout the film, there are hints of such play with representational modes, which make the violation of the distinct depiction of the world of the device and that of the diegesis part of the character's own growing instability. In this way, *Brainstorm* proposes an emerging materiality of representation that puts it in conversation with another release of the same year, David Cronenberg's study of the new flesh of images, *Videodrome* (1983).[13]

The most appealing and deliberately flashy stylistic conceit of the film is found in the sequences that represent the brain recordings. These sequences also incorporate inserts of the characters while wearing the helmet in standard diegetic presentational mode, which provides a clear contrast between the two presentational modes. As figured in the film, the brain recordings are mostly an embodiment of the ostensible camera-operator; it is not the artist nor the fabricator of the work that manufactures the experience but the documentarian, who has recorded their own experiences, including that of death, for another to share. The presentation of the device—in order to pitch its commercial potential—to a corporate board room gives the spectator an opportunity to see a contrasting visual experience; the demo reel presented to the board includes POV shots of a rollercoaster, sweeping vistas of dramatic landscapes, and a speeding truck edited to make it look like the truck is sailing off a cliff and starting to fly. Throughout this scene,

13 Interestingly, when *Silent Running* was first released, critic William Johnson noted the significance of shifting perspective with POV inserts of the anthropomorphic drones in that earlier film. See Johnson, "Silent Running, Douglas Trumbull."

Figure 2.7, 2.8, 2.9: Boardroom demonstration of the brain recording device.

there are shots of the board room members sporting the headgear and responding to what they are "seeing"—which, of course, is what we are seeing, too—their bodies swaying with the camera movement in the demo reel as though they are transported elsewhere. As corporate head Alex (Cliff

Robertson) declares, the technology demonstration showed "the potential for travel and education and news ... You've blown communication as we've known it right out of the water, you know that, don't cha?"

Even if the newness of a technological advance is the focus of the story, the representation of the device draws from cinema conventions; there is editing in the recorded experiences for visceral effect as well as dramatic and sometimes unmotivated shifts to third person and away from the subject's POV, suggesting that the interior mind's eye is rather like film montage. The recordings are stored on what looks like film or tape reels and shelving. The brain-recording tapes are edited with procedures that appear to be very much like analogue film editing. Lingo is also used to make the recordings seem like those of a film production; for instance, when the researchers get an adjustment they wish to keep, they say "slate it" with what looks like a clapperboard. As such, *Brainstorm* figures the technological innovations depicted via abundantly familiar mechanisms and practices, making its futuristic technologies a future of the past.

At the time of *Brainstorm*'s theatrical release, in 1983, the initial reception almost uniformly mentioned the production complications that resulted from the death of one of its stars, Natalie Wood, before shooting had completed. Distributor M.G.M. wanted to shut down the production and claim the insurance money following her tragic death. Trumbull fought this and eventually won the right to complete the film, though the additional financial investment required was substantial. The trade and popular press covered the production situation relatively closely with headlines like "Brainstorm May Never Be Finished," "The Battle to Save the Film," and "Cameras Roll Again on Natalie's Last Film."[14] When released, the story of *Brainstorm* was also the story of Wood, of her having drowned under suspicious circumstances, as well as of the production troubles that followed. Among the re-edits, Trumbull removed potentially insensitive scenes, such as a love scene between Wood and Walken in a canoe and a shot of their son drowning.[15] For some, the production challenges and compromises were evidence of a subpar work. Critic Bernard Drew concluded that the final result was "jumbled, full of missing scenes, confused and a total fiasco ... that should have never been released."[16]

The box office returns of the film echoed that damning comment. The film grossed just over 3 million US dollars in domestic rentals that year,

14 Scott, "*Brainstorm* May Never Be Finished," p. 14; Scott, "The Battle to Save the Film," p. 25; and Pollock, "Cameras Roll Again on Natalie's Last Film," p. 57.

15 Thomas, "Natalie Wood's Final Film to be Released Friday," p. 35.

16 Drew, "'Brainstorm' a Confusing Jumble," p. 35.

Entertainment
In theatres, Argentines cry for their country
GIOVANNI'S
NAC ORCHESTRA
EDUARDO MATA
YO YO MA
BRITANNIA
ANGELO My Love
The Tin Flute
Tomorrow
VERTIGO
Entertainment
Movies make hookers look respectable
Carson gets catty with wife
Dead Zone
The Right Stuff
La Traviata
Merry Christmas, Mr. Lawrence
Zelig
Mr. Mom
Jedi
Tom Cruise All The Right Moves
Brainstorm
Osterman Weekend

Figure 2.10: Movie listing page, *Ottawa Citizen*, October 20, 1983.

which did not cover its production budget.[17] It also had to come up against fierce competition by its fourth week of release, as shown in the October 20, 1983, newspaper movie pages for Ottawa (Ontario, Canada), listings that were typical of North American cities at that moment: two James Bond movies (Irvin Kershner's *Never Say Never Again* and John Glen's *Octopussy*), two Tom Cruise movies (Michael Chapman's *All the Right Moves* and Paul Brickman's *Risky Business*), Richard Marquand's *Return of the Jedi*, which would be the biggest revenue-generating film of the year, as well as new films by Woody Allen (*Zelig*), David Cronenberg (*The Dead Zone*), Peter Greenaway (*The Draughtsman's Contract*), Nagisa Ôshima (*Merry Christmas, Mr. Lawrence*), Franco Zeffirelli (*La Traviata*), and Sam Peckinpah (*The Osterman Weekend*). Lawrence Kasdan's *The Big Chill* was also a growing sleeper hit at the time and Philip Kaufman's *The Right Stuff* was about to open. Most distinctively, in this outrageously crowded field—a slate that any distributor would do anything to avoid when releasing films today—arrived the rerelease of what was at the time a virtually forgotten film, but one that, from that weekend on, would grow in critical stature, *Vertigo* (Alfred Hitchcock, 1958).

17 "Big Rental Films of 1983."

Among this slate of entertainment offerings in October 1983, *Brainstorm* notably presents a direct comment on technological experience. Other such gestures are found in *The Osterman Weekend*, which integrated video surveillance into its tale of conspiracy, and *Zelig*, which presented the appearance of aged film stock to suggest the veracity of its fictional tale. But, *Brainstorm* makes this aspect central to its story and to its circulation.[18] The print advertisements capture this technological focus. One observer, in *Film Comment*, felt that the movie was "hampered by ads that looked as if they were pushing the newest Sony Walkman."[19] The advertisements were in fact echoes of promotional material for other films that claimed to offer a novel spectatorial experience and so tried to picture it, as with Cinerama from the early 1950s. This facet is key to the entire film and to its production context, namely *technological reflexivity in the pursuit of a reimagined motion picture experience.*

A straightforward reflexive reading could seize upon the movie's complicated production, which has Trumbull, the technological innovator, trapped by corporate logic, and fold it into *Brainstorm*, which is itself a story about technological innovators trapped by corporate logic. However, to access the material and discursive conditions that produced *Brainstorm*'s realization of an immersive technological future, there are three layers of reflexivity: 1) the film's own status as a container and a transmitter of experience and historicity; 2) the film's relation to a specific and actual technological innovation; and 3) the technological conditions under which *Brainstorm* ultimately circulated, which shaped the way the film was experienced by movie-goers and structured the two distinct representational modes of brain-recording and diegesis described above. Jay David Bolter and Richard Grusin's work on "remediation," which they define as the "mediation of mediation,"[20] proposes that two logics shape the relation between form and content of new media: "media transparency"—that is, a sense of immediacy— and a heightened presence and appearance of media—which they call "hypermediacy." These logics are linked in such a way that a transparent and ostensibly invisible medium can be so only insomuch as another exaggerates its presence. Or, as they write, "[t]ransparency needs hypermediacy."[21] *Brainstorm*'s three

18 Though not explicitly about a technological experience, *The Dead Zone* establishes two representational modes in a way that is similar to *Brainstorm*. Where the latter has a distinct style for the brain recordings, the former marks the psychic episodes of the lead character, as he transfers into the future or past experiences of a person with whom he has physical contact, with dramatic music and sudden POV camerawork.

19 Kilday, "Ninth Annual Grosses Gloss," p. 64.

20 Bolter and Grusin, *Remediation*, p. 55.

21 Bolter and Grusin, *Remediation*, p. 84.

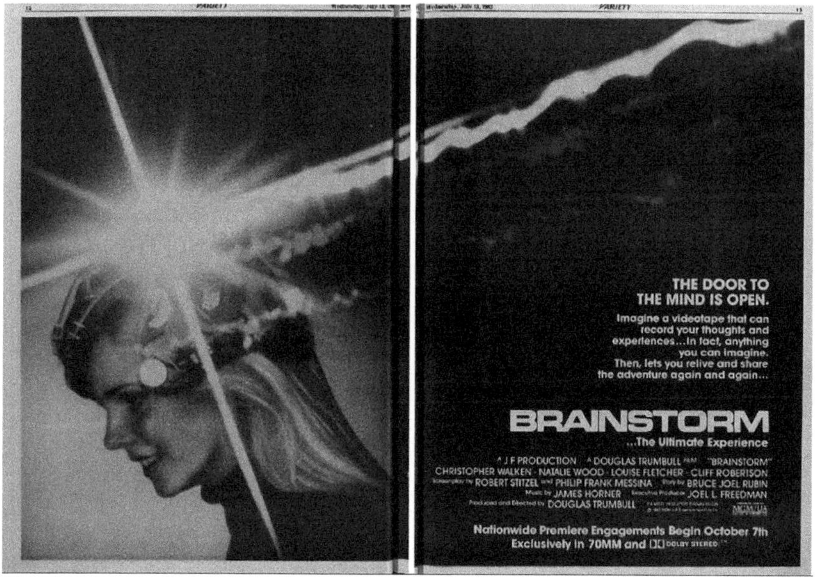

Figure 2.11: *Brainstorm* advertisement, *Variety*, July 13, 1983.

layers of technological reflexivity operate in a comparable fashion, essentially making one media technological assemblage and its presumed immersive experience seem to be a notable advancement by pushing another into the background.

For the first reflexive layer, consider the film's oblique cinematic appeal to sensory transport and intersubjectivity. Though it is presented as a futuristic device in the movie, the virtuality of the recorded and shared experience sounds very much like the way moving images and other media have always worked and been comprehended. A copy of the print advertisements says as much: "The door to the mind is open. Imagine a videotape that can record your thoughts and experiences ... in fact, anything you can imagine. Then, lets you relive and share the adventure again and again."[22] When audiences see *Brainstorm* today, the most immediate parallel that can be made for the recording and playback device as it appears in the film is virtual reality; while precursors existed, Jaron Lanier's popularization of the term and concept was still a couple of years away when the film came out, making such connections anachronistic. At the time, in 1983, the immediate imagistic reference for the device in *Brainstorm* was videotape, a contemporaneous idea about a "future" personalized recording and moving image storage. Near newspaper movie listing pages of the time, including the illustration

22 "*Brainstorm*...The Ultimate Experience," pp. 12–13.

shown above, were pages of advertisements for videocassette recorders (VCRs). The *Brainstorm* ad's characterization—record thoughts and experiences, anything you can imagine, relive them, and share them—could have described film, or essentially any other medium, but resonated specifically in the early 1980s with home video.[23]

The measures or discourses of progress and improvement do not wash away the root impulse to document and circulate expression, whatever the media format. Cultural artifacts are time capsules, freezing the look and the feel of a moment, allowing disparate populations to tap into sensibilities and meanings, later and elsewhere. *Brainstorm*'s casting, for instance, constitutes a fascinating intersection of career paths, one that can be accessed to some degree these many decades later, even though, as critic Janet Maslin put it, the real star of the film is the helmet.[24] Oscar winners Cliff Robertson, as a military-industrial exploiter, and Louise Fletcher were basically at their acting plateaus when the film was released. Fletcher was cast in a role that conventionally would have been awarded to a man. Instead, viewers were offered something rare in American cinema: a smart, tough, female, middle-aged scientist, who was the co-worker, collaborator, and mentor of Christopher Walken's lead. Watching the film today, we could even argue that Sigourney Weaver's performance in *Avatar* (James Cameron, 2009) is a direct descendant of Fletcher's performance as Lillian. Walken's star was continuing to rise—*The Dead Zone*, in which he also starred, was in theatres when *Brainstorm* was released—and Natalie Wood was at the tail-end of her career: she was already a face from the past before filming started. Baked into this filmic artifact, then, is a snapshot of the temporal fluidity and career narratives of these performers.

Wood died in November 1981. Her death was followed by intense press coverage. This final film of hers wasn't released until two years later, essentially transporting the audiences of 1983 back to the sensation of the deceased's performance, a phenomenon that was observed by and advanced in the public discourse of the time. The *Miami Herald*, for instance, called the film a "ghoulish attraction," and some accused M.G.M. of exploitive treatment for the purpose of promotion, including the dedication to the

23 Further to this association with home video, *Brainstorm* contrasted its lackluster box-office with strong home video sales. As *Variety* reported, it "shipped gold, meaning orders topped US$1,500,000 or 37,500 units before the film ever reached the homevideo stores." "New Lease On Life For Pix Via Video," p. 44. As such, it was part of a then relatively recent observation that theatrical flops could be video hits.

24 Maslin, "Screen," p. C17.

actress that appeared in the credits.[25] Almost to a one, articles and reviews addressed the film's representation of a far-out futuristic journey beyond life, then discussed Wood's performance seen two years following her demise, as though this movie was a document that allowed viewers to visit a person who was no longer living, which of course is the primary diegetic concern of *Brainstorm*. Though these two observations ought to be linked—*Brainstorm*'s fictional exploration of a device that captures and shares the experience of dying and the actual way that *Brainstorm* was taken as a moving image vehicle that captured and shared impressions of those who were no longer alive—they were not. This parallel relation asks us to think not only about the sci-fi technology of the movie but also about the fundamental of media ontology: to what extent do media make it possible for us to be with the absent? With its two representational modes, including one that draws attention to itself as a fictional technology of immersion and intersubjectivity, the film distances those qualities from the actual media form being consumed—that is, a theatrical motion picture feature and its similar appeal to experiential expansion. To use the functionalist terms discussed in the opening of this essay, the "improved" fictional immersive experience in the film becomes the measure that produces the "lesser" experience of conventional filmic realism and buries the operation of extant media as container and transmitter of ideas and sensations.

The second layer of technological reflexivity of the film is *Brainstorm*'s status as an historical artifact that sits at a crucial turning point in a parallel endeavor of Trumbull's: the development of a heightened immersive cinematic experience, first called Super 70 in the mid-1970s, then Futurex, and finally Showscan, in 1980. In her work on high frame rate (HFR) films, which includes a synthesis and commentary on Trumbull's format experiments, Julie A. Turnock explains that "HFR can be used as needed to smooth out the motion action sequences, without making slower paced or close-up shots seem strange."[26] Showscan was to be a scalable theatre-specific environment with a screen that occupied the audience's field of vision with specially prepared films shot, and was to be projected, with a HFR of sixty frames per second (FPS) in 70mm format. Doing so would produce intense clarity and sensation of presence.[27] Despite this advantage, Marc Mancini described Showscan as "reminiscent of television" with its look of video, a comment that prefigured the "soap opera look" comments made later about HFR

25 Pollock, "Director Had to Brainstorm to Finish Film," p. 45.
26 Turnock, "Removing the Pane of Glass," p. 43.
27 See Krohn and Lond, "Showscan."

films like *The Hobbit: An Unexpected Journey* (Peter Jackson, 2012) as well as various motion-smoothing settings on contemporary television sets.[28]

Shorts and proof of concept films in Showscan made by Trumbull and his associates reportedly wowed exhibitors and investors. Most ambitiously, *Brainstorm* was to be partially shot and presented in Showscan. Specifically, the sequences that represented the helmet recording device were to be filmed at a high frame rate, so audiences would see the film at the usual twenty-four FPS, then the helmet parts would appear with unusual clarity and dimensionality at sixty FPS. Interestingly, Turnock cites Trumbull saying later that, in the end, an alternation of frame rates might be one of the better uses for an eventual adaptation of HFR, such that only certain sequences in a film would be at a high frame rate and others would be filmed at the standard rate, rather than what Ang Lee did with the entirety of *Billy Lynn's Long Halftime Walk* (2016) or what Peter Jackson did with *The Hobbit*.[29] *Brainstorm* would effectively have been the mainstream calling card and industry-wide publicity for the Showscan process; it would have been what I described elsewhere as a "technological tentpole."[30] In December 1980, *Variety* reported that they were set to shoot in Showscan with M.G.M.'s approval.[31] But, as it turned out, the idea proved too big-budget for all investors interested in *Brainstorm*. Without the exhibition infrastructure needed for such a presentational mode, the Showscan aspect of the enterprise was too risky to pursue.

I began by describing *Brainstorm* as a mini-manifesto for a technological aesthetic. In this, it equally pointed the way beyond the Hollywood blockbuster franchise even as that aesthetic and economy were being fortified. A recognition that competition with home entertainment and the related deterioration of the theatrical environment would necessitate a reimagining of movie-going began to develop in the 1980s, and then happened in earnest in the 1990s with the megaplex theatre.[32] Trumbull was an early voice commenting on the demise of the theatrical experience, saying cinema was, as he put it, "multiplexing itself to death" by not paying attention to

28 Mancini, "Shades of '39," pp. 2–3. Also see Turnock, "Removing the Pane of Glass."

29 Turnock, "Removing the Pane of Glass," p. 43.

30 See Acland, *American Blockbuster*, p. 12. Interestingly, another technological auteur who had been advocating for HFR films, James Cameron, opted for an alternation between frame rates for his blockbuster *Avatar: The Way of Water* (2022), though some critics noted that these decisions appeared to be unmotivated at times. See Kenigsberg, "'Avatar' and the Headache of High Frame Rate Filmmaking."

31 "'Show-Scan' Device Set for MGM Pic," p. 5.

32 For more on multiplexing, see Acland, *Screen Traffic*.

spectatorial conditions, reluctant to innovate, and allowing degradation of the movie theatres.[33] In *Boxoffice* magazine, he argued further that the coming of high-definition television (HDTV)—this is in 1986, many years before that home format had any major consumer impact in Canada and the United States—should push exhibitors to invest in a dramatically different image process and setting and not tinker with minor adjustments.[34] In this context, he promoted Showscan. Though Showscan proved unviable for feature film, Trumbull forged ahead and launched it as a site-specific alternative format. After the multi-year struggle to complete *Brainstorm*, and within weeks of its release, Trumbull publicly declared he didn't want to direct any more studio features.[35] From this point on, he concentrated more intently on films for rides and site-specific installations, some in the Showscan format and later in IMAX when he oversaw that company's rapid expansion in the 1990s.

Technological capability is only ever one component in media history. The great inventor tale is what we see with Walken's and Fletcher's characters in their laboratory; it's Douglas Trumbull, pushing us for decades to rethink and reconsider what we understand of the moving image experience. The next feature in the history of a media form is its roll-out, scalability, and transformation into a workable format. Showscan, closed off from feature filmmaking, began another path as a location-based ride experience. Trumbull launched an alternative ride-film exhibition chain, the ShowBiz Pizza Parlor, with over a hundred sites planned and financial support from Robert Brock of the Brock Hotel Corporation and of the Chuck E. Cheese franchise.[36] *New Magic* and *Big Ball* (Douglas Trumbull, 1984) were twenty-two-minute films that played in those micro-theatre locations. A deal with Plitt Theatres for the creation of thirty Showscan sites over two years followed.[37] These Pizza Parlors may not have had the compactness of the recording-playback apparatus of *Brainstorm*, but they echoed each other's intent to document, magnify, and share overwhelming and immersive moving image sensation.

33 Trumbull, quoted in Bukatman, *Matters of Gravity*, p. 82.

34 Trumbull, "Coming Soon to a Theatre Near You," p. SW6.

35 Trumbull, quoted in Farber, "Brainstorming," p. 77. For an analysis of one such work, Trumbull's IMAX short *Back to the Future: The Ride* (1991), placed in the context of a wider industrial shift, see Ndalianis and Balanzategui, "'Being Inside the Movie'."

36 On this, see "Trumbull's Special Effects to Unspool via Pizza Parlors," p. 14; Krohn and Lond, "Showscan," p. 10; and Wharton, "Showscan Scores with DC Exhibs," p. 4 and p. 37.

37 See "Plitt Theaters, Trumbull to Form Showscan Corp. for 70m 'Environment'"; and "Plitt Theatres and Trumbull Form Showscan Film Corp." Brock, at the time, backed away from his ownership. See "Brock Sells Off Its 50% of Showscan to Plitt."

Let me return to the newspaper movie listings from October 1983. As much as it is a record of the current cinema in theatrical release at that moment in time, it is also a document of built spaces and technological facilities and, in this way, it is an artifact of infrastructure; it represents a moment in the mutable history of the relations among producers, distributors, and exhibitors; it outlines a stage or platform for an encounter between audiences, their hometown, and internationally circulating moving image culture. It is a concrete example of these relations and of the way they are situated among other appeals and offerings, for example the rising tide of VCR sales. When Trumbull worked to build an alternative format, venue, and experience for Showscan with ShowBiz Pizza Parlor, this was the media configuration, the material and discursive infrastructure to which he was responding, an infrastructure against which technological innovation would be noted and to which it would be applied.

The first layer of technological reflexivity involves the root element of mediated experience that is the subject of *Brainstorm*, which is also offered more generally by film, video, and media and which was given a sharpened focus with the paratextual commentary on Natalie Wood's death. The second layer is about the instantiation of technological innovation; this film, which is about a new format for hyper-fidelity in mediated experience (the helmet apparatus), additionally involves paratextual attention to the development of a new format for hyper-fidelity in a mediated experience (Showscan). An additional third layer involves the actual presentational format of the completed work of *Brainstorm*. When it was clear that a HFR Showscan version of the film would not work to signal the shift between the helmet-recorded sequences and the diegetic story, some other distinctive mode of representational and experiential aesthetics had to be settled upon. In the end, the alternation between the film world and the brain recordings is signaled with a shift from 35mm to 70mm, from 1.66:1 to 2.21:1 screen ratio, and from mono sound to Dolby stereo sound. The theatrical release of the film presented the diegetic real life of the drama in a more conventional audiovisual format, and then switched to the more expansive 70mm widescreen stereo for the helmeted sequences. For 35mm prints and for the eventual DVD release, the ratios were 1.85:1 and 2.35:1.[38] This means that for the 70mm release, not only was the image not as elongated as subsequent versions, but the effect of switching aspect ratios was more

38 See Howson, "Douglas Trumbull," p. 114.

Figure 2.12, 2.13: Approximation of the alternation between POV brain recording in 70mm and third person diegetic world in 35mm.

dramatic; ultimately, one rarely has an opportunity to see the film today as it was originally released in 1983.[39]

This particular material component overdetermines *Brainstorm*'s iteration of immersive media experience. The *Brainstorm* thesis, if you will, is that the representational real is less overwhelming for the senses than the proxy recorded experience. The diegetic technology of the helmet marks the experiential immediacy of shared brain activity. That record, that shared memory and that stored experience, are represented in *Brainstorm* by technological spectacularization and sensory expansion. The entire

39 Screenings today take place at highly specialized venues. I was able to take advantage of one such opportunity in August 2022 at the Museum of the Moving Image, in Queens, New York, where *Brainstorm* appeared in a series called "See it Big: 70mm!" As the first scheduled screening occurred shortly after Trumbull's passing, it became a commemorative event, with some of his family members in attendance.

situation—everything you are watching—is, of course, a staged dramatic record: the film audience isn't really putting on and taking off headgear as the characters are. The curious element here is that everything in and around the movie points to media technology; it is the characters' focus, it is the obsession of corporate and military agents in the movie, it is part of the film's publicity, and it is the most extravagant element in the work's construction. More than this, we are supposed to share the characters' experience of the device, which itself is strategically referenced throughout the film; we see images of the characters experiencing immersive sequences that we, the audience, see, and we are told, addressed directly by the characters, to consider the experience that is about to happen or that has just unfurled. The film proposes that conventional realist narrative form is outside of the story world; it is a transparent representational mode in mono and on 35mm. The only way helmet tech can appear and be distinct from that transparent mode is by featuring or pointing to a more pronounced appearance of cinema technology, situating the large and the loud with the representational fidelity of the brain recordings. The headgear is equated with the technology of an immersive (or, to be precise, a "more" immersive) cinematic spectacle, captured on 70mm widescreen and in Dolby stereo sound. The quotidian—the romantic and corporate intrigue—of the movie is the realist technological invisibility of 35mm gauge and mono sound.

This sign of future media as the experience of immersivity—70mm and stereo sound—is the material heart of the film as initially released. The year before, in 1982, exhibitors had unusual successes with 70mm releases; there were apparently more 70mm prints in circulation than ever before. *Boxoffice* referred to 1983 as the "70mm summer."[40] An estimated 425 sites in Canada and in the US had been equipped to show 70mm, returning exhibitors close to the peak adaptation in the late 1960s.[41] The conversion—remember, exhibitors could not be counted on to convert to Showscan, but they did invest in 70mm facilities—was to handle a record number of prints, about 300, surpassing the record 282 from the previous year. There were nine "summer" titles in 70mm: *Return of the Jedi*, *Krull* (Peter Yates, 1983), *Blue Thunder* (John Badham, 1983), *The Keep* (Michael Mann, 1983), *WarGames*, *Superman III* (Richard Lester, 1983), *The Star Chamber* (Peter Hyams, 1983), *Spacehunter: Adventures in the Forbidden Zone* (Lamont Johnson, 1983), which was in 3D, and *Brainstorm*, in late September. This growth in the

40 "Studios, Exhibitors Prepare for 70mm Summer," p. 37.

41 See "Studios, Exhibitors Prepare for 70mm Summer."

70mm infrastructure continued over the next couple of years. One source even says that *Brainstorm* opened in 170 theatres in 70mm, which means that roughly 40% of all American and Canadian 70mm-equipped sites at that moment were showing the film.[42]

Brainstorm accentuated its technological particularity with ticket pricing that was greater than the usual admission price. A premium theatrical experience needed a premium price to match. The experience of the new media format in *Brainstorm*, the fictional near-future experience recorder, relied upon an existing and at the time rejuvenated infrastructure of 70mm. This is the moment when the imagination hit the material; it collided with the actuality of existing economic and social conditions, that is with the material and discursive conditions that made one media situation appear more completely immersive than another. Think of it this way: every new media form and innovation is a chronicle of existing, residual, and persisting ideas about materials for media, technology, and social relations, and its dialectic relation with other organizations of media materials, practices, and ideas moulds how the new form is understood and experienced. Turnock makes a related observation, pointing out how FPS is currently an importation from the earlier celluloid format into the digital, which reinforces existing ideas about the measures of technological change and improvement.[43] In *Brainstorm*, the representation of the new relies upon other media figured as videotape, film reels, and physical splicing of tape for image and sound editing. The brain recording episodes, designed to construct an immersive experience of heightened individualized POV, still lapse into third person, editing, and montage, again returning to a cinematic language shared with the "real" in the film story. For example, the final image that appears as the credits begin to roll places our reunited couple in a memory bubble—for eternity, presumably—as though in the brain recording device. These violations of the premise of the fictional brain recording device notwithstanding, the *alteration between formats*—from 35mm to 70mm, from a smaller to a wider aspect ratio, from mono to stereo sound, and from exclusive third-person observational storytelling to mostly POV—provides a heightened awareness of the mechanics of the media form as well as of its reliance upon a residual historical mode of figuring media. This intersection of material and discursive formations, and the layers of technological reflexivity that elaborates

42 Kilday, "Ninth Annual Grosses Gloss," p. 64.

43 See Turnock, "Removing the Pane of Glass."

different aspects of greater and lesser immersive experience, produce the conditions under which *Brainstorm* can represent an idea of immersive media. *The media experience, then, is that of becoming immersed rather than immersivity itself.*

Acknowledgment

I thank the organizers of "Immersivity and Technological Innovation" in October 2020: Philippe Bédard, Alanna Thain, and Carl Therrien. Douglas Trumbull participated in that conference and gave a keynote address, and I was honored that he attended my presentation of this research. I also thank Quinn Valencourt for his research assistance.

Bibliography

Acland, Charles R. *Screen Traffic: Movies, Multiplexes, and Global Culture.* Durham, NC: Duke University Press, 2003.

Acland, Charles R. *American Blockbuster: Movies, Technology and Wonder.* Durham, NC: Duke University Press, 2020.

Bartlett, Rhett. "Douglas Trumbull, Visual Effects Maestro on '2001: A Space Odyssey', Dies at 79." *The Hollywood Reporter,* February 8, 2022. https://www.hollywoodreporter.com/movies/movie-news/douglas-trumbull-dead-2001-space-odyssey-1235089009/.

"*Big Ball* and *New Magic.*" *Springfield News-Leader,* March 17, 1984: 48.

"Big Rental Films of 1983." *Variety,* January 11, 1984: 13 and 80.

Bolter, Jay David, and Richard Grusin. *Remediation: Understanding New Media.* Cambridge, MA: MIT Press, 1998.

"*Brainstorm*...The Ultimate Experience," *Variety,* July 13, 1983: 12 and 13.

"*Brainstorm*...The Ultimate Experience," *Variety,* October 19, 1983: 10 and 11.

"Brock Sells Off Its 50% of Showscan to Plitt." *Variety,* May 29, 1985: 7 and 27.

Bukatman, Scott. *Matters of Gravity: Special Effects and Supermen in the 20th Century.* Durham, NC: Duke University Press, 2003.

Drew, Bernard. "'Brainstorm' a Confusing Jumble." *Salinas Californian,* September 29, 1983: 35.

Farber, Stephen. "Brainstorming." *Film Comment* 19, no. 5 (September–October, 1983): 76–82.

Gregory, Lon B. "Creativity, Involvement Keynote New Look, Name at IFPA Conference." *Business Screen* 29, no. 7 (November, 1968): 45, 46, 48.

Howson, Simon. "Douglas Trumbull: The Quest for Immersion." *Metro Magazine* 169 (2011): 112–114.

"In 2001, We Reached the Stars." *IEEE Spectrum*, August, 1984: 1.

Johnson, William. "Silent Running, Douglas Trumbull." *Film Quarterly* 25, no. 4 (1972): 52-56.

Kenigsberg, Ben. "'Avatar' and the Headache of High Frame Rate Filmmaking." *New York Times*, January 2, 2023. https://www.nytimes.com/2023/01/02/movies/avatar-the-way-of-water-high-frame-rate.html.

Kilday, Gregg. "Ninth Annual Grosses Gloss." *Film Comment*, March–April, 1984: 62–66.

Krohn, Bill, and Harley W. Lond. "Showscan: A New Type of Exhibition for a Revolutionary Film Process." *Boxoffice*, February, 1984: 10–11.

Mancini, Marc. "Shades of '39." *Film Comment* 21, no. 5 (September–October, 1985): 2–4.

Maslin, Janet. "Screen. 'Brainstorm,' Discovery Goes Awry," *New York Times*, September 30, 1983: C17.

Ndalianis, Angela. *Neo-Baroque Aesthetics and Contemporary Entertainment*. Cambridge, MA: MIT Press, 2005.

Ndalianis, Angela, and Jessica Balanzategui. "'Being Inside the Movie.' 1990s Theme Park Ride Films and Immersive Film Experiences." *The Velvet Light Trap* 84 (September 2019): 18–33.

"New Lease on Life for Pix Via Video." *Variety*, September 5, 1984: 44.

"*New Magic*." *Springfield News-Leader*, February 27, 1984: 50.

Pierson, Michele. *Special Effects: Still in Search of Wonder*. New York, NY: Columbia University Press, 2002.

"Plitt Theaters, Trumbull to Form Showscan Corp. for 70m 'Environment'." *Variety*, December 1984: 7 and 31.

"Plitt Theatres and Trumbull Form Showscan Film Corp." *Boxoffice*, March 1985: 104.

Pollock, Dale. "Cameras Roll Again on Natalie's Last Film." *Los Angeles Times*, February 15, 1982: 57.

Pollock, Dale. "Director Had to Brainstorm to Finish Film." *Miami Herald*, September 24, 1983: 45.

Prince, Stephen. *Digital Visual Effects in Cinema: The Seduction of Reality*. New Brunswick, NJ: Rutgers University Press, 2012.

Rogers, Ariel. "'You Don't So Much Watch It as Download It': Conceptualizations of Digital Spectatorship." *Film History: An International Journal* 24, no. 2 (2012): 221–234.

Scott, Vernon. "The Battle to Save the Film." *Windsor Star*, October 21, 1983: 25.

Scott, Vernon. "*Brainstorm* May Never Be Finished." *Telegraph-Forum*, March 26, 1982: 14.

"'Show-Scan' Device Set for MGM Pic." *Variety*, December 17, 1980: 5.
"Studios, Exhibitors Prepare for 70mm Summer." *Boxoffice*, June 1983: 37.
Studlar, Gaylyn. "Trumbull on Technology." *Spectator* 3, no. 1 (Fall 1983): 7.
Thomas, Bob. "Natalie Wood's Final Film to be Released Friday." *Salinas Californian*, September 29, 1983: 35.
Trumbull, Douglas. "Coming Soon to a Theatre Near You." *Boxoffice*, March 1986: SW6.
"Trumbull's Special Effects to Unspool Via Pizza Parlors." *Variety*, September 21, 1983: 14.
Turnock, Julie A. "Before Industrial Light and Magic: The Independent Hollywood Special Effects Business, 1968–75." *New Review of Film and Television Studies* 7, no. 2 (2009): 133–156.
Turnock, Julie A. "Removing the Pane of Glass: *The Hobbit*, 3D High Frame Rate Filmmaking, and the Rhetoric of Digital Convergence." *Film Criticism* 37, no. 3 (2013): 30–59.
Turnock, Julie A. *Plastic Reality: Special Effects, Technology, and the Emergence of 1970s Blockbuster Aesthetics*. New York, NY: Columbia University Press, 2015.
Wharton, Dennis. "Showscan Scores with DC Exhibs." *Variety*, April 4, 1984: 4 and 37.

About the Author

Charles R. Acland is Distinguished University Research Professor in the Department of Communication Studies, Concordia University, Montreal. He edited the collection *Residual Media* (University of Minnesota Press, 2007) and he is author of *American Blockbuster: Movies, Technology, and Wonder* (Duke University Press, 2020).

3. Dual Cognitive Approach

Eugene Kukshinov

Abstract: This book chapter explores the concept of immersion and its relationship with technology, narratives, and human cognition. Immersion has historically been associated with cognitive absorption in various narrative forms, but the emergence of virtual reality (VR) shifted its focus towards technological experiences. However, technology alone does not guarantee immersion, as individual experiences can vary. The chapter introduces a dual cognitive approach to immersion, differentiating between sensory and mental immersions, particularly through the manifestations of presence and narrative engagement. The chapter examines the nature, contributing factors, and potential effects of these immersive states, and explores their interaction in media such as video games. This cognitive framework aids in addressing issues in presence studies, immersive storytelling, VR research, and video game analysis.

Key words: immersion, presence, narrative engagement, cognition, perception, imagination

Introduction

Historically, researchers applied the concept of immersion to describe some sort of cognitive *absorption* or *blending*, which was long before virtual reality (VR) and contemporary media technologies appeared.[1] They applied immersion to mental engagement with paintings, music, literature, movies, or other forms of narratives.[2] However, the development of VR technologies changed the whole concept of immersion. In fact, in the 1990s, only VR

1 See Collins, *The Act of Musical Composition*; Green et al., "Transportation Across Media"; and Martínez, "Storyworld Possible Selves and the Phenomenon of Narrative Immersion."
2 See Grau, *Virtual Art*; and Green et al., "Transportation Across Media."

Bédard, Philippe, Alanna Thain, and Carl Therrien, eds. *States of Immersion Across Media: Bodies, Techniques, Practices*. Taylor & Francis Group, 2025
DOI 10.5117/9789048558766_CH03

technologies earned the label "immersive" while some vehicle simulators did not, even though today they would certainly qualify. Eventually, immersion became a more widespread term, but it still often refers to technology, or rather to the technological quality of simulation media such as VR.[3]

At the same time, technology is not necessarily a definitive factor in immersive experiences, because qualities such as a bigger screen or a higher fidelity of representation "might make it easier for users to focus and to keep their attention on the representation, but this does not necessarily mean that users will feel more present in the environment portrayed."[4] In other words, technology can be designed to be immersive, but it does not guarantee the production of immersion. Different people might have unique experiences with the same technology, or even more than that, the same technology might both evoke and disrupt immersion.[5]

Therefore, it makes sense to distinguish "immersion" as a psychological process from "*immersiveness*" as "the medium's ability to afford the psychological process of immersion."[6] Technology may be designed to be more immersive (i.e., to simulate as many sensorial cues or representations of some aspect of reality as possible to make the experience indistinguishable from this part of reality). It can even have as a primary goal to be immersive, but immersion per se is not an inevitable outcome of it. Just putting on a head-mounted display (HMD) is not equal to immersion.[7]

Nevertheless, immersion, as a psychological state, may also be a product of interaction with narratives, but not in the way that VR scholars understand it. Immersion into narratives is a different psychological process compared to immersion in the simulation that VR or similar technologies afford. This chapter offers a dual, human-centred cognitive approach to immersion that aims to differentiate "sensory" and "mental" immersions or, more specifically, their major manifestations: presence (often defined as the purpose of VR and related technologies) and narrative engagement. William R. Sherman and Alan B. Craig (2002) were among the first to draw this conceptual line, using the terms "mental" immersion and "sensory/physical" immersion when referring to different types of immersions.[8] They noted that physical

3 See Cummings and Bailenson, "How Immersive Is Enough?"

4 Calleja, *In-Game*, p. 33.

5 See Evans, "Where Is the Virtual Self?"

6 Riecke and Schulte-Pelkum, "An Integrative Approach to Presence and Self-Motion Perception Research," p. 205.

7 See Shen et al., "Effects of Level of Immersion on Virtual Training Transfer of Bimanual Assembly Tasks."

8 Sherman and Craig, *Understanding Virtual Reality*, p. 9.

immersion is neither necessary nor desired when reading books.[9] There are other dualistic approaches to immersion, many of which differentiate immersion in video games as psychological and sensory/perceptual, diegetic and situated, sensory/challenge-based and imaginative, or technological and phenomenological immersions, among others.[10]

In conceptualizing immersions, it is important to understand the differences in the experiences obtained through these psychological states, rather than the specific terms used to name them. The distinction between sensory and mental immersions is preferred because it a) incorporates a human-centred approach rather than a technocratic one; b) clearly illustrates the nature of the illusions/experiences on which these types of immersions rely; and c) overlaps with other terms used in the field. Specifically, this differentiation also aligns with the major manifestations of immersion in media, namely presence as a perceptual illusion of non-mediation and narrative engagement as a form of mental and emotional connection with stories and their characters.[11]

In this chapter, I will first review the dual approach to cognition that helps to distinguish the two sides of immersive experiences by showing how presence and narrative engagement can develop. Then, I will describe the differences between these two immersive states in terms of their nature, contributing factors, and potential effects. Finally, I will examine how presence and narrative engagement can interact with each other in media that naturally afford both immersions, such as video games. This dual cognitive framework of immersion can help to address several issues in the presence studies, such as the "book problem,"[12] as well as guide immersive storytelling that aims to produce both sensory and mental immersion. It can also help in VR studies by addressing the technocratic approach, and in video game research by considering the issues in studying its effects.

Dual Approach to Cognition: Perception and Imagination

Perception and imagination are interconnected cognitive processes. Mental imagery relies on real visual input from a scene and on object recognition

9 Sherman and Craig, *Understanding Virtual Reality*, p. 384.

10 See Carr, "Space, Navigation, and Affect"; Ermi and Mäyrä, "Player-Centred Game Design"; Therrien, "Immersion"; and Veale, " 'Interactive Cinema' Is an Oxymoron, but May Not Always Be."

11 See Lombard and Ditton, "At the Heart of It All."

12 See Gysbers et al., "Exploring the Book Problem"; and Schubert and Crusius, "Five Theses on the Book Problem."

mental subsystems to function, but mental images are also necessary for visual recognition to support vision, which was a primary function of imagination from an evolutionary perspective.[13] In other words, imagination relies on perception and/or memory in terms of prior experiences and representations as necessary elements of the relational imagining process.[14] Additionally, imagination itself assists perception in terms of cognitive and meaning-making processes.

While these processes intersect, perception and imagination are functionally distinct. Their difference can be well described by the possibilistic model of consciousness, which represents the quantum view of cognition and suggests that perception and imagination mutually define what is real and what is not.[15] Reality is always in a state of possibility for us, as we only ever observe it partially. As an example, Kieron P. O'Connor and Frederick Aardema (2005) put forward the idea that even if we may not see the corner of a chair, but the absence of this view does not deter our belief in the chair's solidity.[16]

In this model, "perception operates outwards, and deals with what is there, while imagination is inward-looking and deals with what is not there," and we do not question fully the existence of what is not there.[17] As O'Connor and Aardema (2012) stated: "[I]magination is unbounded in its possibilities in that it does not depend on feedback from reality or physicality to be established. [...] [P]erception, however, depends crucially on the boundaries and laws of probability."[18] In other words, perception deals with what is there or with whatever we can perceive, and imagination reproduces what is not there, or *possibilities* as relationships between memories, representations, and/or ideas. Thus, what we imagine are possibilities, not images or words.[19]

As a result, mental immersion into narratives that rely on imagination inevitably stimulates possibilities that can lead, for instance, to anticipation that drives emotions and greater imagining. Possibilities in imagination are

13 See Barkowsky, "Modeling Mental Spatial Knowledge Processing."

14 See Craver-Lemley and Arterberry, "Imagery-Induced Interference on a Visual Detection Task"; and Thornton and Mitchell, "Consistent Neural Activity Patterns Represent Personally Familiar People."

15 See O'Connor and Aardema, "The Imagination"; and O'Connor and Aardema, "Living in a Bubble."

16 O'Connor and Aardema, "The Imagination", p. 242.

17 O'Connor, Aardema, and Pélissier, *Beyond Reasonable Doubt*, p. 69.

18 O'Connor and Aardema, "Living in a Bubble," p. 233.

19 See Morton, *Emotion and Imagination*.

not bounded by the temporal limits of narrative consumption, meaning that some side of narrative engagement can proceed long after we have watched a movie or read a book.[20] It is different for presence, because it depends on the exact perceptual moment and sensorial input, which usually should be realistic, i.e., it should represent or rather simulate some aspect of reality. Thus, for presence to appear, the perceived should be credible to be lived-in, but for mental immersion it is the opposite:

> In the imagination, possibilities do not need to be arrived at by a linear transport system since rather they need to be lived-in before becoming credible. Once lived-in, a coherent context can be inferred later. In fact, one could say in imagination, multiple contexts are grouped together "out of context" exactly in order to arrange a personally liveable reality. The person needs to be emotionally connected in order for imaginative possibilities to feel real.[21]

As a result, stimulating mood and imagination, narratives can lead to absorption and immersion in a lived-in reality, in a mentally simulated one. As well as that, it is not a semantic formality to call presence the *perceptual* illusion of non-mediation. It implies certain mechanisms and processes that media users engage in to achieve it. Books are often more immersive than movies because they include more possibilities for the imagination to simulate in the viewers' mind, but VR requires more aspects of reality to be reproduced to be deemed "immersive." At the same time, VR is not the only technology that affords presence.

In the (para)social contexts, the difference between immersions and corresponding cognitive processes is also evident. Social presence results from a mediated communication with a real person, a perceptually involved process that, to some extent, requires imagination to complete the "picture" of the partner, for instance, in a phone conversation.[22] In imagined parasocial interactions (whether with characters or real people), perception assists imagination in simulating the interaction, mostly in the mind. Overall, immersion is an umbrella concept that encompasses several psychological states that rely on different aspects of cognition to gain genuine experiences from media interactions.

20 See Martínez, "Storyworld Possible Selves and the Phenomenon of Narrative Immersion."
21 O'Connor and Aardema, "Living in a Bubble," pp. 233–234.
22 See Kukshinov, "Visual Familiarity as a Factor of Social Presence Represented Via Non-Verbal Communication in Audio-Mediated Channel."

Different Immersive States: Presence and Narrative Engagement

Presence and narrative engagement are immersive psychological states that represent at least two ways in which identities and experiences from media interaction can be blurred or blended.[23] For the purposes of explicating the current analytical framework, the sense of presence will be viewed as the primary form of sensory immersion, and narrative engagement will represent mental immersion (into narratives).

There are certain media that typically afford either presence or narrative engagement. Using Gonzalo Frasca's distinction, I would define them as simulation and narrative media.[24] Simulation media, such as VR and social media, can afford presence by recreating activities, environments, and/or social entities supported by external sensory cues. Narrative-based (and more representational) media, on the other hand, rely on imagination for mental immersion to occur.

Presence is a psychological state or a perceptual illusion of non-mediation and non-simulation, which means that when we feel presence while interacting with media, we do not recognize the technology's role in our experiences on some level.[25] Perception is a primary cognitive process for presence, meaning that its intensity may increase with the number of simulated aspects of reality (such as fidelity and details), and it depends crucially on the level of certainty that simulated environments provide. However, flaws in the simulation can break presence and stimulate imagination.

One of the core features of presence that highlights its opposition to mental immersive processes is that it is based on active involvement with simulated processes. As Luciano Floridi stated, presence includes "the capacity of acting and interacting (the agent can be the receiver of an action and respond to it accordingly) with the environment, both pragmatically (by doing or changing things) and epistemically (for example, by observing things locally)."[26] However, observing a location from a first-person perspective is not the same as watching a movie and imagining the possibility of a location, in which we do not act as it is imagined.

Narrative engagement is not a form or part of presence, but rather a form of immersion.[27] It is a mental immersion into narratives, such as those found in

23 See Martínez, "Storyworld Possible Selves and the Phenomenon of Narrative Immersion"; and Snodgrass et al., "A Formal Anthropological View of Motivation Models of Problematic MMO Play."

24 See Frasca, "Simulation Versus Narrative."

25 See Lee, "Presence, Explicated."

26 Floridi, "The Philosophy of Presence," p. 662.

27 See Martínez, "Storyworld Possible Selves and the Phenomenon of Narrative Immersion."

books, movies, and musical compositions. Its intensity, in contrast to sensory immersion, depends heavily on the framework of possibilities provided by the narrative for the imagination to simulate (e.g., through suspense, mysteries, etc.). Narrative engagement is a broader phenomenon that can include transportation, identification, and other (parasocial) relationships with characters as key elements of narratives.[28]

The following subsections describe the major differences between the sense of presence and narrative engagement. They highlight how these psychological states can appear, what factors contribute to them, what effects they may or may not produce, and how they are cognitively processed. More specifically, narrative transportation is sometimes referred to as the state of absence to contrast it with the state of presence.

Differences in Focus: Absence vs. Presence

While presence is a result of a perception of some externally simulated environment that we all potentially can observe in the same manner within a shared physical environment, the world(s) we imagine in our heads while reading a book, for instance, is or are not the same as any other person would imagine.[29] While mental immersion into narratives (such as movies, book/novels or music) is a product of mental simulations, "presence is about the present, the here and now."[30] Being a product of internal (mental) simulation, mental immersion can cause narrative consumers to lose track of time while being transported.[31] As a result, the transportation side of narrative engagement is frequently measured through observing latencies in reaction or attention to the content (i.e., the plot) of the narratives.[32] In that sense, mentally immersive state (in terms of transportation) in contrast to presence is more likely to be described as "absence."

In other words, mental and sensory immersive states represent different types of attentional focus, external for presence, and internal for "absence" as an expression of narrative engagement. While "presence and absence can

28 See Busselle and Bilandzic, "Fictionality and Perceived Realism in Experiencing Stories"; Cohen, "Parasocial Break-Up from Favorite Television Characters"; and Giles, "Parasocial Relationships."

29 See Waterworth et al., "Presence."

30 Waterworth et al., "Presence," p. 37.

31 See Green et al., "Transportation across Media."

32 See Bezdek and Gerrig, "When Narrative Transportation Narrows Attention"; and Green and Brock, "The Role of Transportation in the Persuasiveness of Public Narratives."

both be seen as absorption states, the former [is] based around the current perceptual flow, the latter around imagined events and situations."[33] Thus, mental immersion can include a complete state of absorption in an internal world detached from the external perceptual flow. While presence can coexist with the perceptual flow (if it is relevant) and an outside world, it is impossible to feel present in the world around us while being lost in a fantasy.[34]

It is crucial to acknowledge that presence is bounded by the current perceptual flow. Immersion into narratives can appear and evolve to some extent beyond the narrative consumption. For instance, it is possible to revisit former parasocial relationships with characters from a favorite TV show or book.

The Origin Issue, or the "Book Problem"

It is true that there is a spectrum of immersive experiences that media can provide. The boundaries are blurred as well because media are convergent themselves (e.g., video games or news), and because imagination and perception are inseparable, they coexist with each other. That is why it might be difficult sometimes to exclude media with rich sensorial output from association with sensory immersive experiences, as with movies or TV shows.[35] Nevertheless, narratives are generally designed to mentally stimulate consumers, while some of them are more stimulating and immersive than others are. In the end, narratives are not designed to induce presence.

There is an issue in the presence scholarship called "the book problem." It is defined as a "problem" because presence scholars do understand that books and text-based narratives can be immersive, but these media do not fit into the usual technological affordances (such as those of VR), levels of realism and sensory output that are supposed to be "immersive" in terms of presence or sensory immersive state. In fact, scholars sometimes even define computer or television screens as low-immersive technologies. But, depending on the context, the usual screen (e.g., the screen that simulates a flat surface) can provide an absolute state of sensory immersion or a perfect perceptual illusion.[36]

33 Waterworth et al., "Presence," p. 37.
34 See Waterworth and Waterworth, "Focus, Locus, and Sensus."
35 See Bracken, "Presence and Image Quality"; and Kim and Biocca, "Telepresence via Television."
36 See Horovitz, "LG End-of-the-World Video Prank Mean but Effective."

Some scholars define book reading as low-immersive, or even totally non-immersive media.[37] It is "true" if one assumes that presence can describe each experience of interacting with media. But textual and/or oral narratives are potentially one of the most immersive media, in mental terms. Reading can be more immersive than watching a movie, in terms of mental immersion, or narrative transportation for which it was designed. As Melanie C. Green, Sheryl Kass, Jana Carrey, Benjamin Herzig, Ryan Feeney, and John Sabini stated:

> The "canned" imagery provided by a video presentation could reduce transportation because it limits the imaginative investment that the viewer is able to provide. A written version allows a reader to participate more fully in creating a mental image of the story. [...] A print presentation may also provide more direct insight into the thoughts and emotions of the characters. An author can directly report a character's thoughts, whereas a film director must convey the same information through behavior.[38]

So, a book includes more possibilities for the imagination to be simulated, although it does not mean that every book (or every movie, and any other media) is necessarily immersive. It depends on multiple factors (such as relevance or emotional connection) that, in most cases, involve imagination as a mediator.

Sometimes, "mirror neurons" are used to explain some physical sensations that may appear during movie watching.[39] Due to mirror neurons, when we observe someone in a movie, we can recognize the actions of others, we can understand the emotions of other.[40] We may even mimic some of their emotions, but we do not literally feel ourselves *being* the character. Instead, we more often *relate* to the situation, to the character, to their feelings (if we care enough, of course, because we do not care about every character we see), and thus sometimes feel something (like being afraid) *for* the character. Even "identification," one of the most common forms of parasocial relationships, is not about *being* the character: it is about adopting a character's motivation, feelings, and so forth. In fact, identification is more often about our ideal Self interacting with characters rather than about an actual Self.

37 See Gysbers et al., "Exploring the Book Problem"; and Martingano et al., "The Limited Benefits of Using Virtual Reality 360° Videos to Promote Empathy and Charitable Giving."

38 Green et al., "Transportation Across Media," pp. 517–518.

39 See Grodal and Kramer, "Empathy, Film, and the Brain."

40 See Iacoboni, "Imitation, Empathy, and Mirror Neurons"; and Kohler et al., "Hearing Sounds, Understanding Actions."

Finally, technologies that generate perceptual sensations, such as VR, can be distracting in terms of linear narrative flow and concentration on the storyline and, as a result, its users can fail to recall plot details.[41] This is why emotional content (as in movies) is better suited to screen displays than to VR.[42] We do not act in a movie, novel, or comic; instead, we relate to what we observe. Any realistic movie, abstract cartoon, or music composition includes a story to which we can potentially relate, and as a result of consuming this meaningful media content, we can formulate authentic emotions and relationships. Presence is a product of media *context* or simulation (such as a chat room, a video call, or a virtual environment [VE]) in which we can *act/participate* but traditional media are representational, while simulation is not. Narratives are sequences of signs but simulations that afford presence are systems of signs that reproduce some behavior.[43]

Differences in Factors: What Contributes to Immersion

The feeling of presence is heavily reliant on the realism of the perceptual output, particularly in terms of image quality, screen size, visual feedback, spatialized sound, update rate, and anything else that allows technology to more completely simulate and represent most aspects of reality (or its cues) in a perceptually coherent manner.[44] On the other hand, narrative engagement/transportation does not necessarily require realistic (in terms of details and of social realism) content to stimulate the imagination and, therefore, immersion. In fact, more abstract or fictional content can also stimulate the imagination.[45] This is because the incompleteness of fictionality provides fewer references to real-life experiences that can be made and therefore more possibilities to imagine them.[46] Cartoons, for example, can be highly

41 See Pressgrove and Bowman, "From Immersion to Intention?"; and Szita, Gander, and Wallsten, "The Effects of Cinematic Virtual Reality on Viewing Experience and the Recollection of Narrative Elements."

42 See Baños et al., "Immersion and Emotion."

43 See Frasca, "Simulation Versus Narrative."

44 See Bracken, "Presence and Image Quality"; Cummings and Bailenson, "How Immersive Is Enough?"; Hou et al., "Effects of Screen Size, Viewing Angle, and Players' Immersion Tendencies on Game Experience"; Kobayashi, Ueno, and Ise, "The Effects of Spatialized Sounds on the Sense of Presence in Auditory Virtual Environments"; and Welch et al., "The Effects of Pictorial Realism, Delay of Visual Feedback, and Observer Interactivity on the Subjective Sense of Presence."

45 See Goldstein, "The Pleasure of Unadulterated Sadness"; and Wolf, "Abstraction in the Video Game."

46 See Juul, *Half-Real*.

immersive if one can relate to their story or characters. Additionally, mental immersion does not require large screens because the mental simulation does not take place on a screen.[47]

In fact, it is the "not there" that often stimulates mental immersion via anticipation of the possible worlds. Anticipation embraces the possibilities that narrative can provide within mental simulations that constitute "an indispensable preview of the character and potential consequences of yet-to-be-experienced events."[48] In contrast to presence, which is based on our perception of external stimuli, mental immersion is about what we are about to imagine and is not necessarily tied to what we are currently perceiving.

Narratives are designed to reveal their structure in the imagination, absorbing us so that we do not actually think about them, but rather live or feel them through. Thus, narrative consumers are invited to feel; they need to be emotionally connected to narratives for imaginative possibilities to feel real.[49] Narratives should also be relevant to them in order to achieve this emotional connection.[50] Therefore, mental immersion is about the emotional connection and imagined relationships between the Self, the characters, their actions, and/or the events. Narrative engagement essentially constitutes itself within the relationship between emotions and imagination, because imagination expresses and responds to emotions, while narratives appeal to, stimulate, change, and maintain them.[51] As a result, emotional regulation, attachment, and/or arousal are not only major predictors of mental immersion into narratives, but also major effects of it.

Differences in Effects: Senses vs. Emotions

Differences in effects produced by presence and narrative engagement can be seen as differences between internally appearing *emotions* and *senses* as cues that map and connect to the external world.[52] Sensorial reproduction of the realistic experiences is the main (or maybe even the only) purpose of presence as a psychological state. In other words, via presence, individuals can obtain (intentionally or not) "real," or non-simulated, experiences from

47 See Appel and Mengelkamp, "Watching Videos on a Smartphone."
48 Miles et al., "Not So Close Encounters of the Third Kind," p. 558.
49 See Gorton, *Media Audiences*.
50 See Bae, Lee, and EunGyuhl Bae, "Emotional Engagement with the Plot and Characters."
51 See Mar et al., "Emotion and Narrative Fiction"; and Morton, *Emotion and Imagination*.
52 See Rouby, Fournel, and Bensafi, "The Role of the Senses in Emotion."

the simulated contexts, for example, when replicating a combat situation (or other training situations) or social circumstances of a meeting via various perceptual cues. More specifically, presence is crucial for virtual reality exposure therapy (VRET).[53] It is a popular and a widespread method of treating a range of phobias, including social phobia, arachnophobia, acrophobia, agoraphobia, aviophobia, and also Post-Traumatic Stress Disorder (PTSD). It can also be used for stress recovery: by simulating nature and its natural sounds that, via presence, can cause calming effects.[54] These treatments are potentially effective because they are experienced as if non-simulated by the patience involved in those activities.

The nature of the narrative engagement impact can be best described by the narrative structure that predetermines the emotional effect.[55] Characters are fundamental semiotic elements of the narrative: they act according to their motives while changing the plot, which is important because the character and the action are "built up" into the narrative development.[56] Thus, being transported into the narrative means relating to the characters in some way.[57] Within parasocial relationships, audiences usually make moral judgements about certain characters. Predictably, we enjoy these moments when we support the protagonists' actions, as well as when we dislike the antagonists', or rather the characters' positions with which we identified.[58] In contrast to presence, which relies on our exact experiences and self-references, mental immersion and parasocial relationships are often based on the possible selves that our self-concept includes.

Another difference in effects produced by presence and narrative engagement can be demonstrated by the challenges of VR and of immersive technologies to produce empathy, at least in the way it is currently done. As a meta-analysis of recent studies showed (2022), VR affords emotional reaction towards the perceived, but not the feeling or understanding of

53 See Parsons and Rizzo, "Affective Outcomes of Virtual Reality Exposure Therapy for Anxiety and Specific Phobias."

54 See Annerstedt et al., "Inducing Physiological Stress Recovery with Sounds of Nature in a Virtual Reality Forest."

55 See Porter et al., "Re(de)fining Narrative Events Examining Television Narrative Structure"; and Wirth and Schramm, "Media and Emotions."

56 See Eder, Jannidis, and Schneider, *Characters in Fictional Worlds*; and Martin, *Recent Theories of Narrative*.

57 See Busselle and Bilandzic, "Fictionality and Perceived Realism in Experiencing Stories."

58 See Raney, "Media Enjoyment as a Function of Affective Dispositions Toward and Moral Judgment of Characters."

others' feelings.[59] It happens for various reasons: first, being positioned in a first-person perspective, or 1PP, for a short period of time is not enough to experience the struggles of other people for yourself (which is also an ethical issue); second, 1PP is not equal to identification, which is a mental relationship with others.

Finally, because mental immersion relies on imagination to simulate experiences, it does not necessarily depend on the actual sensorial moment of media use. Thus, one of the definitive features of TV show narratives is the continuity of the storyline: being transported in a TV series usually means that we care about characters, about their lives and their problems week after week, season after season.[60] At the same time, "long-running story arcs may deliver emotional cues along the way, but only a faithful viewer will experience the full emotional impact of a well-crafted story arc."[61] Therefore, only very strong emotional ties usually encourage viewers to anticipate each upcoming episode of a certain TV show, i.e., not each story element (when present) can transport one into it. Overall, there are multiple differences that exist between mental and sensory immersive states. A few of them can be seen in Table 1.

Table 1. Key differences between sensory and mental immersions

	Sensory immersion (as presence)	Mental immersion (via narratives)
Origin	Perceptual illusion of the external mediation/simulation (i.e., of the *shared* environment)	Mental simulation in imagination (i.e., in an *individual* mind)
Attention/ focus	External	Internal ("absence")
Main factors	Certainty: realism, details, fidelity, etc.	Uncertainty and individual relevance
Effects	Sensorial experiences of non-mediated activities	Emotion regulation, parasocial relationships, etc.
Source	Simulation media	Narrative media
Media richness	For mediated communication: video → audio → text (as the least immersive)	Textual narrative can be (or canbe not)the most immersive
Temporality	Happens at the (perceptual) moment of media interaction	Can last after the media interaction

59 See Martingano et al., "The Limited Benefits of Using Virtual Reality 360° Videos to Promote Empathy and Charitable Giving."

60 See Porter et al., "Re(de)fining Narrative Events Examining Television Narrative Structure."

61 Gorton, *Media Audiences*, p. 80.

Dual Immersion: The Case of Video Games

Video games are a great example of a medium that demonstrates that there are (at least) two sides to immersive experiences, because they relate to both narrative and virtual simulation, meaning that they incorporate virtually represented narratives.[62] In other words, video games are narratives that you must participate in to experience them.[63] Therefore, they contain the potential to promote both types of immersion, i.e., presence by participating in a simulation and narrative engagement. Laura Ermi and Frans Mäyrä also distinguished "challenge-based" immersion, which is closely related to presence (as activity-based feeling) and, in part, to the state of flow.[64]

Many video games scholars suggest that mental and sensory immersions are essentially combined during the play. That means, for instance, that being a character in a video game implies "physically enacting a fantasy role and receiving visceral feedback that corroborates the fantasy."[65] Overall, it seems like role-playing is one of the major ways to psychologically combine both immersive experiences. Jamie Banks basically distinguished various relationships of players with their digital representations and characters within the games' narratives.[66] For instance, "Other"-relationship incorporates mental immersion and relationship with the character who acts by themselves; "Me"-relationship is more associated with avatar and a sense of being present despite the mediation process. Finally, "symbiotic" relationship, as a combination of the sensory and mental immersive experiences, involves players using their avatars to interact with the characters, to enact an ideal Self or alternate personas (e.g., brave, strong, happy, social, or confident), or to have the character as a partner in play.

Another reason why mental and sensory immersion can coexist in the video games context is that while players *participate* in the narrative, storylines in video games are designed/coded to follow players (not vice versa). In game design, it is also a widespread practice to encourage *curiosity* and to produce *suspense* as states of uncertainty that stimulate imagination. These states are tied to the "decisions" that players make, which helps to tie

62 See Frasca, "Simulation Versus Narrative"; and Grodal and Kramer, "Empathy, Film, and the Brain."

63 See Domsch, *Storyplaying*.

64 See Ermi and Mäyrä, "Player-Centred Game Design."

65 Isbister, *How Games Move Us*, p. 102.

66 See Banks, "Object, Me, Symbiote, Other."

sensorial involvement with storytelling.[67] Nevertheless, video games still often require cutscenes that offer a linear sequence to deepen the story's immersion because it is not entirely possible for these different cognitive states to always be in a symbiotic relationship.[68]

In fact, there is an experiential spectrum of play possibilities because some video games are more narrative-based while others are more simulation-based, thus stimulating one side of immersion more than the other. Specifically, there is an experiential difference between video games with more complex worlds (with themes and characters), story sequence, and third-person perspective (3PP) versus realistic simulations games with first-person perspective (1PP).[69] More specifically, 1PP stimulates spatial presence, while 3PP is more suitable for observing the way the story develops.[70]

Quite often, those psychological states can contradict each other. For instance, Bimal Balakrishnan and S. Shyam Sundar suggested that narrative transportation negatively affects possibilities for action, vection, and reality judgement that pertain to presence.[71] In a similar manner, Despain claimed that some players can be annoyed if they feel storytelling is getting in their way.[72]

Mental immersion and attention to the story development might be interrupted by unsolicited sensorial input. It is nonetheless important to acknowledge that video games are designed to combine those feelings, which cannot be said about traditional linear narratives, because they only support mental immersion, thus the experience can be interrupted. It may happen due to the manipulative affordance of video games, because "the physical interactions with the interface work as a constant call-back to the real world, which inhibits users from becoming deeply immersed in a drama [...] that requires the development of a story line in users' mind."[73]

Even though mental and sensory immersive states can sometimes coexist, it is important to distinguish them, because they are distinct phenomena

67 See O'Connor and Aardema, "Living in a Bubble"; and Roth, Vorderer, and Klimmt, "The Motivational Appeal of Interactive Storytelling."

68 See Vosmeer and Schouten, "Interactive Cinema."

69 See Grodal and Kramer, "Empathy, Film, and the Brain."

70 See Havranek et al., "Perspective and Agency During Video Gaming Influences Spatial Presence Experience and Brain Activation Patterns"; and Tamborini and Skalski, "The Role of Presence in the Experience of Electronic Games."

71 Balakrishnan and Sundar, "Where Am I?"

72 See Despain, *Writing for Video Game Genres.*

73 Oh, Chung, and Han, "The More Control, the Better?," p. 87.

represented by unique and standalone scholarships that can be integrated in different ways, as in studies of parasocial interaction and of social presence.[74] It is also important to understand the distinctions between these different types of immersive experiences in order to better understand the range of effects that media interactions can have on us.

Discussion and Conclusion

Immersion is not a feature of technology; it is a psychological state of blurring/blending of the actual and mediated experiences. There are at least two forms of media immersions that differ in terms of how they appear, how they evolve and what effects they may produce. It is crucial to acknowledge those differences, especially when media is designed to produce them both (as in video games) or when media producers or scholars merge them into a single instance.

Most of the time, sensory and mental immersions are afforded by different media. Sensory immersion is often afforded by technologies that can provide realistic and high-quality perceptual output, such as virtual reality or large screen displays. These media are designed to simulate real-world experiences as closely as possible and can produce a feeling of presence in the simulated environment. Mental immersion, on the other hand, is more closely tied to the content of the media and can be achieved through narratives that stimulate the imagination and engage the emotions. Even though TV has more perceptual output than a book, or the same story that is verbally narrated, it is still designed as a narrative, meaning that it is designed to follow the story mentally and emotionally. A book can be more mentally immersive than a movie because it allows for more possibilities for the imagination to simulate the story. The *mind*, rather than external stimuli, in this case, is the immersive "technology." We can still be immersed in and enjoy a movie with a small screen or poor quality, especially if we care about the story, characters, or plot. Mysteries, suspense, and sudden endings can stimulate the imagination and take advantage of its ability to consider multiple possibilities.

While some media can be easily distinguished as either simulation(al) (e.g., a chatroom, a VE, etc.) or representational (or narrative-based), it is not always that clear for others. For instance, a flat screen and an HMD are just technologies through which it is possible to access content. Through

74 See Kim and Song, "Celebrity's Self-Disclosure on Twitter and Parasocial Relationships."

both, it is possible to watch both a linear usual video and a 360-degree one, even though one type of video is more appropriate for one of the media technologies, and vice versa.

Nevertheless, applying a narrative to a 360-degree video observed via a VR headset can still be confusing for users because the technology designed for sensory immersion is accessing a narrative designed to stimulate mental immersion, and simply does not allow users to follow the storyline.[75] That is why it makes sense that "immersive" storytelling today is based on "non-immersive" or *extractive* content constituted via information, as in news or nonfiction.[76] From a different perspective, the flat screen is a more appropriate medium for consuming (linear) narratives as it allows easier transportation and emotional connection. A 360-degree video of some space (accessed via a VR headset) is not an interactive experience in a strict sense but is a simulation of a user standing and being able to observe the location while possibly feeling a physical presence in it. A flat screen with some video can also be a pure simulation but in very specific circumstances, as in the case of a simulation of a catastrophe seen from a screen masked as a "window." In this case, both content and medium are designed to simulate a real situation, not to tell a story.

In the end, this dual approach is, in part, an analytical oversimplification that attempts to distinguish immersive experiences that are often viewed from a single point of view. For instance, many presence researchers ignore the "book problem" and apply their frameworks to studying narratives. But, if we do not distinguish between different types of immersion and the experiences they relate to, it becomes difficult to accurately study the effects of these experiences. For example, immersive technologies such as VR are not designed to produce empathy. Immersive technologies can vividly inform about a situation or a location, but they do not directly promote emotional relationships by design.

It is a fundamentally different experience to relate to something that we have observed in our mind versus to actually do it ourselves. This dual human-centred approach can help us to look differently at the video games play process and at the potential effects of play on players, because in such circumstances they are dynamically immersed in one of the possible ways (or even in both). This may cause a unique combination of effects on the players' self-concept, attitude, and/or behavior. Specifically, there are different

75 See Pressgrove and Bowman, "From Immersion to Intention?"

76 See Green et al., "'You Wouldn't Get that from Watching TV!'"; and Humphreys, von Pape, and Karnowski, "Evolving Mobile Media."

relationships that players have with avatars as their digital representation, and with characters as elements of the narrative, which can intersect in video games even if they remain distinct entities. Therefore, we identify with characters and embody avatars. There is no identification with the avatar, because the avatar is us. It may help us to feel a presence in the virtual world, but it is not a part of the story with which we formulate an emotional connection. We do not adopt the avatars' motivations or desires, it's the characters that have them. They represent two sides of the video games that produce different immersive states.

Bibliography

Annerstedt, Matilda, Peter Jönsson, Mattias Wallergård, Gerd Johansson, Björn Karlson, Patrik Grahn, Åse Marie Hansen, and Peter Währborg. "Inducing Physiological Stress Recovery with Sounds of Nature in a Virtual Reality Forest: Results from a Pilot Study." *Physiology & Behavior* 118 (2013): 240–250.

Appel, Markus, and Christoph Mengelkamp. "Watching Videos on a Smartphone: Do Small Screens Impair Narrative Transportation?" *Media Psychology* 25, no. 5 (2022): 653–674.

Bae, Hyuhn-Suhck, Doohwang Lee, and Rosie EunGyuhl Bae. "Emotional Engagement with the Plot and Characters: A Narrative Film on Hearing-Impaired Sexual Assault Victims." *Narrative Inquiry* 24, no. 2 (2014): 309–327.

Balakrishnan, Bimal, and S. Shyam Sundar. "Where Am I? How Can I Get There? Impact of Navigability and Narrative Transportation on Spatial Presence." *Human-Computer Interaction* 26, no. 3 (2011): 161–204.

Banks, Jaime. "Object, Me, Symbiote, Other: A Social Typology of Player-Avatar Relationships." *First Monday* 20, no. 2 (February 4, 2015). http://dx.doi.org/10.5210/fm.v20i2.5433.

Baños, Rosa María, Cristina Botella, Mariano Alcañiz, Víctor Liaño, Belén Guerrero, and Beatriz Rey. "Immersion and Emotion: Their Impact on the Sense of Presence." *Cyberpsychology & Behavior* 7, no. 6 (2004): 734–741.

Barkowsky, Thomas. "Modeling Mental Spatial Knowledge Processing." In *Spatial Processing in Navigation, Imagery and Perception*, edited by Fred Mast and Lutz Jäncke, 67–84. Boston, MA: Springer, 2007.

Bezdek, Matthew A., and Richard J. Gerrig. "When Narrative Transportation Narrows Attention: Changes in Attentional Focus During Suspenseful Film Viewing." *Media Psychology* 20, no. 1 (2017): 60–89.

Bracken, Cheryl Campanella. "Presence and Image Quality: The Case of High-Definition Television." *Media Psychology* 7, no. 2 (2005): 191–205.

Busselle, Rick, and Helena Bilandzic. "Fictionality and Perceived Realism in Experiencing Stories: A Model of Narrative Comprehension and Engagement." *Communication Theory* 18, no. 2 (2008): 255–280.

Calleja, Gordon. *In-Game: From Immersion to Incorporation*. Cambridge, MA: MIT Press, 2011.

Carr, Diane. "Space, Navigation and Affect." In *Computer Games: Text, Narrative and Play*, edited by Diane Carr, David Buckingham, Andrew Burn, and Gareth Scott, 59–71. Cambridge, MA: Polity, 2006.

Cohen, Jonathan. "Parasocial Break-Up from Favorite Television Characters: The Role of Attachment Styles and Relationship Intensity." *Journal of Social and Personal Relationships* 21, no. 2 (2004): 187–202.

Collins, Dave. The Act of Musical Composition: Studies in the Creative Process. London: Routledge, 2016.

Craver-Lemley, Catherine, and Martha E. Arterberry. "Imagery-Induced Interference on a Visual Detection Task." *Spatial Vision*, no. 2 (2001): 101–119.

Cummings, James J., and Jeremy N. Bailenson. "How Immersive Is Enough? A Meta-Analysis of the Effect of Immersive Technology on User Presence." *Media Psychology* 19, no. 2 (2016): 272–309.

Despain, Wendy. *Writing for Video Game Genres: From FPS to RPG*. Boca Raton, FL: CRC Press, 2009.

Domsch, Sebastian. *Storyplaying: Agency and Narrative in Video Games*. Berlin: De Gruyter, 2013.

Eder, Jens, Fotis Jannidis, and Ralf Schneider. *Characters in Fictional Worlds: Understanding Imaginary Beings in Literature, Film, and Other Media*. Berlin: De Gruyter, 2010.

Ermi, Laura, and Frans Mäyrä. "Player-Centred Game Design: Experiences in Using Scenario Study to Inform Mobile Game Design." *Game Studies* 5, no. 1 (October, 2005). https://www.gamestudies.org/0501/ermi_mayra/.

Evans, Simon. "Where is the Virtual Self? Virtual Worlds and the Self as a Cyborg." *Annual Review of CyberTherapy and Telemedicine* 14 (Summer, 2016): 51–57.

Floridi, Luciano. "The Philosophy of Presence: From Epistemic Failure to Successful Observation." *Presence* 14, no. 6 (2005): 656–667.

Frasca, Gonzalo. "Simulation Versus Narrative. Introduction to Ludology." *The Video Game Theory Reader*, edited by Mark J. P. Wolf and Bernard Perron, 221–235. London: Routledge, 2003.

Giles, David C. "Parasocial Relationships." In *Characters in Fictional Worlds: Understanding Imaginary Beings in Literature, Film, and Other Media*, edited by Jens Eder, Fotis Jannidis, and Ralf Schneider, 442–459. Berlin: De Gruyter, 2010.

Goldstein, Thalia R. "The Pleasure of Unadulterated Sadness: Experiencing Sorrow in Fiction, Nonfiction, and 'in Person'." *Psychology of Aesthetics, Creativity, and the Arts* 3, no. 4 (2009): 232–237.

Gorton, Kristyn. *Media Audiences. Television, Meaning and Emotion*. Edinburgh: Edinburgh University Press, 2009.

Grau, Oliver. *Virtual Art:From Illusion to Immersion*. Cambridge, MA: MIT Press, 2003.

Green, David Philip, Mandy Rose, Chris Bevan, Harry Farmer, Kirsten Cater, and Danaë Stanton Fraser. "'You Wouldn't Get that from Watching TV!' Exploring Audience Responses to Virtual Reality Non-Fiction in the Home." *Convergence* 27, no. 3 (2021): 805–829.

Green, Melanie C., and Timothy C. Brock. "The Role of Transportation in the Persuasiveness of Public Narratives." *Journal of Personality and Social Psychology* 79, no. 5 (2000): 701–721.

Green, Melanie C., Sheryl Kass, Jana Carrey, Benjamin Herzig, Ryan Feeney, and John Sabini. "Transportation across Media: Repeated Exposure to Print and Film." *Media Psychology* 11, no. 4 (2008): 512–539.

Grodal, Torben, and Mette Kramer. "Empathy, Film, and the Brain." *Recherches sémiotiques / Semiotic Inquiry* 30, no. 1–3 (2010): 19–35.

Gysbers, Andre, Christoph Klimmt, Tilo Hartmann, Andreas Nosper, and Peter Vorderer. "Exploring the Book Problem: Text Design, Mental Representations of Space, and Spatial Presence in Readers." In *Proceedings of the Seventh International Workshop on Presence*, 13–20. Valencia: Technical University of Valencia, 2004.

Havranek, Michael, Nicolas Langer, Marcus Cheetham, and Lutz Jäncke. "Perspective and Agency during Video Gaming Influences Spatial Presence Experience and Brain Activation Patterns." *Behavioral and Brain Functions* 8, no. 34 (July 19, 2012). https://doi.org/10.1186/1744-9081-8-34.

Horovitz, Bruce. "LG End-of-the-World Video Prank Mean but Effective." *USA Today*, September 5, 2013. https://www.usatoday.com/story/money/business/2013/09/05/lg-electronics-end-of-world-prankvertisement-viral-ad-social-media/2771335/.

Hou, Jinghui, Yujung Nam, Wei Peng, and Kwan Min Lee. "Effects of Screen Size, Viewing Angle, and Players' Immersion Tendencies on Game Experience." *Computers in Human Behavior* 28, no. 2 (2012): 617–623.

Humphreys, Lee, Thilo von Pape, and Veronika Karnowski. "Evolving Mobile Media. Uses and Conceptualizations of the Mobile Internet." *Journal of Computer-Mediated Communication* 18, no. 4 (2013): 491–507.

Iacoboni, Marco. "Imitation, Empathy, and Mirror Neurons." *Annual Review of Psychology* 60 (2009): 653–670.

Isbister, Katherine. *How Games Move Us: Emotion by Design*. Cambridge, MA: MIT Press, 2016.

Juul, Jesper. *Half-Real: Video Games between Real Rules and Fictional Worlds.* Cambridge, MA: MIT Press, 2011.

Kim, Jihyun, and Hayeon Song. "Celebrity's Self-Disclosure on Twitter and Parasocial Relationships: A Mediating Role of Social Presence." *Computers in Human Behavior* 62 (2016): 570–577.

Kim, Taeyong, and Frank Biocca. "Telepresence via Television: Two Dimensions of Telepresence May Have Different Connections to Memory and Persuasion." *Journal of Computer-Mediated Communication* 3, no. 2 (1997): JCMC325.

Kobayashi, Maori, Kanako Ueno, and Shiro Ise. "The Effects of Spatialized Sounds on the Sense of Presence in Auditory Virtual Environments: A Psychological and Physiological Study." *Presence: Teleoperators and Virtual Environments* 24, no. 2 (2015): 163–174.

Kohler, Evelyne, Christian Keysers, M. Alessandra Umilta, Leonardo Fogassi, Vittorio Gallese, and Giacomo Rizzolatti. "Hearing Sounds, Understanding Actions: Action Representation in Mirror Neurons." *Science* 297, no. 5582 (2002): 846–848.

Kukshinov, Eugene. "Visual Familiarity as a Factor of Social Presence Represented via Non-Verbal Communication in Audio-Mediated Channel." PhD diss., Temple University, 2021.

Lee, Kwan Min. "Presence, Explicated." *Communication Theory* 14, no. 1 (2004): 27–50.

Lombard, Matthew, and Theresa Ditton. "At the Heart of It All: The Concept of Presence." *Journal of Computer-Mediated Communication* 3, no. 2 (September, 1997). https://doi.org/10.1111/j.1083-6101.1997.tb00072.x.

Mar, Raymond A., Keith Oatley, Maja Djikic, and Justin Mullin. "Emotion and Narrative Fiction: Interactive Influences Before, During, and After Reading." *Cognition & Emotion* 25, no. 5 (2011): 818–833.

Martin, Wallace. *Recent Theories of Narrative.* Ithaca, NY: Cornell University Press, 1986.

Martínez, M. Angeles. "Storyworld Possible Selves and the Phenomenon of Narrative Immersion: Testing a New Theoretical Construct." *Narrative* 22, no. 1 (2014): 110–131.

Martingano, Alison Jane, Sara Konrath, Adam Brown, and Evan Henrtize. "The Limited Benefits of Using Virtual Reality 360° Videos to Promote Empathy and Charitable Giving." *Nonprofit and Voluntary Sector Quarterly* (October 5, 2022). https://doi.org/10.1177/08997640221125804.

Miles, Lynden K., Brittany M. Christian, Nomita Masilamani, Luiza Volpi, and C. Neil Macrae. "Not So Close Encounters of the Third Kind: Visual Perspective and Imagined Social Interaction." *Social Psychological and Personality Science* 5, no. 5 (2014): 558–565.

Morton, Adam. *Emotion and Imagination.* Cambridge, MA: Polity, 2013.

O'Connor, Kieron, Frederick Aardema, and Marie-Claude Pélissier. *Beyond Reasonable Doubt: Reasoning Processes in Obsessive-Compulsive Disorder and Related Disorders*. Hoboken, NJ: John Wiley & Sons, 2005.

O'Connor, Kieron P., and Frederick Aardema. "The Imagination: Cognitive, Pre-Cognitive, and Meta-Cognitive Aspects." *Consciousness and Cognition* 14, no. 2 (2005): 233–256.

O'Connor, Kieron P., and Frederick Aardema. "Living in a Bubble: Dissociation, Relational Consciousness, and Obsessive Compulsive Disorder." *Journal of Consciousness Studies* 19, no. 7–8 (2012): 216–246.

Oh, Jeeyun, Mun-Young Chung, and Sangyong Han. "The More Control, the Better? The Effects of User Control on Movie Trailer Immersion and Enjoyment." *Journal of Media Psychology. Theories, Methods, and Applications* 26, no. 2 (2014): 81–91.

Parsons, Thomas D., and Albert A. Rizzo. "Affective Outcomes of Virtual Reality Exposure Therapy for Anxiety and Specific Phobias: A Meta-Analysis." *Journal of Behavior Therapy and Experimental Psychiatry* 39, no. 3 (2008): 250–261.

Porter, Michael J., Deborah L. Larson, Allison Harthcock, and Kelly Berg Nellis. "Re(de)fining Narrative Events Examining Television Narrative Structure." *Journal of Popular Film and Television* 30, no. 1 (2002): 23–30.

Pressgrove, Geah, and Nicholas David Bowman. "From Immersion to Intention? Exploring Advances in Prosocial Storytelling." *Journal of Philanthropy and Marketing* 26, no. 2 (2021): e1689.

Raney, Arthur A. "Media Enjoyment as a Function of Affective Dispositions toward and Moral Judgment of Characters." In *The Routledge Handbook of Emotions and Mass Media*, edited by Katrin Döveling, Christian von Scheve, and Elly A. Konijn, 166–178. New York, NY: Routledge, 2010.

Riecke, Bernhard E., and Jörg Schulte-Pelkum. "An Integrative Approach to Presence and Self-Motion Perception Research." In *Immersed in Media: Telepresence Theory, Mesurement & Technology*, edited by Matthew Lombard, Frank Biocca, Jonathan Freeman, Wijnand IJsselsteijn, and Rachel J. Schaevitz, 187–235. Cham: Springer, 2015.

Roth, Christian, Peter Vorderer, and Christoph Klimmt. "The Motivational Appeal of Interactive Storytelling: Towards a Dimensional Model of the User Experience." In *Interactive Storytelling. Second Joint International Conference on Interactive Digital Storytelling. ICIDS 2009*. Guimarães: Lecture Notes in Computer Science, December, 2009.

Rouby, Catherine, Arnaud Fournel, and Moustafa Bensafi. "The Role of the Senses in Emotion." In *Emotion Measurement*, edited by Herbert L. Meiselman, 65–81. Sawston: Woodhead Publishing, 2016.

Schubert, Thomas, and Jan Crusius. "Five Theses on the Book Problem: Presence in Books, Film and VR." In *PRESENCE 2002: Proceedings of the Fifth International Workshop on Presence*, 53–59. Porto: Universidad Fernando Pessoa, 2002.

Shen, Songjia, Hsiang-Ting Chen, William Raffe, and Tuck Wah Leong. "Effects of Level of Immersion on Virtual Training Transfer of Bimanual Assembly Tasks." *Frontiers in Virtual Reality* 2 (May 20, 2021). https://doi.org/10.3389/frvir.2021.597487.

Sherman, William R., and Alan B. Craig. *Understanding Virtual Reality: Interface, Application, and Design*. San Francisco, CA: Morgan Kaufmann, 2002.

Snodgrass, Jeffrey G., H. J. François Dengah, Michael G. Lacy, and Jesse Fagan. "A Formal Anthropological View of Motivation Models of Problematic MMO Play: Achievement, Social, and Immersion Factors in the Context of Culture." *Transcultural Psychiatry* 50, no. 2 (2013): 235–262.

Szita, Kata, Pierre Gander, and David Wallsten. "The Effects of Cinematic Virtual Reality on Viewing Experience and the Recollection of Narrative Elements." *PRESENCE: Virtual and Augmented Reality* 27, no. 4 (2018): 410–425.

Tamborini, Ron, and Paul Skalski. "The Role of Presence in the Experience of Electronic Games." In *Playing Video Games: Motives, Responses, and Consequences*, edited by Peter Vorderer and Jennings Bryant, 225–240. New York, NY: Routledge, 2006.

Therrien, Carl. "Immersion." In *The Routledge Companion to Video Game Studies*, edited by Mark J. P. Wolf and Bernard Perron, 451–484. New York, NY: Routledge Taylor & Francis Group, 2014.

Thornton, Mark A., and Jason P. Mitchell. "Consistent Neural Activity Patterns Represent Personally Familiar People." *Journal of Cognitive Neuroscience* 29, no. 9 (2017): 1583–1594.

Veale, Kevin. "'Interactive Cinema' Is an Oxymoron, but May Not Always Be." *Game Studies* 12, no. 1 (2012). https://gamestudies.org/1201/articles/veale.

Vosmeer, Mirjam, and Ben Schouten. "Interactive Cinema : Engagement and Interaction." In *International Conference on Interactive Digital Storytelling*. Cham: Springer, 2014.

Waterworth, Eva L., and John A. Waterworth. "Focus, Locus, and Sensus: The Three Dimensions of Virtual Experience." *CyberPsychology & Behavior* 4, no. 2 (2001): 203–213.

Waterworth, John A., Eva L. Waterworth, Giuseppe Riva, and Fabrizia Mantovani. "Presence: Form, Content and Consciousness." In *Immersed in Media: Telepresence Theory, Measurements & Technology*, edited by Matthew Lombard, Frank Biocca, Jonathan Freeman, Wijnand IJsselsteijn, and Rachel J. Schaevitz, 35–58. Cham: Springer, 2015.

Welch, Robert B., Theodore T. Blackmon, Andrew Liu, Barbara A. Mellers, and Lawrence W. Stark. "The Effects of Pictorial Realism, Delay of Visual Feedback, and Observer Interactivity on the Subjective Sense of Presence." *Presence: Teleoperators & Virtual Environments* 5, no. 3 (1996): 263–273.

Wirth, Werner, Tilo Hartmann, Saskia Böcking, Peter Vorderer, Christoph Klimmt, Holger Schramm, Timo Saari, Jari Laarni, Niklas Ravaja, Feliz Ribeiro Gouveia, Frank Biocca, Ana Sacau, Lutz Jäncke, Thomas Baumgartner, and Petra Jäncke. "A Process Model of the Formation of Spatial Presence Experiences." *Media Psychology* 9, no. 3 (2007): 493–525.

Wolf, Mark J. P. "Abstraction in the Video Game." In *The Video Game Theory Reader*, edited by J. P. Mark Wolf and Bernard Perron, 69–88. London: Routledge, 2013.

About the Author

Eugene Kukshinov is a post-doctoral researcher at the HCI Games Group at the University of Waterloo, who focuses on media psychological research. The key area of his expertise involves conceptualization of presence and explication of other immersive states. He has done research on VR, video games, Zoom, and other media.

4. Emotional and Immersive Experiences in Films and Games: Looking into the Perceptual Model and Striving for a Motivational Theory

Maxime Deslongchamps-Gagnon

Abstract: To what extent does immersion contribute to the emotional experience of fictional works? Could it be that emotions are essential ingredients to immersive experiences? How should we distinguish these two states of mind? Many scholars have modelled the viewer's experience of emotion after perception and have connected it with immersion, suggesting that fictional works are similar to illusions. Game studies have mostly turned to motivational theories of emotion and immersion, drawing attention to how both states require investment of the self. This chapter examines both the perceptual model and the motivational account with the intent of laying an appropriate groundwork for understanding the conceptual connection between emotion and immersion in the context of engaging with fictional works.

Keywords: emotion, investment, video game, fiction, fear, guilt

One common assumption about emotions is that they are indicative of a lack of perspective, whether they are caused by works of fiction or not. Emotions are said to denote biases and partiality, not critical distance and considerations. When we experience them, we see the world from a very specific angle, without being able to account for what may be relevant to our understanding of the situation. In art or media analysis classes, we have learned to set our emotions aside, so they do not interfere with the cold, calculating, and all-encompassing judgment that is privileged instead. This is also what expressions such as "love is blind" or "blind rage" suggests. The

Bédard, Philippe, Alanna Thain, and Carl Therrien, eds. *States of Immersion Across Media: Bodies, Techniques, Practices*. Taylor & Francis Group, 2025
DOI 10.5117/9789048558766_CH04

emotional person is not perceiving the world as it truly is. Hence, following this line of thought, emotions are sources of misguidance of which we should be wary.

This widespread assumption—notably in the history of Western philosophy—is also tied to another common idea about how emotions operate when consuming fictional works.[1] Why do we care about characters and feel invested in their fate? One popular explanation is that we lose awareness, for a moment, of the fictional status of the represented events. It is as if we were *there* with the characters, thanks to a spell that makes fiction and reality blur together. We may call this spell "immersion" or "presence," which is also indicative of a lack of perspective. When we are immersed, it feels like we are transported from our seats to another world. We are engaged with the work of fiction to such an extent that we forget about our surroundings. Only the represented events seem to exist, just like only the object of our love or anger seem to exist when we are stirred by an emotion. It is not surprising that the aforementioned suspicions about emotions have also been raised with regards to immersion, fearing we may fail to distinguish illusions from reality—Marie-Laure Ryan calls this the "Don Quixote syndrome."[2] We seemingly tend to believe that these two states—emotion and immersion—are deeply related.

In this chapter, I want to examine this relationship closely. To what extent does immersion contribute to the emotional experience of fictional works? Could it be that emotions are essential ingredients to immersive experiences? How should we distinguish these two states of mind? Here, I want to focus on both single-player video games, since the media is often characterized as valuing immersion above any other type of experience, and Hollywood movies, since film studies have produced far more developed contributions on emotions than game studies have.[3] Unsurprisingly, many scholars have modelled the viewer's experience of emotion after perception and have connected it with immersion, suggesting that fictional works are similar to illusions (both are deceptive and captivating) and that artists act as illusionists (both go through their bag of tricks to manipulate the audience's emotions or perception). This perceptual model appears clearly in Torben Grodal and Jonathan Frome's respective works, which I will present in the first part of my demonstration. I will highlight how this model frames fear as a paradigmatic type of emotion and why this framing is limiting by

1 See Solomon, "The Philosophy of Emotions."

2 Ryan, *Narrative as Virtual Reality*, p. 99.

3 See Lankoski, "Player Character Engagement in Computer Games," p. 292.

arguing that it cannot account for the whole range of emotions. We will see that the perceptual model, in its basic or classic form, is unconvincing, which explains why it has not been fully adopted by game studies. I will therefore turn to motivational theories of emotion and immersion, drawing attention to how both states are also viewed as going hand-in-hand, and how this close connexion is yet to be developed fully.

If We Are Afraid of the Fictional Monster, Why Don't We Flee It?

Frome has introduced the perceptual model of emotion through the famous paradox of fiction, which may be summarized as follows:

1. Emotions are beliefs.
2. But we do not believe that fiction is reality.
3. Yet we have emotions towards fiction.[4]

Considering these statements, how is it possible for anyone to experience emotions towards fictional beings and events? Drawing on the work of Grodal, Frome offers what he calls the "hybrid illusionist theory," according to which emotional activity and evaluation of reality status belong to separate mental systems.[5] The author illustrates his point with the Zöllner illusion, an optical illusion depicting parallel lines which appear to converge (fig. 4.1).[6] Even if we prove with a measuring tool that the lines crossing the image are at equal distance from each other, the illusion nevertheless persists before us. Thus, activities of the mind are not unified: perception and belief operate at different levels. The illusion maintains its effect on our perception even though we do not regard what we see as true. To dissolve the paradox of fiction, Frome argues that emotions are akin to perceptions rather than beliefs.

Indeed, it seems like we find the same duality in the emotional experience of fiction. To use Frome's example, the viewer reacting with fright to the *Alien* (Ridley Scott, 1979) (or any fictional monster) does not run for their life out of the theatre because the content is processed by separate mental systems. One of these responds to the illusion produced by images and sounds, whereas the other remains unaffected by the deception: "Some of the viewer's mental

4 Frome, "Reality, Representation, and Emotions Across Media." The paradox of fiction was first developed in Radford and Weston, "How Can We Be Moved by the Fate of Anna Karenina?"

5 See Grodal, *Moving Pictures*.

6 Frome, "Reality, Representation, and Emotions Across Media," p. 14.

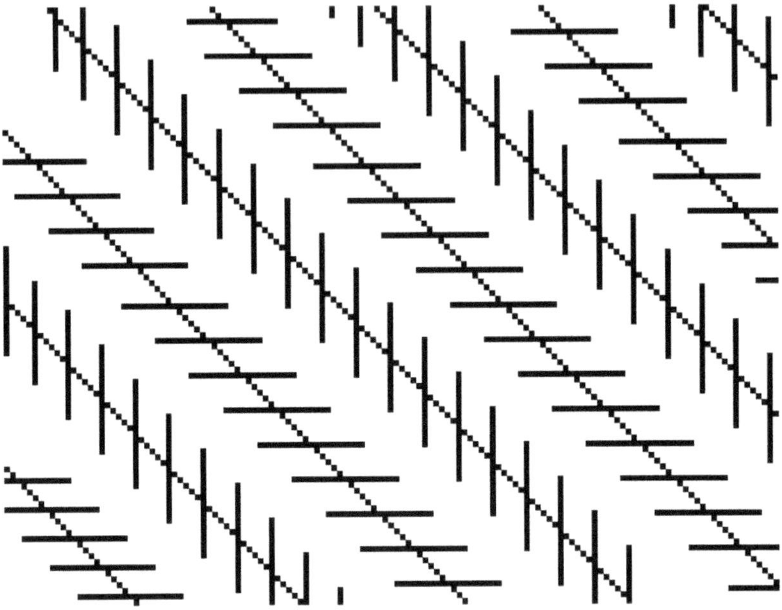

Figure 4.1: The Zöllner illusion.

systems respond to the alien as if it is a real danger, raising her heart rate and increasing her adrenaline. Other mental systems, such as the ones responsible for running and hiding, recognize that the alien is fictional and thus not a real danger that she must avoid."[7] According to Frome, visual media must create convincing illusions that lure the audience's senses to elicit "prototypical" emotions. How so? The author suggests that fiction must replicate the experience of reality, implementing the laws of physics, high-resolution sensory information, interactivity, and proprioception.[8] Illusions are thus enhanced with various technological devices (a large screen, 3D effects, sound spatialization, etc.). Although Frome acknowledges that no media is able to perfectly emulate all of these characteristics, resulting in "partial" emotions in the audience, the underlying technological determinist discourse hints that emotional and immersive experiences are achieved the same way: by maintaining an illusion of reality.[9]

7 Frome, "Reality, Representation, and Emotions Across Media," p. 15.

8 Frome, "Reality, Representation, and Emotions Across Media," p. 16.

9 It is not exactly clear what Frome means by "partial." Are emotions towards fictional objects partial in nature, in intensity, or in internal activity? The hybrid illusionist theory seems to answer these three possibilities in the affirmative. It is not clear whether Frome implicitly refers to Kendall Walton's concept of "quasi-emotion." See Walton, *Mimesis as Make-Believe*.

Before commenting on this issue, it is important to note that Grodal offers a slightly different version of the hybrid illusionist theory. Supporting his claims with neuroscience research, evolutionary psychology, and theories of embodied cognition, Grodal describes mental systems that are not disconnected, but operate together at different speeds. In the first step of the process, a local evaluation system based on perception automatically detects auditory and visual stimuli, triggering emotional responses; in the second step, a global evaluation system determines the reality status of the perceived information.[10] This is why the viewer (or game player) is afraid of the fictional monster, but does not actually try to flee it. A local evaluation alerts the viewer at the sight of what looks like a real danger and a global evaluation corrects the misperception. In other words, the slower, colder mental system defuses the false alarm.[11]

Grodal explains this sequence of internal events in more detail with the aid of the "PECMA flow," a biocerebral information processing model, which refers to how filmic experiences bring together "perception," "emotion," "cognition," and "motor action":

> During film viewing, the "stupid" robot neurons in the visual cortex are bombarded with often strongly emotion-evoking images, and it is only further along in the PECMA flow that a special brain mechanism will evaluate the reality status of the images in order to control our responses to the images according to their reality status. [...] Contrary to the common credo that fiction demands a suspension of disbelief, it actually demands a modification of belief or a suspension of belief so that film viewing does not produce full-scale illusions. Our brains were not constructed for film viewing, for a world in which it is possible to be exposed to very salient and complex moving images; thus even as we watch fictional films it remains true that seeing is believing, because to believe incoming information is, as previously mentioned, the default mode and to disbelieve demands a special effort.[12]

The PECMA flow model illustrates how bottom-up mental processes, which are triggered by sensory inputs, precede top-down concept-driven mental

10 Grodal, *Moving Pictures*, pp. 33–34.

11 This dual process is akin to what psychologist Daniel Kahneman presents in his influential work, *Thinking, Fast and Slow*. Hence, critics of Kahneman's theory may teach us a thing or two about the hybrid illusionist theory.

12 Grodal, *Embodied Visions*, p. 154.

processes in the experience of emotions. This leads Grodal to assert that "seeing is believing." From an evolutionary perspective, it makes perfect sense. Emotions must precede global evaluations of reality status for adaptive reasons. The animal that has to deliberate on the dangerousness of a situation before reacting with fear is likely to live a short life and have a poorly functioning genetic legacy.[13] Our hunter-gatherer ancestors had to preemptively respond to apparent predators to ensure their survival. Fear of the fictional monster is evidence of this adaptive function. We are predisposed to react to danger, whether or not it is real.

The dual process described by Grodal implies that emotions are not fully perceptual: disbelieving in the existence of the fictional monster, even though it "demands a special effort," is enough to dissipate fear until the next illusion. According to the author, emotions are rather both perceptions *and* beliefs. If this is true, we might still conceive the emotional experience of fiction as a conflict between one part of the mind which sustains the illusion on the screen and another part of the mind which keeps trying to dispel it. Although Grodal offers a different flavour of the hybrid illusionist theory, emotion and immersion are still understood as working hand-in-hand. Grodal notably establishes a link between the two through the concept of mental simulation. When mentally simulating the protagonist's situation, the viewer perceives it in the first-person. As stated by the author, this "immersed" position opens up the range of emotions to be experienced, something that traditional observer theories have neglected by assuming a distanced, third-person perspective.[14] For instance, when the gigantic boulder starts rolling towards Indiana Jones after he has gotten his hands on the golden idol in the *Raiders of the Ark* (Steven Spielberg, 1981), the immersed viewer feels fear for herself through mental simulation, while the observer would merely be afraid *for* the life of the protagonist.[15] For Grodal, the former description of fear corresponds better to our actual embodied experience of the scene.

13 For more information about how emotions and evolution are related, see LeDoux, *The Emotional Brain*.

14 Grodal, *Embodied Visions*, p. 182. The author responds mostly to Noël Carroll's account of the paradox of fiction and of the resulting conceptualization of fear. See Carroll, *The Philosophy of Horror*.

15 This is similar to the common distinction between empathy and sympathy. According to Stephen Darwall, "[e]mpathy consists in feeling what one imagines [the other person] feels," while "[s]ympathy [...] is felt as from the perspective of 'one-caring'." Darwall, "Empathy, Sympathy, Care," p. 261. Likewise, Amy Coplan claims that empathy is feeling *with* another while sympathy is feeling *for* another. Only the first emotional disposition requires perspective-taking and results in sharing the other's experience. See Coplan, "Empathic Engagement with Narrative Fictions," pp. 144–145.

The author's explanation is elegant and has the advantage of not resorting to technological determinism, which creates a hierarchy between media (video games supposedly elicit emotions of a greater "prototypicality," to use Frome's term, than literature, photography, and more, thanks to their mimetic mode of interactivity and tactile feedback, which emulates the actional experience of reality that is lacking in "traditional" media) and a hierarchy between eras (high-budget works of today supposedly create "more emotional" experiences than works of the past). However, it still depends on the idea that emotions respond above all to sensorial inputs, leaving many unanswered questions about the scope of the perception analogy, including the following: is the perceptual model able to account for the audience's experience of many other emotion types beside the immersed fear for oneself? Taking this question into consideration is important in order to develop a theory of emotion rather than a single account of a particular emotion. Theorizing how we come to feel emotions when viewing films or playing video games, our solution should aim for generality, be applicable to any or at least most emotion types, or else it will present an incomplete and unpredictable system. "A theory of emotion needs to be a theory of *emotions*," as Robert Zajonc, Sheila T. Murphy, and Marita Inglehart once claimed.[16] Thus, we will understand fear for oneself as a paradigmatic case on which the perceptual model is based and compare it to another type of emotion to test its validity.

From Fear for Oneself to Guilt about Oneself

When Grodal refers to the immersed fear for oneself, he characterizes it as "naïve," experienced in first-person, and mirroring the character's own fear through mental simulation. This emotion is contrasted with fear for others, which is cognizant and felt in third-person, with the awareness that it is a fictional character who is in danger. To show the limits of the perceptual model and the previous dichotomy, I want to discuss guilt, a first-person emotion that does not necessarily rely on mental simulation since it is self-directed. The object of guilt is usually one's own fault in a situation that is controllable and thus could have been avoided, which differentiates it from shame and its attribution of uncontrollability.[17] Even when we are feeling guilty for an action we did not perform, such as a parent for the

16 Zajonc, Murphy, and Inglehart, "Feeling and Facial Efference," p. 412.

17 On this, see Tracy and Robins, "Appraisal Antecedents of Shame and Guilt," pp. 1343–1344.

misbehaviour of their child, we share a sense of responsibility for it due to exercising indirect control over it. We thus ask ourselves what we should have done to avoid this turn of events. Guilt is always about oneself.

Because it is a reflexive or self-conscious emotion, meaning that "[i]n the face of transgression or error, the self turns toward the self—evaluating and rendering judgment—,"guilt is not likely to respond to illusions rendered by fiction works.[18] Feeling guilt implies recognizing one's own mistakes, preventing the misattribution of the character's faults to oneself. This may be one reason why guilt is not part of conventional experiences of fiction. No illusion is deceptive enough to make the audience of a narrative film believe that the wrongdoing depicted is the result of their own actions. For instance, when the superhero lives with the guilt of having failed to save one of their relatives from death, the viewer does not share their feelings, but merely has sympathy for them—just compare how Peter Parker feels about Uncle Ben's death in *Spider-Man* (Sam Raimi, 2002) with how the audience feels about it. While mental simulation may be involved in such a situation, it would still not be enough to stir some sort of perceptual guilt in which one's own fault is mistaken with the character's. In order to elicit this emotion, fiction works have to directly address the audience, for example, through a political commentary or by making them reflect on their previous desires or emotions.[19] The ensuing guilt would be cognitive by originating from a heightened awareness of a real fault instead of stemming from an obnubilated perception.

What is interesting about video games is that they are able to organize the experience of guilt more easily than other media, thanks to their unique structure.[20] Jesper Juul has wisely remarked that "the fundamental difference between tragedy in games and stories is that in stories we never feel responsible for failure and suffering; in games, we do."[21] Over the past decades, many narrative video games have aimed to elicit guilt in players in response to what they do during gameplay—be warned that I am about to spoil the plots of *The Walking Dead*'s first season (Telltale Games, 2012), *Shadow of the Colossus* (Team Ico, 2005), and *Life Is Strange*'s second episode

18 See Tangney and Dearing, *Shame and Guilt*, p. 2.

19 Carl Plantinga considers the latter case of guilt to be "meta-emotion." Plantinga, *Moving Viewers*, pp. 162–164. For Plantinga, this type of emotional experience can be found in Alfred Hitchcock's works such as *Strangers on a Train* (1951), in which the viewer is inclined to feel bad for experiencing satisfaction at the death of Miriam, Guy's fiancée, a despicable character who nevertheless meets an undeserved death.

20 This observation is further developed in Isbister, *How Games Move Us*, pp. 8–10.

21 Juul, *The Art of Failure*, p. 112.

of the first season (Dontnod Entertainment, 2015). Think of how you may have reacted after making the young Clementine cry for prioritizing survival goals over moral ones in *The Walking Dead*, for destroying sublime stone colossi to bring back to life a princess only to learn that the ritual has released a demonic spirit in *Shadow of the Colossus*, for not being able to save Kate, a student victim of bullying, from her suicide attempt in *Life is Strange*, and for unjustly killing, letting die, or stealing from innocent characters by uncritically following video game conventions (e.g., *Spec Ops. The Line* [Yager, 2012]; *Undertale* [Toby Fox, 2015]) or by not playing well enough (*Papers, Please* [Lucas Pope, 2013]; *This War of Mine* [11 bit studios, 2014]). In all these cases, we may feel guilty because we are made aware of the nature of our actions as players; as the ones responsible for the resulting events. This realization could not possibly be the fruit of sensory perception. Guilt demands a more elaborate mental activity, in which the self is brought into focus and assessed, and a mode of thinking that crosses boundaries between fiction and reality, given that the player understands *they* are in control of both themselves and, to a certain extent, their character.

One could say that some of these games create an illusion of choice, hence of fault, by virtue of orchestrating scripted situations in which it is impossible to steer the narrative in one direction or the other. For instance, in *Shadow of the Colossus*, the only way to finish the game is to eliminate all sixteen of the titular creatures. But their immensity, their majesty, and their half-human, half-animal appearance inspire a form of respect close to the sublime. These giants move slowly in desolate places without paying attention to the player-character until they are attacked. When they are slain, the player is not rewarded with resources or points. Only a cutscene showing the creature agonizing and a funereal music occur. Shadows then escape from the fading stone bodies and fly through the player-character, who loses consciousness before waking up in the sanctuary, ready to move on to their next victim. Going through this sequence over and over again, players realize they are not pursuing an admirable quest as suggested at the beginning of the game and may at that point experience remorse.

Now, if the player feels guilty even when they exercise no control over the story, is it not proof that this emotion is perceptual in nature? One could claim that the player feels it in a similar way a subject would see an illusion: by being tricked. According to this view, the player experiences guilt even if they know they are not to blame, just like viewers become afraid of the monster even though they know it is fictional. This would lend credence to Grodal and Frome's models, since the explanation of immersed fear responses applies to guilt, encompassing two emotions which appear *prima facie* different in nature. But I think this view is misleading. Generally speaking, once we learn that we have no real responsibility for what is happening in a given game, our guilt dissipates and we remain unresponsive to the next "illusion," as we are now aware of the true structure of the game, over which we exercise no influence. We might instead feel a different emotion about the game's attempt at instilling a feeling of guilt, from annoyance at the unfair accusation to appreciation for its audaciousness. We may also go through guilt because we are reacting to our previous emotions or desires, such as feeling bad for initially having had fun killing the colossi. In any case, a clash between perception and belief would not occur, contrary to what Frome and Grodal describe. Our guilt would be less capricious than sensory perception: once we know the outcome has been predetermined, we would have no reason to feel this particular way.

Such an account of guilt does not correspond at all with the earlier description of fear for oneself. In order to extend the scope of their theory of emotions, advocates of the perceptual model would need to go much further to demonstrate how that could be possible. Alternatively, they could acknowledge this incompatibility and limit the scope of their account to a smaller class of emotions. I have raised doubts about the feasibility of the first option, but choosing the second option also comes with a flaw. Psychologists and philosophers of emotion have called into question the supposed disunity of the category of emotion implied by the popular conception of basic emotions, which are biologically "hardwired" and triggered by specific stimuli (fear being one typical example).[22] Basic emotions are usually distinguished from nonbasic emotions, which are cognitively complex, meaning that they depend on learned concepts and on what one thinks in a given situation (as is typically the case for guilt). Julien Deonna and Fabrice Teroni identify two problems regarding this conception: "First, does the word 'emotion,' as an umbrella term covering both kinds of state, correspond to a coherent and unified class of mental states? Second, and directly related to this, how are we to understand the relation between these?"[23] If fear is analogous to perception and guilt is cognitive, the difference between the two is one of nature. Film and video game research aiming for a theory of *emotions* would have to reunite the category by demonstrating how these two emotions share the same form and underlying mechanisms in the

22 See Ekman and Cordaro, "What Is Meant by Calling Emotions Basic."

23 Deonna and Teroni, *The Emotions*, p. 21. Psychologists have also raised criticisms against the conception of basic emotions. See Ortony, "Are All 'Basic Emotions' Emotions?"

reception of fiction. As of now, we should be careful not to conclude hastily that the experience and inner workings of fear for oneself apply equally to other types of emotions such as guilt.

Comparing Perception with Emotion and Turning to an Alternative View

Another way of examining the perceptual model, beside raising the previous counterexample, is to highlight relevant similarities and differences between perception and emotion. The attractiveness of the perceptual model certainly lies in its phenomenological dimension. Emotions seem akin to perceptions in the sense that they are unmediated "impressions" of the world. They vary according to the way the world appears to oneself, without demanding any sort of reasoning or any kind of intellectual effort. This is why they are signs of passivity: we do not choose to have an emotion, just as we do not choose to see what is right in front of us. We undergo both experiences passively.[24] Emotions and perceptions are also points of contact with the world and a point of view on it. They give access to its properties, whether it is the redness of an apple or the fearsomeness of a creature. The resemblance between emotions and perceptions contains implications for the emotional experience of fiction. One of them is that we fear the fictional monster because it appears to the consciousness in a direct, instantaneous fashion—something to which Frome and Grodal draw attention. Another is that emotional experiences depend on the way the work presents itself. In other words, it is caused entirely by the properties of the perceived work. This implication is more problematic since it creates the image of a completely passive audience, deprived of discernment—an image that comes with moral concerns. Before addressing this issue, I want to point out differences between emotions and perceptions.

Deonna and Teroni have identified many key distinctions between the two, including the following, which is especially relevant to the experience of fiction: "[W]hile perception provides us with information about our environment through mechanisms that are relatively impervious to our motivational states, emotions, by contrast, are highly responsive to our desires, concerns, and character traits."[25] Indeed, perception operates in an

24 Although action-based theories of perception exist, we should limit our attention to the classic model of perception as implied with the illusion analogy used by Frome.

25 Deonna and Teroni, "In What Sense Are Emotions Evaluations," pp. 23–24.

autonomous way (as Frome has shown with the Zöllner illusion), meaning the redness of an apple, to stick with this example, is accessible as long as the organ of sight remains intact. What one thinks about apples or what prior attitudes one has towards fruit does not change anything about their sensory perception. But emotions "latch onto other psychological states."[26] In particular, their occurrence is prepared by motivational states. For instance, if we react to the hero's fate or to the result of a match, it is because we first value them. We are not mere witnesses of a situation; we desire a specific state of affairs in which the hero has good fortune or in which we are the winner of the match. The same is true for guilt. If we feel bad for hurting fictional beings, it is because we have no desire to hurt them in the first place. Many specialists of emotions assert that motivational states, such as care and concern, are necessary to the occurrence of any emotion and there are many proponents of this conception in both film and game studies.[27] I will freely rely on their work to sketch the outlines of this other account of emotion, which problematizes the perceptual model.

Considering the role of motivation in emotion, the illusion metaphor seems to propose an insufficient answer. It is not enough for the artist to create illusions to manipulate emotions. Rather, they have to make their audience care about what is happening. Here, care and concern are not to be understood as emotional occurrences, but rather as emotional dispositions, according to Prinz: "care is not an emotion, but rather a disposition to experience different emotions. If you care about someone, you will be delighted by her achievements and distressed by her suffering."[28] Before attempting to provoke any emotional response, the artist has to tap into their audience's caring disposition. Goals and conflicts are one way to achieve this. Video games give players something to care about when they implement goals to accomplish. They may then harness the player's caring disposition by offering resistance to their actions, making the given goals more difficult to attain. Let us take as an example permanent death systems, in which player-characters have only one life. When the most powerful player-character dies in a tactical role-playing game such as *Darkest Dungeon* (Red Hook Studios, 2015) or *XCOM. Enemy Unknown* (Firaxis Games, 2012), the game may not be over, but it is hard for the player not to feel pain given

26 Deonna and Teroni, "In What Sense Are Emotions Evaluations," p. 25.

27 On the underlying motivational states of emotions, see Düringer, "Caring"; Frijda, *The Laws of Emotions*; and Roberts, *Emotions*. On this conception in film and game studies, see Leino, "Emotions in Play"; Perron, "Emotions in Video Games"; and Plantinga, *Moving Viewers*.

28 Prinz, *The Emotional Construction of Morals*, p. 84.

that they have invested many hours into it and that their chance of success just diminished significantly. Hollywood movies also establish goals that are met with resistance to consolidate the viewer's caring disposition, as Plantinga observes: "With regard to the sympathetic emotions, the 'concern' part of the audience's concern-based construal arises from the gap between the desired goals for the character and the actual, probable, or possible outcomes that conflict with those goals."[29] Conflict, whether in films or video games, possesses a dramatic structure directing our motivation, which in turn prepares us to feel emotions towards certain states of the narrative or game. We usually experience the most intense emotions when the conflict reaches its peak, during the narrative climax or the final boss fight, not because of better illusions, but because of the stakes at play and because our motivational level has reached its highest point.

Nor is the perception model able to explain why emotionless experiences happen. Sometimes, we prevent ourselves from reacting to a film that is too "sentimental" or "cheesy" in its depiction of a dramatic situation. At other times, we play a game that takes so long to beat that we just want to reach its conclusion and get rid of it. These examples suggest a lack of motivation (we are engaged with the work at the most minimal level) and a lack of passivity (we are unwilling to cooperate with it from an emotional standpoint). The classic perceptual model cannot give an account of this sort of experience since it suggests that emotions are simply caused by the configuration of "illusions" or, to put it differently, by how fiction comes into sight. Concepts such as care and concern allow for a better understanding of the emotional experience of film viewing and game playing. Audiences react to moving images because they first get attached to something: how the narrative will end, how the level is beatable, or how the character will make it through. Emotional occurrences arise from an antecedent. To work their magic, film and games require a certain degree of previous investment from their audience.

Bringing Back Investment to the Mix

Despite the previous critique of the perceptual model of emotion, we could still be satisfied with the idea that emotion operates on a motivational basis and that immersion operates on a perceptual basis. However, immersion has also been conceived on the premise that it is driven by motivation. Consider the differences between immersion as transportation and

29 Plantinga, *Moving Viewers*, p. 93.

immersion as absorption, which has been made by Gordon Calleja in his literature review in game studies.[30] Understood in terms of transportation, immersion is the feeling of being in the fictional world. It is better described as the phenomenological effect of perceiving fiction from the inside. The perceptual model does not apply as well to absorption, a psychological process in which information becomes easier to assimilate as the player invests themselves in their activity. Absorption is not the experience of being *there*, but the experience of having the mind filled with the fictional world. It is my understanding that game scholars have mostly developed absorption accounts of immersion in which emotional "investment" or "engagement" is involved in the process of constituting the immersive experience. Thus, they have brought immersion and emotion into a close conceptual proximity, although in a different manner than film studies.

For instance, Emily Brown and Paul Cairns have categorized participants' discourse in semi-structured interviews on an immersion scale divided into three phases of intensity: engagement, engrossment, and total immersion.[31] Engagement involves an initial effort by the player to learn to play the game and seek self-improvement. This is followed by engrossment, the result of an increased and prolonged engagement, which coincides with a heightened disposition to experiencing emotions: "At this level of immersion due to the time, effort and attention put in, there is a high level of emotional investment in the game."[32] The intensification of these two absorption phases finally leads to total immersion, in which the player feels like they are inhabiting the game world and are cut off from their immediate environment. Following Brown and Cairns, immersion is a process which mobilizes significant effort from the self and entails emotional readiness.

In a similar vein, Calleja attempts to encompass both transportation and absorption with the concept of "incorporation," but the author's involvement model mostly emphasizes absorption. The model illustrates six dimensions of involvement: kinaesthetic, spatial, shared, narrative, ludic, and affective.[33] At a low level of involvement, such as in the early moments of a game, play entails conscious and fragmented attention to the different dimensions. For example, players need to figure out the effect of each button on the controller (kinaesthetic dimension), make a mental map of places (spatial), remember names and characters' purposes (narrative), be familiar with how mechanics and

30 Calleja, *In-Game*, pp. 25–32.
31 Brown and Cairns, "A Grounded Investigation of Game Immersion."
32 Brown and Cairns, "A Grounded Investigation of Game Immersion," p. 1299.
33 See Calleja, *In-Game*, p. 38.

rules interact with each other (ludic), know how to communicate efficiently with their teammates (shared), and give way to their feelings (affective). When fully involved, players do not have to pay conscious attention to each of these dimensions anymore. Thanks to their sustained effort, the six of them have been internalized in such a way that they appear fluid and globalized. At this point, players have finally reached the complete realization of immersion. Akin to what is presented by Brown and Cairns, the affective dimension, which encompasses "various forms of emotional engagement," is here one of the building blocks of a fully immersive experience.[34]

Carl Therrien follows a similar path when discussing emotions in immersive experiences of films and video games.[35] For the author, immersion involves psychoaffective dynamics of investment and return, during which the experience of unpleasant emotions is accepted knowing that fictional institutions promise subsequent "affective gratifications." For instance, the audience endures the suspense caused by a ticking bomb scenario or the frustration resulting from repeated failures of a level because they expect ensuing euphoric emotions from the happy ending or the mastery of the game. According to Therrien, what they invest may be akin to an intellectual effort, such as the mental reconstruction of a puzzle film's narrative, or may be of a more affective nature, such as the feeling of suspense or frustration described above.[36] While this account may not entirely amount to absorption, it is still close to the previous authors' idea of immersion by virtue of characterizing an energy-intensive experience in which the audience spends temporal, attentional, intellectual, or affective resources in order to feed their immersive state. Such a conceptualization departs from the classic perceptual model because senses, as autonomous organs, are not prepared by any sort of investment or effort.

While all the aforementioned authors claim that emotional investment is constitutive of immersion, only Therrien provides an explanation of the former psychological process. The investment metaphor is indeed compatible with the motivational aspect of emotion, since care and concern can be seen as forms of internal resources expenditure. The viewer's concerns take shape

34 Calleja, *In-Game*, p. 44. That being said, the affective dimension occupies an odd place in Calleja's model. One could wonder how this dimension stimulates the same mental processes as the five others. Examples provided by the author, such as the relaxation elicited by peaceful environments in *World of Warcraft* (Blizzard Entertainment, 2004) or the adrenaline rush triggered by an unpredictable wave of zombies in *Left 4 Dead 2* (Valve, 2009), seem to bear no link to any sort of learning process specific to the affective dimension. See Calleja, *In-Game*, pp. 142–145.

35 Therrien, "Illusion, idéalisation, gratification."

36 Therrien, "Illusion, idéalisation, gratification," p. 256 and pp. 266–267.

as they attempt to make sense of the protagonist's inner life, pay attention to narrative details, and anticipate plot twists. The player's caring about the success of a goal increases as they work towards it. In everyday life, being concerned about one's job, child, or partner may indeed be exhausting. We see here that emotions come at a certain price, not necessarily in a negative sense, but in the sense that one has to first give something of oneself in order to be affected by the world. That being said, to obtain a motivational theory of emotions, we still need a finer description of what it is like to be emotionally invested in films and games, a more systemic understanding of what resources are mobilized and how they are related, an exploration of other emotional dispositions involved in this process beside care and concerns, and a clarification of whether these two dispositions are necessarily indicative of immersion as absorption. Despite not answering these questions, I hope to have laid essential foundations for such work.

Conclusion

The classic perceptual model of emotion is not perfectly adequate to describe the emotional experience of films and games. While fear for oneself appears as one of the best examples for supporting the perception analogy, its peculiarities cannot be ascribed to other emotions such as guilt, since we do not feel guilty for characters' faults as we may feel afraid for ourselves when characters are braving dangers. Motivational accounts seem more promising to explain how emotions occur in audiences, that is by stimulating cares and concerns instead of coming up with sophisticated illusions. Such accounts of emotion are very similar to how absorption has been characterized in game studies. In both conceptualizations, investment of the self is a prerequisite to achieving emotional or absorbed states of mind. Nevertheless, film and game studies still have to explore this connexion in greater detail.

Bibliography

Brown, Emily, and Paul Cairns. "A Grounded Investigation of Game Immersion." In *CHI EA '04: Extended Abstracts on Human Factors in Computing Systems*, 1297–1300. New York, NY: Association for Computing Machinery, 2004. https://doi.org/10.1145/985921.986048.

Calleja, Gordon. *In-Game: From Immersion to Incorporation.* Cambridge, MA: MIT Press, 2011.

Carroll, Noël. *The Philosophy of Horror, or, Paradoxes of the Heart*. New York, NY: Routledge, 1990.

Coplan, Amy. "Empathic Engagement with Narrative Fictions." *Journal of Aesthetics and Art Criticism* 62, no. 2 (Spring, 2004): 141–152.

Darwall, Stephen. "Empathy, Sympathy, Care." *Philosophical Studies* 89 (March 1998): 261–282.

Deonna, Julien A., and Fabrice Teroni. *The Emotions: A Philosophical Introduction*. London: Routledge, 2012.

Deonna, Julien A., and Fabrice Teroni. "In What Sense Are Emotions Evaluations." In *Emotion and Value*, edited by Sabine Roeser and Cain Todd, 15–31. Oxford: Oxford University Press, 2015.

Düringer, Eva-Maria. "Caring." In *Evaluating Emotions*, edited by Eva-Maria Düringer, 89–116. New York, NY: Palgrave Macmillan, 2014.

Ekman, Paul, and Daniel Cordaro. "What Is Meant by Calling Emotions Basic." *Emotion Review* 3, no. 4 (October, 2011): 364–370.

Frijda, Nico. *The Laws of Emotions*. London: Psychology Press, 2017.

Frome, Jonathan. "Reality, Representation, and Emotions Across Media." *Film Studies: An International Review* 8, no. 1 (Summer 2006): 12–25.

Grodal, Torben. *Embodied Visions: Evolution, Emotion, Culture, and Film*. New York, NY: Oxford University Press, 2009.

Grodal, Torben. *Moving Pictures: A New Theory of Film Genres, Feelings, and Cognition*. New York, NY: Oxford University Press, 2002.

Isbister, Katherine. *How Games Move Us: Emotion by Design*. Cambridge, MA: MIT Press, 2016.

Juul, Jesper. *The Art of Failure: An Essay of the Pain of Playing Video Games*. Cambridge, MA: MIT Press, 2013.

Kahneman, Daniel. *Thinking, Fast and Slow*. New York, NY: Farrar, Straus, and Giroux, 2011.

Lankoski, Petri. "Player Character Engagement in Computer Games." *Games and Culture* 6, no. 4 (June, 2011): 291–311.

LeDoux, Joseph. *The Emotional Brain: The Mysterious Underpinnings of Emotional Life*. New York, NY: Simon and Schuster, 1996.

Leino, Olli Tapio. "Emotions in Play: On the Constitution of Emotion in Solitary Computer Game Play." PhD diss., University of Copenhagen, 2010.

Ortony, Andrew. "Are All 'Basic Emotions' Emotions? A Problem for the (Basic) Emotions Construct." *Perspectives on Psychological Science* 17, no. 1 (2022): 41–61.

Perron, Bernard. "Emotions in Video Games: Are You Concerned?" In *Video Games and the Mind: Essays on Cognition, Affect, and Emotion*, edited by Bernard Perron and Felix Schröter, 189–209. Jefferson, IN: McFarland, 2016.

Plantinga, Carl. *Moving Viewers: American Film and The Spectator's Experience.* Berkeley, CA: University of California Press, 2009.

Prinz, Jesse J. *The Emotional Construction of Morals.* Oxford: Oxford University Press, 2007.

Radford, Colin, and Michael Weston. "How Can We Be Moved by the Fate of Anna Karenina?" *Proceedings of the Aristotelian Society* 49 (1975): 67–93.

Roberts, Robert C. *Emotions: An Essay in Aid of Moral Psychology.* Cambridge, MA: Cambridge University Press, 2003.

Ryan, Marie-Laure. *Narrative as Virtual Reality: Immersion and Interactivity in Literature and Electronic Media.* Baltimore, MD: Johns Hopkins University Press, 2001.

Solomon, Robert C. "The Philosophy of Emotions." In *Handbook of Emotions*, edited by Michael Lewis, Jeannette M. Haviland-Jones, and Lisa Feldman Barrett, 3–16. New York, NY: Guilford Press, 2008.

Tangney, June Price, and Ronda L. Dearing. *Shame and Guilt.* New York, NY: Guilford Press, 2002.

Therrien, Carl. "Illusion, idéalisation, gratification. L'immersion dans les univers de fiction à l'ère du jeu vidéo." PhD diss., Université du Québec à Montréal, 2011.

Tracy, Jessica L., and Richard W. Robins. "Appraisal Antecedents of Shame and Guilt: Support for a Theoretical Model." *Personality and Social Psychology Bulletin* 32, no. 10 (October, 2006): 1339–1351.

Walton, Kendall. *Mimesis as Make-Believe.* Cambridge, MA: Harvard University Press, 1990.

Zajonc, Robert B., Sheila T. Murphy, and Marita Inglehart. "Feeling and Facial Efference: Implications of the Vascular Theory of Emotion." *Psychological Review* 96, no. 3 (July, 1989): 395–416.

About the Author

Maxime Deslongchamps-Gagnon is a lecturer in the video games production program at the University of Québec in Abitibi-Témiscamingue. He holds a PhD in Film Studies from the University of Montréal. He has published on video games and their connections to virtue ethics, emotions, aesthetics, walking simulators, and films.

5. Interaction and Immersion Logics in *Encounters*: Coalescence of Bodies in a Liquid Space

Claire Chatelet and Mathieu Pradat

Abstract: *Encounters* is a free roaming collective experience in virtual reality. Participants are engaged in different ways and at different stages during the production process and the reception of the work. Based on analysis of this original project, how the complex relationships between spaces (physical, virtual, scenic, representational, narrative), time (real, narrative) and bodies (lived, simulated) determine a dynamic, spatialized and "embodied" narrative are examined. This involves questioning the concept of "immersion," not only in its theoretical presuppositions, but above all in its practical applications and, thus, in its relation to multiples issues within artistic creation (technological, corporeal, affective, narrative, etc.). Also, it shows that the concept of "border-contact" from Gestalt Therapy could assist the understanding of specific immersive experiences.

Keywords: virtual reality, Gestalt theory, water

From Immersion to Immersions

The metaphor of immersion—that is, immersion in a liquid—is systematically used in the rhetoric of manufacturers who develop new tools of diffusion (virtual reality headsets, mixed-reality glasses) and in the cultural and media discourses that accompany the creations conceived for these devices, as if the technology did not only predetermine, but also homogenize the receptive experience. The multiple aesthetic potentialities offered by these artworks, the diversity of the proposed setups, the disparity of the involved "users-spectators"—and thus their reception—indeed seem to dissolve in

Bédard, Philippe, Alanna Thain, and Carl Therrien, eds. *States of Immersion Across Media: Bodies, Techniques, Practices*. Taylor & Francis Group, 2025
DOI 10.5117/9789048558766_CH05

Figure 5.1: © La prairie productions, SoWhen? and Serendipity Films

an immersivity which appears constituent, both as an aesthetic principle of the artworks and as an "horizon of expectation" (*horizon d'attente*[1]) for the public. Yet, immersion is difficult to apprehend as an aesthetic concept, to say the least, and probably even more so from an ontological point of view.

1 The concept of "horizon of expectation" was coined by Hans R. Jauss for literary theory. See Jauss, *Pour une esthétique de la réception*. Unless otherwise noted, all translations are our own.

Can it only be defined or at least circumscribed? Doesn't the polysemy of the term require taking its plural nature into consideration? Not a single immersion, therefore—because there could not be a definitive and unique form of reaching out to the public—, but various *immersions*, whose nature should be specified each time: psychological, physical, emotional, sensorial, perceptual, or even fictional, technological, or narrative. For some researchers, such as Mathieu Triclot, the very notion of "immersion" constitutes "an epistemological obstacle to thinking the relation to the images and the technical devices that generate them," because by replacing "more ancient other modes of conceptualization of media forms," the metaphor of immersion might occlude "the ways of tangible engagement with the apparatus itself and the making of a specific regime of perception."[2] In the field of video games, Gordon Calleja, relying in particular on the work of Katie Salen and Eric Zimmerman,[3] insists on the double meaning of the term: immersion can be apprehended as an *absorption* and as a *transportation*. Calleja refers here, on the one hand, to the fact that the player is involved in the medium, and, on the other hand, to their *sensation of being present* in a different space.[4] Instead of the ambiguous concept of "immersion," he favours the concept of "incorporation," which allows us to consider this double phenomenon:

> Incorporation thus operates on a double axis: the player incorporates (in the sense of internalizing or assimilating) the game environment into consciousness while simultaneously being incorporated through the avatar into that environment. [...] We can thus conceive of incorporation *as the absorption of a virtual environment into consciousness, yielding a sense of habitation, which is supported by the systematically upheld embodiment of the player in a single location, as represented by the avatar.*[5]

Incorporation, in this perspective, is considered as an intensification of the "internalized" engagement manifesting itself following multiple more or less imbricated dimensions: kinaesthetic, spatial, shared (in the case of setups soliciting a collective reception), narrative, affective, or ludic engagement. However, for Calleja, it is the spatial and kinaesthetic dimensions that appear essential in the experiential phenomenon of incorporation.

2 Triclot, "L'immersion n'existe pas."

3 Salen and Zimmerman, *Rules of Play.*

4 Calleja, *In-Game*, pp. 167–179. It can be noted that Richard Gerrig also uses the concept of "transportation" in the literary field to analyze the experience of narrative worlds. See Gerrig, *Experiencing Narrative Worlds.*

5 Calleja, *In-Game*, p. 169, original emphasis.

On the artists' side, we note that the notion of "immersion" is also subject to caution. For example, plastician Grégory Chatonsky criticizes the aesthetics of the absolute that underlies the concept of immersion: "What we question is not only the immersive aesthetic, it is also the immersive ontology, an ontology of the absolute that makes us believe that we could have access to the real as such, that we could get rid of this inextricable relation between the two, the thing and the object [...]."[6] Artist-researcher Liam Jarvis insists on the manifest contingency of the immersive experience: "[I]mmersion through an aesthetic experience is contingent—we are already 'in' a world, much as we are already bound 'within' a body. The specific 'in' of immersion hints at a desire to reach something beyond or extranormative."[7] In a general way, to apprehend the immersion starting from the technological dimension of the equipped body only and from the "explorable" or interactive nature of the virtual environment proposed via the head-mounted display occludes the genuine quality of the experience. What is the real engagement of the physical body? What about the spatiality of the body and the perception of space in these devices? Also, what are the emotional and affective implications of the "body-subject" (*corps propre*), as this body, according to Maurice Merleau-Ponty, is what reveals and at the same time supports the subjectivity, and is proved to be the principle of all perception?[8] If, like William James, we consider that aesthetic emotions are essentially bodily,[9] and if we follow another representative of the pragmatist philosophy, John Dewey, in saying that "the scope and content of the relations measure the significant content of an experience,"[10] then it is important to grasp the artistic experiences that the devices of virtual reality or mixed reality offer us, according to the relations that they engage between the body (perceiving and lived) and the spaces (real, virtual; corporal, pericorporal, emotional; and yet, scenic, representational, narrative). It is indeed in this confrontation of the body (in movement) with spaces with multiple dimensions that the possible configurations of a significant immersion are drawn.

Initiated at the end of 2019, during the *Novembre Numérique* (Digital November) event of the French Institute and the Universidad de las Americas (UDLA) in Quito, *Encounters* was further developed during the summer

6 Chatonsky, "L'immersion comme ontologie et esthétique de l'absolu."
7 Jarvis, *Immersive Embodiment*, p. 73, emphasis added.
8 Merleau-Ponty, *La phénoménologie de la perception*, pp. 114–179.
9 See Shusterman, "The Pragmatist Aesthetics of William James," pp. 347–361.
10 Dewey, *L'art comme expérience*, p. 95.

of 2020 at the VRHAM! festival in Hamburg, and during a multi-user residency in Avignon in the spring of 2021. A fourth creative residency was held in October 2022, in Kaohsiung, as part of the selection of the project for the Taiwan XR residency. Part of the elaboration of the project was thus realized *in public* and *with the public*. The approach during these residencies was mainly focused on the work of staging, correlated to that of the user experience. Mathieu Pradat and his team proceeded by iterations from observations that aimed to examine how different audiences adopt an immersive narrative. The experimental sessions with the public were recorded and questionnaires were given to the participants at the end of the experience. Filmed interviews, allowing for more open exchanges, were also set up. The aim of these sessions, questionnaires, and interviews was to use these multiple feedback materials to identify what each person understood and felt, and thus adapt the project for a better reception. In this chapter, we will therefore examine the double dynamic of immersion that *Encounters* proposes, namely a poietics and aesthetic dynamic, to use the terminological distinction made by Paul Valéry.[11]

Encounters as a Setup and as an Experience

Encounters is a thirty-two-minute virtual reality experience that is still in a production phase. Four parts of eight minutes follow one another—"The Crowd," "Birds," "Colossus," and the "Epilogue"—and constitute an accelerated approach to the trajectories of humanity, like a shortcut of the Anthropocene. Participants live the experience equipped with stand-alone virtual reality headsets with open-back audio headphones. They roam freely, barefoot, on a horizontal waterproof platform of 10x5m, which can accommodate up to eight participants. Water that is 3mm deep spreads over the substrate. A screen with video-mapping, suspended above the pool, reflects what the participants see in the virtual world. In the virtual universe, users are visible through avatars that are graphically close to the fictional characters of this world, inducing a voluntary blurring between the entities represented.[12] As for the sound environment, four loudspeakers on stands placed on each corner of the set emit ambient sounds. These sounds come from recordings made in real time via microphones placed in ecosystems

11 Valéry, "Discours prononcé au Deuxième Congrès international d'esthétique et de science de l'art."

12 However, there is a detail that distinguishes them: only virtual entities wear a hat.

directly threatened by rising water levels (for example in Camargue, a region located in the south of France). The experiment is a fiction but, in addition to the sound recordings produced in these fragile places, it relies on two other documentary materials: the real-time positioning data of planes, container ships, and satellites in the China Sea, off the coast of Shanghai, in "Colossus"; and the movements and voices of the participants of the experiment in "The Epilogue." It should be noted that the voice of the participants is subject to a singular treatment, which makes it unintelligible during the first three chapters, thus hindering communication before "The Epilogue."

From a narrative standpoint, the different parts of the narration are interspersed with an interlude of rain (seen in the virtual space), which refers as much to the birth of life on earth as to the imminent and inescapably destructive rise of the waters. In the first chapter, "The Crowd," characters spring from each of the participants as if they were their doubles. Seemingly refusing any form of exchange, these humanoid entities quickly turn away and gather among themselves. In "Birds," small, sublime birds are transformed by the actions of the public into greedy fowls that grow, burp, and fart. They end up poisoning themselves in a pantagruelic atmosphere. In the third chapter, "Colossus," users in turn become immense, without immediately realizing this change in scale. It is the landscape and the objects that surround them that allow them to apprehend their new size. 12,000m tall, they move above an abstract and infinite ocean where tiny container ships sail. In the sky, airliners, at eye level, are perceived as mere buzzing mosquitoes. Planes, boats, and satellites are diverted one towards the other by the users' actions and they end up exploding through the body envelopes of their virtual avatars. All the interactions of the participants lead in the end to a saturation, in visual and sonic terms, of the different worlds discovered successively in the chapters. This phenomenon aims to implicitly show that each of our actions constantly builds and rebuilds the world around us and has a real impact. Finally, "The Epilogue" constitutes a moment of hope during which each participant sees a virtual character trying to enter into relationship with them by a kind of childish mimicry. The participants are invited to follow the movements of the characters and to respond to them. They then understand that they can leave an imprint of their passage in the world of *Encounters* and thus call out to other participants from a distance in time.

From the point of view of interactivity, a double graphic treatment responds to the presence of users. A blue envelope draws the volumetric contours of all the characters, animals, and objects (the avatars of the participants, the humanoid entities, the birds, the ships, etc.). The aspect of the envelope

Figure 5.2: The narrative progression of the first three chapters functions according to the same principle of passage from the one to the many, then to the countless. When the saturation is total, the water takes over again. La prairie productions, SoWhen? and Serendipity Films.

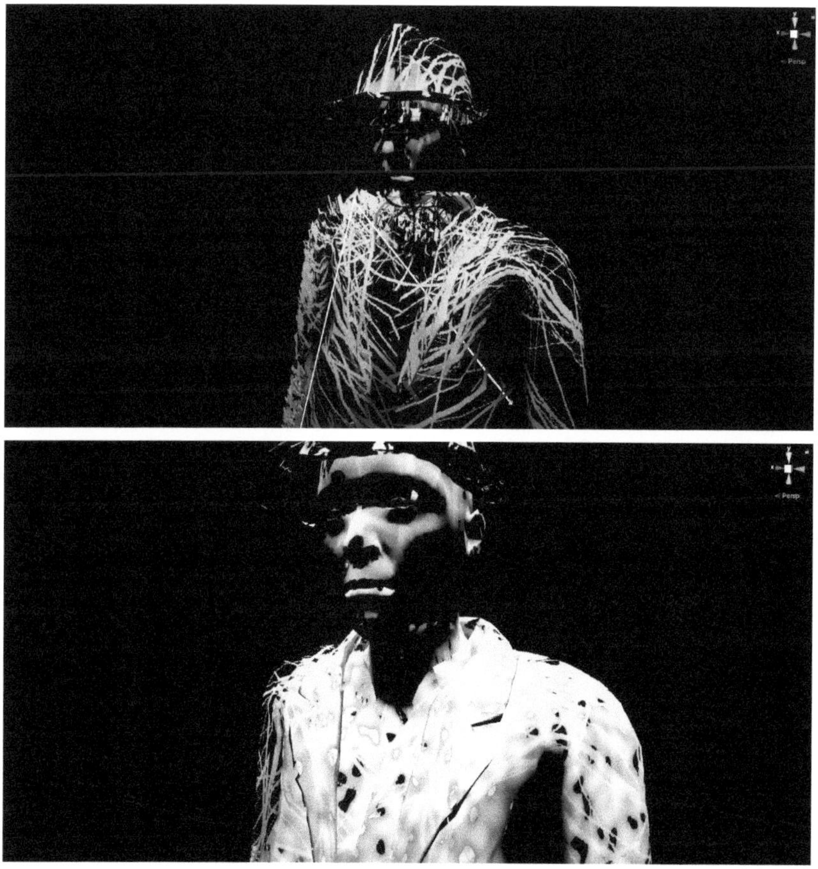

Figure 5.3, 5.4: The blue envelope is a function of the distance. The operation is dynamic and dual. © La prairie productions, SoWhen? and Serendipity Films. Shader artist: Florent Brillet.

Figure 5.5: The blue envelope is a function of the distance. The operation is dynamic and dual. © La prairie productions and SoWhen? and Serendipity Films Shader artist: Florent Brillet.

varies according to the position of the users, which is calculated by a tracking system. The level of detail increases as the participants move closer to the characters and objects and decreases as they move away from them. Moreover, each entity comprises an inner material made of white particles. This material reacts to the level of movement. Conversely, after a few seconds of immobility, the inner matter of the characters or of the user's avatar fades. The visual perception of the environment thus depends on the position of the users in the space and on their level of movement in time. The public evolves in a dynamic, vibrant, and reactive universe which transforms itself from them, and in real time, throughout the experience.

The Immersive Logics of *Encounters*: Formativity, Performance, and Contact

The double involvement of the spectators-users in the work, at the moment of its conception and at the moment of its reception, is a distinctive feature of *Encounters*. The public fully participates in the "instauration" of the project, to take up the notion developed by René Passeron following Étienne Souriau,[13] but more generally in its "formativity." We refer here to the "operative" conception of the form as a process, as developed by Luigi Pareyson in the 1950s.[14] Seeking to focus "on the making more than on

13 Souriau, *L'instauration philosophique*; Passeron, *Recherches poétiques*, p. 14 and following.

14 Pareyson, *Esthétique*.

the contemplating," the philosopher considers the artistic experience as a complex dynamic which exceeds the cleavages between invention and production, between form and content, between finished work and work in becoming, and between creator and spectator, to propose "an aesthetics of the formativity" where the form is rather apprehended as "a result, or better the success of a process of formation," and the artwork like a "formed form" and a "forming form."[15] Although the use of this theory can seem anachronistic, considering the contemporary technological context, it nevertheless seems enlightening to us when examining the experiences of virtual reality, as long as one integrates the spectators-users as stakeholders of the dynamic process of formation of the work, and thus here, of its actualization. The immersion that *Encounters* proposes should not only be considered as a simple reception modality, but also as a formative operation, while constituting a performative experience. Experimentally involved in the conception, then sensorially and corporally incorporated into the setup, and cognitively immersed into the fiction, the participants are indeed "immersants," in the sense that Liam Jarvis, following the artist Char Davies, uses this term; that is to say, they are subjects who perceive, experiment, and *perform*—and we could add: subjects who, at the same time, "form."[16] Immersion in the project *Encounters* is therefore a setting in space, a setting in motion, and a setting in action. It implies a practice in which "the practitioner sets up in the sense of action: they do not only operate an apparatus foreign to them, they become apparatus."[17] And the practitioner also becomes an interface, via their body. They are thus as much an "incorporated interface," a "body-operator," as a body-living and a body-perceived in a multidimensional space where real and virtual intermingle.[18]

"As soon as we arrive, the first characters we see [in the VR headset] are the ones that are supposed to represent our classmates, so we were touching

15 Pareyson, *Esthétique*, pp. 73–109.

16 "[T]he immersant [is] both performer and perceiver, and experimenter-subject. The immersed spectator is audience to themselves from within their unrehearsed act of performance," explains Liam Jarvis. Jarvis, *Immersive Embodiment*, p. 77. Char Davies coined the term in the 1990s and theorized it in several papers, such as Davies, "Virtual Space."

17 Mahé, "Les pratiqueurs," p. 126. Jean-Paul Fourmentraux and Samuel Bianchini have previously used the term "practitioner" to define the spectator of interactive installations. See Fourmentraux and Bianchini, "Médias praticables." The term seems particularly adapted to qualify the status and the role of the participants in the experiences of Free-Roaming Virtual Reality, in particular because it convokes the idea of agency.

18 On the notions of "embedded interface" and "body-operator", see Chatelet, "The Body at Work."

each other," explains Baya, a young participant in an experimentation session during the creation residency in Avignon.[19] She continues: "[T]hen, others are added, so we no longer know who is who."[20] The blurring of the borders voluntarily established between the space of the representation and the physical space, notably by the fact that the representation of the androids and of the avatars embodying the participants is graphically similar, and the fact that water is present both virtually and in reality, imply, as the testimonies clearly show, that the body, although destabilized, appears to be an essential reference point for situating oneself in this unique environment. It is important to situate one's "own body" or "body-subject," according to Merleau-Ponty's concept, but also to situate the bodies of others in order to apprehend space. Jade, another participant, explains: "I didn't dare go too far into the dark or get too close because I didn't know from which distance I was going to touch the person."[21] In *Encounters*, we experience the phenomenological affirmation that the perception of space is in and through our bodies. "The body is our general medium for having a world," as Maurice Merleau-Ponty reminds us.[22] In *Encounters*, the body becomes this "storm center" constitutive of our relationship to the environment, as William James describes it: "The world experienced comes at all times with our body as its centre, centre of vision, centre of action, centre of interest."[23]

The body involved in the project is also a machinic body, because it is technologically equipped. It is not, however, reduced to the head-mounted display alone. Through its complex device that makes water a decisive parameter of perception, *Encounters* seeks to distinguish itself from a majority of virtual reality projects that grant vision a central place. Falling within the enactive perspective of art, *Encounters* is not only about seeing, but above all about doing and feeling in order to see.[24] This is why it was conceived as a collective kinaesthetic and sensory experience. If the water felt by the participants with their bare feet, the sounds heard in their headphones, the images seen on the screen on their head-mounted display, and the touch of the other participants felt on their bodies constitute

19 La prairie productions, "*Rencontres*—récit d'une création avec et en public." These testimonies and the following ones come from the recordings made during the creation residency of 2020, at Le Grenier à Sel, in Avignon.
20 La prairie productions, "*Rencontres*—récit d'une création avec et en public."
21 La prairie productions, "*Rencontres*—récit d'une création avec et en public."
22 Merleau-Ponty, *La phénoménologie de la perception*, p. 171.
23 James, *Essai d'empirisme radical*, pp. 137–138.
24 See on these notions, Noë, "Experience and Experiment in Art"; and Noë, *Art as Enaction*.

exteroceptive stimulations, the setting in movement of the "practitioners," for whom the water incites a certain slowness, situates them in a proprioceptive perception, which corresponds to their exteroceptive perception, thus placing them in a real wandering, redoubled—or "translated"—via their avatar in the virtual environment. The process of performative corporeality induced by this "technaesthesic experience"[25] thus implies a plurality of bodies. Following Don Ihde, we could consider, for example, a "body one," the "phenomenological body" engaged in "microperception," and a "body two,"[26] culturally constructed and engaged in "macroperception."[27] But it seems difficult to distinguish them strictly in *Encounters*, because we are more in a logic of reciprocal interactions, where each body ultimately turns out to be the product of technological mediations. What we must also remember is that the mediation of touch is essential here: "I was trying to touch to see if she [Baya] was there," Amine said after the immersive experience, "and it was another person ... When Philemon touched me, I had a little jolt ... It was weird ... We were really in there."[28] Philemon, another participant, added: "You are under the impression of being in the experience, because these are not the same sensations as usual, in particular because we walk barefoot in the water."[29] To touch is to "come into contact with someone or something,"[30] and one of the issues of *Encounters*, in terms of its narrative and staging, is precisely to work on the notion of "contact" as a link, an edge, and a border. In this perspective, water is an essential element.

According to Gestalt Therapy—which is inspired by the great philosophical currents of the twentieth century, such as phenomenology, pragmatism, and existentialism—psychology consists in observing "what happens at the *border-contact between the individual and the environment*."[31] Frederick S. Perls indeed considers that: "[i]t is here, at the boundary between the two-limit that psychological phenomena happen. Our thoughts, actions, behaviour, emotions are our way of interacting with and experiencing these borderline events."[32]

25 We borrow this expression from Edmond Couchot. See Couchot, *Images*, and *La technologie dans l'art*.
26 Ihde, *Bodies in Technology*, p. 69 and following.
27 Idhe, *Technology and the Lifeworld*, p. 29 and following.
28 La prairie productions, "*Rencontres*—récit d'une création avec et en public."
29 La prairie productions, "*Rencontres*—récit d'une création avec et en public."
30 "Toucher."
31 Perls, *Manuel de Gestalttherapie*, p. 34; our emphasis.
32 Perls, *Manuel de Gestalttherapie*, p. 34.

It seems fruitful to us to import the concept of "border-contact" into the artistic field in order to apprehend more finely the immersive logics at work in *Encounters.*[33] In this context, the notion has a rich polysemy: the "contact-border" first refers to the link between the real and the virtual created through the familiar sensation of water on the skin, which summons the memory of the "practitioners" inside the virtual experience. It then refers to the correspondence between real and virtual, which is created through the real sensory restitution of the virtual representation seen inside the headset, that is to say the same shallow water that they really go through. Finally, it evokes the frontier between the material and the spiritual to which walking on water, commonly known as a miraculous journey on liquid matter, refers. Beyond the question of immersion, the sensoriality of water on the skin gives the "practitioners" an anchorage. It even amplifies their perception of the "border-contact" by giving it a materiality. Water thus offers a concrete support to their engagement in the environment of the experience: it simultaneously crosses the real, virtual, and imaginary dimensions of the relations they maintain with the multiple spaces of the experience. Physicality thus gives consistency to the fictional world of *Encounters*, making the immediately felt physical sensations interact with the technologically remedied sensations of the experience. In the "theory of contact," Gestalt Therapy distinguishes between the notions of "pre-contact," "contact" (or "making contact"), "full contact," and "post-contact."[34] To schematize, the first refers to the phase of emergence of the need; the second to the encounter with the object; the third to the actual experience; and the last to the assimilation of the experience. We could then interpret these different phases in the light of the experience proposed in *Encounters*. The pre-contact phase would be the discovery of the apparatus: the pool envisaged from its perimeter, the public looking not only at the participants already in immersion, but also at the screen above the pool, giving access from the outside to a vision of the virtual world. The contact phase, envisaged as a transition phase, could be both the vision of the virtual world through the screen and the preparation to the experience, with the various related thresholds: the removal of the shoes, the washing of the feet, the installation of the headset. The full

33 During the project's development phase, Mathieu Pradat wondered whether it was possible to use the concepts of Gestalt Therapy, originally conceived from a therapeutic perspective, to imagine an artistic experience such as *Encounters*. In this section, we propose to retrace this questioning and expand it.

34 See Masquelier-Savatier, *La Gestalt-Thérapie*, pp. 61–86.

contact phase could correspond to the experiment run, and in particular to the phases of interaction with the characters. The post-contact phase, finally, could find an echo in the project's fourth chapter, which has a reflective dimension. In this chapter, we meet another self, similar to us, and we are led to reflect on the effect of our actions on the world. The goal here is indeed to assimilate the experience. The exit from the experiment, with the removal of the headset, the putting on of one's shoes, and the leaving of the place after possibly wandering around the basin, could also be apprehended as a phase of "post-contact."

From the narrative standpoint, these concepts also seem relevant. The first chapter thus presents a situation of pre-contact with a perceptual exploration at a distance from the characters, before the participants are allowed to instigate a first relation with them. The phase of making contact is found when the users try to face the characters, while they evade and seem to flee a little further. We can note that these two sequences of pre-contact and making-contact take place at a distance: users are spectators of a 360-degree space, where an action is being prepared and in which they have not yet taken part, or where an action is taking place while they momentarily remain simple witnesses. The full contact phase is the narrative moment when users carry out an action, a gesture, or pronounce a word. It is the moment of action, beyond the perceptual and spatial exploration that the experience implies. The graphic treatment of the experiment gives legibility to these phases of pre-contact and making contact, in particular. This is particularly the case with the blue envelope of the fictional characters, which densifies little by little, at a short distance, when the full contact phase is about to occur. On the contrary, the white particles, which are like a visual sounding board of the movements of the characters, inform us of the phases of pre-contact and making contact as well as of the post-contact phase. Generally speaking, we note that immersive experiences possess a singularity in their succession of *exploration* and *action* or, in other words, in the alternation between phases of pre-contact/contact and phases of full contact/post-contact.

A second concept originating from Gestalt Therapy also seems enlightening for our project, both for its development and for its analysis: that of "field," which led to the founding idea of "organism/environment field." According to Perls, we are part of a field and our behaviour is always the result of a global field including ourselves and our environment.[35] Kurt Lewin, one of the major contributors to "field theory," defines it as "the totality of coexisting

35 Perls, *Manuel de Gestaltthérapie*, p. 33.

facts which are conceived of as mutually interdependent."[36] The field is therefore not given once and for all, but because it is part of a dynamic co-evolutionary process, it appears as a potentiality that is actualized. Moreover, the field does not exist by itself, it always proves to be in relation with something. Jean-Marie Robine explains:

> To install an "organism/environment field" is to emphasize that there is an articulation between the organism and its environment, indicated by the slash. The place of this experience between the two poles of the field is called the "contact-border," since it is the locus of the movements of differentiation and integration that animate the field, at the same time unifying it into a totality and limiting it by tracing its own contours.[37]

Thanks to the "field theory" of Gestalt Therapy, we realize that the virtual cannot be detached from the real. Working on the association of the two terms, on their synergy, was precisely one of the creative tracks for the *Encounters* project. Importing these concepts from Gestalt Therapy to the field of virtual reality allows to consider that users, by integrating the field of the virtual during an experience, see that their behaviour is really affected. They respond to audiovisual stimuli via their head-mounted display by acting on the virtual environment. But the simulation only produces the expected real effect if the user can act on this virtual environment. It is on this principle that virtual reality therapies are used, for example, in the treatment of vertigo or arachnophobia.[38] We are reminded that:

> During cognitive behavioural therapy, the reference psychotherapeutic technique in the treatment of anxiety disorders, exposure, must be progressive, hierarchical, and long enough to be therapeutic and to allow anxiety to be reduced by half, thus leading to the phenomenon of habituation with extinction of the fear response. Traditionally, exposure is carried out either in reality (*in vivo*) or in imagination (*in imagino*).[39]

VR technologies seem an interesting alternative for conventional therapies, because they operate with *presence*: beyond exposition, there is an

36 Lewin, *Field Theory in Social Science*, p. 240.

37 Robine, "Le fond du champ," p. 198.

38 See Bouchard et al., "Efficacité d'un traitement d'exposition en réalité virtuelle pour le traitement de l'arachnophobie chez l'enfant."

39 Levy, Rautureau, and Jouvent, "La thérapie par la réalité virtuelle dans la prise en charge des troubles anxieux," p. 662.

embodied action. Thus, the virtual reality experience ultimately remains anchored in reality. In other words, the *in imagino* works at the same time as the *in vivo*.

The virtual world could therefore be described, following the definition of Gestalt Therapy, as a world where potentialities constantly cease to be potentialities, only to be actualized in an accelerated time, a time distinct from the temporal unfolding of the non-virtual world, in which potentialities very often remain latent. Here, we find the Deleuzian distinction between virtual and actual, and between real and possible, later renewed by Pierre Lévy, who explains that the virtual is not opposed to the real:

> The word "virtual" is derived from the Medieval Latin *virtualis*, itself derived from *virtus*, meaning strength or power. In scholastic philosophy the virtual is that which has potential rather than actual existence. The virtual *tends* toward actualization, without undergoing any form of effective or formal concretization. The tree is virtually present in the seed. Strictly speaking, the virtual should not be compared with the real but to the actual, for virtuality and actuality are merely two different ways of being.[40]

In the virtual world, the characters around us engage us more than in the real world, and space is constantly changing. Walls can come down to afford a change of view. This is the case in Benjamin Nuel's *Isle of the Dead* (2018), where we go from a mundane apartment to the open sea in a matter of seconds, or in Jennifer Brea and Amaury La Burthe's *Unrest VR* (2017), where a patient's room turns into a lush forest. The constitutive immateriality of the space induces a change of its status. The potentialities induced in the environment are in fact actualized in the virtual, accelerating the slow, uncertain chronology of the real in order to give place to a vibrant world where space as field is no longer simply a decor, a secondary element for the narrative action and the characters. It rather constitutes a character of the narrative in its own right, evolving according to a dramatic course and a dynamic component of the experience, revealing it, through work on the rhythms, the allusions, the ellipses, which one could bring closer to the work of editing in cinema. Placing users in the field of the virtual is equivalent to positioning them in a world where the potentialities of the real, by being accomplished in the virtual, are based on an increased possibility of actions. They thus experience condensation of amalgamated realities.

40 Pierre Lévy, *Becoming Virtual*, p. 23.

Virtual reality does not hack the brain. It hacks perception. It tries to immerse the perceptive organs in a different sensory "bath," but the immersion is never complete because the brain keeps its reference points and remains anchored in the real. Therefore, it does not lead to a switch from one state to another, or to an effective cut-off from reality, but rather to a cohabitation of two perceptual states that play precisely on their articulation. If this coexistence of states, or better, this redoubling, is perceived as a limit, it seems to us, on the contrary to, that it constitutes a richness. It is in this sense that we can apprehend the virtual experience, thanks to Gestalt Theory, as an edge or a surface of contact which offers the remarkable possibility to be *here* and *elsewhere* at the same time. Because the singularity of the virtual resides in this constitutive duality, the term "immersion" thus seems to characterize less the virtual experience itself than the general phenomenon of virtualization at work in our digital societies.

By working on this dual relationship between the real and the virtual, *Encounters* ultimately seeks to be faithful to the promise suggested in its title, namely to amplify the quantity of our meaningful actions in relation to the world. A true "encounter" does indeed change the way we see and act. This is undoubtedly an underlying desire in any encounter, but it is even more so in the context of a project that is an ecological fable, with which the author wishes to show, through simulation, that another future is possible and that it is up to us to make it happen.

Bibliography

"Toucher." *Centre National de Ressources Textuelles et Lexicales*, 2012. https://www.cnrtl.fr/definition/toucher.

Bianchini, Samuel, and Jean-Paul Fourmentraux. "Médias praticables. L'interactivité à l'œuvre." *Sociétés* 2, no. 96 (2007): 91–104.

Bouchard, Stéphane, Julie Saint-Jacques, Geneviève Robillard, and Patrice Renaud. "Efficacité d'un traitement d'exposition en réalité virtuelle pour le traitement de l'arachnophobie chez l'enfant. Une étude pilote." *Journal de Thérapie Comportementale et Cognitive* 17, no. 3 (October, 2007): 101–108.

Calleja, Gordon. *In-Game: From Immersion to Incorporation.* Cambridge, MA: MIT Press, 2011.

Chatelet, Claire. "The Body at Work: Sensory-Motoric and Emotional Partitions of Film." In *Expanded Senses*, edited by Marc Ries and Bernd Kracke, 371–384. Bielefeld: Transcript Verlag, 2015.

Chatonsky, Grégory. "L'immersion comme ontologie et esthétique de l'absolu." *Grégory Chatonsky*, January 2012. http://chatonsky.net/immersion/.

Couchot, Edmond. *Images. De l'optique au numérique.* Paris: Hermès, 1988.

Couchot, Edmond. *La technologie dans l'art. De la photographie à la réalité virtuelle.* Nîmes: Jacqueline Chambon, 1998.

Davies, Char. "Virtual Space." In *Space: In Science, Art, and Society*, edited by Francois Penz, Gregory Radick, and Robert Howell, 69–104. Cambridge: Cambridge University Press, 2004.

Dewey, John. *L'art comme expérience.* Paris: Gallimard, 2005.

Gerrig, Richard J. *Experiencing Narrative Worlds: On the Psychological Activities of Reading.* New Haven, CT: Yale University Press, 1993.

Ihde, Don. *Technology and the Lifeworld from Garden to Earth*, Bloomington: Indiana University Press, 1990.

Ihde, Don. *Bodies in Technology.* Minneapolis, MN: University of Minnesota Press, 2002.

James, William. *Essai d'empirisme radical* [1906]. Marseille: Agone, 2005.

Jarvis, Liam. *Immersive Embodiment: Theatres of Mislocalized Sensation.* London: Palgrave Macmillan, 2019.

Jauss, Hans Robert. *Pour une esthétique de la réception.* Paris: Gallimard, 1978.

La prairie productions. "*Rencontres*—récit d'une création avec et en public." *Vimeo*, December 10, 2021. Video, 34:06. https://vimeo.com/655259571.

Lévy, Pierre. *Becoming Virtual: Reality in the Digital Age.* New York, NY: Plenum Trade, 1998.

Levy, Fanny, Gilles Rautureau, and Roland Jouvent. "La thérapie par la réalité virtuelle dans la prise en charge des troubles anxieux." *L'Information Psychiatrique* 93 (2017–2018): 660–663.

Lewin, Kurt. *Field Theory in Social Science: Selected Theorical Papers.* New York, NY: Harper & Brothers, 1951.

Mahé, Emmanuel. "Les pratiqueurs." In *L'ère post-média. Humanités digitales et cultures numériques*, edited by Jean-Paul Fourmentraux, 117–136. Paris: Hermann, 2012.

Masquelier-Savatier, Chantal. *La Gestalt-thérapie.* Paris: Presses universitaires de France, 2015.

Merleau-Ponty, Maurice. *La phénoménologie de la perception* [1945]. Paris: Gallimard, 2001.

Noë, Alva. "Experience and Experiment in Art." *Journal of Consciousness Studies* 7, no. 8–9 (2000): 123–135.

Noë, Alva. *Art as Enaction.* Cambridge, MA: MIT Press, 2006.

Pareyson, Luigi. *Esthétique. Théorie de la formativité.* [1954] Paris: Rue d'Ulm, coll. Æsthetica, 2013.

Passeron, René. *Recherches poétiques. Tome 1.* Paris: Klincksiek, 1975.
Perls, Frederick S. *Manuel de Gestaltthérapie* [1973]. Paris: ESF, 2003.
Robine, Jean-Marie. "Le fond du champ. En arrière-plan du concept." *Cahiers de Gestalt-thérapie* 22, no. 1 (2008): 197–210.
Salen, Katie, and Eric Zimmerman. *Rules of Play: Game Design Fundamentals.* Cambridge, MA: MIT Press, 2004.
Shusterman, J. R. "The Pragmatist Aesthetics of William James". *British Journal of Aesthetics* 51, no 4 (2011): 347–361.
Souriau, Étienne. *L'instauration philosophique.* Paris: Félix Alcan, 1939.
Triclot, Mathieu. "L'immersion n'existe pas." In *L'image virtuelle,* edited by Valentina Tirloni, Paris: L'Harmattan, 2012. https://shs.hal.science/halshs-01666832/document.
Valéry, Paul. "Discours prononcé au Deuxième Congrès international d'esthétique et de science de l'art." In *Variété IV,* 235–265. Paris: Gallimard, 1938.

About the Authors

Claire Chatelet is a Lecturer in Audiovisual and New Media at the Université de Montpellier Paul-Valéry, and a member of RIRRA21. Her research focuses on the generation of images by artificial intelligence, connected audiovisual writings, and the aesthetic challenges of immersive technologies. She also publishes her own photographs and AI images (https://www.instagram.com/claire_chatelet/).

Mathieu Pradat is a director and architect who works in the field of virtual reality and cinema. His projects have been selected in many festivals. He is a 2022 laureate of the Villa Albertine and has since 2022 been a member of the Open Documentary Lab of the Massachusetts Institute of Technology. He created La prairie productions in 2010.

6. Facing Oneself: Out-of-Body Experience, Immersion, and the Remediation of Cinema through Virtual Reality[1]

Pietro Conte

Abstract: Cinema and virtual reality have explored different strategies to visualize so-called out-of-body (and, more precisely, near-death) experiences, where the traditional first-person perspective ends up coinciding, paradoxically enough, with a seemingly disembodied gaze. The observers (in cinema) or users (in immersive virtual environments) get the impression that they are perceiving their own gazeless body, and they therefore feel embodied in a body that is, and at the very same time is not, their own. I maintain that this phenomenon is to be regarded as emblematic of the experience of bilocation that defines presence in virtual environments, thereby arguing against the myth of "total," "absolute," or "complete" immersion.

Keywords: bilocation, first-person shot, subjectivity, death, phenomenology

Evolution or Revolution?

The debate within media studies on the nature, potential, and limits of virtual reality technologies has developed through two fundamental

1 This chapter was written within the framework of the research project "AN-ICON. An-Iconology: History, Theory, and Practices of Environmental Images". The project received funding from the European Research Council (ERC) under the European Union's Horizon 2020 research and innovation program (grant agreement no. 834033 ANICON), and is hosted by the Department of Philosophy "Piero Martinetti" at the University of Milan (Project "Departments of Excellence 2023–2027" awarded by the Italian Ministry of Education, University, and Research).

Bédard, Philippe, Alanna Thain, and Carl Therrien, eds. *States of Immersion Across Media: Bodies, Techniques, Practices*. Taylor & Francis Group, 2025
DOI 10.5117/9789048558766_CH06

contrasts. The first is the opposition between utopists and apocalyptists. While the former emphasize the possibilities offered by the new medium in every field of human knowledge (or even of non-human knowledge—one need only think of the issue of the application of artificial intelligence to the avatars populating immersive environments), the latter point to the risk of "disconnecting" from everyday physical reality. The second contrast, which concerns us more here, is that between continuists and discontinuists.

Oliver Grau may be counted among the staunchest champions of the thesis that virtual reality is only the last stage—for the moment—in a gradual evolution that began with the frescoes enveloping diners in Pompeiian villas and that unfolded across the centuries through the search for increasingly complex immersive effects. New technological discoveries were constantly put to the service of this pursuit, enabling the creation of the dizzying Baroque ceilings covered in *trompe-l'œil* paintings, the eighteenth- and nineteenth-century dioramas and panoramas, magic lanterns, phantasmagorias, and stereoscopes, to mention only some of the stages in a journey that reached another crucial turning point with cinema.[2] While avoiding the risk of a linear, deterministic teleology, Fabienne Liptay and Burcu Dogramaci, too, have argued that immersion may embrace "any act or experience of plunging into something, without necessarily applying to computer-generated virtual environments."[3] From a different perspective, Werner Wolf, Walter Bernhart, and Andreas Mahler came to a similar conclusion by taking playing a board game or reading a book as different, yet equally typical immersive experiences.[4]

By setting out from a critical reading of the continuist hypothesis, one of the leading contemporary phenomenologists, Lambert Wiesing, has instead sided with those who believe that the new medium has engendered nothing less than "images of a new kind [*neuartige Bilder*]."[5] According to this German philosopher, virtual reality should be understood as an umbrella term encompassing all kinds of digital simulative and interactive media (including traditional video games or by now outdated virtual social worlds such as *Second Life* [2003]) through which the generated objects act as if they have "a substance and the properties of a substance."[6] The

2 See Grau, *Virtual Art*.
3 Liptay and Dogramaci, *Immersion in the Visual Arts and Media*, p. 1.
4 Wolf, Bernhart, and Mahler, *Immersion and Distance*, p. 4.
5 Wiesing, *Artificial Presence*, p. 89.
6 Wiesing, *Artificial Presence*, p. 100.

peculiarity of the kind of images that only the new media can generate—a peculiarity that entails a genuine "qualitative jump"[7] with respect to all previous image-making techniques—would lie in the fact that, for the first time in history, it has become possible to modify these images and interact with them in real time. By adopting the spirit and terminology of Edmund Husserl's phenomenological tradition, Wiesing has identified the real *differentia specifica* of virtual technologies as being the possibility of transforming a phantasy object (*Phantasieobjekt*) into an image object (*Bildobjekt*) that is visible, shareable, and constantly open to manipulation: "The viewer gains access to the imaginary image object."[8]

The thesis of a radical discontinuity between old and new media, however, increasingly seems to entail a rather naive enthusiasm that tends to replace the analytical rigour of a historical-theoretical proposal such as Wiesing's—however debatable it may be—with a whole range of bombastic slogans and rash rhetorical arguments. For instance, Chris Milk has famously described virtual reality (here to be taken in the narrow sense of the word to designate immersive digital environments accessible through head-mounted displays) as the "last medium" and "ultimate empathy machine":[9]

> Virtual reality [is] the first medium that has actually interfaced on a truly human level with our human senses—two of them right now, eyes and ears. Ultimately, what we're talking about is a medium that disappears, because there is no rectangle on the wall, and there is no page you're holding in your hand. It feels like real life. [...] Now, you're confined to a couch or you're confined to a 10-by-10 rug and you can't smell things, you can't taste things, you can't feel things. But you have to figure, if you scale the technology out, those will come eventually.[10]

By marking a caesura compared to traditional screen-based media, which are deemed to be rapidly growing obsolete, virtual reality would provide access to images whose hyper-mediated character is increasingly obliterated in favour of an apparent non-mediatedness.[11] The medium would tend to become imperceptible. This interpretation is not unreasonable in itself, but it is frequently turned into a far more radical claim: under the rallying cry:

7 Wiesing, *Artificial Presence*, p. 97.
8 Wiesing, *Artificial Presence*, p. 95.
9 Milk, "How Virtual Reality Can Create the Ultimate Empathy Machine."
10 Milk, "Virtual Reality Is 'The Last Medium'."
11 On immediacy (and on its paradoxical aspects in relation to its logical opposite, hypermediacy), see Bolter and Grusin, *Remediation*; and Pinotti, "Self-Negating Images."

"The Information Age is dead, welcome to the Experiential Age!," immersive virtual reality is celebrated for the *absolute* immediacy that it is believed to ensure (if not now, at any rate in some unspecified future), and which was regarded as unattainable by previous media apparatuses.

The main target of this frequently propagandistic narrative is quite clear: cinema. In the context of the opposition between continuists and discontinuists just outlined, the debate has indeed come to focus on the relationship between the medium of film, which is considered as by now outdated, and that of virtual reality. Significantly, Milk trained as a director, just like Alejandro González Iñárritu, who, at the seventieth Cannes Festival, in 2017, presented what remains the most famous VR installation, *Carne y Arena*, a six-and-a-half-minute solo experience conceived to make the user feel what it is like for Central American migrants to attempt to illegally cross the border between Mexico and the United States. Rather than discussing the work itself, which has already been the subject of intense debate among scholars from different scientific fields, I here wish to concentrate on the way in which Iñárritu advertised it by employing highly rhetorical strategies designed to draw a clear opposition between cinema and virtual reality.[12]

John Bailey, the then director of the Academy of Motion Picture Arts and Sciences, described *Carne y Arena* as an experience capable of opening up "new doors of cinematic perception," thus suggesting that the Mexican director's work, however innovative, still fell within the framework of traditional film-making.[13] By contrast, Iñárritu himself repeatedly stressed what he regards as unavoidable mediatic differences. In an interview for the *Los Angeles Times*, for example, he stated:

> The language of film gives you a partial reality, a fragmented reality and you have to interpret it. If I shoot two guys in a restaurant talking, you have to imagine the other 360 degrees. [...] In virtual reality, I have to give you a multidimensional thing. I have to build the whole thing. It ends the dictatorship of the frame. And the viewer is not passive. You are absolutely actively deciding where to go. I can try to manipulate you, but I will have no control. In cinema, I have the control.[14]

12 See, among others, D'Aloia, "Virtually Present, Physically Invisible"; Montani, *Tre forme di creatività*, pp. 132–138; and Raessens, "Virtually Present, Physically Invisible."

13 Bailey, "The Academy's Board of Governors Awards an Oscar to Alejandro G. Iñárritu's *Carne y Arena* Virtual Reality Installation."

14 Iñárritu, "How a Migrant Woman's Death Influenced *Carne y Arena*."

In *Carne y Arena*'s press kit, Iñárritu was even more categorical, unequivocally arguing that "while both are audiovisual, VR is all that cinema is not, and vice versa; the frame is gone and the two-dimensional limits are dissolved."[15] A similar argument has been made by immersive journalism pioneer Nonny de la Peña, who explicitly declared her ambition to put people directly "on scene," so as to make them live an experience in the first person, without the constraints imposed by traditional framing devices.[16] Such claims, whose eminently rhetorical nature has already been highlighted in various critical analyses,[17] can easily lead to the idea that sooner or later traditional images will be replaced by a *complete* immersion and a *perfect* symbiosis between human beings and technology. The "total cinema" that André Bazin still described in 1946 as a myth now seems to be within reach and increasingly virtual reality tends to be emphatically described as the realization of an immersive dream worthy of Adolfo Bioy Casares's 1940 novel *The Invention of Morel*.[18]

Despite authoritative warnings such as by Janet H. Murray, who explicitly urged scholars not to buy into new forms of "magical thinking" and to acknowledge that virtual reality is but "a medium of representation that the brain will process in its appropriate cultural context, just as it has learned to process speech, writing, photography, or moving images,"[19] the idea of a "complete immersion" is finding more and more supporters. Phenomenologically speaking, this notion is open to many different interpretations, since it equivocally refers to either ontological or psychological aspects, from a looming osmotic fusion between the world of images and physical reality to a loss of critical distance in the confrontation with pictures; from the blurring, or even disappearing, of the threshold "between the representational world

15 Iñárritu, *Carne y Arena*. Similar words are used by Jay David Bolter and Richard Grusin: "In painting and photography, the user's point of view was fixed. In film and television, the point of view was set in motion, but it was the director or editor who controlled the movement. Now computer animation can function like film in this respect, for it too can present a sequence of predetermined camera shots. However, the sequence can also be placed under the viewer's control, as it is in animated computer video games or virtual reality." Bolter and Grusin, *Remediation*, p. 28.

16 De la Peña, quoted in Gayet, "'With the XR medium we want to put people on scene'."

17 See Conte, *Unframing Aesthetics*; Dalmasso, "The Body as Virtual Frame"; Kubiński, "Immersion vs. Emersive Effects in Video Games"; and Murphy, "Virtual Reality and the Rhetoric of Empathy."

18 Bazin, "The Myth of Total Cinema"; and Bioy Casares, *The Invention of Morel*. For a detailed analysis of this novel in relation to immersive virtual environments, see Casetti and Pinotti, "Post-Cinema Ecology."

19 Murray, "Virtual/Reality," p. 20.

and the actual world",[20] to the impression of being teleported into an artificial, simulated universe. Ultimately, what we are dealing with is a "mythical media topos,"[21] an almost oneiric vision that resurfaces again and again in the founding myths of both cinema and virtual reality, extending—just to name a few—from the "stereokino" yearned for by Sergej M. Eisenstein in 1947[22] or the "cinema of the future" described by Morton L. Heilig in 1955,[23] to present-day notions such as that of "ultimate display," first proposed in 1965 by Ivan Sutherland, the inventor of a device (the so-called "Sword of Damocles") which is widely regarded as the first example of virtual reality.[24] Originally introduced in the late 1920s by Saint-Pol-Roux, the concept of an "organic" cinema capable of overcoming the boundaries imposed by the screen surface, thus blurring the threshold between viewers and images,[25] has been recently reformulated in terms of "embodiment" by cyberspace pioneer Randall Walser:

> Whereas film is used to show a reality to an audience, cyberspace is used to give a virtual body, and a role, to everyone in the audience. Print and radio tell, stage and film show; cyberspace embodies. […] A spacemaker sets up a world for an audience to act directly within, and not just so the audience can imagine they are experiencing an interesting reality, but so they can experience it directly. […] The filmmaker says, "Look, I'll show you." The spacemaker says, "Here, I'll help you discover."[26]

This leads us to the key point that is either implicitly or explicitly made in many analyses of the relationship between "old" and "new" media: cinema *forces us to imagine*, whereas virtual reality *enables us to experience*. The verbs used are in themselves meaningful: the activity of imagining that the film medium requires viewers to engage in is described as a "constraint," a limit imposed by the very nature of the medium. By contrast, the (supposedly) direct experience of an immersive environment is praised as an advantage—or, rather, as *the* advantage—that is "made possible" by virtual reality. The shift in the mode of use, through the transition from the conventional figure of the film-viewer (and, before that, of the viewer

20 Murray, *Hamlet on the Holodeck*, p. 103.
21 Schweinitz, "Totale Immersion, Kino und die Utopien von der virtuellen Realität," p. 139, my translation.
22 Eisenstein, "On Stereocinema."
23 Heilig, "The Cinema of the Future."
24 Sutherland, "The Ultimate Display."
25 Saint-Pol-Roux, *Cinéma vivant (1925–1930)*, p. 60.
26 Walser, "Spacemakers and the Art of the Cyberspace Playhouse," pp. 60–61.

of a painting or sculpture) to that of the explorer or even designer of an environment-image, is seen as marking the crucial leap from cinema to post-cinema.

Yet, it was not all that long ago that cinema itself was criticized not because of the space it *lends to* imagination but, on the contrary, because of what it *takes away* from imagination. For instance, in the 1910s and 1920s, a leading art historian and theorist such as Konrad Lange did not hesitate to condemn motion pictures and the film medium, accusing them of being so mimetic and lifelike that they hampered the imaginational process of "animating the inanimate," which he regarded as a necessary condition for any authentic aesthetic experience.[27] By contrast, cinema today (just like television) is viewed with growing dissatisfaction for the very opposite reason, namely for its alleged incapacity to give the viewer a genuine, all-round, multisensory experience—in brief, an experience that can be fully assimilated to that of the life-world.

It seems clear that in this clash between continuists and discontinuists one often tends to assume a dichotomous attitude which favours either one or the other extreme. On the one hand, there is a propensity to downplay the differences that undoubtedly exist between cinema and virtual reality; on the other, there is a tendency to overstress the contrast between the two media and to extol the more recent medium to the detriment of the "older" one. The thesis I wish to advance here instead is that avoiding Manichaean attitudes would be of much help in the debate on the nature of virtual reality as a *medium*. The advent of a new technology always entails an absorption and a remediation of previous technologies, but in turn it is prefigured by the latter. In the case we are discussing, the medium of virtual reality is foreshadowed

> not only by the narratives through which cinema has envisaged it in the digital age, but also by the visual and linguistic forms that cinema has adopted in prefiguring it and by the kind of experience it has consequently shaped. In cinema, a series of "figures of immersion" already appeared [...] that entail a redefinition of the sense of distance *vis-à-vis* the image in view of an *immediate* physical and emotional enjoyment of the image itself. [...] In genuine VR experiences, in turn, we can notice a constant negotiation between different paradigms that were dominant at different stages in the history of cinema.[28]

27 Lange, *Die Bewusste Selbsttäuschung als Kern des künstlerischen Genusses*, p. 10. Also see, in particular, Lange, *Das Kino in Gegenwart und Zukunft*.

28 Grossi, "Il visore immaginario," pp. 84–85, original emphasis, my translation.

A Matter of Point of View

If each medium is incorporated and assimilated into other media, it is only through this process of "positioning" with respect to previous dispositives that a representational technology comes to define itself. In the case under discussion, the question arises to what extent virtual reality can be said, on the one hand, to mark a radical break with cinema, as the advocates of the discontinuist paradigm maintain, while, on the other hand, it continues to rely heavily on syntactic structures, stylistic elements, and narrative techniques that are typically cinematic. Therefore, understanding the novelty and power of the new medium means at the same time grasping its limitations and the way it does not *replace*, but rather *reconfigure* the "older" medium. This is all the more true when the analysis shifts from the material condition and uses of virtual reality to its associated discursive, cultural, and political practices, which often insist on a techno-deterministic reading of immersion as such. To debunk the rhetorical strategies that continue to hamper a critical discussion of virtual reality, further reflection is needed on *both* the continuities and ruptures that characterize its relationship with cinema and cinematic experiences.

Now, one of the main features that are deemed to distinguish virtual reality from other immersive media is no doubt the possibility it grants users to constantly change the framing by moving their head. Certainly, the possibilities offered by a VR environment that only provides users with three degrees of freedom (corresponding to the three directions in which they can move their head) are limited—and limiting—compared to those offered by an environment that instead allows six degrees of freedom (i.e., one in which sensors—accelerometers, gyroscopes, and magnometers—also enable the mapping of bodily movements). The sense of embodiment of the experiencer of *Clouds over Sidra* (Gabo Arora and Barry Pousman, 2015), which features the use of a recording of pre-rendered environments, will be different from, and inferior to, that of the experiencer of the aforementioned *Carne y Arena*: while in the former case one can only look around and one is bound to the original shots of the 360-degree camera (much as in the virtual tours of apartments offered by cutting-edge real estate agencies), in the latter one can freely move about and explore the surroundings. In both cases, however, the rhetoric of "freedom" exploited to celebrate the advent of virtual reality by propagating the myth of its ability to replace physical reality clashes with all the technical constraints that affect even the most sophisticated immersive environment, which can allow the user to perform only a limited number of actions—namely those correlated to some relevant depicted features of the scene. As Matthias Wittmann puts it, "despite the new navigation options" offered by virtual reality, the user "can perform zoom-ins, axis jumps, and other viewpoint changes only provided that these have been planned in advance" by the programmers as a significant part of the experience.[29] Nevertheless, the fact remains that the choice is left to the user what part of the environment to view by framing a given part of the scene. In cinema, by contrast, this choice is always made by the person operating the mechanical eye of the camera and the viewers are such precisely because they can only accept the director's "cut"; if they do not, they will necessarily have to abandon the world of the image, as is the case when we shut our eyes before a horror scene or focus for an instant on the glowing "emergency exit" sign of the cinema.

On the technical, aesthetic, and semiotic levels, the specific features of virtual reality can therefore better be understood by considering the ways in which the new medium pays tribute to the old, while at the same time revisiting it and dialectically entering into competition with it. In particular, the fluid point of view ensured by the VR experience can—or, rather, must—be interpreted with reference to the traditional filming device of the subjective shot and its post-cinematographic hybrid, namely the first-person shot.[30] Indeed, it is tempting to define virtual reality as an absolutizing of the subjective shot; as its sublimation into the experience of a seamless world algorithmically generated in accordance with the direction of the user's gaze. Virtual reality would thus amount to "the presentation of first-person experiences through the use of a head-mounted display and headphones that enable users to experience a synthetic environment as if they were physically there."[31]

This apparently straightforward definition in fact betrays a misunderstanding of virtual reality's potential, which is not at all limited to a sort of juxtaposition between the concept of the subjective shot and that of virtuality. Indeed, the subjective experience is only *one possibility* disclosed by the virtual interface—a possibility which has been the rhetorical focus of countless narratives seeking to celebrate the new medium's capacity to put ourselves in other people's shoes by empathetically adopting their different point of view.

While in merely *syntactical* terms, there is no doubt that VR technology, by virtue of its very structure, is based on the perspective embodied by the

29 Wittmann, "Vom Ende der 'Suture'," my translation.
30 On this, see Eugeni, "Remediating the Presence."
31 Mateer, "Directing for Cinematic Virtual Reality," p. 14.

user's gaze, things are quite different from both a *semiotic* and a *phenomenological* standpoint, for a subjective shot is never merely the outcome of the positioning of the subject of the scopic action (be it the film viewer or the experiencer of a digital environment) within the diegetic universe.[32] Virtual audiovisual enunciation undoubtedly expresses an intentional act of the perceiving subject, but this does not mean that any VR experience can be regarded as a "subjective" one, unless we are to use this expression in such a broad sense as to make it misleading. Rather, the construction of the diegetic world must primarily be understood as the outcome of a subject-viewer's perception and only secondarily as—*possibly*—also coinciding with the perception of a subject actually operating within the diegetic world. The transition from the first to the second sense in which subjectivity can be understood in relation to virtual reality is made possible only by the use of particular solutions provided by the storytelling or by the way in which the immersive environment is built through the interface. An immersive VR experience may therefore be characterized by the adoption of a first-person point of view, but also by a third-person or—more rarely—second-person point of view.[33]

So, while in virtual reality the digital universe is generated (or rather, actualized) starting from the gaze of the experiencer, who represents the visual centre of that universe, this does not mean that we can always speak of a subjective point of view. Moreover, while it is true that the first-person shot, as a "semiotic figure," presents certain differences compared to the classic device of the subjective shot,[34] it is equally true that virtual reality *remodels* the cinematic sequence based precisely on the subjective point of view by evoking it as a specific semantic element.

In order to clarify this point, I am now going to focus on an apparently marginal case that is nonetheless paradigmatic in its capacity to shed light on the remediating relations between cinema's expressive forms and stylistic features and those of virtual reality: the case of the visualization of an out-of-body (and more specifically a near-death) experience. This example is of particular interest since it implies the use of peculiar strategies designed to ensure that the viewer or the experiencer interprets a given scene as expressing not merely a generically subjective perspective, but rather a subjective perspective whereby the framed object enigmatically overlaps with the body of the framing subject. Thus, one faces the paradoxical situation in which the

32 On this, see Bédard, "La machine subjective?"
33 See Larsen, "Virtual Sidekick"; and Sherman and Craig, *Understanding Virtual Reality*, p. 309.
34 Eugeni, "Remediating the Presence," p. 209.

subjective perspective comes to coincide with a seemingly *disembodied* gaze in such a way that the observers get the impression that they are perceiving their own now gazeless body, and therefore feel embodied in a body that is, and at the very same time is not, their own.

Remediating the First-Person Shot

In cinema, the first way to visually render the dissociation between the body and the gaze is provided by the most classic strategy on which the first-person point of view is based, namely the use of the shot/reverse shot through two separate and specular frames: a first shot showing a character looking at something, and a second one showing what the character sees (or, more rarely, vice-versa). This is the simplest case, which does not differ syntactically from any other ordinary subjective shot. It is to be found, for instance, in a blockbuster such as *Ghost* (Jerry Zucker, 1990), precisely in the scene where Sam (Patrick Swayze) is killed during a walk in the company of his girlfriend Molly (Demi Moore). The viewer does not immediately realize what has happened, since after the murder the camera continues to film the protagonist (or rather his phantasmatic alter-ego—but the viewer does not know this yet—) as he chases the criminal in an effort to apprehend him. When he fails to catch him, Sam walks back and sees a desperate Molly kneeling on the street floor. He approaches her to find out the reason of her despair and only then does he see himself—his own lifeless body—resting on the ground. As usual, the two frames that make up the subjective shot first show the viewing subject and then the object viewed, but, in this case, the latter coincides with the (body of the) former (fig. 6.1).

Immediately after this scene, *Ghost* offers a second strategy for visualizing an out-of-body experience: the *simultaneous* appearance of viewer and viewed *within the same frame* (fig. 6.2). This occurs only through a special effect of superimposition, since it would be otherwise impossible for the film medium to show the perceiving subject and the object perceived within a single frame. Juxtaposition and a skilful use of transparency enable the director, Jerry Zucker, to simultaneously show Sam's lifeless body and his soul, which has inherited that body's gaze.[35]

Then, there is a third way to make the out-of-body experience visible cinematically, namely by showing not the *detachment* but the *detaching*

35 As far back as 1983, Francis Ford Coppola's *Rumble Fish* had offered an example of both strategies for the visualizing of the out-of-body experience that *Ghost* deployed in 1990.

Figure 6.1: *Ghost* (Jerry Zucker, 1990)

of the soul from the body—not the outcome, but the process leading to the disembodiment. Rather than use two separate frames of the same subject (shot/reverse shot) or "split" the subject within the same frame (special effect), this third strategy is based on the camera's movement. Such a visualization technique can certainly be described in words, but given its intrinsically motive nature it cannot be reproduced through static images, unlike the two previous instances: single stills may indeed provide a general idea of the process by reproducing certain moments of the represented event, but in no way do they represent an analogon of the process itself, which remains an eminently *filmic* operation, that is, one specifically pertaining

Figure 6.2: *Ghost* (Jerry Zucker, 1990)

to the cinematographic medium as such and essentially belonging to its "language."

The movement of the camera to convey the feeling of one's self leaving their body can either seamlessly lead from a shot to a reverse shot through a 180 degree rotation or come into play only after the "break" of a reverse shot by moving back and up in what is known as a depth dolly. The former technique is exemplified by the last scene in Brian De Palma's *Carlito's Way* (1993), where the dying protagonist, played by Al Pacino, is laid out on a stretcher and taken away from the station where he has just fallen victim to an ambush. The camera first adopts the Hispanic gangster's point of view by showing the neon lights on the ceiling; then a continuous 180-degree rotation leads it to film the face of the dying Carlito perpendicularly from above (fig. 6.3). The overall effect enables the viewer to feel like they are being lifted up by the camera's motion, so as to identify with the protagonist's departure from their own body. A very similar technique is used, for the same purpose, in Gaspar Noé's highly experimental *Enter the Void* (2009).

The other possibility that cinema has to visually suggest the dissociation between body and gaze in the out-of-body experience instead involves two distinct stages. It *first* exploits the shot/reverse shot technique and *then* the camera's movement, which comes into play to hint at the self's departure from the body to which it was hitherto confined. This visual strategy is *semantically* equivalent to the one proposed in *Carlito's Way* and *Enter the Void*, yet it is *syntactically* different. It was adopted by Douglas Trumbull in a masterpiece of sci-fi cinema, *Brainstorm* (1983), which foreshadowed the plot, poetics, and technique of *Strange Days* (Kathryn Bigelow, 1995); unlike the latter, it is rarely mentioned in studies on immersiveness, even though

Figure 6.3: *Carlito's Way* (Brian De Palma, 1993)

the American director himself has explicitly described it as "an immersive movie about an immersive technology."[36]

The story revolves around the invention of a headset that allows wearers to record their own experiences and emotions so as to enable other persons to relive them. In one of the film's key scenes, the protagonist, Lillian (Louise Fletcher), suffers a fatal heart attack and decides to record the last moments of her life. Her friend and colleague Michael (Christopher Walken) finds the tape and sets out to go through something that, by definition, can only be experienced in the first person: death as "the ultimate experience"—a significant expression used on a poster advertising the film to describe death as much as the movie itself. It is at this point that the camera, masterfully

36 Trumbull, "Immersive Cinema." For more on *Brainstorm*, see Charles Acland's chapter in this volume.

Figure 6.4: *Brainstorm* (Douglas Trumbull, 1983)

handled by Trumbull, first focuses on the equipment used by Lillian a few moments before her death (shot) and then on Lillian herself (reverse shot); then it progressively moves away from her lifeless body through a retrograde motion (fig. 6.4). The viewer is thus led to interpret the first frame as a subjective shot taken from Lillian's point of view when she was still alive and the second frame as another subjective shot, but this time from the perspective of Lillian's soul after it has left her body.

The analysis carried out so far shows how the visual rendering of the out-of-body experience, and more specifically of the near-death experience, can be understood as a figure of immersivity on which cinema has focused through the use of the subjective shot, applied in different ways (shot/reverse shot, camera motion, or a combination of the two). In order to get the viewers involved and give them the feeling that they are having a first-person experience, cinema employs a *dispositif* that can never be reduced to a merely syntactical construct, since it also brings into play the semantic and phenomenological sphere of the attribution of meaning in and through the very act of viewing.

There is one thing that cinema cannot do, however, *in principle*, that is, for reasons pertaining to the intrinsic features of the medium: it cannot show the viewer their *own* out-of-body experience. The subjective shot always refers to an intradiegetic character who is not—and cannot be—the viewer, not even in the exceedingly rare cases (ranging from *Lady in the Lake* [Robert Montgomery, 1947] to *Hardcore Henry* [Ilya Naishuller, 2015]) in which the film is entirely shot in the first person. When I watch a movie, subjective shooting allows me to identify with one or more protagonists, yet always with *someone else*. Clearly, even in the paradoxical case in which a character whose perspective the viewer adopts were to find themselves before a mirror in the intradiegetic world, the reflected image would be that of the character themselves or—a paradox within the paradox—of the camera (as in the aforementioned virtual tours of for sale apartments, where I find myself gazing at an environment from a 360-degree perspective, yet when I look down I do not see my legs, but rather the camera that has been placed at the centre of the room during the filming).

In remediating the *dispositif* of subjective shooting, virtual reality seems in a better position than film when it comes to letting us experience in the first person the process of leaving a body that we perceive *as our own*. In a 2017 experiment, a team of researchers headed by Mel Slater tested the capacity of an immersive VR environment to awaken in users the feelings of: 1) finding themselves embodied in a vicarious body capable of replacing their physical one; and 2) having an out-of-body experience through this virtual body perceived as their own.[37]

The first point pertains to the body ownership illusion, which is to say the fact that participants feel ownership of—and agency over—a digital body, even though they know that it is only an image. This is done by way of motion capture devices and a haptic suit which allow the avatar to move in sync with the physical body and to react to tactile stimuli. While the subject remains perfectly aware that they are dealing with an avatar, they cannot avoid perceiving this vicarious body as their own.[38] This is all the more so if, as in the particular case we are discussing, the virtual alter ego is placed before a mirror and casts a shadow upon the wall, so that the user perceives its movements as being in sync with their own, thereby identifying even more with the digital double.

During the second phase of the experiment, the participant's point of view is progressively shifted back and up, according to the ascending and retrograde motion imposed by the experimental setting. Some virtual spheres that were striking the digital body in the first phase, synchronically eliciting a haptic reaction from the physical body thanks to the electrical muscle stimulation granted by the haptic suit, now follow the same movement and rotate around the space surrounding the centre of visual perception, while the body that the user previously "possessed" remains motionless below, clearly visible from the new raised perspective (fig. 6.5). When the centre of visuo-tactile perception is moved towards the ceiling of the room, the virtual reality experience offers the potential for the subject

37 Bourdin et al., "A Virtual Out-of-Body Experience Reduces Fear of Death."

38 This has been proven by a whole range of experimental studies inspired by the phenomenon of the rubber hand illusion, originally described in 1998 by Matthew Botvinick and Jonathan Cohen. See Botvinick and Cohen, "Rubber Hands 'Feel' Touch That Eyes See."

Figure 6.5: Out-of-body experience in an immersive virtual environment (Bourdin et al., "A Virtual Out-of-Body Experience Reduces Fear of Death.").

of the experiment to feel like they have left their own body, and thus find themselves personally having an out-of-body experience. This would result in reduced death anxiety, as the authors of the experiment argue on the basis of the answers given by participants to a specifically designed questionnaire.

Here, I wish to stress two defining aspects of this immersive strategy. On the one hand, the use of subjective shooting in virtual reality seems to enable users to experience their *own* self leaving their *own* body, thereby overcoming the intrinsic limitation of the cinematic medium. On the other hand, however, virtual reality itself cannot be said to have achieved that "perfect," "total," or "complete" illusion of living a true out-of-body experience which cinema could not convey. On closer inspection, the disembodied perception that the immersive environment should give access to is never really disembodied, since it is still—and necessarily—bound to a specific technological setting, purposedly designed to allow only very limited freedom. In other words, the way the installation is construed *orients* the whole experience from the outset. Regardless of how intense the user's feeling of embodiment and sense of presence in the virtual world may be, an enunciator is still there and it still manifests itself in the use of typically cinematic devices and visual strategies, such as the "ascensional" movement of the camera "dragging" the self upwards and leaving the experimental subject no other option than to look around. This, in turn, contrasts with the perceptual information provided by the physical body, which makes itself felt in the "here" of the real world. Roland Barthes's observations about the intrinsically double nature of cinematic experience could therefore be suited for describing virtual reality experience as well: it is as if we had "two bodies at the same time: a narcissistic body which gazes, lost, to the engulfing mirror, and a perverse body, ready to fetishize not the image but precisely what exceeds it."[39]

Conclusion

Moving beyond sterile contrasts, while at the same time avoiding undue conflation between the two media, it must be argued that immersive virtual reality remediates cinema (as well as television) insofar as "it depends on the conventions and associations of the first-person point of view or subjective camera."[40] This chapter has developed this main argument by concentrating on out-of-body experiences. Although they might first appear as an all-too-specific case of consequently limited interest, they instead represent an emblematic example of the immersive experience, both in cinema and in virtual reality.

Consider film watching first. Autoscopic phenomena can be described in terms of "cinematic experiences of a body without a screen."[41] Conversely, the traditional film experience requires the viewers to move out of themselves in order to interact—thanks to the camera's movement—with the virtual characters inhabiting the image world: by adopting the filmic strategy of the subjective or point-of-view shot, the cinematic *dispositif* seeks to transform the "camera-eye" into a "camera-I."[42] This resembles an out-of-body experience insofar as it implies a sense of disrupted self. Unlike genuine out-of-body experiences, however, cinema-goers are asked to identify with an individual—the fictional character—*other* than themselves, and thus with a body *other* than their own. In the case of the film experience, then, the out-of-body experience comes to be "more of an extended 'inside-of-the-other-(body)-experience'."[43]

Virtual reality, instead, is hailed as granting us the possibility to live a genuine out-of-*our-own*-body experience. Thanks to its capacity to teleport us to a digital "elsewhere," the new medium would let us fully immerse into a simulated environment where the detachment of our own self from our

39 Barthes, "Leaving the Movie Theater," p. 421.
40 Bolter and Grusin, *Remediation*, p. 48.
41 Gerner and Guerra, "On the Cinematic Self," p. 91.
42 Sobchack, *The Address of the Eye*, pp. 200–202.
43 Gerner and Guerra, "On the Cinematic Self," p. 91.

own body could be artificially induced. Yet, despite all the rhetoric of perfect immersion and total illusion, the promise of a fully disembodied perception *within* the virtual diegesis is not—and cannot be—really fulfilled, since it clashes with the still-embodied perception of our physical body *outside* the virtual diegesis. Rather than misconceiving the idea of "telepresence" as if it implied the perfect illusion of inhabiting the virtual world while forgetting physical reality, it must be acknowledged that virtual reality users—in this respect similar to film viewers—have the feeling of *both* "being there" (in the mediated environment) and "staying here" (in physical reality).[44] This phenomenon of "bilocation" implies a particular kind of dissociation that could be fruitfully compared to the phenomenological concept of *Ichspaltung*, that is, of the "split" between an imagining I and an imagined I, which is the hallmark of all phantasy experiences. This can also allow us to retain the notion of the out-of-body experience as a useful tool for describing the specific kind of experience made possible by immersive virtual environments—not, however, as a completely disembodied perception, but rather as the continuous negotiation between the bodily self and its virtual alter ego. Therefore, the new interactive, 360-degree digital environments cannot be said to erase (nor even to be aimed to erase) the previous technologies of the subjective camera and first-person shot; rather, they simultaneously leech off and celebrate these technologies, sublimating them and enabling them to continue to exist, dialectically, as both technical and cultural points of reference to measure the potential, effects, and successes of virtual reality as well as its limitations and even failures.

Bibliography

Bailey, John. "The Academy's Board of Governors Awards an Oscar to Alejandro G. Iñárritu's *Carne y Arena* Virtual Reality Installation." *A.Frame: The Digital Magazine of the Academy*, October 27, 2017. https://www.oscars.org/news/academys-board-governors-awards-oscarr-alejandro-g-inarritus-carne-y-arena-virtual-reality.

Barthes, Roland. "Leaving the Movie Theatre." In *The Art of the Personal Essay: An Anthology from the Classical Era to the Present*, edited by Phillip Lopate, translated by Richard Howard, 418–421. New York, NY: Anchor Books, 1995.

Bazin, André. "The Myth of Total Cinema." In *What Is Cinema?*, edited and translated by Hugh Gray, 23–27. Los Angeles, CA: University of California Press, 1967.

44 See Pinotti, "Staying Here, Being There."

Bédard, Philippe. “La machine subjective? Les appropriations cinématographiques des dispositifs immersifs contemporains.” *Canadian Journal of Film Studies* 28, no. 1 (2019): 66–92. https://doi.org/https://doi.org/10.3138/cjfs.28.1.2019-0012.

Bioy Casares, Adolfo. *The Invention of Morel.* New York, NY: NYRB Classics, 2003.

Bolter, Jay David, and Richard Grusin. *Remediation: Understanding New Media.* Cambridge, MA: MIT Press, 1998.

Botvinick, Matthew, and Jonathan Cohen. “Rubber Hands ‘Feel’ Touch That Eyes See.” *Nature* 391, no. 756 (February 19, 1998). https://doi.org/10.1038/35784.

Bourdin, Pierre, Itxaso Barberia, Ramon Olivia, and Mel Slater. “A Virtual Out-of-Body Experience Reduces Fear of Death.” *PLoS ONE* 12, no. 1 (January 9, 2017). https://doi.org/10.1371/journal.pone.0169343.

Casetti, Francesco, and Andrea Pinotti. “Post-Cinema Ecology.” In *Post-Cinema. Cinema in the Post-Art Era*, edited by Dominique Chateau and José Moure, 193–217. Amsterdam: Amsterdam University Press, 2020.

Conte, Pietro. *Unframing Aesthetics*. Milan: Mimesis International, 2020.

D’Aloia, Adriano. “Virtually Present, Physically Invisible: Virtual Reality Immersion and Emersion in Iñárritu’s *Carne y Arena*.” *Senses of Cinema* 87 (June, 2018). http://sensesofcinema.com/2018/feature-articles/virtually-present-physically-invisible-virtual-reality-immersion-and-emersion-in-alejandro-gonzalez-inarritus-carne-y-arena/.

Dalmasso, Anna Caterina. “The Body as Virtual Frame: Performativity of the Image in Immersive Environments.” *Cinéma & Cie* 32 (2019): 101–119.

Eisenstein, Sergej M. “On Stereocinema.” In *3D Cinema and Beyond*, edited by Dan Adler, Janine Marchessault, and Sanja Obradovic, 20–59. Toronto: Public Books, 2013.

Eugeni, Ruggero. “Remediating the Presence: First Person Shots and Post Cinema Subjectivity.” In *Rimediazioni. Immagini interattive*, edited by Tiziana Migliore, 210–214. Ariccia: Aracne, 2016.

Gayet, Mathieu. “ ‘With the XR Medium We Want to Put People on Scene.’ Nonny de la Peña (Emblematic Group).” *XRMust*, June 8, 2021. https://www.xrmust.com/xrmagazine/nonny-de-la-pena-emblematic-group/.

Gerner, Alexander, and Michele Guerra. “On the Cinematic Self: Cinematic Experience as ‘Out-of-Body’ Experience?” In *Altered Self and Altered Self Experience*, edited by Alexander Gerner and Jorge Gonçalves, 85–106. Norderstedt: Books on Demand, 2014.

Grau, Oliver. *Virtual Art: From Illusion to Immersion*. Cambridge, MA: MIT Press, 2003.

Grossi, Giancarlo. “Il visore immaginario. La nascita della realtà virtuale nello spirito del cinema.” *La Valle dell’Eden* 33 (2018): 77–85.

Heilig, Morton L. "The Cinema of the Future." In *Multimedia: From Wagner to Virtual Reality*, edited by Ken Jordan and Randall Packer, 239–251. New York, NY: Norton & Company, 2002.

Iñárritu, Alejandro González. *Carne y Arena. Press Kit.* Milan: Fondazione Prada, 2017.

Iñárritu, Alejandro González. "How a Migrant Woman's Death Influenced *Carne y Arena*." Interview by Carolina A. Miranda. *Los Angeles Times*, November 28, 2017. https://www.latimes.com/entertainment/arts/miranda/la-et-cam-alejandro-inarritu-lacma-20171127-htmlstory.html.

Kubiński, Piotr. "Immersion vs. Emersive Effects in Video Games." In *Engaging with Video Games: Play, Theory, and Practice*, edited by Dawn Stobbart and Monica Evans, 133–141. Oxford: Inter-Disciplinary Press, 2014.

Lange, Konrad. *Die Bewusste Selbsttäuschung als Kern des künstlerischen Genusses.* Stuttgart: Veit & Comp., 1895.

Lange, Konrad. *Das Kino in Gegenwart und Zukunft.* Stuttgart: Enke, 1920.

Larsen, Mads. "Virtual Sidekick. Second-Person POV in Narrative VR." *Journal of Screenwriting* 9, no. 1 (2018): 73–83.

Liptay, Fabienne, and Burcu Dogramaci. *Immersion in the Visual Arts and Media.* Leiden: Brill Rodopi, 2016.

Mateer, John. "Directing for Cinematic Virtual Reality: How the Traditional Film Director's Craft Applies to Immersive Environments and Notions of Presence." *Journal of Media Practice* 18, no. 1 (2017): 14–25.

Milk, Chris. "How Virtual Reality Can Create the Ultimate Empathy Machine." March 2015. Video, 10:16. https://www.ted.com/talks/chris_milk_how_virtual_reality_can_create_the_ultimate_empathy_machine.

Milk, Chris. "Virtual Reality Is 'The Last Medium'." Interview by Eric Johnson. *Re/Code*, October 1, 2015. https://www.vox.com/2015/10/1/11619088/virtual-reality-is-the-last-medium-says-filmmaker-and-vrse-ceo-chris.

Montani, Pietro. *Tre forme di creatività. Tecnica, arte, politica.* Roma: Cronopio, 2017.

Murphy, Dooley. "Virtual Reality and the Rhetoric of Empathy." In *A Companion to Motion Pictures and Public*, edited by Mette Hjort and Ted Nannicelli, 488–508. Hoboken, NJ: Wiley-Blackwell, 2022.

Murray, Janet H. *Hamlet on the Holodeck: The Future of Narrative in Cyberspace.* New York, NY: Free Press, 1997.

Murray, Janet H. "Virtual/Reality: How to Tell the Difference." *Journal of Visual Culture* 19, no. 1 (2020): 11–27.

Pinotti, Andrea. "Self-Negating Images: Towards An-Iconology." *Proceedings* 1 (2017): 1–9. https://doi.org/10.3390/proceedings1090856.

Pinotti, Andrea. "Staying Here, Being There: Bilocation, Empathy, and Self-Empathy in Virtual Reality." *Bollettino filosofico* 37 (2022): 142–162.

Raessens, Joost. "Virtually Present, Physically Invisible: Alejandro G. Iñárritu's Mixed Reality Installation *Carne y Arena*." *Television & New Media* 20, no. 6 (July 7, 2019): 634–648.

Saint-Pol-Roux. *Cinéma vivant (1925–1930)*. Mortemart: Rougerie, 1972.

Schweinitz, Jörg. "Totale Immersion, Kino und die Utopien von der virtuellen Realität. Zur Geschichte und Theorie eines Mediengründungsmythos." In *Das Spiel mit dem Medium. Partizipation—Immersion—Interaktion*, edited by Britta Neitzel and Rolf F. Nohr, 135–152. Marburg: Schüren, 2006. https://doi.org/10.25969/mediarep/14265.

Sherman, William R., and Alan B. Craig. *Understanding Virtual Reality: Interface, Application, and Design*. San Francisco, CA: Morgan Kaufmann, 2002.

Sobchack, Vivian. *The Address of the Eye: A Phenomenology of Film Experience*. Princeton, NJ: Princeton University Press, 1992.

Sutherland, Ivan. "The Ultimate Display." *Proceedings of IFIP Congress* 2 (1965): 506–508.

Trumbull, Douglas. "Immersive Cinema." Keynote paper presented at the *Immersivité et innovations technologiques/Immersivity and technological innovations* virtual conference. Université de Montreal and McGill University, October 26–30, 2020.

Walser, Randall. "Spacemakers and the Art of the Cyberspace Playhouse." *Mondo 2000* 2 (1990): 60–61.

Wiesing, Lambert. *Artificial Presence: Philosophical Studies in Image Theory*. Stanford, CA: Stanford University Press, 2010.

Wittmann, Matthias. "Vom Ende der 'Suture.' Zeit- und Raumformen im 360°-Film." *Erzählen im 360°-Film* (2020). https://www.360storytelling.org/de/zeit-und-raumformen-im-360-film.

Wolf, Werner, Walter Bernhart, and Andreas Mahler. *Immersion and Distance: Aesthetic Illusion in Literature and Other Media*. Amsterdam: Rodopi, 2013.

About the Author

Pietro Conte is an associate professor in the department of philosophy at the University of Milan. His research focuses on illusion, hyperrealism, immersion, and the multifarious practices of un-framing, a thematic cluster he has explored in the monographs *Unframing Aesthetics* (2020) and *In carne e cera. Estetica e fenomenologia dell'iperrealismo* (2014).

7. The Time it Takes to Tend Virtual Succulents: Immersing Oneself in a Playthrough of *Viridi*[1]

Gabriel Tremblay-Gaudette

Abstract: This chapter looks at the independent video game *Viridi*, an indoor gardening simulation in which the player has to tend a pot of succulents. This rather unique video game premise offers such an apparent lack of challenge that it might seem unlikely to engender a state of immersive play. However, as we shall see, a closer look at the game's modalities reveals that the gradual establishment of a particular form of immersion is made possible through two specific factors: an engagement and commitment to the game building up over time, and a hermeneutic investment in the game's practice.

Keywords: immersion, video game, *Viridi*, zen, plant media

The video game distribution platform Steam, through its interface, informs players of how much time they have dedicated to a given game since its installation. I have never fully understood why keeping track of the amount of time spent on a game could be considered valuable information by a player. Who could gain any sense of pride from having spent dozens—if not hundreds—of hours in a virtual territory? The number of hours played over time can become impressive—or worrisome—for some players. Maybe it has to do with the immersive pull of a game.

The mechanics of some games and what they offer to players can more readily explain why the number of hours played can sometimes reach three

1 I am grateful to Holly Vestad, a truly skilled reader and botanist, for her insights on the first draft of this chapter.

Bédard, Philippe, Alanna Thain, and Carl Therrien, eds. *States of Immersion Across Media: Bodies, Techniques, Practices*. Taylor & Francis Group, 2025
DOI 10.5117/9789048558766_CH07

or even four digits—think of the ever-expanding Massively Multiplayer Online Role-Playing Games (MMORPGs) like *World of Warcraft* (Blizzard Entertainment, 2004) or of games without a clear end goal like *Stardew Valley* (ConcernedApe, 2016). Yet, at the time of writing this chapter, I have spent more than two hundred hours "in" the video game that will soon be discussed in the next few pages, even though it is a game that… well, that doesn't ask of its players to *do* much in terms of gameplay. As we will see, this surprisingly high figure of time played has a lot to do with how I came to consider what I qualify as "playing" this game.[2] In this chapter, I will perform what can loosely be called an "auto-ethnography" of my time spent playing *Viridi* (Ice Water Games, 2015).[3] After exploring the features of the game, I will attempt to pinpoint the specific forms of engagement (which can lead to a state of immersion) that can arise—or rather, into which one can sink—while playing *Viridi*.[4]

A Fistful of Succulents

In 2015, Ice Water Games released *Viridi* (fig. 7.1). On the company's website, *Viridi* is dubbed a "gardening sim."[5] It could adequately be described as a mix between a botanical simulation game and an idle game, except for the fact that there are very few management options and few tangible rewards or resources to reallocate in this simulation, as is often the case in—if not the sole purpose of— idle games. Perhaps a more apt categorization would

2 My tentative use of the word game, highlighted by the scare quotes surrounding it, echoes an ongoing debate about the definition of what a "true" game is, could, or should be. I will not venture deeply in this conversation here. For further discussion on this topic, see Consalvo and Paul, *Real Games*; and Myers, *Games Are Not*.

3 I hesitate to use the term "auto-ethnography" as this could lead the reader to believe my endeavour was thorough or organized and that it was accomplished while following strict protocols. The opposite is true; I mostly played the game for its own sake, without considering the possibility that I could be thinking hard and long about it at some point in the future. What follows is a recollection, to the best of my knowledge and capacity, of the global experience that was playing *Viridi* for hundreds of hours over a long period of time.

4 I fully acknowledge that the structure of my demonstration amounts, in a sense, to putting the cart before the horse, since I will only be narrowing down my own take on immersion *after* I will have examined the game that will serve as its case study. I felt it was necessary, and perhaps more convincing for the reader, to follow the same process I went through (which consisted in playing the game *first*, before broadening my conceptualization of immersion by its analysis) in order to explain my experience. Thankfully, readers expecting thorough examinations of the concept of immersion will find many other useful chapters in this anthology.

5 "*Viridi*."

Figure 7.1: Presentation of *Viridi* on Ice Water Games's website

be to describe it as a "zen game," but, as Ian Bogost warns, that term is frowned upon in Japan due to its imprecision.[6] In a sense, *Viridi* does not fit neatly into any known video game category—or, at the very least, no quick classification can explain what is going on when you "play" it.

Viridi is a Latin word; specifically, it is the dative masculine singular form of *viridis*, which derives from the verb *vireo* (to grow) and which, in its adjectival form, roughly means "green," "fresh," "lively." Interestingly, this word, which describes both an action and a state, is also fit to sum up *Viridi*'s gameplay: when playing it, you don't get to *do* much; the focus is instead on appreciating the virtual space of the game. This virtual space is a territory that the player owns and develops. They can choose to nurture it to its full capacity with maximum efficiency if they are so inclined, but the goal of the game goes beyond growing plants. Instead, its aim, the way I see it at least, is to make the player become aware of *how to properly inhabit*, rather than occupy—a term that connotes unwarranted ownership—or manage—again, connoting a form of control over—the virtual space. I will discuss this further below.

When a player starts *Viridi*, they are first asked to choose one pot among thirteen different options (fig. 7.2). This choice is of a cosmetic nature: it

6 Bogost, *How to Do Things with Video Games*, p. 93.

Fig. 7.2: The first step. Choosing a pot.

is inconsequential from a performance standpoint as none of the pots have a direct influence on the future plants' growth. This purely aesthetic decision is common with many other aspects of the game and it is possible at all times to select a different pot without any cost or penalty. Then, the player must choose a starter kit of plants—all of them succulents, a family of plants known for their retaining of water, slow growth, and little upkeep. Once the player chooses a starter kit, the selected plants are dispersed in the pot at random. The player always has the option to move them around at will and to add new seeds, either by buying them through the in-game store at a very low cost or by waiting for a free random seed, delivered to players every Friday.[7] Yet again, the player's initial choice of a starter kit (and the subsequent addition of new seeds) is based on sensual (qualitative) considerations rather than on productive (quantitative) ones.

After these basic decisions are made, the player can click on each plant to check on it.[8] Their neutral state is "sated"; after a couple of hours or days, the plant becomes "thirsty" and needs to be watered by "spritzing," an action that can be accomplished by clicking on a button on the lower right corner of the game's interface. If the player neglects to do so for a prolonged period

7 On the PC version of the game, the cost of a seed ranges between nine cents for common plants, such as *Cheiridopsis denticulata*, to thirty-nine cents for rarer plants, such as *Sedum morganianum*. On the game's mobile version, bundles of seeds are available for ninety-nine cents. The number of seeds in each bundle varies according to the plant's rarity.

8 For the purposes of this chapter, I will limit my description of the game's mechanics to its PC version. There is also a mobile version of the game, but, as will become clear in what follows, I do not think that the intermittent, brief sessions a player usually commits to with a mobile game are best suited for the gaming experience I am focusing on here.

Figure 7.3: An unattended pot in which all the plants are parched.

of time, the plant becomes "parched" and its appearance changes: it wilts and its colours become less vivid (fig. 7.3). This situation can be rectified in two clicks. An overzealous player can also get carried away and water a plant more than it needs: the plant then becomes "overwatered" at first and then "waterlogged," states which can only be reversed by waiting until the plant absorbs the excess of water. If one chooses, it is possible to name each plant in order to personalize it.

After a while, weeds grow. The player has the option of removing them by dragging the cursor over a patch of weeds: an "X" then appears, and clicking on it instantly removes the weeds, which is signified both by their sudden disappearance from the visual environment and by a short and somewhat brutal sound effect, which momentarily interrupts the soothing, synthy, and smooth soundtrack of the game, which normally plays in a loop. After a longer while, the plants reach a state of maturity: they eventually become flowers and "go to seed." There is no option to gather seeds from these flowers; the only thing left to do is to "throw [them] in the garden," as the game states it, but not before being invited to commemorate the plant's full life cycle (fig. 7.4).

Aside from exploring the pot by rotating it and zooming in and out of it, and apart from labelling each plant with a personalized name, the only other action the player can do is to focus on an individual plant or on the only other living inhabitant of the virtual territory, a snail named "Thunderbolt." The player can click on any plant and, after a few seconds, five horizontal parallel lines will appear. Soon afterwards, dots will spread over the lines: they form a musical partition, which means that the player is now singing to the plant. This action can put the plant in various states, such as "vibrant,"

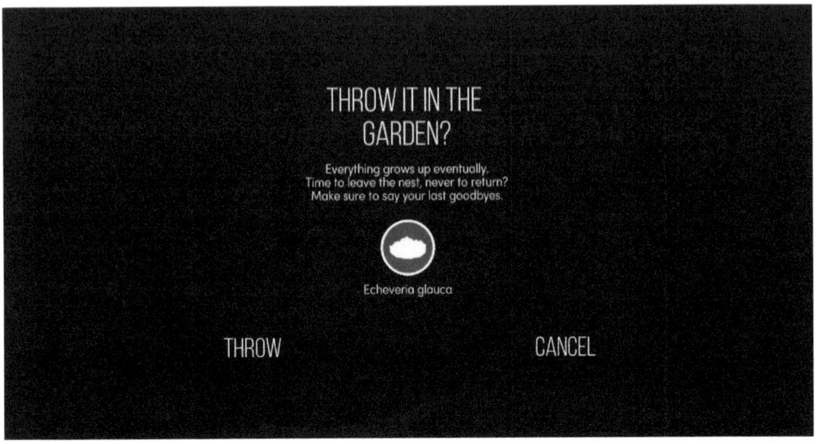

Figure 7.4: The game's screen is activated by the player's decision to throw a plant in the garden.

"vivacious," "vital," and so on (fig. 7.5). As for the snail, it circles the pot and is described as "wow cute" (fig. 7.6). The player can also water it and sing to it.

And... that's about it. The game does not explicitly establish any conditions for "winning"; thus, the performance of the player cannot be evaluated in terms of "victory" or "failure" (although one might *certainly* feel a certain degree of failure if one consistently under- or overwaters one's plants). *Viridi* is thus deprived of many essential components that normally characterize games, as stated in Katie Salen and Eric Zimmerman's famous definition of the term as "a system in which players engage in an artificial conflict, defined by rules, that results in a quantifiable outcome."[9] There is no conflict, the rules are implicit, and the quantifiable outcome is never specified by the system. The objective and, in a sense, the purpose of *Viridi* is simply to spend some time tending to your pot of succulents, possibly removing weeds from it and singing to it. Engaging with this virtual territory amounts to spending time lying back and appreciating it; you don't play towards a goal, you play for... well, for *something else*, which is left to the player's judgment.

How could one become engaged, let alone immersed, in such a limited video game experience? Intuitively, one could jump to the conclusion that this game is extremely uninteresting: its narrow possibilities of active engagement should indeed result in a limited involvement of the player, leading to their disengagement from the game. According to Emily Brown and Paul Cairns, there are three stages that lead to videoludic immersion: first there is *engagement*, characterized by a minimal investment of time, effort, and concentration in a game; then, when a game is sufficiently

9 Salen and Zimmerman, *Rules of Play*, p. 80.

Figure 7.5: By singing to a plant, the player can bring it into a "vibrant" state.

Figure 7.6: The snail, named "Thunderbolt," in action.

well-built to generate emotional reactions from its player, this is followed by *engrossment*; finally comes *total immersion*, which is akin to feeling that one is "present" within the game space.[10] However loosely defined these terms are, we can establish that the path towards immersion is a gradual process that starts with the player's engagement. But, as Carl Therrien explains, "[t]he path towards ludic gratification relies on the acquisition of a specific competence in the execution of the primary actions, and also involves—according to the previous experience of the player—more or

10 Brown and Cairns, "A Grounded Investigation of Game Immersion," quoted in Arsenault and Picard, "Le jeu vidéo entre dépendance et plaisir immersif," pp. 4–5.

less sustained efforts."[11] Few such efforts are required to play *Viridi* and its learning curve is so flat it almost seems nonexistent. Therrien also adds that the potential of gratification derived from a video game experience could be correlated with the potential of action within the fictional universe that it offers the player:

> There are few games that model all of the trivial fictive actions, since this implementation would only require a basic participation that would correspond to the minimal condition of the cybertextual work: the user must accomplish a simple gesture that requires no coordination effort or cognitive challenge other than the comprehension of the fictive situation. In sum, the gratification potential of such a modality remains limited.[12]

Following Therrien's philosophy, the probability of experiencing sustained engagement in *Viridi* which could evolve towards engrossment and total immersion, should be very low: its gameplay, which is solely based on a very trivial set of actions, does little indeed to engage the player's videoludic activity. However, in my case, quite the opposite was true, as is attested at the very least by the considerable amount of time I spent playing *Viridi*. In order to understand how and why I became immersed in this particular game, further investigation of its mechanics is required.

Thinking Long and Hard about *Viridi*

I have a few hypotheses regarding why my immersion into *Viridi* was—and still is—so strong, which mostly have to do with my "professional" disposition as a semiotician. I devised these hypotheses by asking myself a set of questions about the game.

The first question that came to my mind was: what is the role of the weeds? What is their purpose? Their sudden emergence in the "field of play," my virtual pot of succulents, can be interpreted as an invasion of the video game territory by a foreign agent. "Ah," I thought at first, "perhaps they are the opponents in the artificial conflict that I assume always occurs in a game!" Indeed, if there is one thing I have learned after years of playing video games, it is that whenever there is an intruder in the virtual space in which my character finds itself, or, in other words, if an "alien" shows

11 Therrien, "Illusion, idéalisation, gratification," p. 159, my translation.
12 Therrien, "Illusion, idéalisation, gratification," p. 269, my translation.

up on my screen, I must almost always eliminate it. In *Viridi*, the game informs me that I can remove the intruder by clicking the visual sign of the "X" button or the "Pluck weeds" text that appears at the top right corner of the screen. But the text is merely a designation of the action the player can take; strictly speaking, it is not an injunction, thus it cannot be considered an instruction the player has to follow automatically.

Why, then, do I remove the weeds? Do I do it solely because it is a possibility provided by the game? Is it a simple procedural mechanism of a recurring feedback loop implemented into the game design to bring me back to my pot and give me something else to do aside from planting, watering, and throwing away succulents? Will ripping away the weeds allow for the roots of my succulents to expand and grow faster? (And if so, why should a plant growing faster towards its eventual removal from my pot be considered a *good* thing?) Is the game setting a trap for the player through its suggestion to remove the weeds? After all, maybe the weeds are helpful, or even necessary, for the succulents to thrive; maybe they are a natural part of their ecosystem and I should leave them be.

One thing is certain: from a strictly videoludic standpoint, the weeds can be described as an obstacle for the player's experience in the sense that they block visual access to the succulents. It is thus preferable to clear the territory of them so as to locate the succulents more easily. Moreover, the potential positive or negative effects of removing the weeds are never established within the game, since it does not state clear objectives. The effects of the weeds' presence are rather difficult to infer from the gameplay itself: the growth feedback is too slow to allow a discernable difference between removing the weeds or leaving them be. If the player likes the visual effect of blades of grass surrounding their succulents, they can decide that the weeds are, in fact, an important addition to the pot. I think of these weeds as inhabitants of the virtual space I tend to: I can either choose to make them vanish in order to keep my space clean or accept their presence as rightful dwellers of the pot; after all, they showed up there on their own.

Another feature of the game that elicits consideration is the option it gives the player to sing to a plant. As is the case for the removal of weeds, it is nearly impossible to determine with any assurance if singing to my plant results in growth acceleration. In this particular case, I chose to conduct a little online research. In 2016, Wolfgang Stuppy claimed that there was no scientific consensus on the subject of singing to real plants.[13] Attempts to elucidate this mystery date as far back as Charles Darwin, who carried

13 Stuppy, quoted in Nicholls, "Can Your Plants Really Hear You If You Sing to Them?"

out an experiment but reached inconclusive results. A later study from the Royal Horticultural Society came to the conclusion that speaking to tomato plants can slightly increase their growth rate, but the exact reason why has not yet been established.[14]

If there is no valid explanation of the positive effect of a human's vocal activity on plants, there are some hypotheses about its effect on their gardener. Firstly, talking to a plant makes one more aware of its existence and prevents gardeners from forgetting to water it for a long stretch of time, thus increasing the survival rate of the gardener's botanical companion. Secondly, this active *engagement* with plants leads to an additional gratification derived from this hobby, thus leading to an overall increase of the positive effects of gardening on the gardener, namely relaxation and stress decrease.

As passive as zooming in on a plant in *Viridi* and leaving my cursor there for a few minutes or hours may seem, it is nonetheless a form of engagement with the game, one that adequately corresponds to its slow rhythm but that nevertheless is more akin to spectatorship than playing. Coincidentally, Roger Odin explains that in order to immerse oneself in the activity of filmviewing, one has to achieve what he calls a *mise en phase*, (roughly translatable as "phasing") which refers to "[t]he process that leads me to *vibrate* to the rhythm of what the film is giving me to show and to hear. The phasing is a modality of the affective participation of the spectator in the film."[15] In what is most likely a coincidence, if the player chooses to sing to one of their plants in *Viridi*, they can put the plant in a state that the game calls "vibrant"; perhaps this enables a form of synchronicity between the virtual succulent and the player, as both achieve a *mise en phase*.

Talking or singing to plants is a common cliché in fiction. According to the web's leading authority in this field, *TV Tropes*, "[p]eople talking to their plants in fiction is often used to show how boring they are. A common shtick, especially in cartoons, is for the plant to die of boredom as a result."[16] And, in effect, passively playing *Viridi* the way I did—launching the game and briefly tending to the plants, then leaving the game open in the background, filling whichever room my computer was in with both the relaxing music emanating from its speakers and the visually pleasing spectacle of the camera hovering slowly above my pot—might be described as excruciatingly *boring* by some people. Perhaps some of you reading this might think exactly so. In that sense, *Viridi* can be considered the antithesis of what is

14 See Vanderlinden, "It's True—You Should Really Talk to Your Plants."

15 Odin, *De la fiction*, p. 38, my translation, my emphasis.

16 "Talking to Plants."

traditionally considered a "good" video game, which usually provides the player nearly incessant stimulation, retroaction, engagement, reward, and a well-paced difficulty curve that should progress at a forty-five degree angle between anxiety and boredom, according to Mihály Csíkszentmihályi's first theory of "flow."[17] In the frenetic contemporary society we live in, boredom is perceived negatively. It is often avoided at all costs, almost as if it were forbidden by decree: external stimulation is ever present, content is pushed to us via notifications, there is no time to rest or to stand still, and the dreadful "fear of missing out" is never far. *Viridi* stands as a remedy against this pressure of the external world to be doing something, anything, at all times. In the first paragraph of the game's description on its official website, we are informed that "[p]layers can enjoy a moment of blissed out plant tending when they need a break from work, to wind down after a long day, or instead of checking those [expletive] emails."[18]

I found myself doing exactly this more and more. After selecting a plant to sing to, I would leave the game open in the background of my computer, the discrete soundtrack reassuring me that it was on and that the plants were still there, growing slowly on another software, reachable at will by quickly pressing ALT-TAB on my keyboard, always ready to offer me a more relaxing form of procrastination than checking social media. What is more, my pot followed me during my travels to France, in 2018, when I carried out this amateur auto-ethnographical experience: it helped me calm down after a long day of attending a conference in a castle in Normandy; it complemented the decoration of the living room in my cramped living quarters in Paris; and it was a life-saver during a train ride between Lyon and Paris disrupted by annoying passengers who would just not shut up.

Weeks after my introduction to the game, by which point I had already racked up dozens of hours of play, something unexpected happened. As mentioned in the introduction, in order to launch *Viridi*, one must go through Steam's interface, which stores all of one's games in a library. This interface does more than just keep track of the amount of time played; it also lists a series of achievements. These achievements are usually offered to the player as a means to extend their involvement after reaching the ending of a game; there is thus a replayability value inherent in accomplishing all the achievements listed in Steam. They often amount to nothing more than secondary objectives one can collect in a virtual trophy case, but they are nonetheless important for some players.

17 Csíkszentmihályi, *Flow*, p. 39.
18 "*Viridi*."

In order to understand why this list of achievements was of interest to me, it is perhaps worthwhile to consider the category of player to which I likely belong. In 1996, Richard Bartle published the visionary article "Hearts, Clubs, Diamonds, Spades: Players Who Suit MUDs." While initially aimed at describing play behaviours in multi-user-dungeons (MUDs), a near-extinct text-based game genre, Bartle's taxonomy can nowadays be applied to many video games. This taxonomy comprises four categories of players, which can be summarized as "achievers," "explorers," "socializers," and "killers." An easy way to remember them is to consider suits in a conventional pack of cards:

- Achievers are Diamonds (they always seek treasures);
- Explorers are Spades (they dig around for information);
- Socializers are Hearts (they empathize with other players);
- Killers are Clubs (they hit people with them).[19]

I identify with the "diamond" type: I am precisely the type of player who can become interested, and even obsessed, with completing all the achievements offered by a game. So, naturally, I explored the full list of achievements possible in *Viridi*. It turns out they are a bit disappointing: almost all of them consist of making a specific succulent reach maturity and flower, which makes the achievements quite redundant. But I did notice some achievements pertaining to throwing mature plants in the garden. The first two have rather innocuous names: "Leaving home to face the world" is unlocked after the player throws out their first plant, and "Pleasant little garden" after throwing out ten. The next two are a bit more troubling: throw away fifty plants and you achieve the status of "Small ecological disaster," while the final possible goal can be reached after the player throws away one hundred plants. This ultimate achievement is titled, rather bluntly: "You are really good at growing plants; maybe you should take it easy!" What does it mean to have the game politely suggest, by its ultimate paratextual achievement, that its player should stop playing it?

Maybe it is because we erroneously tend to conflate "virtual" and "digital" with "immaterial." The objects that appear as pixels on the screen might not be tangible to people, as they do not occupy a physical volume in front of them, but any document that is stored in "the cloud" has a material existence somewhere on a server room; any digital information has its corresponding material "weight" and energetic cost. As Jussi Parikka explains, in *A Geology of Media* (2015), "[i]ncreasing bandwidth, more efficient processors, data-heavy digital design practices, and the sheer increase in the number of

19 Bartle, "Hearts, Clubs, Diamonds, Spades."

computational events tell the story of the upward curve in absolute energy consumption of the supposedly immaterial matrix of computation."[20] When I participate in this collective story of computation by playing *Viridi*, leaving the program open on my computer and displayed on my screen while I read at the other end of the room, as I tend to do, I am expending resources. Nevertheless, as inconsequential as they might be in the grand scheme of things, the energetic resources that are consumed when I make my virtual territory of succulents appear on my screen are indeed real. Even knowing this, I cannot bring myself to stop playing the game. The best explanation for my continued interest in *Viridi* might therefore lie in the specific ways it enticed a state of immersion in me.

Immersing Oneself in *Viridi*

Therrien suggests that immersion should not be confused with immediacy, virtual illusion, or cognitive adhesion to a fictional world, since it is both none and some of these factors all at once. He also states that there is another element we tend to neglect while considering videoludic immersion: the affective component of playing a video game. "It is obvious," Therrien claims, "that the interactive posture offered by a video game increases the emergence of emotions directed at their own performance for the gamer, emotions such as shame or pride,"[21] adding that, "[i]n a video game, the nature of the interactions with a virtual character and the duration of play—often more than a dozen hours for the average player—can generate a particularly strong bond with it."[22] In other terms, emotional engagement in a game is another pathway towards engrossment, and eventually immersion; beyond or besides the high scores, the achievements and the successes, a player can develop an affective attachment to a game experience.

Even though there is indeed an implicit avatar who carries out the action that I, the player, command it to do in *Viridi*, it might seem like a bit of a stretch to call my succulents "characters." But I did spend more than a hundred hours with them; I gave them names and I saw them evolve because I made them grow. Even though these actions did not require a colossal effort, I do take a certain amount of pride in what I achieved in my virtual territory. I also spent some time engaging virtually with my plants by

20 Parikka, *A Geology of Media*, p. 137.

21 Therrien, "Illusion, idéalisation, gratification," p. 261, my translation.

22 Therrien, "Illusion, idéalisation, gratification," p. 290, my translation.

re-arranging them in my pot: this manipulation can be considered relevant when examining the immersive potential of *Viridi*.

Nick Yee established a model of motivations for playing multiplayer online games. Even though *Viridi* is a single-player experience, it is worthwhile to consider this model since it is broken down into three components: achievement, social, and immersion. This last category is broken down into four sub-categories: discovery, role-playing, customization, and escapism. Customization is described as the player "[h]aving an interest in customizing the appearance of their character."[23] In *Viridi*, tending to our succulents can be seen as amounting to this task, since we strive to keep our plants healthy, we can move them around the pot to display them in different configurations, and we can choose to remove the weeds from our pot in order to achieve a more visually ascetic result. Though not as spectacular or intense as the other forms of motivations for play listed by Yee, customizing our succulent garden in *Viridi* still is a form of immersion in play.

Let us consider my playing of *Viridi* through different immersion taxonomies to find out if any of them can directly apply to my experience. One of the earliest models devised to distinguish different forms of immersion comes from Ermi and Mäyrä.[24] It is broken down into three forms, namely sensory, challenge-based, and imaginative immersions. Sensory immersion is "related to the audiovisual execution of games. [...] [D]igital games have evolved into audiovisually impressive, three-dimensional, and stereophonic worlds that surround their players in a very comprehensive manner."[25] Challenge-based immersion is "the feeling of immersion that is at its most powerful when one is able to achieve a satisfying balance of challenges and abilities. Challenges can be related to motor skills or mental skills such as strategic thinking or logical problem solving, but they usually involve both to some degree."[26] Imaginative immersion is "the area in which the game offers the player a chance to use [their] imagination, empathise with the characters, or just enjoy the fantasy of the game."[27] *Viridi* lends itself to all three of these forms of immersion: from a sensory perspective, the game is visually and auditorily pleasant and soothing; though not especially challenging, it solicits some forms of mental skill, as we will see in the following pages; and finally, there is room for emphasis on the plants/

23 Yee, "Motivations of Play in Online Games," p. 772–774.
24 Ermi and Mäyrä, "Fundamental Components of the Gameplay Experience."
25 Ermi and Mäyrä, "Fundamental Components of the Gameplay Experience," p. 6.
26 Ermi and Mäyrä, "Fundamental Components of the Gameplay Experience," pp. 6–7.
27 Ermi and Mäyrä, "Fundamental Components of the Gameplay Experience," p. 7.

characters. However, none of these categories can satisfyingly and precisely pinpoint the specific immersion experience that playing *Viridi* generates.

Let us now consider the taxonomy developed by Gordon Calleja.[28] Calleja lists six types of possible player involvement (called "incorporations" by the author) within a video game: kinaesthetic, spatial, shared, narrative, affective, and ludic. I will leave aside the shared and narrative elements of this list, since *Viridi* is not a narrative-driven game and offers no multiplayer functionalities. With regards to kinaesthetic involvement, Calleja states the following: "The focus on control and affordances for movement designed into game environments means that understanding an experience such as kinaesthetic involvement requires a simultaneous consideration of the visual, audio, and haptic output along with the disposition to act on the part of the player."[29] Even though the range of movements and actions in *Viridi* is very limited, the visually and aurally soothing qualities of the game's experience can increase its potential for immersion.

As for spatial involvement, Calleja's considerations rely heavily on the idea of movement through space for the sake of discovery and exploration:

> Digital games and virtual worlds are particularly adept at facilitating spatial exploration that enables players not only to project their imagination into the represented landscapes but also to traverse them. [...] Contemporary digital games offer players geographical expanses to inhabit, interact with, and explore. Certain game worlds invite players into picturesque landscapes which give the impression that they extend as far as the eye can see.[30]

Again, *Viridi* fares poorly in that category as it invites the player to visit and revisit the same space, which can be encompassed in a single screen (zooming in on the pot or a specific plant indeed constitutes a limited form and range of exploration). However, the attentive reader might recall that I mentioned that I would circle back to the issue of inhabiting a virtual territory. To this effect, Lucas distinguishes four types of inhabitant in a video game space: the opportunist inhabitant, who "does without"; the slit inhabitant, who "does not"; the developing inhabitant, who "does in"; and the building inhabitant, who "does."[31] Following this taxonomy, *Viridi*'s player

28 Calleja, *In-Game*.
29 Calleja, *In-Game*, p. 70.
30 Calleja, *In-Game*, pp. 73–74.
31 Lucas, "Les figures de l'habitant dans les mondes virtuels," my translation.

is a developing inhabitant: they "seize the grips that are offered to them in order to act on the environment and shape an interior space."[32] Thus, the involvement in the virtual world of *Viridi* does not rely at all on a desire for exploration, but rather on a desire to cultivate, customize, revisit, and maintain a very small patch of land, the pot. In other words, it is what we "do in" *Viridi* that generates engagement with the game in terms of spatial involvement and inhabiting.

Calleja's ludic involvement revolves around the concepts of rules, goals, and rewards.[33] As we have seen, *Viridi* does not state explicit rules or goals through its game system and the secondary rewards provided through Steam's achievements culminate in a mild scolding of the player. Even my facetious explorations of the affordances of the game, which sought to maximize the potential for ludic involvement within *Viridi*, produced musings that, however satisfying they might have been for me, cannot be deemed intrinsic to the standard gameplay provided by the game.

What, then, about Calleja's last category, affective involvement? The author surely states that:

> The player's active input creates the potential for a more intense emotional experience, whether satisfying or frustrating, than non-ergodic media provide. The cognitive, emotional, and kinesthetic feedback loop that is formed between the game and the player makes digital games a particularly powerful medium for affecting players' moods and emotional states.[34]

However, Calleja stresses that the potential for affective involvement with a game lies both in its appealing aesthetic qualities (which *Viridi* provides through its soothing, luscious, and elegant formal qualities) and in the potential for excitement through a quick feedback loop (quite the opposite from what is experienced when playing *Viridi*).

Let us now briefly turn to another taxonomy, elaborated by Olivier Caïra, to consider the multiple dimensions of ludic engagement which can in turn lead to immersion.[35] The MÉDIAL model, built as both an improvement and an expansion of Calleja's model, is an acronym for its different components: *Moteur* (Motor), *Épreuve* (Challenge), *Diégèse* (Diegesis), Intrigue, *Autrui* (Other), and *Ludologie* (Ludology). I want to focus on the most significant

32 Lucas, "Les figures de l'habitant dans les mondes virtuels," my translation.
33 Calleja, *In-Game*, pp. 147–156.
34 Calleja, *In-Game*, p. 135.
35 Caïra, "Les dimensions multiples de l'engagement ludique."

addition that Caïra's model presents: the ludological angle. Caïra states: "On the table or in front of a screen, alone or in mass practice, passionate players develop reflexive and increasingly erudite thinking."[36] If Caïra approaches this reflexive thinking more in terms of the game's performance (for instance in improving strategies, in exploiting game mechanics, or in editorial choices by its creators) or of commercial considerations, it seems to me that the deep dive into *Viridi*'s peculiarities that I accomplished while playing the game and later while writing this contribution fall into a ludological form of engagement. In other words, playing a video game does not start or stop when one pushes keys or buttons; the intellectual engagement with a game is an integral part of the play experience, which can lead to engrossment and immersion.

Even though the brunt of this chapter attempted to establish the degree to which I became intellectually engaged with *Viridi*, I should not minimize the affective involvement that resulted, in my case, in hundreds of hours of play. To that effect, Lennart E. Nacke and Craig A. Lindley state: "[Donald Arthur] Norman makes a clear distinction between emotion and affect, defining emotion as consciously experienced affect, which allows us to identify who or what caused our affective response and why. Affect on the contrary is defined as a discrete, conscious, subjective feeling that contributes to and influences players' emotions."[37] Emotions can be intense and encompassing, whereas affect is a constant state which, while not as spectacular and enthralling as an emotional reaction to a game, can become deeply linked to a game experience. I went back again and again to *Viridi* to feel something that will be described more plainly in the following conclusion.

Conclusion

I would like to offer two thoughts to address the immersive power of a game like *Viridi*, in the hopes of adequately pinpointing its specific qualities.

Firstly, *Viridi* is a game that does not offer an artificial conflict, that does not state clear rules of play, and that does not steer its player toward a precise goal. It offers something different: an encounter with "nature," albeit simulated. Alenda Y. Chang opens her book *Playing Nature* (2019)

36 Caïra, "Les dimensions multiples de l'engagement ludique," my translation.

37 Nacke and Lindley, "Affective Ludology, Flow, and Immersion in a First-Person Shooter." The authors refer to Donald Arthur Norman's works. See Norman, *Emotional Design*.

by relaying a common prejudice against video games, which is that "whether played on computers, consoles, mobile devices, or arcade cabinets, [video games] are reviled as the most unnatural of offenders against environmental communion."[38] She then offers a definition of what an "environmental game" would be like: "a game's 'environmentality' does not at all depend on its possessing an overt environmentalist rhetoric, but rather follows from its modelling of ecological states and relations, however admirable or terrible that modeling may be."[39] As we've seen, the accuracy of *Viridi*'s botanical modelling might not be optimal, but it does require that the player return to the game often in order to supervise and appreciate the slow growth of their plants. It is also perhaps for that exact reason that *Viridi* strips its gameplay of any elements that would distract from the activity at the core of the game: tending to plants and establishing a mostly contemplative, mutually beneficial relation with the succulents *over time*.

Secondly, considering the remarks that opened this chapter, I would argue that any attempt at a taxonomy of immersion should include an important parameter: *temporal* involvement. Obviously, there can be no direct correlation established between the amount of time spent in a game and its eventual affective immersion potential. One can traverse Jason Rohrer's *Passage* (2007) in a matter of minutes and still emerge deeply affected by this experience, and I once managed to clear a level of Devolver Digital's *Hotline Miami* (2012) in a brief but exhilarating encounter that I would describe as one of the most intense and immersive videoludic experiences of my life. Still, it is not irrelevant to consider how a game offered an experience that drew in a player so intensely that they invested what amounts to days, if not weeks, of playing it. I would also suggest that Caïra's model could include a seventh category, a specifically *hermeneutic* form of engagement, which has less to do with the gameplay's affordances or even with its potential narrative content but rather with the expansive considerations about a game's significance and signification for a player. Several sets of factors could potentially trigger an immersive state while playing *Viridi*, but, in my case, it was the interpretive endeavour surrounding the game's meanings that established the prolonged engagement in which I found myself, which then led to immersion.

In the end, my main drive to come back again and again to *Viridi* might have been linked to an affective form of immersion. What this game achieved

38 Chang, *Playing Nature*, p. 2.

39 Chang, *Playing Nature*, p. 21.

through its bare-bones gameplay is the creation of a state of relaxation through prolonged contact with "nature." Ian Bogost writes:

> As designers and players, we must reject the principle of engagement. Relaxation and reflection arise from constrained environments in which the senses are de-emphasized and focused rather than escalated and elevated. Video games may often overwhelm and titillate our senses, but relaxation comes instead from withdrawal and placidity. To relax through a game requires abandoning the value of leaning forward and focusing on how games can also allow players to achieve satisfaction by leaning back.[40]

It is precisely by leaning back that I managed to "vibrate" to *Viridi*, as Odin would put it. The non-frenetic pace of the game allowed me to engage *intellectually* with its components. Playing *Viridi* became a relaxing endeavour where I could let my thoughts about the game sink in in a soothing setting. This form of engagement led to immersion, even though it bypassed the step of engrossment, or rather, the engrossment was achieved through affect rather than through emotions.

In her manifesto "Rambunctious Games" (2019), Chang declares that "one way to mitigate games' considerable antienvironmental tendencies would be to encourage games that alter the normal palette of possible actions or redefine success in terms of affective change, cognitive development, or even failing to enjoy oneself."[41] Perhaps one can enjoy *Viridi* because it is, in a way, impossible to "fail" at it, which leaves more room for mental musing and reflexive thinking. Who or what grows more when we become immersed in *Viridi* and decide to inhabit a virtual environment instead of conquering it—the plants, or oneself?

Bibliography

Arsenault, Dominic, and Martin Picard. "Le jeu vidéo entre dépendance et plaisir immersif. Les trois formes d'immersion vidéoludique." In *Actes du colloque Homo Ludens. Le jeu vidéo. un phénomène social massivement pratiqué*, 1–16. Trois-Rivières: Congrès de l'ACFAS, 2008.

Bartle, Richard. "Hearts, Clubs, Diamonds, Spades: Players Who Suit MUDs." *MUD*, August 28, 1996. http://mud.co.uk/richard/hcds.htm.

40 Bogost, *How to Do Things with Video Games*, p. 95.

41 Chang, "Rambunctious Games," p. 70.

Blizzard Entertainment. *World of Warcraft*. Blizzard Entertainment, 2004.

Bogost, Ian. *How to Do Things with Video Games*. Minneapolis, MN: University of Minnesota Press, 2011.

Brown, Emily, and Paul Cairns. "A Grounded Investigation of Game Immersion." In *CHI EA '04: Extended Abstracts on Human Factors in Computing Systems*, 1297–1300. New York, NY: Association for Computing Machinery, 2004. https://doi.org/10.1145/985921.986048.

Caïra, Olivier. "Les dimensions multiples de l'engagement ludique." *Sciences du jeu*, no. 10 (2018). https://doi.org/10.4000/sdj.1149.

Calleja, Gordon. *In-Game: From Immersion to Incorporation*. Cambridge, MA: MIT Press, 2011.

Chang, Alenda Y. *Playing Nature: Ecology in Video Games*. Minneapolis, MN: University of Minnesota Press, 2019.

Chang, Alenda Y. "Rambunctious Games: A Manifesto for Environmental Game Design." *Art Journal* 79, no. 2 (2019): 68–75.

ConcernedApe. *Stardew Valley*. ConcernedApe, 2016.

Consalvo, Mia, and Christopher Paul. *Real Games: What's Legitimate and What's Not in Contemporary Video Games*. Cambridge, MA: MIT Press, 2019.

Csíkszentmihályi, Mihály. *Flow: The Psychology of Optimal Experience*. New York, NY: Harper Collins, 1990.

Ermi, Laura, and Frans Mäyrä. "Fundamental Components of the Gameplay Experience: Analysing Immersion." *Worlds in Play: International Perspectives on Digital Games Research* 37, no. 2 (2005): 37–53.

Ice Water Games. *Viridi*. Ice Water Games, 2015.

Lucas, Jean-François. "Les figures de l'habitant dans les mondes virtuels." *Sciences du jeu*, no. 10 (2018). https://doi.org/10.4000/sdj.1353.

Myers, David. *Games Are Not: The Difficult and Definitive Guide to What Video Games Are*. Manchester: Manchester University Press, 2017.

Nacke, Lennart E., and Craig A. Lindley. "Affective Ludology, Flow, and Immersion in a First-Person Shooter: Measurement of Player Experience." *Loading… The Journal of Canadian Game Studies Association* 3, no. 5 (2009). https://doi.org/10.48550/arXiv.1004.0248.

Nicholls, Henry. "Can Your Plants Really Hear You If You Sing to Them?" *BBC Earth*, January 18, 2016. http://www.bbc.com/earth/story/20160118-can-your-plants-really-hear-you-if-you-sing-to-them.

Norman, Donald Arthur. *Emotional Design: Why We Love (or Hate) Everyday Things*. New York, NY: Basic Books, 2004.

Odin, Roger. *De la fiction*. Louvain-la-Neuve: De Boeck Supérieur, 2000.

Parikka, Jussi. *A Geology of Media*. Minneapolis, MN: University of Minnesota Press, 2015.

Salen, Katie, and Eric Zimmerman. *Rules of Play: Game Design Fundamentals.* Cambridge, MA: MIT Press, 2004.

"Talking to Plants." *TV Tropes.* https://tvtropes.org/pmwiki/pmwiki.php/Main/TalkingToPlants.

Therrien, Carl. "Illusion, idéalisation, gratification. L'immersion dans les univers de fiction à l'ère du jeu vidéo." PhD diss., Université du Québec à Montréal, 2011.

Vanderlinden, Colleen. "It's True—You Should Really Talk to Your Plants." *The Spruce,* February 17, 2022. https://www.thespruce.com/should-you-talk-to-your-plants-3972298.

"*Viridi*: A Quiet Place for You to Return to." *Ice Water Games.* http://www.icewatergames.com/viridi.

Yee, Nick. "Motivations of Play in Online Games." *Journal of CyberPsychology and Behavior* 9, no. 6 (December 1, 2006): 772–775.

About the Author

Currently an independent researcher, **Gabriel Tremblay-Gaudette** was previously a professor at TÉLUQ and lecturer at UQAM, UQAT, Paris 3, Paris 8, RIT, and WVU. Founder of the Collectif Obèle, a research-creation initiative whose projects interpolate video games and literature, he is working on a hypertextual non-fiction essay on cephaloclastophilia.

Section 2

Finding New Frames

8. The Many Births of Virtual Reality

Philippe Bédard

Abstract: Since the revival of virtual reality (VR) in the second decade of the twenty-first century, the technology has been used by a series of preexisting media (cinema, video games, theatre, dance, etc.), all while searching for an identity of its own. Following André Gaudreault and Philippe Marion's concept of the "double birth of media," this chapter analyses the many births of VR: first as a technology, and second, eventually, as a medium with its own language, generic conventions, and institutional recognition. More specifically, while this chapter is interested in applying the "double birth" model to VR, it also looks at the way artists working in the cinematographic tradition have appropriated this immersive technology. Finally, it looks forward at what a *medium* of VR might look like that is not limited to preestablished media.

Keywords: technology, medium, intermediality, genealogy

Virtual Reality (VR) is a great many things. What it is definitely not, at the time of writing, is either dying or already dead, as so many clickbait headlines would have us believe. In fact, since VR's revival in the second decade of the twenty-first century,[1] tales of its demise have been both ceaseless and greatly exaggerated.[2] In this chapter, I want to refocus our attention; not on VR's death, but rather on its recent rise from the dead and on the period

1 The precise date of this revival is up for debate, but we can point to the release of two major consumer VR devices as a good point for marking VR's official rise from the dead. Those are the Oculus Rift and the HTC VIVE, which were released on March 28 and April 5, 2016, respectively.

2 See Collins, "VR Headsets Are Dying a Lonely Death"; to name but a few of the articles predicting VR's impending downfall, see Mandt, "VR Can Be a Mainstream Tech, but It's not Ready Yet"; Roose, "This Should Be VR's Moment"; or Topolsky, "The Virtual Reality Dream Is Dying," For a counterargument to these, see Parlock, "Stop Saying Virtual Reality Is Dying."

Bédard, Philippe, Alanna Thain, and Carl Therrien, eds. *States of Immersion Across Media: Bodies, Techniques, Practices*. Taylor & Francis Group, 2025
DOI 10.5117/9789048558766_CH08

of gestation it has undergone since. What makes me describe the current rise in VR's popularity as a "revival"? When and in what manner was it first born? And, finally, what can we learn from the forces currently shaping VR's renaissance in the 2010s and 2020s?

In order to answer these questions, this chapter adopts a genealogical approach, the general goal of which is to highlight some of the forces currently shaping the identity of VR as an up-and-coming medium. Doing so requires that we first define what VR is, a task that will have us go over the tumultuous history of this technology in search of an identity. Following André Gaudreault and Philippe Marion's concept of the "double birth" of media,[3] I then approach VR from two distinct perspectives: firstly, from the point of view of its technologies and their affordances and, secondly, from the perspective of intermediality. While the former allows us to understand how VR might be said to have failed the first time—as it did in the 1990s—the latter gives us the outlook needed to tackle the intermedial nature of contemporary virtual reality, which currently borrows much from pre-existing media, as it has yet to coalesce into its own standardized medium. To help us navigate this overview of VR's many births and its failure to rise to the occasion thus far, I will make frequent mention of the way virtual reality has been depicted in science-fiction media. While we wait for a paradigmatic example for VR, fictional examples remain the best way to evoke the potential and shortcomings of "actual" virtual reality.

Ultimately, while my intermedial analysis focuses on similarities between current VR experiences and the earlier medium of cinema, my purpose in this chapter is to highlight the importance of going beyond formal similarities and of considering, instead, the experiential characteristics of each medium. Looking at the way cinema has shaped the formal characteristics of contemporary VR is a useful point of departure, but a similar approach could also tease out VR's genealogical antecedents in video games, theatre, dance, and so on. Importantly, the lack of historical perspective makes analyzing this still emerging medium a challenging affair. It is also what makes it so important a task. And while I cannot claim to predict what identity the medium of VR might make for itself, I hope that studying part of its genealogy and formation will help inform our understanding of this up-and-coming medium.

3 Gaudreault and Marion, *The End of Cinema?*

What is VR?

In his doctoral dissertation, Tobias Bowman proposes a "technical and social history of Virtual Reality in the United States of America, from 1965 to 2005."[4] As this time frame suggests, virtual reality has a relatively long history, marked by technical innovations of varying complexity and efficacy.[5] To avoid confusion and focus exclusively on the genealogical ancestors of current VR offerings, Bowman proposes the following definition of VR: "A technology which uses a Head-Mounted Display (HMD) and motion tracking to provide a point of immersive interface between a person and a digital computer-generated graphical environment, in order to create a sense of presence in that environment."[6] Contrary to most definitions of VR, Bowman's avoids any misdirection by getting to its underlying principle: the technology, as well as its affordances. With this material and experiential definition of VR in hand, the author is able to distinguish the current trends in immersive technologies from other inventions that simply muddy the waters. In fact, Bowman suggests that semantic confusion concerning the meaning of "VR" was a significant problem throughout the 1980s and 1990s, when the moniker was used to describe a great many things related to computers which had little to do with the kinds of immersive experiences we typically associate with the term VR. Glancing at definitions of virtual reality from the period of its emergence, in the 1980s and 1990s, what might we glean about its untimely downfall in the late 1990s and about its recent revival in the 2010s?

As the story goes, the expression "virtual reality" was coined by Jaron Lanier, a programmer who helped develop one of the first commercial-grade virtual reality devices in the late 1980s.[7] A futurist and a dreamer, Lanier has since come up with as many as fifty definitions of virtual reality, each more whimsical than the last. These definitions paint the picture of an absolutely fantastical technology: "the technology of noticing experience itself," "a medium that could convey dreaming," "[t]he science of comprehensive

4 Bowman, "Fate Amenable to Change."

5 This history appears long because we are still in the infancy of the medium, some forty to fifty years into the lifespan of VR. While still relatively young in comparison to cinema and other media, VR seems to be undergoing a prolonged "early history" phase.

6 Bowman, "Fate Amenable to Change," p. 24.

7 Lanier co-founded VPL Research Inc. in 1984. His claim on the expression "virtual reality" is recounted by several authors in their survey of the origins of VR. See Bolter and Grusin, *Remediation*; Manovich, *The Language of New Media*; Ryan, "Immersion vs. Interactivity"; and Ryan, *Narrative as Virtual Reality*, pp. 48–85.

illusion," "[a] hint of the experience of life without all the limitations that have always defined personhood," and finally, "[a] shared, waking state, intentional, communicative, collaborative dream."[8] An overview of the literature at the time shows, however, that Lanier was not the only one to cloud the meaning of "VR" with unnecessarily vague or ambitious definitions.[9] As these examples demonstrate, at the time of VR's initial development at the end of the twentieth century, there was no single, clear-cut definition of "virtual reality."

Amid the terminological confusion that accompanied the development of virtual reality in the 1990s, Bowman suggests that "much of VR's linguistic and semantic identity came from its futuristic depictions when the 'brand' was launched in the 1980s."[10] Indeed, one could argue that VR's identity was shaped in no small part by depictions in science-fiction media, including cyberpunk novels like William Gibson's *Neuromancer* (1984) and Neal Stephenson's *Snow Crash* (1992) and *The Diamond Age* (1995), as well as films such as *Brainstorm* (Douglas Trumbull, 1983), *The Lawnmower Man* (Brett Leonard, 1992), and *Strange Days* (Kathryn Bigelow, 1995). That being said, it is the Holodeck—as represented in *Star Trek: The Next Generation* (1987–1994)—that has consistently served as one of the most stimulating visions of virtual reality's possible futures.[11]

As I have demonstrated elsewhere, the Holodeck produces a simulated reality that allows one (or many) to inhabit detailed fictional worlds, replete with virtual characters and objects that appear completely real and tangible.[12] As represented in the television series, this simulation is total, in that it allows users to interact with the virtual world and its inhabitants in much the same way they would under normal circumstances (assuming one were to be transported to that particular place and time). And as Jay David Bolter puts it, in respect to the development of VR in relation to films and television: "The ultimate goal remains the *holodeck*—a virtual environment that embodies a seamless story-world and enables one human to be present

8 These are the twelfth, third, forty-seventh, fiftieth, and forty-eighth definitions, respectively. See Lanier, *Dawn of the New Everything*.

9 See Bolter and Grusin, *Remediation*, p. 22; Bricken, "Virtual Worlds"; Ditlea, "Another World," p. 112; and Ryan, "Immersion vs. Interactivity," p. 110.

10 Bowman, "Fate Amenable to Change," p. 250.

11 This intuition was strengthened in late 2022 when the rapid development of AI text-to-image generators (DALL·E 2, Midjourney, Stable Diffusion, etc.) was often described as a stepping stone towards Holodeck-style simulators.

12 Bédard, "La machine subjective?"

(and two or more humans to be copresent) in it."[13] Since the 1980s, then, the Holodeck has remained the public face of virtual reality.

Whatever happened to VR?

As an example of what VR might be one day, the Holodeck has also consistently served as a reminder of the shortcomings of existing VR solutions. After a handful of attempts at commercializing virtual reality devices in the late 1980s and early 1990s,[14] the technologies being developed suffered from the comparison to the Holodeck and other representations of VR that emerged at the time. As Marie-Laure Ryan writes in *Narrative as Virtual Reality*:

> An imagination easily aroused is usually quick to feel let down, and as Lanier observed in late 1998, the Holodeck analogy turned out in the long run to be a double-edged sword for the PR of VR: "As for the waning of virtual reality from public attention, 'I bear some of the blame for it. I always talked about virtual reality in its ultimate implementation and when that didn't happen, interest declined. Because everyone wanted the Holodeck from *Star Trek*, virtual reality couldn't fulfill its promise so quickly.'"[15]

When considering what might have led to VR's death in the eyes of the public in the 1990s, one need not look any further than this divide between fiction and reality. Indeed, already in 1999, Ellen Strain suggested that "the immense amount of speculation surrounding VR makes it necessary to differentiate between current VR applications and what I like to call 'virtual' virtual reality—in other words, the goggle-and-senso-sheath technologies

13 Bolter, "Transference and Transparency," p. 23, original emphasis.

14 Two examples of these attempts are Lanier's VPL, which marketed HMDs and datagloves as human-computer interfaces to the enterprise sector, and Jonathan Waldern's Virtuality, which offered virtual reality games in arcades.

15 Ryan, *Narrative as Virtual Reality*, p. 51. Lanier's original quote comes from Ditlea, "False Starts Aside, Virtual Reality Finds New Roles." Speaking about Jonathan Waldern's Virtuality and its attempts at commercializing VR arcade machines in the 1990s, Lanier has also been quoted as saying: "The problem is that they were a little unrealistic about whether they were ready for primetime. I think people spent their dollar and were disappointed with the level of graphics that were possible at the time." Donovan, *Replay*, pp. 252–253.

which seem virtually real in popular discourses but *which have yet to enter the realm of actuality*."[16]

The issue, then, was a semantic one, as Ryan surmised, and as Bowman elaborates, calling VR "oxymoronic by name and nature, meaning different things to different people, dividing the academic community, and preventing any specialist language, typology, or central concept to be defined."[17] In other words, VR in the 1990s lacked a *paradigm*. Bowman borrows this idea of the aparadigmatic nature of VR from Peter G. M. Swann and Tim P. Watts, who characterize VR's failure to rise to the occasion in the 1990s as a "lack of coordination of expectations and vision between diverse technology developers and diverse users" rather than "an issue of technological failure or lack of potential demand."[18] Considering the state of affairs at the turn of the millennium, Swann and Watts conclude: "VR remains in a pre-paradigmatic stage and until that stage is passed diffusion will remain slow. VR lacks a coherent vision and is, in a sense, a victim of its own ubiquitous potential."[19]

What VR failed to do in the 1990s, then, was not so much to develop useful tools and applications, but rather to coalesce into a clear and identifiable entity.[20] Since the first musings in university and military research labs in the 1960s and through the rest of the twentieth century, as it developed into commercial applications, VR had been coming into its own as a set of technologies. In the mid-1990s, however, it failed to crystallize around a

16 Strain, "Virtual VR," p. 10, emphasis added. In a chapter entitled "Virtual Reality as Dream and as Technology," Ryan likewise writes: "When virtual reality technology burst into public view in the early 1990s, it was less through a revolutionary computer system than through a grand flourish of rhetoric. The idea of VR sprang fully formed from the brain of its prophets, and it was presented by the media to the general public as being in a state of perfect implementation," to which she adds, "[l]ittle did many of us realize in the early 1990s that their enthusiastic and quite precise descriptions of VR applications were largely castles in the air, and how much separated 'real' VR from the virtual brand." Ryan, *Narrative as Virtual Reality*, p. 48.

17 Bowman, "Fate Amenable to Change," pp. 272–273.

18 Swann and Watts, "Visualization Needs Vision," p. 42.

19 Swann and Watts, "Visualization Needs Vision," p. 42. Brenda Laurel gives a more pragmatic reason for VR's demise in the 1990s: "VR as entertainment was going to be extremely difficult to 'monetize.' There was one really gnarly problem with VR as location-based entertainment, what amusement-park people would call 'throughput.' To get anything out of a VR experience, a person reasonably needs five minutes—probably longer—even to get a taste of the world, never mind having a dramatic experience in it." Laurel, *Computers as Theatre*, p. 185.

20 According to Bowman, "[i]t might be fair therefore to argue that VR's failure was not one of technological performance but of establishing what it was good for, or what its purpose was and instead publicly demonstrating what it was not good for." Bowman, "Fate Amenable to Change," pp. 268–269, original emphasis.

single paradigmatic organization of technologies, affordances, and practices. This was the first death of VR. This death was obviously not final, however, as evidenced by the continued presence of VR in laboratories and training facilities—whether for industry or military applications—as well as by the commercial revival it underwent in the second and third decades of the twenty-first century.

The "Double Birth" Hypothesis

> More than any other art, the cinema has died repeatedly and with great regularity over the course of its relatively brief ... existence.
>
> —Stefan Jovanovic[21]

To understand what VR went through at the time of its original demise in the 1990s—as well as what we seem to be witnessing since its revival in the mid-2010s—we must consider the early life of media, from their invention to their crystallization into an institutionally recognized medium. One of the most notable examples in this regard is that of early cinema and of its transformation from a technical apparatus first born in the mid-1890s into the medium we know today. André Gaudreault and Philippe Marion's depiction of the "double birth" of cinema serves as a particularly stimulating point from which to consider the transformations undergone by media as they come into being.

In their book *The End of Cinema* (2015), Gaudreault and Marion present the idea that cinema was not invented all at once in 1895, but also some years later, in the 1910s.[22] The authors explain their reassessment of the established narrative of the invention-birth of cinema by alluding to the shift in paradigm that occurred in the 1910s when the "cinema of attractions,"[23] prevalent in the early period of the cinema gave way to a media different from those that had previously existed. When it was first invented as a technology for recording and reproducing moving images, cinema was not yet cinema *per se*, but rather a protean set of uses of a particular technology. This is what the authors call the "kinematograph," in reference to the

21 Jovanovic, "The Ending(s) of Cinema."

22 Gaudreault and Marion, *The End of Cinema?*, p. 34.

23 "Cinema of attractions" is an expression coined by Tom Gunning to describe the tendency in films of the late 1890s and early 1900s to foster an aesthetic of astonishment that relished the spectacularity of its apparatus and of the experiences it afforded. See Gunning, "The Cinema of Attractions," p. 61.

technical apparatus born at the turn of the century. In the early years of the invention, then, the kinematograph was not a medium onto itself, but rather a technology that grafted its capabilities (recording and reproducing moving images) onto other pre-existing cultural series. It is not until the 1910s, then, that a clearer and unique identity is discerned for cinema, in what the authors describe as a paradigm shift "of such extent that we would be justified in seeing the two entities as completely separate."[24] As they make clear:

> Changes of paradigm do not necessarily occur at the same time as the invention of new techniques, and the sudden availability of a new technology does not necessarily revolutionize the surrounding culture or the behavior and activities of the various cultural agents who latch onto it. Nor does it make it possible to accede immediately to a new cultural, artistic, or media order. Nascent media take their first steps by reproducing in a rather servile manner the other media from which they are to greater or lesser degrees derived. This was the case with cinema.[25]

This was the case with cinema, and so is it the case with VR as well. The example of cinema's institutionalization in the 1910s can indeed serve as a stimulating lesson for understanding VR's downfall in the 1990s as a result of its aparadigmatic nature, as well as for analyzing the current transitional period in virtual reality's life. In both cases—VR in the 1990s and since the 2010s—it must be made clear that virtual reality has remained in the pre-paradigmatic stage of a technology not yet reborn with a clear identity as a medium.

In both periods of virtual reality discussed in the present chapter, as well as in the case of early cinema discussed by Gaudreault and Marion, we are faced with a period in the life of a medium that predates its institutionalization, when it is still "a kind of hodgepodge of other expressive forms."[26] While looking at the second birth of cinema as a fully formed medium will certainly help us understand what VR failed to do in the 1990s and what it could possibly achieve in the 2020s and beyond, I suggest that analyzing the period of turmoil that preceded cinema's institutionalization is a better way of understanding the fluid and intermedial nature of a proto-media.

24 Gaudreault and Marion, *The End of Cinema?*, p. 34.
25 Gaudreault and Marion, *The End of Cinema?*, p. 106.
26 Gaudreault and Marion, *The End of Cinema?*, p. 105.

Intermedial Meshing

> To me [...][,] the computer looks more each day like the movie camera of the 1890s: a truly revolutionary invention humankind is just on the verge of putting to use as a spellbinding storyteller.
>
> —Janet H. Murray[27]

As Gaudreault and Marion demonstrate, "a medium *is always born twice.*"[28] "Its first birth," they explain, "takes place when a technological innovation is used to give new life to existing cultural practices and series, under whose authority this technology places itself."[29] This is the period from which VR failed to evolve in the 1990s and in which we find ourselves again today, in the 2020s. Considering history is slated to repeat itself—as Janet H. Murray opines in the above-quoted epigraph—one of the most important lessons we can take away from this first birth of cinema (as a technology born under the auspices of existing media), comes from its interaction with the media formations that came before it.[30] In the case of early cinema, it has been demonstrated on many occasions that the kinematograph was not simply "photography but moving," but rather the product of intermedial exchanges with a great many cultural series: "theater, fairy plays, photography, magic lanterns," and so on.[31] As a sort of proto-medium in the process of crystallizing its identity, early cinema was indebted to many media of varying origins. The same can be said of VR in the years since its revival in the mid-2010s. If, as Gaudreault and Marion argue, "a medium's first years of life, before it takes on a conscious identity, are decisive and shape it once it has 'grown up,'"[32] what are the genealogies that are currently shaping virtual reality? In the remainder of this chapter, I highlight ways in which VR seems to be taking inspiration from cinema, among other media, in order to form its own, as yet inchoate, media identity. Seeing what VR has taken from cinema and, more importantly, how it differs from that earlier medium, can help us understand what identity VR might eventually craft for itself.

27 Murray, *Hamlet on the Holodeck*, p. 9.

28 Gaudreault and Marion, *The End of Cinema?*, p. 107, original emphasis.

29 The quote continues: "A second birth occurs when the expressive resources made possible by a medium, or more precisely those of a technological apparatus that has become a medium, achieve institutional legitimacy and work toward establishing the specificity of these resources as the norm." Gaudreault and Marion, *The End of Cinema?*, p. 107, original emphasis.

30 Gaudreault and Marion, *The End of Cinema?*, p. 113.

31 See Dulac, Gaudreault, and Hidalgo, *Companion to Early Cinema*, p. 6.

32 Gaudreault and Marion, "Measuring the 'Double Birth' Model against the Digital Age," p. 159.

VR Films: An Oxymoron?

> VR is utterly a first-person point-of-view medium.
> —Brenda Laurel[33]

In her interpretation of a quote from Jaron Lanier that echoes Tobias Bowman's definition of VR, Marie-Laure Ryan underlines some of the affordances of the technology: "the experience of being inside a computer-generated world involves three distinct components: a sense of being surrounded, a sense of depth, and the possession of a roving point of view. Each of these dimensions improves on, or remediates, a variety of earlier technologies."[34] As Ryan demonstrates, each of these components has a genealogy that can be traced back to earlier media formations: the sense of being surrounded goes back to the panorama and cyclorama, the sense of depth is engrained into perspectival representations and stereoscopes, while the movement of a point of view is part and parcel of films and video games.[35] "Here again," the author posits of the latter, "VR represents the ultimate achievement in the history of what Jay Bolter and Richard Grusin call 'point of view technologies.'"[36] As Ryan suggests, then, Bolter and Grusin present virtual reality as "a remediation of the subjective style of film, an exercise in identification through occupying a visual point of view."[37]

If we understand this identification to be the primary identification with the camera, as Christian Metz defined it, then myriad VR experiences certainly continue the filmic tradition of allowing spectators to

33 Laurel, *Computer as Theatre*, p. 184.

34 Ryan, *Narrative as Virtual Reality*, p. 53. She continues: "For a body to enter into a world, this world must be fully spatial. Lanier, in his pioneering pronouncements about VR, describes the VR experience as follows: 'When you put [VR glasses] on you suddenly see a world that surrounds you—you see the virtual world. It's fully three-dimensional and it surrounds you, and as you move your head to look around, the images that you see inside the eyeglasses are shifted in such a way that an illusion is created that while you're moving around the virtual world is standing still.' " Ryan, *Narrative as Virtual Reality*, p. 53, emphasis added. Lanier's quote originally comes from Zhai, *Get Real*, p. 176.

35 Ryan, *Narrative as Virtual Reality*, pp. 53–54.

36 Ryan, *Narrative as Virtual Reality*, pp. 53–54. See also Bolter and Grusin, *Remediation*, p. 162.

37 Bolter and Grusin, *Remediation*, pp. 165–166. The authors continue by stating: "In our media-saturated culture, we can hardly avoid regarding virtual reality in this way, and that is perhaps why Hollywood filmmakers have offered us so many films that incorporate or refer to virtual reality." Bolter and Grusin, *Remediation*, p. 166.

identify with the point of view of the camera and feel as though they were seeing the filmic world directly. "Virtual reality films" or "360-degree VR" are terms used to describe experiences wherein a HMD can be used to view moving images that seem to be unbounded by the typical cinematic frame as they span the entirety of a sphere that encircles the viewer. Although often limited by the graphical capabilities of current displays, VR films benefit from the expansive field of view and from the stereoscopic effect of the display, which give shape and volume to the world represented on screen. VR films like *Clouds over Sidra* (Gabo Arora and Chris Milk, 2015), *Borderline* (Assaf Machnes, 2018), *Dinner Party* (Angel Manuel Soto, 2018), *Rooms* (Christian Zipfel, 2018), and *21–22 Alpha* (Thierry Loa, 2019) take part in this lineage by presenting spaces that the viewer can explore visually through the point of view of the camera.

If, however, we take the identification of which Bolter and Grusin speak to be Metz's secondary identification with the characters in the story, then VR films have certainly taken part in this "remediation of the subjective style of film." As I have demonstrated in a past essay, a particularly prevalent tendency in early VR productions was to emulate the first-person perspective of narrative films by placing the spectator in the body of a character in the narrative.[38] Works that fit this description include *The Doghouse* (Johan Knattrup Jensen, 2014), *Notes on Blindness* (Arnaud Colinart, Amaury La Burthe, Peter Middleton, and James Spinney, 2016), *I, Philip* (Pierre Zandrowicz, 2016), and *Miyubi* (Félix Lajeunesse, 2017), among countless others. This remediation of cinema's subjective point of view can be attributed to the affordances of the technical apparatus of virtual reality, which differs fundamentally from that of cinema in that the former allows users to control the point of view. However, VR filmmakers have taken this affordance one step further by putting spectators "in a character's shoes."[39]

This proclivity towards first-person VR films recalls an example favoured by Bolter and Grusin. In their introduction of virtual reality—and of their

38 Bédard, "La machine subjective?"

39 One particularly telling quote to that effect can be found in the digital press kit for *The Doghouse*: "You sit down, put on the glasses and headphones, and instantly you establish a connection between the perceived and the perceiver [...]. We call it First Person Reality, because we put the audience even before the first row, inside the character [...]. In the Doghouse you experience, what the actor experiences. As if you were there. You see, what the actor sees, and you hear, what the actor hears. So you, by the help of the actor, become a character in the film." *The Doghouse*, Electronic Press Kit.

concept of remediation—the authors refer to *Strange Days* and its so-called "wire," a contraption used to record and represent the lived experience of an individual in all its sensorial and affective complexity. As represented in Bigelow's film, namely as entirely first-person long takes, the experiences afforded by the wire appear entirely unmediated. Bolter and Grusin actually present the relation between VR and the wire from *Strange Days* wire in reciprocal terms, going as far as to say: "The wire is just a fanciful extrapolation of contemporary virtual reality," and, conversely, "contemporary virtual reality is, like the wire in *Strange Days*, an experiment in cinematic point of view."[40]

Using *Strange Days* to represent VR's remediation of cinematic strategies brings several issues to the fore, not the least of which is the fact that "point of view technologies," as Bolter and Grusin call them, can scarcely be limited to the medium of film anymore, especially since the advent of 3D video games. The problem with comparing VR to the way the wire functions is that the latter, while fully immersive, is also an essentially passive experience, much like cinema. The technical apparatus of VR, as I have demonstrated here, allows the viewer to shift their point of view. This affords them a degree of agency which, as minimal as it might be in some cases, sets it apart from the cinematic tradition. As Torben Grodal suggests, at a base level interactivity consists in having the ability to "change the visual appearance of a computer screen," the zero degree of which I take to mean the power to affect the point of view.[41] In *Strange Days*, as in the medium of cinema, the spectator cannot affect what is being seen. In virtual reality experiences—even those closest to the lineage of cinema—the ability to affect the representation by virtue of one's own movements hints at other sources of influence. In other words, while there might indeed be a family resemblance between early "VR films" and certain *formal* characteristics of cinema (point of view, "camera" movement, editing, etc.), the *experiential* aspects of virtual reality force us to recognize the difference between the passivity of the filmic spectator and the activity of those who take part in VR experiences.

40 Bolter and Grusin, *Remediation*, p. 4.

41 Grodal, "Stories for Eye, Ear, and Muscles," p. 142. Bolter and Grusin propose a similarly minimal approach to interaction when they describe a user's influence on a computer animation sequence: "viewing can be interactive, although the interaction may be as simple as the capacity to change one's point of view." Bolter and Grusin, *Remediation*, p. 28.

Conclusion: VR's (Yet to Be) Media Identity

> With VR, you're just not limited to reality [...]. You're only limited by imagination. The most exciting stuff is yet to come.
> —Yelena Rachitsky, executive producer of experiences at Oculus[42]

If the limited interactivity of VR films precludes me from placing virtual reality solely in the lineage of cinema, it is because VR borrows as much—if not more—from other media, including video games, theatre, dance, painting, and many more. Myriad VR experiences also currently challenge any attempt to neatly trace a media genealogy. To name but a few:

- *The Machine to be Another* (*TMTBA*) (BeAnotherLab, 2014–present) invites two users to exchange bodies and see the world from someone else's perspective (including seeing themselves from the other's point of view). It borrows more from performance art, guided meditation, and psychological experiments;
- *Eve, Dance is an Unplaceable Place* (Compagnie Voix and Omnipresenz, 2019) is an offshoot of *TMTBA* that invites three spectators to take part in a dance choreography that takes place inside three distinct virtual environments as well as in front of a live audience. It is a child of dance, but also video art, and cyborg art;
- *Taste of Hunger* (Christoffer Boe and David Adler, 2020) presents the viewer with a looping, deconstructed narrative they must reconstruct by walking around a circular space. It descends from sculpture and other such spatial media, as well as from anamorphoses and GIFs;
- *Agence* (Pietro Gagliano, 2020) puts the user in control of a small planet inhabited by five creatures powered by artificial intelligence, which makes it entirely impossible for any two experiences to be alike. Its clearest forebearers are Darwinian evolutionary models, behaviourism, and recent developments in AI.

In all these experiences—as well as in countless others I have not yet had the pleasure of trying myself—the intermedial porosity of contemporary virtual reality is made even more apparent than in the VR films discussed previously. This, of course, is without considering the myriad other applications VR has found in recent years, from serving as a tool for digital sculpting and painting to social applications that have offered new ways

42 Rachitsky, quoted in Soloski, "Theater's Next Act?"

to connect with strangers, friends, and loved ones, especially during the isolation brought about by COVID-19 and global lockdowns.

The constant proliferation of VR in all shapes and forms is a proof of its unwillingness to die as so many have predicted. It is also a sign of VR's current focus on remediating the media formations that existed when it was initially born. Reborn as a technology in the form of HMDs in the mid-2010s, VR has yet to live through the second birth that comes in the form of institutionalization. While it is impossible to say exactly what identity the medium of VR will take on if and when it succeeds in becoming recognized, as cinema was in the mid-1910s, all we can do as scholars of media—and as scholars of technical innovations especially—is adopt strategies for interpreting the intermedial meshing, or porosity, that accompany media in formation. And while much energy may be spent trying to identify the genealogy of VR by looking at its *formal* qualities for signs of its closest relatives, VR's inherently interactive and immersive persona requires that we pay as much—if not more—attention to its *experiential* characteristics to understand the media identity current projects are driving towards.

Bibliography

Bédard, Philippe. "La machine subjective ? Les appropriations cinématographiques des dispositifs immersifs contemporains." *Canadian Journal of Film Studies* 28, no. 1 (2019): 66–92. https://doi.org/https://doi.org/10.3138/cjfs.28.1.2019-0012.

Bolter, Jay David, and Richard Grusin. *Remediation: Understanding New Media.* Cambridge, MA: MIT Press, 2000.

Bolter, Jay David. "Transference and Transparency: Digital Technology and the Remediation of Cinema." *Intermédialités* 6 (Fall, 2005): 13–26.

Bowman, Tobias. "Fate Amenable to Change: A Technical and Social History of Virtual Reality in the United States of America, from 1965 to 2005." PhD diss., University of Oxford, 2018.

Bricken, Meredith. "Virtual Worlds: No Interface to Design." In *Cyberspace: First Steps*, edited by Michael Benedikt, 363–382. Cambridge, MA: MIT Press, 1991.

Collins, Barry. "VR Headsets are Dying a Lonely Death." *Forbes*, May 4, 2020. https://www.forbes.com/sites/barrycollins/2020/05/04/vr-headsets-are-dying-a-lonely-death/.

Ditlea, Steve. "Another World: Inside Artificial Reality." *PC Computing* 2, no. 1 (1989): 90–112.

Ditlea, Steve. "False Starts Aside, Virtual Reality Finds New Roles." *New York Times*, March 23, 1998. https://www.nytimes.com/1998/03/23/business/false-starts-aside-virtual-reality-finds-new-roles.html.

Donovan, Tristan. *Replay: The History of Video Games*. Lewes: Yellow Ant, 2010.
Dulac, Nicolas, André Gaudreault, and Santiago Hidalgo. *Companion to Early Cinema*. Malden, MA: Wiley-Blackwell, 2012.
Gaudreault, André, and Philippe Marion. *The End of Cinema? A Medium in Crisis in the Digital Age.* New York, NY: Columbia University Press, 2015.
Gaudreault, André, and Philippe Marion. “Measuring the ‘Double Birth’ Model against the Digital Age.” *Early Popular Visual Culture* 11, no. 2 (2013): 158–177.
Grodal, Torben. “Stories for Eye, Ear, and Muscles: Video Games, Media, and Embodied Experiences.” In *The Video Game Theory Reader*, edited by Mark J. P. Wolf and Bernard Perron, 129–155. London: Routledge, 2003.
Gunning, Tom. “The Cinema of Attractions: Early Cinema, Its Spectator and the Avant-Garde.” In *Early Cinema: Space, Frame, Narrative*, edited by Thomas Elsaesser and Adam Barker, 56–62. London: British Film Institute Publishing, 1990.
Jovanovic, Stefan. “The Ending(s) of Cinema: Notes on the Recurrent Demise of the Seventh Art.” *Offscreen* 7, no. 4 (April, 2003). https://offscreen.com/view/seventh_art1.
Lanier, Jaron. *Dawn of the New Everything: Encounters with Reality and Virtual Reality*. New York, NY: Henry Holt and Company, 2017.
Laurel, Brenda. *Computers as Theatre*. Reading, MA: Addison-Wesley, 1991.
Mandt, Neil. “VR Can Be a Mainstream Tech, but It’s not Ready Yet.” *VentureBeat*, August 8, 2017. https://venturebeat.com/2017/08/08/vr-can-be-a-mainstream-tech-but-its-not-ready-yet/.
Manovich, Lev. *The Language of New Media.* Cambridge, MA: MIT Press, 2001.
Murray, Janet H. *Hamlet on the Holodeck. The Future of Narrative in Cyberspace.* New York, NY: Free Press, 1997.
Parlock, Joe. “Stop Saying Virtual Reality Is Dying.” *Forbes*, May 5, 2020. https://www.forbes.com/sites/joeparlock/2020/05/05/stop-saying-virtual-reality-is-dying/.
Roose, Kevin. “This Should Be VR’s Moment: Why Is It Still so Niche?” *New York Times*, April 30, 2020. https://www.nytimes.com/2020/04/30/technology/virtual-reality.html.
Ryan, Marie-Laure. “Immersion vs. Interactivity: Virtual Reality and Literary Theory.” *SubStance* 28, no. 2 (1999): 110–137. https://doi.org/10.2307/3685793.
Ryan, Marie-Laure. *Narrative as Virtual Reality: Immersion and Interactivity in Literature and Electronic Media*. Baltimore, MD: Johns Hopkins University Press, 2001.
Soloski, Alexis. “Theater’s Next Act? A Show That’s All in Your Head.” *The New York Times*, July 8, 2020. https://www.nytimes.com/2020/07/08/theater/virtual-reality-the-tempest.html.
Strain, Ellen. “Virtual VR.” *Convergence* 5, no. 2 (Summer, 1999): 10–15.

Swann, Peter G. M., and Tim. P. Watts. "Visualization Needs Vision: The Pre-Paradigmatic Character of Virtual Reality." In *Virtual Society? Technology, Cyberbole, Reality*, edited by Steve Woolgar, 41–60. Oxford: Oxford University Press, 2002.
Topolsky, Joshua. "The Virtual Reality Dream Is Dying." *The Outline*, October 25, 2018. https://theoutline.com/post/6443/virtual-reality-dream-is-dead-hype-oculus-rift-facebook-playstation.
Zhai, Philip. *Get Real: A Philosophical Adventure in Virtual Reality*. New York, NY: Rowman & Littlefield, 1999.

About the Author

Philippe Bédard is an independent researcher with a fascination for moving image media and their technologies. As part of his postdoctoral research, he focused on virtual reality, with a focus on the notion of empathy. Finally, he has worked extensively in and around the immersive media industry in Montreal, including work with Xn Québec, Festival du Nouveau Cinéma, the National Film Board of Canada's interactive studio, and SODEC.

9. Seeing by Automobility: Protoautomobility, Immersion, the Traveller-Spectator, and the Phantom Ride, 1897–1912

Michael J. T. Stock

Abstract: This chapter delves into the phenomenon of "phantom ride" films, a popular form of early actuality cinema at the turn of the twentieth century. These films, shot from the front of moving vehicles such as trains, trams, and eventually automobiles, provided viewers with a unique point of view (POV) experience, immersing them in the space traversed by the camera. Drawing from Lynne Kirby's concept of locomotive perception, the author explores how these films, typically associated with train travel, introduced audiences to a form of "protoautomobility" long before automobiles became widely accessible. Finally, through a discussion of notable examples like "Leaving Jerusalem by Railway" and "The Haverstraw Tunnel," the article highlights the immersive and anticipatory nature of these films in shaping viewers' perceptions of mobility.

Keywords: point of view (POV), early cinema, film history, travel films

Fusing the rollercoaster, the locomotive, and the cinema (and anticipating the point of view [POV] of automobility), one of the most popular early actualities at the turn of the century was the "phantom ride" film. Typically lasting anywhere from one to over six minutes, phantom ride films were shot from the front of a moving vehicle (most of them were filmed from a train, but others were shot from trams, trolleys, streetcars, subways, boats, and eventually automobiles). The camera could be mounted to either the front or the back of the moving vehicle. Most often, the vehicle that the camera and cameraman or camerawoman were mounted to remained

Bédard, Philippe, Alanna Thain, and Carl Therrien, eds. *States of Immersion Across Media: Bodies, Techniques, Practices*. Taylor & Francis Group, 2025
DOI 10.5117/9789048558766_CH09

unseen in the film, hence the ghostly or "phantom" nature of these films. The resulting POV ranged from striking to spectacular, putting the spectator in the position of a traveller, immersed in the space that the camera moved through. Aligned seamlessly with the camera, its movement became their movement, transforming the viewer into a *Traveller-Spectator.* This chapter will explore how these phantom ride films, typically associated with train travel, are responsible for introducing the mode of perception equated with automobility long before most filmgoers had even seen an automobile. Building from the influential argument framed by Lynne Kirby in her book *Parallel Tracks: The Railroad and Silent Cinema* (1997) that locomotive perception was protocinematic, this chapter argues that phantom ride films offered cinema's earliest audiences a unique experience of *protoautomobility* long before most people had even seen an automobile (let alone been in one).[1]

Although the automobile's origins date back to the late nineteenth century, up until the introduction of Ford's Model T, in 1913, the experience of automobility was a privilege and a luxury only enjoyed by the elite. As automobile historian Steven Parissien explains, with the Model T, "for the first time, a car cost less than the average annual wage. A vehicle that was now within the reach of almost everyone, the Model T, was effectively the first global car."[2] Most first-generation filmgoers were introduced to automobility in the cinema, catching their first glimpse of an automobile on the movie screen before they began appearing on the street. But, even before that, Traveller-Spectators were introduced to a unique and immersive experience, the phantom ride films, that anticipated the POV they would eventually experience within automobiles.

For Lynne Kirby, it is the paradoxical simultaneity of movement and stillness that bonds the train to the cinema. As she explains in the introduction of her influential 1997 book: "In cinema, instability is built into the basis of the filmgoing experience: the perceptual illusion of movement is tied to the physical immobility of the spectator and to the sequential unfolding of a chain of still images that constitute the basis of every film."[3] Kirby goes on to describe the train as "cinema's mirror image," in the sense that trains were responsible for training locomotive travellers how to eventually become film spectators via their experience of the moving landscape seen

1 Kirby, *Parallel Tracks*.

2 Parissien, *The Life of the Automobile*, p. 14. Tellingly, while in 1908 there were only 190,000 automobiles registered in the United States, by 1916 there were more than 3.6 million registered automobiles; 570,000 of those were Model Ts manufactured that year alone. On this, see Seiler, *Republic of Drivers*, p. 6; and Vinsel, *Moving Violations*, p. 32.

3 Kirby, *Parallel Tracks*, p. 3.

through the train's windows.[4] As Kirby describes it, train rides were in essence a "*protocinematic* phenomenon."[5] She explains:

> In both cases, passengers sit still as they rush through space and time, whether physically and visually, as on the train, or merely visually, as in the cinema. The train would then be cinema's mirror image in the sequential unfolding chain of essentially still images and the rapid shifts of point of view that the train and cinema experiences entail. [...] As a *perceptual* paradigm, the railroad established a new, specifically modern mode of perception that the cinema absorbed naturally. In other words, the kind of perception that came to characterize the experience of the passenger on the train became that of the spectator in the cinema.[6]

Just as the railroad established "a new, specifically modern mode of perception that the cinema absorbed," as Kirby asserts, the phantom ride films introduced viewers to a new mode of perception that would prepare them for their experiences in an automobile. Key to this experience was the unique immersive POV that the phantom ride films offered the Traveller-Spectator, moving through the space that surrounded them. It is precisely this state of immersion that marks these films as striking examples of protoautomobility. No longer limited to merely staring out the window of a train at the panorama of images that flashed by them outside, the Traveller-Spectator of the phantom ride film enjoyed an unprecedented level of immersion in the spectacle.

In this earliest period of the cinema, before narrative had become inextricably linked with the medium, spectators were drawn to the cinema to experience what Tom Gunning famously refers to as "attractions." Gunning describes the "cinema of attractions" as a "harnessing of visibility" and an "act of showing and exhibition," which he links to the avant-garde of the early twentieth century (the Futurists, the Dadaists, and the Surrealists) in its use of the cinema to break with its "enslavement to traditional art forms, particularly theatre and literature."[7] The exhibition of the films amplified this notion of the cinema as an "attraction." As Gunning points out, the largest chain of film theatres before 1906 was *Hale's Tours and Scenes of the World*, where "not only did the films consist of non-narrative sequences taken from moving vehicles (usually

4 Kirby, *Parallel Tracks*, p. 2

5 Kirby, *Parallel Tracks*, p. 2, original emphasis.

6 Kirby, *Parallel Tracks*, p. 2 and p. 7, original emphasis.

7 Gunning, "The Cinema of Attractions," p. 381.

trains), but the theatre itself was arranged as a train car with a conductor who took tickets, and sound effects simulating the click-clack of wheels and hiss of air brakes."[8] The most prominent elements of these attractions were mobility and a setting that recreated the unique mobility afforded to passengers on a locomotive. As Lauren Rabinowitz describes, these simulated train trips or "'movie rides' coordinated sounds, motion pictures, and mechanical movement to reproduce the realism of train travel and to offer virtual travel to remote areas that the railroad had recently opened up for tourism in the United States, Canada and Europe."[9] The popularity of *Hale's Tours* spawned numerous competitors offering similar cinematic travel experiences, not only in the faux setting of a fictional locomotive car, but in similarly constructed touring cars, automobiles, and buses as well as in simulations of air travel by balloon or of water travel by boat.[10] For *Tim Hurst's Auto Tours*, Traveller-Spectators boarded a giant touring car described in one 1906 article as an "imitation sight-seeing bus" that gave filmgoers a tour of New York City:

> The imitation car is suspended in such a manner as to make the effect very realistic. Apparently the vehicle is forcing its way through the labyrinth of traffic, with occasional stops at crossings, pedestrians dodge to the right and left and the lecturer recites the objects of interest along the route. New York City requires a film of 840 feet, and the route starts at the Flatiron building, covers Fifth avenue, crosses to Riverside Drive, returns down Broadway, invades the East Side, and crosses the Brooklyn Bridge. Another film, 800 feet long, pictures Washington in like manner.[11]

A *Scientific American* article published the same year explains the illusion, effect, and process in more detail:

> As the car does not move, it is necessary to have the landscape pass in review, and to the tourists reposing in their comfortable seats,

8 Gunning, "The Cinema of Attractions," p. 383.

9 Rabinovitz, "From Hale's Tours to Star Tours," p. 42. See also Gunning, "The Attraction of Motion"; Harrison, *From Steam to Screen*; Hayes, "Phantom Carriages"; Rabinovitz, "'Bells and Whistles'"; and Slugan, *Fiction and Imagination in Early Cinema*.

10 In Rabinowitz's essays about *Hale's Tours*, she names a number of Hale's competitors: *Tim Hurst Auto Tours, Hurst's Touring New York, Auto Tours of the World, Cessna's Sightseeing Auto Tours, Trolley Car Tours, Palace Touring Cars, Citron's Overland Flyer, A Trip in California Over Land and Sea, Sightseeing in the Principal Cities, White and Langever's Steamboat Tours of the World,* and *Hruby and Plummer's Tours and Scenes of the World.* To date, virtually nothing else has been published regarding these companies.

11 "An Imitation Bus Trip," p. 566.

> there is an ever-changing picture in front of them. This consists of a series of rapidly-made views, secured by covering the route of the alleged trip with a camera mounted on a car. These are subsequently projected on the screen in front of the big auto. It is thus possible not only to make a tour of the city in which the new amusement feature is located, but it is proposed to present a varied programme, so that it will be possible to make similar tours of other cities and localities of interest.[12]

While the novelty in viewing a train film at *Hale's Tours* would have been for filmgoers to compare their simulated experience to what they had no doubt experienced on real train journeys, the experience, in 1906, of a simulated "big auto" would have provided most Traveller-Spectators their first experience of automobility, as only the very rich had had access to automobiles at that point. As Rabinowitz explains, each of these travel ride films acted to "foreground the body itself as a site for sensory experience" and, as a result, "articulated a seemingly contradictory process for the spectator: they attempted to dematerialize the subject's body through its extension into the cinematic field while they repeatedly emphasized the corporeality of the body and the physical delirium of the senses."[13] Of course, it wasn't just the subject's body that a theatre like *Hale's Tours* attempted to dematerialize, but the theatre itself: once the lights dimmed and the film started, the theatre became a phantom. Suddenly, floors, walls, and ceiling became floorboard, doors, and roof. Freed from the static foundations of the built environment, the experience of the phantom ride was an immersion in movement.

While panorama films like *Panorama du grand pris d'un bateau* (Alexandre Promio, 1896) offers the spectator a view of a landscape or a cityscape passing by—a view often compared to the one one would get by staring out of the window of a train—in the phantom ride films, the spectator becomes a traveller who is aligned seamlessly with the camera: its movement becomes their movement. Although the vehicle of movement remains invisible, the view of the setting—train tracks, trolley wires overhead, water around them, waves breaking—hints that the Traveller-Spectator has become fused with a technological means of mobility operating in tandem with the technology of the motion picture camera. Discussing

12 "Brief Notes Concerning Patents," p. 520.

13 Rabinovitz, "From Hale's Tours to Star Tours," p. 42.

the origins of phantom ride films in *The Railway and Modernity* (2007), Patrick Keiller explains:

> One might understand *phantom* merely as a modifier, meaning illusory, in which case railway panoramas can be and sometimes are described as phantom rides, but the term is more specific to a view in a direction close to that of the line of travel. In its strictest sense, a *phantom ride* is a film which looks forward from the front of a moving railway engine, a view then seldom encountered in ordinary experience, even by an engine driver.[14]

Two of the most successful phantom ride films were released in 1897: *Départ de Jérusalem en chemin de fer* [*Leaving Jerusalem by Railway*], a Lumière Brothers production shot by Alexandre Promio, and *The Haverstraw Tunnel*, shot by an uncredited filmmaker and released by Herman Casler and William Kennedy-Laurie Dickson's American Mutoscope and Biograph Company. This pair of films effectively frame both the scope and the limitations of the immersion rendered possible in a phantom ride, offering viewers two distinct and different examples of protoautomobility, which is why they will be discussed at length here.

In the forty-five-second-long *Leaving Jerusalem by Railway*, the camera is mounted to the rear of a train, so our perspective is of the city as we leave it behind. As the film opens, our attention is immediately drawn to three men in suits and hats walking along the railroad tracks. Behind them, the tracks stretch to the horizon just at the left side of the frame. Beyond them, in the distance, we can just make out a tall outcropping of buildings. On the right side of the frame are what appear to be the ruins of a building from some past time. As the camera/train departs, more men appear in the frame, all tipping their hats or waving to the camera as it passes. As they are left in the distance, a second group of men—this time composed of men in suits and a fez—appear in the frame. As the train nears the station, the crowd grows in size and in diversity, including a Franciscan friar, a man with a long sword, and a mixed group of men, women, and children. Leaving the station and the crowd in the distance, the eyes of which are still on the departing train/camera, the train passes a lumber yard, and then the film ends.

The Haverstraw Tunnel is a thirty-five-second film that opens in mid-journey with a POV of train tracks leading through a lush countryside to the vanishing point of the horizon. In this film, the camera is mounted to

14 Keiller, "Phantom Rides," p. 75, original emphasis.

the front of the train, so viewers are given a vantage point on the space in front of them as they move into it. On both sides of the frame, beyond the tracks, a low-rise fence running parallel to the tracks is visible, with fenceposts every six feet or so and telegraph poles every fifty feet or so. Within seconds, the train races past a wide crossing marked by a white wooden fence soon racing towards a bend in the tracks just ahead. The Haverstraw Tunnel itself is only visible shortly before the train enters it. For several seconds, there is complete darkness, then the tracks reappear, glinting in the light of the opening ahead. The light is blinding as the train emerges from the tunnel, now framing the curving tracks ahead, flanked again by telegraph poles. Just as the train comes out of the curve, speeding down a now straight path towards the horizon, a man appears in its path, crossing the tracks just ahead. For a few seconds, the train races towards the man. It seems unlikely that he will get out of the way in time. But he does, and the train blasts past him. In the distance, we can just make out the outline of what may be another figure standing in the middle of the tracks; then the film ends.

The two films are in many ways polar opposites—with trajectories through space and time aimed in opposite directions in both literal and symbolic senses. In *Leaving Jerusalem by Railway*, the camera is locked down on the rear of the train, providing the viewer with a view of Jerusalem as it is left behind—and with it, a rich but volatile mix of cultures that will soon vanish from sight, as the four centuries of rule by the Ottoman Empire disappeared less than twenty years later, after World War I. Situated at the outskirts of Jerusalem, the Traveller-Spectator experiences the city as a place of both ruins and modern expansion, in the form not only of the modern transportation enabled by the train, but of the lumber yard next to it, signifying more building, future expansion, and growth. The speed of the train's movement is slow at first and then speeds up—which corresponds to the natural and necessary movement of a train pulling away from the station, but at the same time signifies the break from the gravity of the train station, of the family members or friends left behind, and indeed, of the city of Jerusalem itself.

In *The Haverstraw Tunnel*, the Traveller-Spectator is aligned with the then-cutting edge of modernity. It offers a POV of not only what lies ahead, but one experienced at a high velocity. Like most actualities, there is no immediate sense of introduction or conclusion to the film. It is rather an excerpt of reality that the Traveller-Spectator is given access to for a few seconds. This is a very different sensation than the Lumière film, which opens just shortly after the train begins to leave the station. Although *The*

Haverstraw Tunnel takes place in the countryside, it is a space that modern technology has taken over—from the fence and the telegraph poles lining the winding tracks to the train and the tunnel themselves—which would have required advanced engineering and explosives to literally blow through the hillside and reshape nature to make way for the path of the locomotive. The train never slows. Not even for humanity. This, of course, provides a powerful metaphor for the specific relationship of the locomotive (and/or of modernity) and humanity that it is bearing down on at a high velocity, and the implication of the cinema in the act, for not only does the sequence create the spectacle, but also exploits the potential for violence and death on screen.

In both cases, the phantom ride offers the Traveller-Spectator a unique immersive experience that allows them to experience two distinct time-spaces at once—both the time-space of the locomotive and the time-space of the environments they are moving through. Beyond the binary of pastness and futurity, the viewer is also immersed in a collision of cultures. In *Leaving Jerusalem by Train*, the train is the shared meeting point of multiple cultures travelling away from the city centre towards what will soon be developed at its edges. While the Traveller-Spectator knows nothing of the denizens of the train in *The Haverstraw Tunnel*, the immersion in the landscape similarly reveals the uneasy balance between nature and technology, reflecting on the train's role in the upsetting of traditional boundaries of rural and urban areas. Immersed in the phantom ride with no sight of either point of origin or point of call, the train serves as an extension of the built environment, redefining the countryside as something that must be moved through between destinations. The precise positioning of the camera that defines the phantom ride—i.e., either looking ahead or looking behind, with its path strictly aligned with the tracks—stresses this point. While a passenger on the train is afforded a view of the countryside that allows them at least the illusion of moving through a pastoral countryside, the viewer of a phantom ride film is always presented with a landscape marked by the constant path of rails and crossties that determines their path ahead (or behind).

Arguably, the most futuristic aspect of the phantom ride in *The Haverstraw Tunnel* is the fenced crossing that appears just seconds into the film. On first analysis, the wideness of the fencing most likely suggests a cattle crossing. But, on closer analysis of the tracks, you can see that the gaps between them have been filled in with slats of wood and with dirt pushed up to an incline on either side of the tracks, in essence forming a sort of bridge over the tracks. Although this makeshift bridge was most likely made to facilitate the crossing of cattle safely across the tracks, it also almost as certainly could soon be used as a crossing for automobiles. In that sense, this phantom ride film is haunted by a spectre from the future: that of the automobile. Like the first roads created by animals picking out the most passable path, here, the pathway for the automobile is already set. The intersection of the paths of the automobile and of the locomotive is inevitable, as is their inevitable collision on a cultural level.

This film is in fact doubly marked by automobility. While the film achieves its stunning spectatorial position by mounting the camera on the front of a speeding locomotive, this perspective is not associated with the locomotive (save perhaps for the engineer). It is rather a perspective that anticipates the unique vantage point of automobility—not only in terms of its linear perspective, with the tracks leading to a vanishing point on the horizon—but also in terms of its enframing—anticipating the windscreen of the automobile. Because we never see the vehicle that transports the Traveller-Spectator in the phantom ride film, the unique perspective they are given in them is the entire field of vision of the path ahead. The original limits of the field are determined by the frame of the camera, which is then experienced by the Traveller-Spectator as the frame of the screen. In 1897, of course, automobiles didn't have windscreens yet. That innovation would not arrive until the twentieth century. The POV of the phantom ride films is thus clearly anticipating what will soon become the POV of automobility, albeit on *roads* that will lead to a vanishing point on the horizon instead of tracks. While the experience of the moving panorama in the late eighteenth century helped train passengers comprehend their new experience of panoramic perception in the nineteenth century, just as surely the experience of the unique perspective of the phantom ride films and the unique immersion they offered helped the first drivers of the twentieth century make sense of the newfound spectacle of automobility afforded by their cars. In that sense, one could argue that the ghost that haunts all of the phantom ride films is actually automobility itself.

The relationship of the frame and camera movement in the phantom ride films is very different than a simple tracking shot or dolly, for in the phantom ride films, camera movement is, as its name suggests, always experienced as a *ride*. Although the Traveller-Spectator may not know what kind of vehicle they are on (hence the "phantom" part), it is an experience of the film as travel, as a specific form of mobility melded to some sort of mechanical vehicle, as opposed to a tracking or a dolly shot in on a space, which is more aligned with vision than the bodily experience of travel. As such, the enframing of this camera movement is experienced as a constantly changing vantage point enabled by our mechanical means of mobility. As Patrick Keiller describes it:

> The combination of the camera's "subjective" view and its conventionally inaccessible position suggests disembodied consciousness, though no one seems to know if this was a factor in the naming of the *phantom ride*, or when the first was first used, and by whom. In any case, in other variations of the moving-camera film the view is similarly oneiric, so that films in which the camera is looking backwards, or is located in a wagon halfway along the train, as well as films photographed from trams and cars in which the forward-facing view was unexceptional, have also been described as phantom rides, though perhaps not at the time they were first exhibited.[15]

This is a far cry from the assumption made by most phenomenological film theory, where the Traveller-Spectator experiences camera movement as an embodied experience within a frameless perspective, somehow disregarding the rectangular boundaries set by our vision, as Vivian Sobchack argues in her influential work on phenomenology and film, *The Address of the Eye* (1992). For Sobchack, "[t]he frame is *invisible* to the seeing that is the film. It is a limit, but like that of our own vision it is inexhaustibly mobile and free to displace itself," it is a "'secret' boundary of the film's act of vision" that "gently peel[s] away."[16] While that may be true of a tracking shot embedded in a narrative film, in a phantom ride film, the abrupt cutting off of vision at the edges of the frame is an integral part of the ride. Here, with velocity and the frame working in tandem and resulting in the erasure of the landscape, it is, as Gunning describes in one of his many crucial essays on early cinema's attractions, as if "an unseen energy swallows space."[17] The frame is crucial in this sensation of "swallowing" or devouring space, as is the hyper-mobility of the engine that drives the Traveller-Spectator's POV; the "unseen energy" is both mobility and the consumption of the space that it enables. As Gunning describes it: "The experience to be reconstituted in these films is the thrill

15 Keiller, "Phantom Rides," p. 75.

16 Sobchack, *The Address of the Eye*, p. 131.

17 Gunning, "An Unseen Energy Swallows Space." The often-cited phrase "unseen energy swallows space" originates from a 1897 *New York Mail and Express* review of the film: "The way in which the unseen energy swallows up space and flings itself into the distances is as mysterious and impressive as an allegory. [...] One holds [their] breath instinctively as [they] [are] swept along in the rush of the phantom cars. [Their] attention is held almost with the vise of fate." See Kemp, *The Biograph Bulletins*, p. 27.

of motion and its transformation of space. Although a large part of their appeal was clearly as a cheap form of vicarious tourism for the masses, the experience was also uniquely cinematic. The sense of penetration of space by the unseen camera gave the spectator an almost uncanny feeling."[18] Petra Löffler echoes this sentiment in her more recent article, "Ghosts of the City: A Spectrology of Cinematic Spaces" (2015), pointing out that in the phantom ride films, "the camera functioned as a seeing machine exploring a permanently changing space where the instability of spatial relations and dimensions revealed the uncanniness of cinematic spacing."[19] This sense of the uncanny that both scholars recognize in phantom ride films is linked to both the "penetration of space" and the "instability of spatial relations and dimensions," and touches on both the erasure of space the Traveller-Spectator experiences and on an as-yet unprecedented entry into the space ahead that anticipates automobility. This sense of the uncanny was seized upon in many of the original reviews of *The Haverstraw Tunnel* in 1897, as in the popular New York magazine, *The Phonoscope*:

> The spectator was not an outsider watching from safety the rush of the cars. [They were] a passenger on a phantom train that whirled [them] through space at nearly a mile a minute. There was no smoke, no glimpse of shuddering frame or crushing wheels. There was nothing to indicate motion save that shining vista of tracks that was eaten up irresistibly, rapidly, and the disappearing panorama of banks and fences. The train was invisible, and yet the landscape [swept by] remorselessly, and far away the bright day became a spot of darkness. That was the mouth of the tunnel, and toward it the spectator was hurled as if a fate was behind [them]. The spot of blackness closed around and the spectator being flung through that cavern with the demoniac energy behind [them]. The shadows, the rush of the invisible force and the uncertainty of the issues made one instinctively hold [their] breath as when on the edge of a crisis that might become a catastrophe. [...] The audience that stood five deep back of the orchestra chairs half reeled as it caught itself. It had been snatched up and rapt away by a phantom train.[20]

The audience's response to *The Haverstraw Tunnel* provides a fascinating counterpoint (and in many ways a polar opposite) to the mythical arrival of

18 Gunning, "An Unseen Energy Swallows Space," p. 363.
19 Löffler, "Ghosts of the City," p. 8.
20 "Life on Canvas," p. 6.

cinema's first train on screen, in the Lumière's *L'Arrivée d'un train en gare de La Ciotat* [*Arrival of a Train at La Ciotat Station*], at the Salon Indien of the Grand Café, in France, two years earlier. According to the famous myth (which is arguably still the most famous myth in film history), as recounted by Gunning, "spectators reared back in their seats, or screamed, or got up and ran from the auditorium (or all three in succession)."[21] Gunning, like many historians, remains skeptical about this actually happening that night, though accounts do exist of similar reactions at subsequent screenings of train arrival films around the globe.[22] For Gunning, "this primal scene at the cinema underpins certain contemporary theorizations of spectatorship. The terrorized spectator of the Grand Café still stalks the imagination of film theorists who envision audiences submitting passively to an all-dominating apparatus, hypnotized and transfixed by its illusionist power."[23] Writing here, in 1989, about the apparatus theory which dominated film theory in the 1970s and 1980s, as evinced by Christian Metz, Gunning is careful to explain that he does not want to deny this "founding myth of the cinema's spectator," but that we should, rather than frame it strictly as fear as Metz does, also examine the attraction of such images.[24] For Gunning, the attraction resides in seeing a still image come to life; in seeing the movement not only of the train but of the film itself. As Gunning describes: "The on-rushing train did not simply produce the negative experience of fear but the particularly modern entertainment form of the thrill, embodied elsewhere in the recently appearing attractions of the amusement parks (such as the roller coaster), which combined sensations of acceleration and falling with a security guaranteed by modern industrial technology."[25] Key to this experience is the now-forgotten experience of the exhibition of these early films—described in detail by film historian Martin Loiperdinger:

> Before projecting a film with the Cinématographe Lumière, the framing had to be adjusted and the film locked into place after it had been inserted, during which one saw a still image, i.e., one projected frame, similar to a slide show. As soon as the framing was correct, the projectionist started

21 Gunning, "An Aesthetic of Astonishment," p. 736.

22 For varying accounts, see Bottomore, "The Panicking Audience?"; Gunning, "An Aesthetic of Astonishment"; Loiperdinger, "Lumière's *Arrival of the Train*"; and Tsivian, *Early Cinema in Russia and Its Cultural Reception*.

23 Gunning, "An Aesthetic of Astonishment," p. 737.

24 Gunning, "An Aesthetic of Astonishment," p. 737.

25 Gunning, "An Aesthetic of Astonishment," p. 743.

> cranking, setting the image on screen in motion. Many commentators described this transition from a still to a continuously moving image as a surprising, even thrilling, perceptual experience.[26]

Loiperdinger goes on to describes the scene as "cinema's founding myth," whether or not it actually happened:

> The cinema's first audiences are interpreted as being unable to distinguish between film image and reality. *Arrival of the Train* is not simply used as an icon of cinema's birth, rather this one-minute film by Louis Lumière stands as a striking example of the manipulative power allegedly inherent in cinema since its beginnings. It serves to illustrate cinema's inherent suggestive forces, elevated to a basic principle. While the fear and panic of the audience facing Lumière's locomotive is retold in the form of an anecdote, its status reaches much higher: reiterated over and over again, it figures as *the* founding myth of the medium, testifying to the power of film over its spectators.[27]

Visually, the two films are far more than an inverted mirror reflection of each other, although that is what they may appear to be at first glance. In the Lumière film, the train is rushing towards a stationary camera set up at a train station. The imminent endpoint that audiences feared was one of impact—the train barrelling into the audience and crashing to a stop. With *The Haverstraw Tunnel*, the train is still the vehicle of mobility, but it is now invisible, and it is the Traveller-Spectator that is moving through a still landscape with no endpoint in sight. This is a vision of pure, unstoppable mobility, especially in theatres like Poli's Wonderland Theatre in New Haven, Connecticut, where the film was shown in a loop as a "continuous performance" for audiences who crowded into the theatre to see this unparalleled cinematic marvel.[28] Film historian Charles Musser points out that this was a commonplace occurrence during the earliest period of cinema, especially before 1900. Musser resists using Gunning's phrase "cinema of attractions" to describe this, though, instead offering up the opposite notion of a "cinema of contemplation" in the case of a

26 Loiperdinger, "Lumière's *Arrival of the Train*," p. 97. One of the most conspicuous of these commentators, as Tom Gunning points out, was no less than George Méliès; see Gunning, "An Aesthetic of Astonishment," pp. 740–741.

27 Loiperdinger, "Lumière's *Arrival of the Train*," p. 92.

28 See the 1897 advertisement of the film published in New Haven, Connecticut's *Morning Journal-Courier*.

film running in a continuous loop for minutes at a time: "This did more than denarrativize individual films: such sustained presentations also encouraged spectators to contemplate and explore the image. As this evidence suggests, one way that early audiences were meant to look at films was not unrelated to the way they were meant to look at paintings."[29] Granted, the examples that Musser cites—*American Falls from Above, American Side* (Thomas Edison, 1896) and *Waterfall in the Catskills* (Thomas Edison, 1897)—suggest a more relaxing form of contemplation (the scenic spectacle of a waterfall) than in the case of a film like *The Haverstraw Tunnel*. But, in either case, running these films in a continuous loop would offer a unique space of contemplation for late nineteenth-century filmgoers, whether reflecting on the continual cascade of a waterfall or on the "unseen energy swallowing space" of a phantom ride. As Musser points out, in the case of *The Wave* (Birt Acres, 1896), where the film featured a violent sea crashing up on shore seen in a loop, "the visceral reaction must have abated. The spectator would gain a sense of mastery of this new medium, settle back into his or her seat and enter a more detached and contemplative state."[30]

Imagine if *The Haverstraw Tunnel* would have been embraced as cinema's founding myth rather than the threat of the train crashing into the Lumière theatre. With the phantom ride, the vantage point offered to the viewer is not simply one of passive traveller, watching the world slide by outside the glass, but rather the POV of the engineer, the pilot (and, anticipating automobility, eventually the person in the driver's seat), and of the experience of continuous travel into an infinitely extending landscape. As Gunning sums up:

> With their front-on viewpoint, the phantom rides provide a unique realization of the fantasy of penetrating a landscape, of chasing the horizon into the depth of an ever-unfolding image. The displacement from the lateral view provided by the train (and the moving panoramas) to a head-on plunge into the center of the image fundamentally transforms the distance (Wolfgang) Schivelbusch described as inherent in panoramic perception.[31]

Interestingly, for Gunning, the potential collision is still in play both as part of the risk and allure of being a Traveller-Spectator. As he goes on to

29 Musser, "A Cinema of Contemplation, a Cinema of Discernment," p. 162.
30 Musser, "A Cinema of Contemplation, a Cinema of Discernment," p. 167.
31 Gunning, "Landscape and the Fantasy of Moving Pictures," p. 57.

explain in his 2010 article, "Landscape and the Fantasy of Moving Pictures: Early Cinema's Phantom Rides":

> The head-on confrontation between the viewer's vantage point and the direction of movement into space evoke the possibility of shock and collision, in contrast to the lateral view of the traditional train rider that created the sense of separation essential to panoramic perception. Nonetheless, a fundamental fissure between viewer and spectacle remains, dependent less on visual viewpoint, yet physically more absolute. As much as the expectation, even the sensation, of collision may be evoked by such film, no collision is ever possible. [...] No physical shock is possible, no meeting between our bodies and the space on screen can occur, however much we may seem to penetrate into it.[32]

For Gunning, the always-anticipated collision or crash that can never happen is also an important part of the phantom status of these films. Like the rollercoaster, this is a carefully controlled experience of moving through space, where the safety of the Traveller-Spectator is an integral part of the ride.

However, this notion is challenged in a film that comes nine years later, at the end of the era of the phantom rides' popularity, which also clearly illustrates how quickly the genre disconnected from its strict identification with the experience of the train, shifting to the perspective of automobility first promised by *The Haverstraw Tunnel*. The 1906 film *A Trip down Market Street*, photographed by Harry J. Miles and co-produced with his brothers, Herbert, Earle, and Joe, provides striking proof of the transformation in transportation already occurring in San Francisco. The film begins at the location of the Miles Brothers' film studio on 1139, Market Street, between 8th and 9th Streets, and ends where the street ends at the Ferry Building. The precise location of the film on the timeline introduces the Traveller-Spectator to another haunting aspect of this particular phantom ride, as the film was shot just four days before the 1906 San Francisco earthquake on April 18, which then resulted in the subsequent San Francisco fire, causing the destruction of 28,000 buildings comprising 80% of the city; more than 3,000 deaths; and leaving over a quarter of a million inhabitants homeless—including, very likely, many of the men, women, and children we see on the phantom ride in this film.[33]

32 Gunning, "Placing the World in a Frame," p. 58.

33 The Miles Brothers studio was destroyed in the fires that consumed much of the city after the earthquake. The survival of the negative is solely due to the fact it was taken by train to the

Shot from the front of a cable car, the twelve-minute phantom ride takes us down the then-thriving Market Street, giving us a view of both the past and the future of the city and of this country as we move through a street crowded with automobiles, horse-drawn carts, carriages, streetcars (both electric and horse-drawn), and pedestrians of all ages darting in and out of traffic, very literally taking their lives into their own hands to cross the street. While licence plates had just been mandated by law in 1905, there are no laws yet in place governing traffic and no police officers to monitor the mayhem.

The film opens on a chaotic view down Market Street. The tracks laid in the wide street mark the middle of it and the frame itself, leading all the way to the horizon, where the pointed shape of the Ferry Building at the street's end is barely visible. The horizon is situated two-thirds of the way up on the screen, so the street, buildings, and sidewalk activity fairly fill the frame. The Traveller-Spectator is greeted by horse-drawn carts both approaching and trotting ahead; strikingly, just a few seconds into the film, one of the horse-drawn carts ahead alters its trajectory from the side of the street to near the centre, riding directly on the rails in front of the trolley. Immediately, the viewer's relationship to the tracks changes; unlike many of the early phantom ride films, this is not a POV strictly aligned with the locomotive, but with the vantage points of all the vehicles that now crowd onto a road where the submerged tracks are only one element. Just fifteen seconds into the film, an automobile slips by on the righthand side of the frame; two seconds later another automobile zips right in front of the trolley, immediately transforming the timeline of the film from the nineteenth century to the twentieth. More cars appear after that, almost in a swarm, shooting past the Traveller-Spectator's phantom ride, darting and crisscrossing just in front of the camera's POV. It is soon apparent that Market Street is actually dominated by the new automobiles, as they clearly outnumber every other form of vehicle on the crowded street. As the phantom ride continues down Market Street, various vehicles appear ahead, also seeming to run along the embedded tracks barely visible in the street: horse-drawn carts, automobiles, and even trucks. The resulting impression on the Traveller-Spectator is that they might be travelling on any one of these vehicles. In this case, the "phantom" part of the phantom

Miles Brothers office in New York the day before the fire. For many years, the film was a powerful curio, uncredited and unsynced to the timeline, until film historians in 2010 identified the date of the film based on a licence plate seen in it, registered in February 1906 by the Reliance Auto Company. Only afterwards were historians able to link the film to the Miles Brothers legacy.

ride shifts; while the Traveller-Spectator is still aligned with the means of mechanized travel, its specific shape seems malleable. Although the trolley follows the tracks leading to the horizon, so is every other vehicle on the road. The Traveller-Spectator is no longer necessarily fused to a locomotive, or even to the street car that the film was shot from, but is rather part of the modern urban ensemble of traffic, which is dominated by the automobile. The automobiles are of course very different. Most notable perhaps is the fact that they are all convertibles, with their soft-tops folded down so the passengers are travelling in the open air and sun. Further contributing to their openness is the fact that none of these cars have windshields, as this is before they were invented.

Clocking in at over twelve minutes, *A Trip down Market Street* is by far the longest of all the phantom ride films. As such, the experience of watching the film is dramatically different from all that preceded it. What begins as a powerful example of a "cinema of attractions" soon transforms into something closer to Charles Musser's "cinema of contemplation," because of its then-unprecedented duration. While the cars whiz by, the horse-drawn carts trot past, and the city's on-foot occupants dart in and out of the seemingly constant stream of multi-vehicular traffic, the street maintains its hold over the bottom two-thirds of the screen, ever so slowly rolling past underfoot, remaining a steady and constant gray slate. The experience of this film is a vision of the future—the normalcy of traffic—dominated by automobiles. The seemingly endless flow of buildings and of people moving up and down sidewalks and back and forth across the street is an image of modernity just at its origins. It is also an image that remains largely unchanged now, almost 120 years later. As a Traveller-Spectator, processing this fact is an inevitable aspect of our process of contemplation and reflection. What was certainly, at the time the film was made, a futuristic vision of both San Francisco and automobility, looking backwards, provides a visceral proof of just how long automobility has been the norm in urban environments. At the same time, it also reminds the Traveller-Spectator just how long the perspective of these urban environments has been aligned with automobility. Although the windshield (or "windscreen," as the British term it) has yet to be invented at this point, the perspective anticipates that vantage point and, indeed, modern viewers looking back at the film will no doubt see the film in that fashion. Already in 1906, we see the clear hierarchy of vehicles on the road, with the automobile, barely a decade into its existence, already ruling the road—with everything and everyone else either left behind it or running to get out of its way. The individualized traveller behind the wheel (or the handful of occupants found in the vehicle) are elevated above the pedestrian, moving at a faster rate than anyone else, in an individualized method of movement that is no longer communal, that is no longer bound to timetables.

For the Traveller-Spectator, the experience of duration is partly responsible for the sensation of automobility, as it recreates in real time the perspective of a car moving down Market Street with no edits and no additional camera movements other than the steady forward momentum of our phantom ride. As Traveller-Spectators, our eyes are free to roam the frame, searching the streets for faces we know, for shop windows that may interest us, or for an inspection of the striking turn-of-the-century architecture of the buildings lining the street that stretch into the sky, beyond the confines of the frame. Eventually, our contemplation of the street scene becomes focused on the building at the far end of Market Street. Although we can barely make out its tower at the beginning of the film, appearing as just a tiny spire in the fog in the distance, it becomes quite striking, several minutes into the film, as it is an isolated tower jutting up into the growing white space of sky that stretches between the buildings flanking either side of the frame. Several more minutes in, it eventually becomes clear that this building is the endpoint of the road, and the destination that we have been travelling towards all along. The Ferry Building, built in 1896 by the State Harbor Commissioners, marked the boundary between city and sea, serving not only as the destination for our phantom ride, but indeed as the destination of commuters who rode the ferry fleets of the Southern Pacific and Key System across the San Francisco Bay and the transcontinental rail lines of the Southern Pacific, Santa Fe, and Western Pacific. When we arrive at our destination, several people on the street in front of the Ferry Building step out of the way, as if to let the phantom ride pass. Instead, it comes to a stop for almost ten seconds, our attention trained on the plaque of engraved stone on the building that reads: "ERECTED A.D. 1896 BY THE BOARD OF STATE HARBOR COMMISSIONERS." Suddenly, our phantom ride does something it's never done before—it slowly turns around a full 180 degree—transforming what was originally the destination of our journey into an all new point of origin. With this dramatic camera move, we are suddenly greeted by a group of young boys, grinning, jumping into the air, and cheering right in front of us; then the film cuts to black, and it is over. The close of the film is both dramatic and stirring, more so for anyone who knows the sad fate of the city just days later, when the Great Earthquake of 1906 destroyed 80% of the city. The jubilance of the boys seems a celebration not only of the virtuoso camera move, but of the presence of the camera itself, a not uncommon greeting in the cinema of attractions. The inhabitants of the film often stare back at us through time, as captivated by the moving presence of the camera (which likely was itself quite a spectacle, strapped onto the front of a cable car) as we are by them. To the Traveller-Spectator, it also seems like a celebration of the fact that the phantom ride will resume once again, that this immersive perspective of automobility accessed through the windscreen is only just beginning.

Bibliography

Bottomore, Stephen. "The Panicking Audience? Early Cinema and the Train Effect." *Historical Journal of Film, Radio and Television* 19, no. 2 (1999): 177–216.

"Brief Notes Concerning Patents." *Scientific American*, June 23, 1906.

Gunning, Tom. "An Aesthetic of Astonishment: Early Film and the (In)Credulous Spectator." *Film Theory and Criticism: Introductory Readings*. 7th ed., edited by Leo Braudy and Marshall Cohen, 736–750. New York, NY: Oxford University Press, 2009.

Gunning, Tom. "The Attraction of Motion: Modern Representation and the Image of Movement." In *Film 1900: Technology, Perception, Culture*, edited by Annemone Ligensa and Klaus Kreimeier, 165–174. Bloomington, IN: Indiana University Press & John Libbey Publishing, 2009.

Gunning, Tom. "The Cinema of Attractions: Early Cinema, Its Spectator and the Avant-Garde." In *Early Cinema: Space, Frame, Narrative*, edited by Thomas Elsaesser and Adam Barker, 56–62. London: British Film Institute Publishing, 1990.

Gunning, Tom. "Landscape and the Fantasy of Moving Pictures: Early Cinema's Phantom Rides." In *Cinema and Landscape: Film, Nation, and Cultural Geography*, edited by Graeme Harper and Jonathan Rayner, 31–70. Chicago, IL: Intellect, 2010.

Gunning, Tom. "An Unseen Energy Swallows Space: The Space in Early Film and Its Relation to American Avant-Garde Film." In *Film Before Griffith*, edited by John L. Fell, 355–366. Berkeley, CA: University of California Press, 1983.

Harrison, Rebecca. *From Steam to Screen: Cinema, the Railways, and Modernity*. London: IB Tauris, 2017.

Hayes, Christian. "Phantom Carriages: Reconstructing Hale's Tours and the Virtual Travel Experience." *Early Popular Visual Culture* 7, no. 2 (2009): 185–198.

"An Imitation Bus Trip." *Automotive Industries*, March 29, 1906: 566.

Keiller, Patrick. "Phantom Rides: The Railway and Early Film." In *The Railway and Modernity: Time, Space, and the Machine Ensemble*, edited by Matthew Beaumont and Michael Freeman, 69–84. Oxford: Peter Lang, 2007.

Kirby, Lynne. *Parallel Tracks: The Railroad and Silent Cinema*. Durham, NC: Duke University Press, 1997.

"Life on Canvas: A Phantom Ride on an Express Train and the Other Remarkable Views Made by the Biograph." *The Phonoscope*, August–September 1897: 6.

Löffler, Petra. "Ghosts of the City: A Spectrology of Cinematic Spaces." *Communication+1* 4, no. 1 (September 2015). http://dx.doi.org/10.7275/R5VM496F.
Loiperdinger, Martin. "Lumière's *Arrival of the Train*: Cinema's Founding Myth." *The Moving Image: The Journal of the Association of Moving Image Archivists* 4, no. 1 (Spring 2004): 89–118.
Musser, Charles. "A Cinema of Contemplation, a Cinema of Discernment." In *The Cinema of Attractions Reloaded*, edited by Wanda Strauven, 159–180. Amsterdam: Amsterdam University Press, 2006.
Niver, Kemp R. *Biograph Bulletins 1896–1908*. Los Angeles: Locare Research Group, 1971.
Parissien, Steven. *The Life of the Automobile: The Complete History of the Motor Car.* New York, NY: St. Martin's Press, 2013.
Rabinovitz, Lauren. "'Bells and Whistles.' The Sound of Meaning in Train Travel Film Rides." In *The Sounds of Early Cinema*, edited by Lauren Rabinowitz and Abraham Geil, 99–125. Bloomington, IN: Indiana University Press, 2001.
Rabinovitz, Lauren. "From Hale's Tours to Star Tours: Virtual Voyages, Travel Ride Films, and the Delirium of the Hyper-Real." In *Virtual Voyages*, edited by Jeffrey Ruoff, 42–60. Durham, NC: Duke University Press, 2006.
Seiler, Cotten. *Republic of Drivers: A Cultural History of Automobility in America*. Chicago, IL: University of Chicago Press, 2008.
Slugan, Mario. *Fiction and Imagination in Early Cinema: A Philosophical Approach to Film History*. London: Bloomsbury Academic, 2019.
Sobchack, Vivian. *The Address of the Eye: A Phenomenology of Film Experience.* Princeton, NJ: Princeton University Press, 1992.
"The Haverstraw Tunnel," *New Haven, Connecticut's Morning Journal-Courier*, December 20, 1897.
Tsivian, Yuri. *Early Cinema in Russia and Its Cultural Reception*, translated by Alan Bodger. Chicago, IL: University of Chicago Press, 1994.
Vinsel, Lee. *Moving Violations: Automobiles, Experts, and Regulations in the United States*. Baltimore, MD: Johns Hopkins University Press, 2019.

About the Author

Michael J. T. Stock is a Professor of Cinema and Media Studies at SCI-Arc, teaching courses on American Film History, World Cinema Film Aesthetics, and Film Genres—including many popular seminars on science fiction, cyberpunk, horror, and animation. His published work reflects these interests, with over a dozen published articles in anthologies and peer-reviewed journals.

10. From Participation to Immersion: Promotional Tactics around Tricked and Dynamic Seats in the Film Theatre

Aude Weber-Houde

Abstract: This chapter examines how advertising posters for cinematographic shows with tricked or dynamic seats presented in theatres explicitly reflect, by their rhetoric, an ideal immersive spectatorial experience. Using advertising materials from 1959 for the film *The Tingler* and the *Percepto* gimmick, and from 2010s by the *D-Box* company, the author studies how the tricked or dynamic seat is sold to the public as a promise to better immerse themselves in the world of a film through the sense of touch. The article also considers how the spectatorial postures conveyed by the advertisements depict immersion quite similarly from one era to another.

Keywords: gimmick, advertisement, spectatorship, exhibition

The seat, as it always is in direct contact with the spectator when they watch a film, is central to the reception apparatus of cinematographic works. Indeed, it is rather rare that we experience films standing up or lying down. However, this piece of furniture is generally considered secondary to cinema's apparatus, since it does not appear to participate in the expressive codes of cinema, unlike the format or ratio of the image or the spatialization of sound, for example. Thus, except when it comes to choosing an optimal seat in the theatre in order to see the screen, to hear the sound effects all around, and not to be invaded by the proximity of neighbours, and except in cases where it might be broken or uncomfortable, the seat—like the film

This article stems from a revised and condensed version of several chapters of the author's doctoral dissertation. See Weber-Houde, "Le siège des sensations." She would like to thank her research adviser, Olivier Asselin, for his generous contribution to the ideas presented here. She is also grateful to Philippe Bédard for his help with the translation of this article.

Bédard, Philippe, Alanna Thain, and Carl Therrien, eds. *States of Immersion Across Media: Bodies, Techniques, Practices*. Taylor & Francis Group, 2025
DOI 10.5117/9789048558766_CH10

theatre itself nowadays—is usually forgotten during the film's screening. Moreover, this particular piece of the cinematographic apparatus has so far received very little attention from scholars. Among the few publications devoted to this subject is Jocelyn Szczepaniak-Gillece's article, "Revisiting the Apparatus: The Theatre Chair and Cinematic Spectatorship" (2016), in which the author studies texts written by architects who designed film theatres and analyzes the way they considered this piece of furniture.[1] Among others, she quotes Ansel M. Moore, an architect who described his vision of the ideal seat in a 1939 *Boxoffice Magazine* article:

> [Ideal modern seating should be] so caressingly comfortable that *the physical is completely forgotten and the illusion created on the screen reigns supreme* [...][.] Every seat is filled by patrons who are *entirely forgetful of surroundings*, which is the *ideal condition of motion picture entertainment reception* [...][.] Were they not comfortably relaxed, even the most elaborate presentation would not hold their interest.[2]

For Moore, this perfect seat was therefore intended to encourage viewers to forget the immediate material and physical reality in which they found themselves.

This ideal has been maintained over time, as evidenced by the design of contemporary theatres. It was from the 1930s, especially in the United States, that the film theatre was gradually configured towards an experience of pure filmic contemplation, as Jocelyn Szczepaniak-Gillece explains:

> More directed focus on the screen and greater comfort became rallying cries for exhibitors, architects, and auditorium interior designers, during and after the Depression, looking to put bodies in the seats and encourage stillness, silence, and theatrical integration after investing in expensive transitions to sound.[3]

The seat, as a privileged piece of furniture in the film theatre, was the subject of sustained attention from designers and architects during this period. For them, the ideal seat must optimize viewing conditions, while minimizing risks of fatigue, noise, and obstacles during the cinematographic screening. In a way, the seat must offer increased bodily comfort in order to make the

1 Szczepaniak-Gillece, "Revisiting the Apparatus."
2 Moore quoted in Szczepaniak-Gillece, "Revisiting the Apparatus," p. 267, my emphasis.
3 Szczepaniak-Gillece, "Revisiting the Apparatus," p. 266.

body of the spectator even more invisible and inaudible to others, but also imperceptible to themself. Thus, in the discourses of architects studied by Szczepaniak-Gillece, an ideal aesthetic program is manifested, suggesting the complete occultation of the physical body of the spectator in favour of a "supreme" screen regime, as suggested by the vocabulary used in Moore's article cited above. In this context, the emphasis is directed on the distant senses (sight and hearing), and the senses of proximity, such as tactility, are evacuated, since they convey the material and physical dimensions and the immediate environment of the theatre, which the audience seeks to forget.

But there are a few cases in the history of cinema where this sensory hierarchy has been questioned. Among these, some rigged or moving seats (the so-called "dynamic seats") have indeed emerged. Their effects differ from one model to another, from simple vibrations to the extreme gyrations more typically found in amusement park rides. However, the promises offered by these different, non-traditional versions of the seat all have the same goal: to intensify the spectator's experience in the theatre by soliciting them in a direct and physical way, through touch. By harnessing the pleasure of visceral experiences conveyed by all the senses, these devices, often referred to as "gimmicks," allow indoor film theatres to imitate amusement park rides and offer an experience in which the viewer's body is readmitted.[4]

In this chapter, I will observe, through the analysis of the type of images and of the vocabulary chosen, how the advertising posters selling some of the cinematographic screenings with rigged or dynamic seats in film theatres explicitly reflect, by their rhetoric, an ideal immersive experience. The advertisements designed around these seats emphasize the creation of a unique connection between viewers and the film's diegetic world, highlighting the bodily and sensory experience thus created. After a short presentation of the intermedial and promotional context in which they emerged, I will study the posters advertising the *Percepto* gimmick, a device with rigged seats designed to accompany the American film *The Tingler* (William Castle, 1959). By focusing on the vocabulary used at the time, I will highlight the "participatory" posture proposed to the audience attending those screenings. I will then be able to compare these advertisements with the posters used to promote the contemporary *D-Box* device in the 2010s. Introduced in 1998 in the province of Quebec, Canada, this dynamic seat technology is designed as an add-on to various large-scale productions, allowing for the integration of proprioceptive sensations into the audiovisual experience.

4 On the concept of "gimmick", see Clepper, "The Rigged House"; Ngai, *Theory of the Gimmick*; and, in French, my PhD dissertation, Weber-Houde, "Le siège des sensations."

The American 1950s: Between "Active" and "Passive" Leisure

It is a well-known fact, in the late 1940s and during the 1950s, film theatres in the USA saw their attendance drop drastically. The massive arrival of television in the American domestic space in the years following the Second World War is generally blamed for this marked decline in film theatre attendance. According to this hypothesis, as televisions became more widespread in American homes, more and more consumers were choosing to spend their evening at home rather than pay for a trip to the theatre. In keeping with this economic logic, then, the small screen became a much cheaper "substitute" for the audience, at least once the consumer had acquired the device.[5]

Douglas Gomery and John Belton have however shown that things were more complex. In fact, the massive arrival of television in homes added to a series of social transformations that had already drastically changed theatre attendance habits. Gomery attributes this drop to profound social changes, such as the now privileged suburban lifestyle and the baby boom, which had a significant impact on the population's tendency to frequent film theatres.[6] For his part, John Belton notes the population's preference for more "active" recreational activities during this decade.[7] According to him, at this particular time, the public was invested ardently in outdoor sports and practised many physical activities.[8] But, another type of entertainment, often forgotten by studies that look at the cinema of the 1950s, occupied the population: at this time, there were records of admissions to amusement parks, fairs, and travelling carnivals.[9] The pleasures associated with frequenting amusement parks were part of a set which explicitly aimed at the physical stimulation of users. For example, the rides offered experiences allowing a mobility of the body and a pleasure based in particular on the *ilinx*, which refers to the search for vertigo and disorientation momentarily destroying the stability of perception, a type of game described by Roger Caillois in his book *Les jeux et les hommes*.[10] Above all, these rides generally place the user at the heart of the device in an "egocentric" position: indeed, they are not facing the ride when they

5 See Gomery, *Shared Pleasures*, p. 83.

6 Gomery, *Shared Pleasures*, pp. 83–88 and p. 230.

7 John Belton sets out the details of the occupations of these new families settled in the suburbs. Belton, *Widescreen Cinema*, pp. 71–74.

8 Belton, *Widescreen Cinema*, p. 77.

9 Belton, *Widescreen Cinema*, p. 77.

10 Caillois, *Les jeux et les hommes*.

experience it, they are within it—as the rides are often designed as cockpits or vehicles.[11]

An analysis of the recreational market published by *Fortune* magazine in 1955 and quoted by John Belton categorizes leisure activities as "passive" or "active" and notes the population's preference for active entertainment, as opposed to more contemplative leisure: "The sharpest fact about the postwar leisure market is the growing preference for *active fun rather than mere onlooking*."[12] Belton also quotes Darryl F. Zanuck, a producer at 20th Century Fox, who mentions the different cycles the film industry went through in an essay titled "Entertainment vs. Recreation" published in 1953 in *The Hollywood Reporter*: "Now, the public is more participation-minded than ever."[13] As the title of his text indicates, Zanuck draws a clear distinction between what he considers to be "entertainment activities" and another type of leisure of a more "recreational" nature: "Entertainment [...] is something others provide for you, while recreation is something you provide in some measure for yourself—something in which you participate."[14]

From Zanuck's comments and from the study of the *Fortune* magazine article mentioned by Belton, we can deduce that the public of the 1950s rather preferred, for its leisure activities, to be involved in an active posture more than a passive one, and that it seemed to have favoured activities of a more physical than contemplative nature, hence the particular choice of the term "participation" used by Zanuck. According to the *Merriam-Webster Dictionary*, "participating" refers to the act of "tak[ing] part" or "to have a part or share in something." Note here that the word implies the idea of a community, of an action exercised in a collective framework, which is a crucial precision in the intermedial context of the time, where film theatres were threatened. But, the term also suggests that the individual who participates takes part and is more directly involved in said activity. Moreover, as Gilles Delavaud and William Paul point out, this term was common in the industry's discourse at the time and was also used by critics.[15] "If there is participation," explains Delavaud, "it must first be understood

11 On this, see the model proposed by Paul Milgram, Haruo Takemura, Akira Utsumi, and Fumio Kishino, which distinguishes between "exocentric" devices and "egocentric" devices. Milgram *et al.*, "Augmented Reality," pp. 282–293. Also see Asselin, "Vers un cinéma neuronal," p. 91; and Asselin and Auger Gosselin, "This Side of Paradise," p. 137.

12 Belton, *Widescreen Cinema*, p. 77, my emphasis.

13 Zanuck, "Entertainment vs. Recreation," quoted in Belton, *Widescreen Cinema*, p. 77.

14 Zanuck, "Entertainment vs. Recreation," quoted in Belton, *Widescreen Cinema*, p. 77.

15 Delavaud, "Le cinéma après la télévision," pp. 58–59; and Paul, "The Aesthetics of Emergence," p. 336.

in the sense that the spectator has the feeling of being part of the image, of being part of the show."[16]

Film theatre managers therefore had to face desertion caused by these multiple factors. They had to tackle this problem by reflecting on the tastes of the audiences they addressed and on the distinction of their offering. Indeed, spectators previously accustomed to patronizing theatres were missing in action, busy as they were with other types of hobbies. In other words, cinema had to compete not only with television, but also with various leisure activities massively adopted during the 1950s. The period of the 1950s was therefore characterized by a series of rapid transformations, which led to theatres whose architecture and layout were completely reconfigured. Varied and heterogeneous, the novelties offered on the market were multiplying: colour images, wide screens covering peripheral vision, stereoscopic images recreating the illusion of depth and relief, and stereophonic and surround sound sought to immerse the viewer in a distinct world, to give them the impression of being at the centre of the fictional space represented, as well as to evoke a physical experience by stimulating their senses in a greater way. In this context, where industrials sought both to renew the theatrical experience and to retain or renew the cinema's audience with teenagers and young adults, advertising strategies also needed to be particularly innovative. This is why, at the time, the rhetoric of advertisements for various technological novelties affecting film theatres suggests that they were participatory events, reflecting Zanuck's vision of them as such.[17]

The Advertising Rhetorics of the 1950s

Advertisements created to promote several new cinematographic devices that appeared during the 1950s—such as the *Cinerama* and *CinemaScope*, but also resurgent technologies, such as stereoscopy, as well as other less widely known technological trials, such as the *Smell-O-Vision*, the *Hypnovista*, and the *Percepto*—were typical of a desire to make the film theatre a recreational place, where the public could enjoy the aforementioned participatory culture.[18] For example, in the case of the *Cinerama*, its now well-known

16 Delavaud, "Le cinéma après la télévision," p. 59, my translation.

17 For more on this, see Belton, *Widescreen Cinema*, p. 77.

18 On these technologies, see Barnier and Kitsopanidou, *Le cinéma 3-D*; Zone, *Stereoscopic Cinema and the Origins of 3-D Film, 1838-1952*; Zone, *3-D Revolution*; and Clepper, "The Rigged House."

advertising posters combine the experience of film viewing with that of an amusement park ride: a group of spectators is seen from behind, sitting in what is supposed to be a film theatre, and appear at the bottom of the poster of the film *This is Cinerama* (Mike Todd *et al.*, 1952), facing a giant image projected on a cinema screen, which occupies most of the space of the poster.[19] The screen presents the point of view of a user sitting in a car, speeding down a roller coaster in an amusement park, or, on other posters, the perspective of a traveller perched on the bow of a boat or sitting on an airplane wing hovering in the air. This juxtaposition of the characters seen from behind with the images on the screen gives the impression that they all take place in the car of an amusement park ride, the subjective point of view of the image shown on screen reinforcing this idea.

On one of the posters of the film *This is Cinerama*, the text featured on the poster suggests the immersive qualities of the device, using capital letters to accentuate the address to the viewer: "Puts YOU in the picture." This second-person formula, coupled with the presence of a visible audience on the poster, indicates that the spectator is considered a central element in this experience. In a way, this advertising seems to introduce the idea that the spectator is now readmitted as an important part of the cinematographic experience, in which they "participate." But even when it comes to the experience of an amusement park in a film theatre, the poster implies that this audience must "participate," albeit in a polite and calm manner befitting the cinema: the spectators on the poster all have their eyes riveted on the screen and they all appear motionless. This is a central condition for the experience: only sight is solicited on the poster, and it consequently suggests an immersive and physical experience that unfolds only on a visual level, for a spectator who remains quietly seated.[20] This paradox is all the more explicit in the advertising text included in a booklet of the program of *This Is Cinerama* (1952), which emphasizes a collective experience with other spectators, in particular through the use of the verb "share":

> You *gasp and thrill with the excitement* of a *vividly realistic ride on the roller coaster...* You *feel the giddy sensations* of a plane flight as you bank and turn over Niagara and skim through the rocky grandeur of the Grand Canyon.

19 See one of the famous posters online at IMDB: https://www.imdb.com/title/tt0045231/mediaviewer/rm2338771969/?ref_=tt_md_4.

20 A critic of *The New York Times* reports, however, that the audience screamed during the roller coaster scenes, as noted in Paul, "The Aesthetics of Emergence," p. 337. Screams are also heard on the film's soundtrack.

> Everything that happens on the curved Cinerama screen is happening to you. And *without moving from your seat*, you *share*, personally, in the most remarkable new kind of emotional experience ever brought to the theatre.[21]

Let me highlight the schism that emerges between the spectatorial posture prescribed by advertising (that of a motionless body, which does not move from its seat), and what is described in the text: "gasping" and "thrilling" with all the excitement of an amusement park ride refers to reactions that involve the whole body and that generate physical stimuli, especially because of movement, or dizziness, and that are related to the vestibular properties of the body. Then, the viewers also have to react to these sensations by expressing themselves loudly (*gasp*). In other words, the program presents in its text an ideal based on a form of compromise: it promotes an attractional experience that suggests a strong physical dimension, which remains, however, only of a visual nature.[22] This makes it compatible with the traditional theatre apparatus, which also remains audiovisual. In that regard, even if the *Cinerama* made use of a purely audiovisual technology, it revealed the transformation of the dominant paradigm of the hierarchy of the senses and contributed to it by its evocation of the ideal experience, in words that describe a very physical experience, a multisensorial one. As we have seen previously, the "active" investment of the spectator in the audiovisual experience of cinema going was particularly sought after at the time. As Belton explains, "[t]he 'participation effect' quite literally meant the creation of a strong sense of *physical* participation."[23]

The evocation of this well-known advertisement allows us to better understand a certain program on the part of the producers of the 1950s era, based on their perception of the audience's expectations: the physical sensations related to movement, typical of attractions and rides, were sought after and were at the heart of this promotional approach, suggesting that the spectator's body was included in the film's spectatorship experience. This formula hit its mark: this direct address to the viewer through the use of the second person, the insistence on realism, and the emphasis on physical sensations conveyed by visual technology gave rise to an advertising rhetoric suggesting an immersive and participatory cinematographic experience, which was quickly imitated and repeated for other films produced with the

21 *This is Cinerama*'s booklet, quoted in Belton, *Widescreen Cinema*, p. 89, my emphasis.
22 The advertisement does not mention the use of surround sound, although it was used.
23 Belton, "Spectator and Screen," p. 243, my emphasis.

Cinerama process, and was also used to promote several other productions and technologies of the time, including the rigged *Percepto* seat.[24]

The Rigged *Percepto* Seat and Its Advertising Program

A few years after the first presentations of the *Cinerama*, in 1959, showman, filmmaker, producer, and master of gimmicks William Castle introduced the "audiovisuotactile" experience *Percepto* to the public during the showing of his film *The Tingler*. *Percepto*, whose name explicitly refers to the idea of perception, was a gimmick based on the rigging of certain seats, which were randomly hidden in the film theatre. They would vibrate in sync with the action presented on the screen during two specific scenes. In addition to this trick, an actress would be hired to play the role of a spectator and "faint," before being escorted out of the theatre on a stretcher. In this experience, the audience would also strongly be encouraged to scream.

The Tingler, a hybrid film working on the borders of science fiction, fantasy, and horror, tells the incredible story of a pathologist, Dr. Chapin (Vincent Price), who performs autopsies on the bodies of prisoners sentenced to death and executed by electric chair. He develops a crazy theory that fear lodges in the spine of human beings in the form of a parasite that feeds on the fears of its host. This grotesque creature, whose appearance is that of a hybrid between a giant earwig and a lobster without claws, is called a "tingler," because of the sensation it causes to its host when they are seized with fright. If properly fed, the tingler can grow impressively, until it crushes the spine of its victim and kills it. The only way to stop the growth of this creature and weaken it is for its host to scream.

The title of the film bears the name attributed to its monstrous creature: it therefore acquires the characteristics associated with it and announces its sensory content. The word "tingler" derives from the verb "to tingle," which evokes a bodily sensation associated with tactility, in addition to being related to anticipation and excitement. Thus, the tactile stimuli felt by the viewers through the rigged seats were linked with the thrill induced by the tingler, as if this creature could cross the screen to reach the theatre's space and the spectator's body. The visceral effects therefore came from the diegetic space of the film and were deployed in the real space of the theatre, through its furniture.

24 For the subsequent advertisements of the *Cinerama* technology, see Belton, "Spectator and Screen," p. 241.

If it induced a tactile stimulus that was then unprecedented in the history of film theatres, the *Percepto* device was not reduced to the simple modification of the seat. It also offered a real staging that took place live in the theatre. This staging was orchestrated like a choreographed performance. At an opportune moment, dictated by the action on the screen, an actress sitting among the audience would suddenly faint. The screening would then be interrupted, in a fake way, while the lights of the theatre would all at once be turned on and while doormen would evacuate the fainted girl, a voice imitating that of the projectionist warning the audience that the tingler was now in the theatre (and no longer in the film). When tension reached its peak, the engines under the seats would be activated and the audience invited to scream.[25] All this staging served to reinforce a lure according to which the fiction world of the film was actualized in the auditorium space: Castle would try to make the audience believe that the monstrous creature that haunts the film had crossed the screen and was now in the theatre, thus undermining the border that supposedly separates the diegetic space of the film from the real space of the auditorium.

In this context, the screams of the audience served to reinforce the illusion. Similarly, producers would allow the spectators to be under the impression they were participating actively in the collective experience of the film, since the device inscribed their presence within a community of spectators. This type of immersion can also be related to what Gordon Calleja calls "shared involvement" in the contemporary context of video games.[26] As such, the visual advertising materials produced to promote *Percepto* were particularly explicit with regard to the participatory and collective spectatorial posture the viewer was encouraged to have toward the film. For the purposes of this analysis, I will now consider a specific lobby card and a poster.[27]

One of the lobby cards (lobby card no. 1 of the set) intended to promote the film *The Tingler* and its gimmick *Percepto* consists of two similar main

25 According to accounts from 1959, the audience screamed with enthusiasm. See Adams, "'Tingler' Sell out at Pilgrim as Kids Come A-Screaming," p. 18; Gardner, "'Tingler' Tailor Made for Vicarious Thrill," p. F6.

26 Calleja, *In-Game*, pp. 93–112.

27 Lobby cards are medium-sized advertising posters (11x14 inches, or 28x36 centimetres), typically distributed in batches of eight different copies. They were very popular during the 1950s. As their name suggests, lobby cards were intended to be displayed in the halls of theatres to announce new films, present or to come. All lobby cards and posters prepared for *The Tingler* are under copyright and cannot be reproduced here, but are available online. See "*The Tingler*: Lobby Card Set." on *TCM Archive*: https://www.tcm.com/tcmdb/title/19997/the-tingler#photos-videos.

illustrations whose mirror effect aims to reflect the analogy between what the characters of the film experience and what is expected from the audience.[28] In the lower right corner, we see the drawing of two important actors in the film, Vincent Price and Pamela Lincoln, respectively playing Dr. Chapin and Lucy Stevens. The first has a worried face; the second, a frightened expression: her mouth is open, giving the impression that she is screaming. The other image is located in the upper right corner and shows a photograph of part of an audience in a film theatre. A yellow circle surrounds the faces of seven spectators on the right and delimits part of the image, as if those characters were illuminated by a projector or by a follow spot. This assumes that the spotlight is not directed toward the stage, toward the actors, but rather toward a part of the audience, which is a very important aspect given the participatory and active dimensions associated with the spectatorial posture.

This representation of the audience on the advertising images of the film corresponds to a rhetorical strategy common at the time. As we have seen in advertisements used to present processes such as the *Cinerama*, it was customary at the time to display the audience as well as its posture or reaction to the film as a model. In the case of this lobby card made for *The Tingler*, the audience is no longer seen from behind, as an indistinguishable and uniform mass illuminated by the screen, as it was the case for *This is Cinerama*. It is rather shown from a frontal view, where the card holders can distinguish the expressive and rather young faces of the members of the audience. In addition, the bodies of the spectators are no longer straight and aligned: they are bent and the overall composition is unbalanced, forming diagonals and subgroups. The position of the hands of the spectators is also notable: they are placed in the foreground, in front of the body or face, as if they were trying to protect themselves, or as if to emphasize their emotion and amplify their cry. It is thus no longer the screen, the image, or the technology that are presented as major, new, and important elements in the advertisement; it is the audience, frightened and screaming, that is at the centre of the composition. Here, it is literally the audience, the body of its members, and their reactions that are spectacularized.[29]

28 Lobby Card no. 1 for the promotion of the film *The Tingler* (William Castle, 1959). "Lobby Card Set." on *TCM Archive*: https://www.tcm.com/tcmdb/title/19997/the-tingler#photos-videos.
29 As André Loiselle points out, this kind of imagery of an audience explicitly horrified by what they are viewing is part of a certain rhetoric associated with horror films. In this genre, it is customary to show the reaction of the spectators as a guarantee of quality, since the audience *seeks* to be frightened. See Loiselle, *Theatricality in the Horror Film*, p. 106.

The text accompanying the two illustrations reinforces this effect. At the very top of the card, occupying its entire width, one can read, in red capital letters: "When the screen screams you'll scream too… if you value your life!" This immediately announces the participatory, performative dimension of the proposed film screening. In other words, it is clearly explained to the audience members what they can expect in terms of experience: the film they are about to see will be powerful enough to make them scream. The phrase "if you value your life" also evokes a death threat to the viewer, which is a constant in Castle's gimmicks. This excerpt also implies an allusion to the scientific hypothesis behind the story told in the film—which is that screaming can free a person from the threat of the monster. In addition to the usual credits, placed in the lower right corner, two short texts are added to introduce the film, each inside a frame. On the left, above Price's head, the card reads: "Percepto! newest and most startling gimmick on the screen!" The text refers to the gimmick as such, but it is strangely associated with the screen, and not with the seat. This displacement already announces the link between the screen and the seat. Furthermore, the typography used for the term *Percepto* is trembling: it suggests a crackling, a vibration, even a shock, which are all phenomenon associated with the screen, but which also announce the shaking of the seat. Thus, already in advertising, the creators evoke the passage of the monster from the film to the theatre.

In the other segment framed, this time with a dotted line, placed obliquely, a text, written in small characters, announces the "guarantees" of the film, a recurrent strategy within the promotional tactics for the film: "GUARANTEED. The Tingler will break loose in the theatre while *you* are in the audience. As you enter the theatre you will receive instructions… how to guard yourself against attack by THE TINGLER!" This box announces several elements of the aesthetic program inherent to the show. Note here the direct address to the spectator, placed at the centre of the action. The text first refers to the irruption of the tingler in the auditorium while the audience attends the performance, thus crossing the boundary between fiction and reality. This is a recurring suggestion on several advertising posters of the time, which assumed the immersion of the audience in the film, or conversely, the "emersion" of the film in the auditorium.[30] Secondly, if the usually "passive" posture of the spectators seems at first expelled in favour of a certain participation on their part, it turns out in fact to be reinforced by the increased discipline imposed by this type of show, which includes the reception of instructions

30 On the concept of "emersion," see Asselin and Auger Gosselin, "This Side of Paradise"; and Paul, "The Aesthetics of Emergence."

when entering the theatre. Indeed, the audience members must perform certain actions from a protocol in order to protect themselves against the attacks of the tingler, an obligation that places them from the outset in a posture of obedience to new rules, to a new decorum.[31] Even if the audience is freed from certain shackles (silence, calm), it is immediately subjected to other rules, which order it to free itself by shouting, suggesting a form of participation that is very controlled, but which also remains consistent with the fictional world of the film, in which the characters must shout to protect themselves from the creature. This participation also implies a form of narrative engagement, since the screams are included in the narrative that unfolds, simultaneously, both in the film and in the theatre.

In the promotional material for *The Tingler*, the focus is on the audience members, on their physical reaction and screams, but the very nature of the monster remains secret, except for the fact that it can break free and attack. The focus is neither on the narrative and on its characters nor on the technological innovations deployed by the film and the gimmick, but rather on their effects on the audience and on their expected reactions. The audience becomes a real advertising "object," a material element in the same way as a bleeding-edge technology might be. In the end, with this lobby card, it is no longer a question of promoting a film or a gimmick, but rather of promoting an extraordinary spectatorial experience, where the spectator's body (their perceptions, sensations, and reactions) is central and included in the course of the film. This reflects the program of the time, where active leisure and "participatory events" were sought.

Several posters were also produced and distributed to promote the film and the *Percepto* gimmick. They reused many of the same iconographic elements and rhetorical strategies found in the lobby cards previously mentioned. On one of the main posters, for instance, it is the theatre seat, an emblem of the *Percepto* device, that is promoted as a major element of the advertising strategy.[32] At the top of the poster reads this text: "Do you have the guts to sit in this chair?" In the centre, a red theatre seat serves as an answer to the demonstrative part of the question.[33] It sits in what

31 A comparison could be made here with Jonathan Crary's description of an "observer." See Crary, *Techniques of the Observer*. On discipline in theatrical projections, see Linda Williams, "Discipline and Fun."

32 See the poster online on *The Grindhouse Cinema Database*: https://www.grindhousedatabase.com/index.php/Percepto.

33 The image of the seat (its style, angle, and design details being identical) seems to come from an advertisement from International Chairs (International Seat Division, Union City Body Company Inc.) published in the *Motion Picture Herald* (July 11, 1959), p. 28.

looks like a bubble with lightning-shaped contours, connoting the shock. It is also the form of speech bubbles used in the rhetoric of comics during exclamations, such as screams. The question is redundant, since it is written twice on the same poster: “Can You Take Percepto?,” then “Can you take it when The Tingler breaks loose?” Here, the word *Percepto*, written in capital letters, as well as the question mark are highlighted using a blue font that seems to sizzle or vibrate, like the effects generated by the gimmick. The seat is designated as potentially threatening, since it takes some courage to sit on it. Its bright red colour and the speech bubble that surrounds it contribute to reinforce this disturbing aspect. The spectator is also made aware of the tactile, physical dimension of the gimmick, thanks to the image of the seat, which is the only object offering direct contact with the spectator’s body in the theatre’s apparatus.

The advertisement therefore focuses on the seat as a horrific element. In doing so, it recalls the metaphor suggested with this rigged seat: it is a recreational, entertaining, and playful version of the electric chair, the one that, in the beginning of the film, is responsible for the execution of a convict and which, in its gimmick version, pretends to electrocute the viewer. Several elements of the graphic evoke this analogy, starting with the electrical connotation inscribed in the image with the idea of discharge or shock. Here, therefore, it is suggested that the theatre seat is a symbolic version of this instrument of death used in some American states and the spectator is invited to take a seat in it, both through challenge and interrogation. This advertising rhetoric of challenging the viewer is also common in the horror genre: it is part of the lineage of midnight ghost shows advertisements that used formulas of this type, but can also be associated with the general promotion of horror films during the 1930s.[34]

On several other advertising images aimed at promoting *Percepto*, we see the language of immersion typical of advertisements of the time (obviously without this specific word appearing), addressing the viewers directly by emphasizing their participatory role. Thus, one can read phrases such as “Filmed in Percepto | amazing new screen technique that makes YOU an actual part of the action!”[35] repeating the same rhetoric I noted earlier,

34 See Berenstein, “‘It Will Thrill You, It May Shock You, It Might Even Horrify You’”; and Kattelman, “Magic, Monsters, Movies.”

35 Teaser no. 210 for the film *The Tingler* (William Castle, 1959), Columbia Pictures Industries, Inc. Source: *Tingler Press Book*, 1959, p. 4. James Welsh Collection SUA-003, Edward H. Nabb Research Center for Delmarva History and Culture, Salisbury University, Salisbury, Maryland.

with advertisements reminiscent of those of *Cinerama*. This is probably the most radical version of this participatory posture of the spectator in a film theatre for that period. But, as I will demonstrate, this type of slogan returned more than fifty years later, using the same terms with little difference.

The Dynamic *D-Box* Seat

Forty years after *Percepto*, the contemporary *D-Box* device, a motion seat, was developed in the Montreal area; it then expanded rapidly from the 2010s in North American film theatres. Unlike *Percepto*, which was only imagined for screenings of the specific film *The Tingler*, this patented technology is designed to accompany various films. With *D-Box*, the motions of the seat are synchronized to the action of the film in order to reproduce the sensations experienced by certain characters.[36] These seats move forward or backward, from side to side, and also up or down. Each seat is equipped with an individual control system allowing the user to adjust the level of intensity of the apparatus. For the purposes of this analysis, I will examine two ads prepared by the company in the 2010s, one of which was also presented by UA Cinemas, a now-defunct Hong Kong-based theatre chain.

Comparing some of the ads created by *D-Box* in the 2010s with the 1959 posters studied previously reveals some striking similarities. In one of the advertisements (fig. 10.1), a *D-Box* seat is illuminated by a light beam, emphasizing that the seat is at the centre of the show, a recurring element that was also found on the lobby cards for *Percepto*. The seat is also at the centre of the composition of the image and elevated above the audience: not only is it an isolated piece of furniture from the rest of the apparatus, but this position also implies a form of luxury entertainment, which literally allows the customer to "rise above the crowd" and to stand out. In contrast, at the bottom of the image, a usual theatre environment is shown, where we see many identical seats, in which spectators with illuminated but indistinguishable faces are seated. This suggests a deviation in spectatorial

36 The haptic track, i.e., the series of seat movements that corresponds exactly to the motion on the screen, is entirely designed for the specific needs of each film. It should also be noted that most films that incorporate such haptic tracks are generally action, horror, or science-fiction films in which movement is omnipresent. For a detailed analysis of the point of view (or to be more exact, of the proprioceptive perception of movement) implied in this apparatus, see Petersen, "The Address of the Ass."

Figure 10.1: Advertisement for *D-Box*. © D-Box Technologies, ca. 2015. All rights reserved. Courtesy of D-Box Technologies. Original in color.

posture, but also in class. Moreover, the cost for occupying such seats is higher than for other "ordinary" seats, as was traditionally the case in theatres, unlike film theatres, where all social classes rubbed shoulders without reserved seats.

As with *Percepto*, *Cinerama*, and several other technological innovations of the 1950s, the texts printed on the posters speak directly to the viewer: they describe the type of experience they are about to have. If, in the 1950s, the audience was addressed in a questioning mode, going so far as to challenge it, in the 2010s, it is conveyed in an imperative manner: "Experience the new way *movies will move you*." (fig. 10.1) These last four words are highlighted, underlining the promise associated with the seat. A semantic shift is made here from the seat to the narrative of the film, as if they were one: it is an effect of metonymy. The word "movement" is understood here both in its literal and physical dimension, but also according to its figurative and psychological meaning, assuming a multiplicity of affects jostling in the audience to the rhythm of the motions of the seat.

The box summons the viewer to act: "Reserve your D-Box motion seat." On the floor, at the bottom of the seat, imperative verbs are legible where the image is enlarged: "jolt," "float," "freeze," "bounce," referring to the various sensations felt by the body in the environment. The strategic positioning of these words on the poster, on the floor and under the seat, further suggests that the movements come from this piece of furniture. By extension, these verbs embody the promise of the advertisement, that of the effects of the seat on the body of the spectator. "Live the action," reads the text on the bottom right. It is no longer a question of seeing, hearing, or feeling, but rather of *living* the action, as if the seat could allow direct and physical access to the experience proposed by the film. Thus, advertising no longer invites the spectator to simply attend a show from their traditional contemplative posture; it summons them to "live the action," in an embodied way. Moreover, there is a particular resemblance between this sentence and the one that appeared sixty years earlier on the *Percepto* advertisements: "Filmed in Percepto | amazing new screen technique that makes YOU an actual part of the action!"[37] This time, though, the viewers are no longer a part of the action; they "are" the action, so to speak.

This enduring rhetoric is also found in other advertisements of the 2010s. For example, in figure 10.2, the words "Live the action" return, accompanied by the famous red seat, now through flames and smoke, which suggest heat and speed, thermal and proprioceptive sensations. Once again, the spectator who occupies this seat is placed at the heart of the action, here represented by physical phenomena. The text below the image describes the effect of the seat: "It creates a unique immersive experience which creates a sensation that you are in the movie!" This time, the word "immersion" appears and is defined as a "sensation" of being in the film by the company. This idea is also explicitly expressed in other advertisements from *D-Box*: "Feel like you're in the movie." Ultimately, between these slogans and the phrase "puts you in the picture" associated with *Cinerama* around sixty years earlier, there is very little difference.

37 Teaser no. 210 for the film *The Tingler* (William Castle, 1959), Columbia Pictures Industries, Inc. Source: *Tingler Press Book*, 1959, p. 4. James Welsh Collection SUA-003, Edward H. Nabb Research Center for Delmarva History and Culture, Salisbury University, Salisbury, Maryland.

Figure 10.2: Advertisement for *D-Box*. © D-Box Technologies, date unknown. All rights reserved. Courtesy of D-Box Technologies.

Between Participation and Immersion: The Ideal Spectatorial Experience in the 1950s and the 2010s

In conclusion, one aspect recurs in advertisements from the 1950s as well as in those from the 2010s: above all, they emphasize the experiential dimension of film screening. These posters thus promote a new form of relationship between spectator and show, in particular through the proximity or even the contact established between the audience, the image, and the sound, expressly through the seat. It is a question of giving the illusion that the film occurs in the theatre, through tactile effects, for example by inducing the thrill produced by the tingler through vibrations of the seat, or by imitating the effects of a chase using the dynamic motions of the latter. This relationship is thus more engaging, more proximate, and sometimes more "participatory," or more "immersive," according to the terms used in each era. Beyond the spectacularization of the auditorium space, which seeks to offer the audience an experience that the small screen and home viewing cannot offer, what the advertisements promoting these rigged and dynamic seating technologies show is a tendency to "attractionalize" the cinema experience, to confer on it certain characteristics specific to amusement parks. In both the cases of *Percepto* and *D-Box*, the seat—a piece of furniture that one usually tries to forget in a traditional film theatre—is a direct, physical point of contact between the spectator's body and the space of the film.

For *Percepto*, the film *The Tingler* offers an ideal spectatorial model. It anticipates its own reception, since it presents a scene taking place in an intradiegetic film theatre. The action in the theatre at the time of the vibration of the seats is therefore the exact copy of the scene taking place at the same time on the screen, like a mirror effect. Thus, the narrative system of the film transposes the fiction into the real theatre, integrating it into its narrative. In a way, the film emerges in the theatre space, just as the tingler crosses the limits of the film: it reaches the real auditorium through the seat, which begins to vibrate, providing a thrill, certainly artificial, to the spectator, but one by which they seek, once again, to blur the boundary drawn between diegetic and real spaces. With *The Tingler*, Castle aims to immerse the audience even more in fiction, which takes place in a real and tangible space, that of the theatre, and no longer only on the screen. The seat and the spectators are therefore integrated into the fiction; far from being concealed and transparent, they become opaque and are, in a way, props and characters of the fictional world that unfolds as much on the screen as in the auditorium space.

As for *D-Box* seats, they are often, on promotional posters of the 2010s, isolated from the rest of the seats, creating a distinction—both physical and of class. In addition, *D-Box* seems to position itself as an individual entertainment rather than inviting the viewer to a collective and shared experience, as was the case with *Percepto*. Viewers are, individually and without regard to other members of the audience, drawn into the action of the film, which takes place only on the screen, not including the auditorium in the fictional device. In other words, the seat only serves as a vehicle toward immersion in the heart of the film, without being integrated into the narrative. Moreover, the auditorium is not a necessity for the device, since, having adapted well to the cinema distribution market, the company *D-Box* also now offers a domestic version of its motion seat.

D-Box is not the only company to offer such motion seating. Other specialized companies, whose advertisements it would be relevant to study to highlight their commonalities, also propose the accumulation of several sensorialities. For example, the motion seats of the companies 4DX and MX4D, respectively marketed by the Korean company CJ 4DPlex and the American company MediaMation, are equipped with bubble dispensers and allow the production of air and wind jets, atmospheric elements such as heat, rain, and snow, flashes, fog, smoke, smells, light, and strobe effects, in addition to stereoscopic diffusion. These dynamic seats are sometimes available in specialized and reserved sections of the theatre, for an additional fee and associating the section with a select area, or in the entire auditorium, depending on the company concerned. The designers of these technologies offering multisensory experiences have understood that the world is not only made of images and sounds, but also of textures, tingling, breezes, icy sensations, heat, jostling, shocks, and touches.

Bibliography

Adams, Marjory. "'Tingler' Sell Out at Pilgrim as Kids Come A-Screaming." *Daily Boston Globe*, September 2, 1959: 18.

Asselin, Olivier, and Louis Auger Gosselin. "This Side of Paradise: Immersion and Emersion is S3D and AR." *PUBLIC. Art | Culture | Ideas*, no. 47 (Spring, 2013): 132–141.

Asselin, Olivier. "Vers un cinéma neuronal. Lunettes, verres de contact, implants." *Cinémas. Revue d'études cinématographiques* 29, no. 1 (Fall, 2018): 89–109.

Barnier, Martin, and Kira Kitsopanidou. *Le cinéma 3-D. Histoire, économie, technique, esthétique*. Paris: Armand Colin, 2015.

Belton, John. "Spectator and Screen." In *Moviegoing in America: A Sourcebook in the History of Film Exhibition*, edited by Gregory A. Waller, 238–246. Oxford, MS: Blackwell Publishing, 2002.

Belton, John. *Widescreen Cinema*. Cambridge, MA: Harvard University Press, 1992.

Berenstein, Rhona J. "'It Will Thrill You, It May Shock You, It Might Even Horrify You.' Gender, Reception and Classic Horror Cinema." In *The Dread of Difference: Gender and the Horror Film*, edited by Barry Keith Grant, 117–142. Austin, TX: University of Texas Press, 2017.

Caillois, Roger. *Les jeux et les hommes. Le masque et le vertige*. Paris: Gallimard, 1958.

Calleja, Gordon. *In-Game: From Immersion to Incorporation*. Cambridge, MA: MIT Press, 2011.

Castle, William, dir. *The Tingler*. 1959; Los Angeles, CA: William Castle Productions and Columbia Pictures, 2018. Blu-ray Disc.

Clepper, Catherine. "The Rigged House: Gimmickry, Exhibition, and Embodied Spectatorship in Mid-Century American Movie-Going." PhD diss., Northwestern University, 2016.

Crary, Jonathan. *Techniques of the Observer: On Vision and Modernity in the Nineteenth Century*. Cambridge, MA: MIT Press, 1992.

Delavaud, Gilles. "Le cinéma après la télévision." In *Le cinéma, et après?*, edited by Maxime Scheinfeigel, 51–61. Rennes: Presses universitaires de Rennes, 2003.

Gardner, R. H. "'Tingler' Tailor Made for Vicarious Thrill." *The Baltimore Sun*. August 16, 1959: F6.

Gomery, Douglas. *Shared Pleasures: A History of Movie Presentation in the United States*. Madison, WI: University of Wisconsin Press, 1992.

Kattelman, Beth A. "Magic, Monsters, Movies: America's Midnight Ghost Shows." *Theatre Journal* 62, no. 1 (2010): 23–39.

Loiselle, André. *Theatricality in the Horror Film. A Brief Study on the Dark Pleasures of Screen Artifice*. London: Anthem Press, 2020.

Milgram, Paul, Haruo Takemura, Akira Utsumi, and Fumio Kishino. "Augmented Reality: A Class of Displays on the Reality-Virtuality Continuum." *Telemanipulator and Telepresence Technologies* 2351 (December, 1995): 282–293.

Ngai, Sianne. *Theory of the Gimmick: Aesthetic Judgment and Capitalist Form*. Cambridge, MA: Belknap Press, 2009.

Paul, William. "The Aesthetics of Emergence." *Film History* 5, no. 3 (September, 1993): 321–355.

Petersen, Christina G. "The Address of the Ass: D-Box Motion Code, Personalized Surround Sound, and Focalized Immersive Spectatorship." *Journal of Film and Video* 71, no. 1 (Spring, 2019): 3–19.

Schaefer, Christopher. "*The Tingler*." *Mondo Cult*, no. 2 (2007): 45–50.

Szczepaniak-Gillece, Jocelyn. "Revisiting the Apparatus: The Theatre Chair and Cinematic Spectatorship." *Screen* 57, no. 3 (October 2016): 253–276.

"*The Tingler.* Lobby Card Set." *Turner Classic Movies*. https://www.tcm.com/tcmdb/title/19997/the-tingler/#photos-videos.

Tingler Press Book, Columbia Pictures Industries Inc., 1959, 6 pages. James Welsh Collection SUA-003, Edward H. Nabb Research Center for Delmarva History and Culture, Salisbury University, Salisbury, Maryland.

Weber-Houde, Aude. "Le siège des sensations. La salle de cinéma en crise et le *gimmick* tactile *Percepto* (William Castle, 1959)." PhD diss., Université de Montréal, 2021.

Williams, Linda. "Discipline and Fun: Psycho and Postmodern Cinema." In *Reinventing Film Studies*, edited by Christine Gledhill and Linda Williams, 351–378. London: Arnold, 2000.

Zone, Ray. *3-D Revolution: The History of Modern Stereoscopic Cinema*. Lexington, KY: University Press of Kentucky, 2012.

Zone, Ray. *Stereoscopic Cinema and the Origins of 3-D Film, 1838–1952*. Lexington, KY: University Press of Kentucky, 2007.

About the Author

Aude Weber-Houde is a professor at the Université du Québec en Abitibi-Témiscamingue (UQAT), Canada. Her research focuses on film gimmicks and advertisement, cinema exhibition history, as well as media and the senses. She has published several articles on touch and polysensoriality, and has coedited the "Monumental and Collective Immersive Viewing Systems" parcours of the TECHNÈS *Encyclopedia of Film Techniques and Technologies*.

11. Surface and Depth: The Picture Plane as a Site of Immersivity in the Digital Age

C. E. Harris

Abstract: This chapter applies the art historical notion of the picture plane to contemporary digital cinema in order to address the emerging spatial complexities which result from the dematerialization of the frame, the camera, and the lens in the digital age. Analyzing instances in which digital films oscillate between visually representing the picture plane as a diegetic surface (e.g., the trope of blood spatter on the camera lens) and visually representing the *dissolution* of this surface (e.g., *trompe l'œil*, negative parallax), it argues that: (1) the picture plane has become an arbiter of cinematic immersion, and (2) its porous, virtual nature in dematerialized digital cinema reconfigures and renegotiates spatial relations on screen, and between screen and spectator.

Keywords: digital cinema, spatiality, materiality, aesthetics, media archeology

This chapter is concerned with the art historical notion of the "picture plane," insofar as it relates to the configuration of immersive spatiality in an increasingly dematerialized digital cinema. The picture plane has largely been framed as a pedagogical concept rather than as a theoretical one, first in terms of how it is taught today, as an entry principle to students of the plastic arts, and second in terms of its historical invention, attributed to the perspectival thinkers of the fifteenth and sixteenth centuries, as an optical tool to aid in the accurate rendering of objects in space as they are transcribed and projected from real-world three-dimensionality into pictorial two-dimensionality. The concept of the picture plane thus already has

Bédard, Philippe, Alanna Thain, and Carl Therrien, eds. *States of Immersion Across Media: Bodies, Techniques, Practices*. Taylor & Francis Group, 2025
DOI 10.5117/9789048558766_CH11

a polyvocal meaning, simultaneously referring to both the physical devices of artistic observation which make this graphic projection possible—Leon Battista Alberti's veil or drawing grid, Albrecht Dürer's draughtsman's net or sheet of glass, or Filippo Brunelleschi's *tavoletta*, for example—and the optical-mental processes that render perspective legible for the spectator.[1]

When it comes to cinematic images, we rarely speak of a picture plane proper; instead, its role seems to have been usurped by theorizations of the frame and of the lens or of the screen and of the spectator's gaze. But today, with the dematerialization of many of cinema's devices and materials and with the rise of films created by increasingly digital means, these elements, their codes and uses, and the relationship between them are shifting. Contemporary digital images can be created—completely or in part—not only without film stock, but also without cameras, lenses, or physical sets, for example. Freed from the physical constraints of the traditional tools of filmmaking, this "dematerialized digital cinema," as it will be termed in this chapter, alternates between upholding and defying the visual and aesthetic codes established through the materiality of analogue cinema. The concept of the picture plane is useful in such instances for unpacking the complex spatiality and immersive impulses of these contemporary digital images, as they re-stage the relationship between the (often virtual) profilmic and the (often virtual) camera lens, between the image and its frame, and between the viewing *dispositif* and the spectator. The picture plane, due to its polyvocality, is a perfect metaphor for the sometimes material, and sometimes dematerialized virtual surfaces so characteristic of the digital age, which separate but articulate: (1) filmed objects and their two-dimensional projections onto our viewing screens; and (2) the spectator's viewing position and the images being viewed. It is a virtual surface which stages images and relations to images in immersive depth. In this light, this chapter proposes an interrogation of the picture plane in the digital age in order to put forward a two-part argument: first, it postulates that the picture plane is more visible than ever, and second, it proposes that the picture plane has become a porous site of circulation. At times, it circumscribes a strict border that separates discrete cinematographic spaces, but it is just as likely to give way to an interpenetration of these spaces and to a circulation between them. This porosity oscillates between visually representing the picture plane as a physical surface in the story-world of

1 These devices are explored in Niels Lindahl Elliot's monograph, *Mediating Nature*. In addition to his chapter on "The Nature of Observation," (pp. 32–53), specific descriptions of the observational instruments of Alberti, Dürer, and Brunelleschi can be found on p. 58 and p. 71.

a film and dissolving this boundary in order to extend the spaces of the story-world until they appear to meet our own. In both cases, the picture plane works as a spatializing agent: when visible, it extends diegetic space, and when dissolved, it organizes spectatorial immersion.

The Picture Plane: From Painting to Cinema

In its historical connotation, the picture plane was a physical tool—a grid, a set of strings, a veil, a pane of glass—that could be erected by artists in front of the real-world object they were depicting on a perpendicular two-dimensional surface, serving as a sort of looking glass, frame, or window to aid in artistic observation toward the accurate construction of geometric perspective. However, the picture plane also represents an immaterial virtual positioning. As William M. Ivins, Jr. explains, for Alberti, it was "a perpendicular cross section of the cone of vision," which revealed that, "pictorially at least, form and position were functions of each other."[2] In this second formulation, the picture plane is not an actual, physical device or surface, but a virtual, imagined one which positions the observer and organizes the act of observation, both during the process of creating an image and during the contemplation of works of art. In the discourse of the ideal spectator position, often invoked in discussions of *trompe-l'œil* art, for example, the picture plane serves as a sort of arbiter of space. How does this apply to cinema?

Traditionally, filmmakers have looked through the camera's viewfinder, which constructs a viewing frame around profilmic objects and figures. Through the use of different types of lenses, which organize the graphical perspective of the three-dimensional profilmic into a two-dimensional image, the depth of the profilmic space can be transposed with various measures of control, with the long lens compressing space but permitting views into a greater distance and the wide lens exploring a smaller range of space in greater depth. The glass pane or veil of the painter's picture plane, in this light, can be simply transposed onto the device of the camera lens as a physical tool for the construction of perspective. However, in terms of image spectatorship, the picture plane becomes (at least, traditionally) immaterial and virtual. Due to the mobility of cinematic images, with their mobile frames and ability to seamlessly change focal lengths even within a single shot, the picture plane qua ideal spectator position fluctuates in terms of its

2 Ivins, "On the Rationalization of Sight," p. 32.

relation to profilmic objects. In some regards, this is a narratological issue, in which the polyvocality of the picture plane arises through cinema's natural capacity for the seamless oscillation along a spectrum of subjectivity and objectivity, between subjective (first-person) and objective (third-person) points of view and the intermediate states in between. The positioning and repositioning of the spectator are constituted as a virtual aspect of projected images. Here, the "virtual" is understood as the real but not actual aspects of the image, as the image's potentiality, which is imperceptible but nonetheless exerts itself upon the spectator.[3] As described by Brian Massumi, "[t]he virtual [...] is inaccessible to the senses. [...] The virtual that cannot be felt also cannot but be felt, in its effects."[4]

Pertinently, Massumi goes on to explain that there are instances in which this virtual is rendered visible: "When expressions of its effects are multiplied, the virtual fleetingly appears. Its fleeting is in the cracks between and the surfaces around the images."[5] For dematerialized digital cinema, this type of multiplication precisely takes place, as the picture plane's functions are subject to a sort of *mise en abyme* effect. During the creation of computer-generated sequences or films, virtual cameras are visually represented on the artist's screen in order to stage myriad simultaneous possible profilmic relations; during the animation process, the simulated camera presents a spatial organization and spectatorship position wholly separate from the computer screen, as artists can temporarily "enter into" the image in order to visualize what future audiences will see and reorganize it accordingly by selecting particular framings or views. This is also an increasingly common operation on set, using previsualization (Previz) technologies. In these instances, the picture plane is both a tool for visualization and, itself, visualized, as a malleable variable of composition. It is neither the screen on which the images are visualized and manipulated, nor the surface of the image itself, but rather an imaginary intermediary between the two. It is a protean, virtual limit—an interface—that negotiates spatial configurations within the image and relations of proximity, distance, and immersivity between the image and the future spectator.

During spectatorship, the visualization of the picture plane is often shifted from the (virtual) camera to the (virtual) lens. The contrast between the immersive *depth* of the digital age (with its renewed predilection for shots

3 This notion of the virtual is borrowed from Gilles Deleuze's work (who, in turn, references the work of Henri Bergson). See Deleuze, *Bergsonism*, pp. 94–98.

4 Massumi, *Parables for the Virtual*, p. 133.

5 Massumi, *Parables for the Virtual*, p. 133.

that move into images along the Z-axis) and the increasing preoccupation with the *surface* of the camera lens—for example, via the common stylistic choice to add in artifacts of lenses that are no longer there, such as lens flares—makes the picture plane perhaps more visible than ever. It draws up new spatial relationships, not only between the image and its creator, but also between the image and the spectator. This highlights what has remained consistent throughout film history, regardless of style, technology, and the variables of film production and projection: that exploring the configuration of space has always been one of the central projects of cinema. In the words of Antoine Gaudin, "[t]he notion of space is consubstantial with the art of cinema."[6] As Tom Conley affirms, "[a] film, like a topographic projection, can be understood as an image that locates and patterns the imagination of its spectators. When it takes hold, a film encourages the public to think of the world in concert with its own articulation of space."[7]

The Paradoxical Materiality of Surface and Substance

In the penultimate sequence of Alfonso Cuarón's *Children of Men* (2006), something curious occurs. As main character Theo seeks cover from gunfire, he narrowly avoids being shot when another character takes a bullet that was meant for him. While he ducks for cover, eight droplets of blood land on the lens of the camera.

This device of blood spatter on the camera lens is not novel. It seems to have first occurred in 1969, in Guy Hamilton's *The Battle of Britain*, and has become a popular gimmick, especially in war and horror movies. What is curious about this particular instance of blood spatter on the camera lens, however, are the circumstances of the effect's creation as well as the philosophical questions that it raises about the materials of cinema and the role of special effects in upholding the spatial and material relations between filmed objects and the filming process.

In a *Cinefex Magazine* breakdown of the film's VFX, Double Negative visual effects supervisor Frazer Churchill explains that "[it] was completely accidental. A fugee extra in the bus was off her mark. She turned to the camera and splashed the lens. We continued the take because daylight was fading fast, and there were so many elements going on it was impossible to

6 Gaudin, *L'espace cinématographique*, p. 5, my translation.

7 Conley, *Cartographic Cinema*, pp. 1–2.

communicate the 'cut' to everybody."[8] This could have been problematic, as the blood droplets landed on the camera lens during a painstakingly choreographed, several-minute-long continuous take that would have been difficult to reset and reshoot, particularly in light of the time constraints that the crew was facing. However, in a 2017 interview with *Vulture.com*, Cuarón explains that the decision to keep the blood spatter for its aesthetic value was made instantly and without question.[9] The only problem became how to deal with the removal of the blood for the duration of the shot without creating a glaringly obvious discontinuity, were it to simply mysteriously disappear. The special effects team solved the problem digitally: "2D artist Niki Wakefield spent a lot of time removing and replacing blood so we could believably animate the drops to thin and roll away as the camera moved around," recounts Churchill, therefore providing a diegetic explanation for the disappearance of the blood droplets.[10] Indeed, as the scene continues, the blood droplets discreetly disappear one by one over the course of roughly a minute and a half, as if by evaporation.

This behind-the-scenes detail of *Children of Men* is pertinent not only for its innovative continuity solution, but especially for the new paradigm of materiality that it represents. Indeed, perhaps the most curious aspect of this anecdote is the fact that the obvious reference to the presence and physicality of the camera lens in the film's story-space was hardly given a second thought (or a diegetic explanation). Instead, concern was focused toward the realistic portrayal and removal of the blood droplets in their transition from materiality to digitality. This is surprising because the blood-spatter-on-the-camera-lens trope has had a controversial history with regard to realist aesthetics and the spectatorial suspension of disbelief, precisely because it can work simultaneously toward greater audience immersion *and* toward a jarring sense of breaking the fourth wall that shakes spectators out of their suspended disbelief. In other words, it can make us feel like we are immersed in the story-world by heightening our perceived proximity to the film action, but it can also pull us out of our immersed state by reminding us of the artifices and devices of the cinema itself. The camera lens that films the sequence, which occupies the nondiegetic space of the filmmakers, is rendered present within the space of the diegesis, visually articulating the limits between the two realms. Furthermore, a series of additional devices is used to add to our sense of the camera's immersivity:

8 Churchill, quoted in Fordham, "*Children of Men*," p. 43.
9 Cuarón, quoted in Riesman, "The Vulture Transcript."
10 Churchill, quoted in Fordham, "*Children of Men*," p. 43.

the shakiness of the handheld camera, the eye-level camera height, and the camera movements which appear to follow characters through the scene. Such devices are often used to make a shot feel subjective, and indeed during this sequence we have the fleeting impression of watching a subjective shot as if from the point of view of some immersed phantom character. However, once we see the blood droplets, the illusion is interrupted: the lens becomes a physical surface, rather than a phantom character. This procedure sets up a spatial continuity which links diegetic action with the nondiegetic world of a film's production and post-production. Material or dematerialized, immersive or jarring, the blood-spatter-on-the-camera-lens trope is a sort of Schrödinger's surface in the constitution of film space: it simultaneously stages a surface which separates spaces, calls our attention to this, and dissolves this separation through its immersive effects.

In order to elucidate this argument, let us consider a series of examples of references to the camera's lens that emerge specifically in films that do not actually use a physical camera for image capture. In the age of dematerialized digital cinema, as the line between film and animation blurs more than ever, there is a persistent trend toward the use of such virtual devices and processes which, by simulating the visual "language" of analogue cinema, highlight the simultaneous absence and presence of cinema's materials.[11]

The Virtual Lens as a Diegetic Object

Computer-animated films have shown a great predilection for referencing camera lenses that are not present by staging them as interdiegetic surfaces with which characters, objects, and substances within the story-world can occasionally interact. As Mike Jones explains, this is a "widespread common practice," which can be "attributed to a desire on the part of the creators to emulate a visual aesthetic that belongs to another form—that of camera-based live-action, subsequently invoking the audience perception and expectations of that very familiar medium."[12] It is a favorite trope of Pixar, for example, whose animators already use it in their 1998 co-production with Disney, *A Bug's Life* (John Lasseter and Andrew Stanton). In a post-credits

11 The topic of virtual film production—its tools and techniques, its aesthetics, its intermedialities and materialities, its effects on the industry, its philosophical stakes—is a site of rich discussion and debate in contemporary film and media studies. See, for example, Bédard, "Virtual Production and the Transformation of Cameras Mechanical, Virtual, and Actual"; Bédard, "Le cinéma sans caméra?"; Crockett, "The 'Camera as Camera'"; and Jones, "Vanishing Point."

12 Jones, "Vanishing Point," p. 242.

sequence, we are shown a "blooper reel." In itself, this device underlines how interested Pixar is in situating their films within a larger conversation about filmic materiality. In the "blooper reel" of *A Bug's Life*, there are several sequences organized around interactions with the "camera," in particular animated characters colliding with the "camera" and leaving a mark on its "lens."[13]

This trope has become ubiquitous in recent computer-animated films, from niche children's movies such as *Surf's Up* (Ash Brannon, Chris Buck, and Henry Yu, 2007), *Trolls* (Walt Dorhn and Mike Mitchell, 2016), and *Trolls World Tour* (Walt Dorhn, 2020) to popular Pixar productions such as *Cars* (John Lasseter, 2006) or *Turning Red* (Domee Shi, 2022), and even including computer-animated sequences from Marvel blockbusters such as *Guardians of the Galaxy, Vol. 2* (James Gunn, 2017), to name just a few. In these films, the virtual surface of the simulated lens is not only acknowledged diegetically, but is increasingly becoming a site of experimentation as well. Renowned for their innovations in the field of photorealistic animation and for their rendering of textures and substances, including difficult-to-render textures like liquids, sand, and glitter, for example, *Surf's Up* and the *Trolls* movies use the simulated lens as a "surface" onto which they can project these difficult-to-render substances in order to show audiences how they pass muster even when viewed up close.[14] Characters, objects, and substances—especially water droplets and glitter—are repeatedly propelled toward the virtual camera "lens" and stick to its "surface," rendering the "lens" temporarily diegetically material.[15] These filmic moments serve several purposes: they break the fourth wall, they nod to earlier filmic materialities, and they have the ludic function of setting up a sort of inside joke with audiences about cinematic reflexivity. However, they also have a spatializing function: first, they extend the diegetic space toward the space of the spectators; then, the virtual surface of the simulated lens is rendered visible in order to mediate and circulate between these two spaces. We no

13 At the time of writing of this chapter, this "blooper reel" can be seen online on the Pixar fandom wiki site, Pixar.fandom.com. See DisneyFan18111928, "*A Bug's Life* Outtakes."

14 These animated films have been the object of presentations at the Special Interest Group on Computer Graphics and Interactive Techniques (SIGGRAPH) annual conference for precisely these sorts of innovations in the rendering of substances. See, for example, Augello et al., "Hair Effects in *Trolls World Tour*"; Bredow et al., "*Surf's Up*"; and Kuan-Ting, "Organizing the Fuzziest Concerts for *Trolls World Tour*."

15 Interestingly, this is the case even in stereoscopic 3D versions of these films; *Trolls World Tour*, for example, contains uses of 3D spurts in which objects spring forth from the screen, alongside instances in which glitter and other substances are made to stick to the intermittently present virtual camera "lens."

longer have the impression of observing from afar; we instead are made to feel that we are located right in front of the profilmic action, involved or interpellated directly by what is happening in it, and able to scrutinize from as close as possible the technological and aesthetic innovations that are being shown off to us.[16] The simulated lens, as an experimental virtual surface, is a key example of the intervention of the picture plane in digital cinema, with its function as arbiter of spatiality and mediator between the spaces it organizes.

Negative Parallax: Dissolving the Virtual Surface

This mediating function is just as conspicuous when the virtual surface of the simulated camera lens dissolves. Just as the picture plane can be represented as a physical diegetic surface, it can also be pierced and rendered as diegetically immaterial—and this, in service of the same objective of organizing digital cinematographic space.[17] Several recent 2D computer-animated films experiment with techniques that work toward this end by referencing and imitating devices native to 3D cinema, for example using *trompe-l'œil* effects or blurring techniques to make objects appear to move too closely to a camera that is not there, exceed their frame, or jump out at the audience. Whereas 3D cinema creates these spurt effects stereoscopically, recent 2D examples often use letterbox framing to do so, playing upon audience expectations about where the frame is located and about where the diegetic space begins and ends. In these examples, objects in the diegetic world seem to pass unobstructed into negative parallax: in other words, they extend out of their frames and appear to enter into our world, into the space that we occupy as spectators.[18] One of the most popular

16 On the topics of the virtual camera and of the construction of space, see Brown, "Digital Cinema's Conquest of Space"; and Jones, "Vanishing Point." For wider discussions on virtual cameras and camera-less cinema in the digital age, see Bédard, "Virtual Production and the Transformation of Cameras Mechanical, Virtual, and Actual"; Bédard, "Le cinéma sans caméra?"; and Crockett, "The 'Camera as Camera'."

17 While this chapter is only concerned with 2D computer-generated films, the author would like to point to Miriam Ross's pertinent work, "Stereoscopic Visuality" (2013). Ross's article, while primarily addressing the optical illusions produced in and by 3D images, raises a relevant discussion about how contemporary cinema constructs space due to and despite the porosity of the screen plane or picture plane. See Ross, "Stereoscopic Visuality."

18 Kristen Whissel defines "negative parallax" images as "emergent images" (images appearing to emerge from the frame), which organize perception "around heightened emotion and sensation." Whissel, "Parallax Effects," p. 235.

examples of this phenomenon is the series of teasers for the *Despicable Me* films, which use the device repeatedly. For example, a teaser trailer for *Despicable Me 2* (2013), entitled "The Minions Announce *Despicable Me 2*," is organized around this visual gag.[19] As the characters perform a song for the audience, the letterbox framing, the frontality of the image, and the direct address of the characters' gaze seem to structure the space into two separate zones: the diegetic world and the separate space which we occupy as spectators. This rigid limit is pierced, however, when a character blows into a party horn which unfurls toward us into negative parallax, revealing that the picture plane of this image is not located where we had anticipated it to be. DreamWorks' *Penguins of Madagascar* (Eric Darnell and Simon J. Smith, 2014) uses the same device during the movie's climax: as mad scientist Dr. Brine releases a horde of mutant penguins on the city of New York, they are surrounded by clouds of a toxic-looking green smoke which billows out over the letterbox at the bottom of the screen, giving the impression of spilling out of the frame toward the audience.[20] In a similar vein, Disney/Pixar's *Turning Red* opens with a sequence that shows a series of photographs of the protagonist as a young child; as the camera settles on the last image, the protagonist comes to life and jumps forward out of the frame of the photograph.

This device operates by subverting our expectations regarding framing: firstly, it induces us to assume that the letterbox framing corresponds to the frame of the image, demarcating the limit between the diegetic and the nondiegetic spaces, only to subsequently reveal that this is not the case and that the limit of the diegetic space is not where we had expected it to be. Secondly, as the action of the sequence extends along the Z-axis, the depth of the film space is exaggerated in both directions, simultaneously toward and away from the audience. It extends backward into the image, as a vanishing point behind the characters, but it also reaches out into the space of the spectators, creating a sense that there is a second vanishing point located behind the viewers from the perspective of the characters. Cinematographic space is elongated and revealed to be malleable rather than fixed. Finally, by breaching the "surface" that would usually be occupied by a camera lens, it admits that there *is* no camera lens there, a fact that typically goes unspoken,

19 At the time of writing this chapter, this teaser trailer is available for viewing on IMDB's website. See "The Minions Announce *Despicable Me 2*."

20 At the time of writing this chapter, this sequence can be viewed on YouTube. See Movieclips, "Penguins of Madagascar (2014). Mutant Penguins Scene (7/10) | Movieclips."

with films preferring to simulate its presence rather than point out its absence. Audience expectations regarding the spatial configuration of these images and our relation to them are subverted. While stereoscopic 3D films have long employed this device of objects or characters bursting forth from the frame, recent dematerialized digital cinema has found creative ways of bringing it to 2D films by animating in visual codes that imitate the gestures of stereoscopy in order to expand cinematographic space beyond the frame and increase circulation between the diegetic and the nondiegetic spaces.

This trope has recently become popular in digital artworks as well, as social networks become sites of circulation for experimental pieces by animators and artists like freelance motion graphics designer Tim Gray, who creates animations that play on the projection of diegetic objects into negative parallax.[21] Gray's work operates by animating digital creatures, which seem to escape from their frame, for example, in mock Instagram posts meant to be viewed on a smartphone and in which characters appear to jump out toward us, escaping the feed of images and posts that we are scrolling through. Whereas in stereoscopic 3D cinema, objects might appear to temporarily occupy the space of the auditorium, these images retain the appearance of residing within our devices. Nonetheless, they work by revealing that the frame and limits of the image do not always reside precisely where we expect them to. These *trompe-l'œil* techniques challenge our expectations about the configuration of digital space, showing us that the picture plane has become flexible, mobile, and porous.

This type of device is indebted to a now classical horror movie trope: the monster that bursts out of a screen to enter into our world. Perhaps the most famous example of this comes from *Ringu* (Hideo Nakata, 1998) and its US remake *The Ring* (Gore Verbinski, 2002), in which the monstrous Sadako (Samara in the remake) is seen in a video and then crawls out of the television screen to attack its spectator. More recently, this also happens in *It. Chapter 1* (Andrés Muschietti, 2017), as the evil clown Pennywise appears in a slide-show projection only to burst out of the screen to attack its viewers. Indeed, the paradoxical role that media devices play both in containing our fears as well as projecting them back onto us is a common reflexive theme in the horror genre, and there is a certain legacy of media panic which has always linked the popularization of new technological forms of

21 Tim Gray's negative parallax animations can be found on his personal Instagram portfolio. See Gray, *@iamtimgray*. He also maintains a web portfolio, dedicated primarily to his freelance film and advertising work. See Gray, "Projects."

entertainment to fears of trickery.[22] From magic lantern phantasmagorias to apocryphal tales of audiences running away in fear of being hit by a train at the showing of the Lumière brothers' *Arrivée d'un train en gare de La Ciotat* (1896) by way of nineteenth-century spirit photography, this preoccupation dates back to historical anxieties about the mystical powers of photography to not only capture up souls and ghostly presences in its images, but also to potentially release them back upon us as they escape from their frames and penetrate into our world. With the shift toward digital media and subsequent reconfigurations of cinematic space, this topic is once again relevant. As audiences seek out greater and greater sensations at the cinema, contemporary films play up the impression that the movie screen is but a thin veil separating our world from some Other Side and that perhaps something could burst through the screen at any moment. This sense of proximity and porosity allows us to feel not only that we are immersed in the image world, but also that images are immersed in our own world.

Now, in the age of Augmented Reality (AR), digital objects are closer than ever to springing forth from the screen into our world. Experimentation with negative parallax in non-stereoscopic films is increasingly informed by this regime of aesthetics, unbounded by the frame and exceeding the screens on which we watch them. They privilege the dissolution of the picture plane's virtual surface, which separates our world from the image world, by extending the diegetic and the extradiegetic spaces into one infinitely deep space. If this circulation between images and our world is a mere *trompe-l'œil*, it reveals that the porosity of the picture plane informs the aesthetics of the images we create and consume as well as the experiences that we seek from them.

The Porosity of the Picture Plane

Experimentation with the organization of cinematographic spaces via the picture plane, as seen in the aforementioned computer-animated films, can be situated within a larger tradition that links cartoons, photography, and cinematographic materiality. In this vein, in her 2019 monograph *Frame*

22 Tom Gunning, for example, takes up this issue through the concept of the uncanny and technological strangeness in "Re-Newing Old Technologies." With regard to horror cinema specifically, much has been written about monsters like vampires as a vehicle for studying the relationship between technology and fear. See for example, Stacey Abbott, *Celluloid Vampires*; and Barbara Laner, "The Vampiric Film."

by Frame: A Materialist Aesthetics of Animated Cartoons, Hannah Frank undertakes an extensive historiography of animated materialities. Here, she refers to an early instantiation of the porous picture plane of camera-less cinema: the cel animations of Chuck Jones. She cites cartoons such as *Duck Amuck* (1953) and *Rabbit Rampage* (1955), in which characters serendipitously discover the limits of their own cartoon world and engage in diegetic meta-explorations of cartoon materiality. At times, they escape the diegetic world and find themselves in the blank void beyond the frame, or they find themselves "tormented by the pencil and paintbrush of an animator," or else the "projector" jams and the "celluloid" of the cartoon breaks, allowing characters to escape an on-coming diegetic danger.[23] Frank understands these gags as a study of animation's relationship to other arts and, ultimately, to our world—from questions of indexicality to the material signifiers of representational and reproductive arts.[24] Indeed, such meta treatments of film's materials are not exclusive to digital cinema. However, this formal experimentation with spatiality via the picture plane, once restricted to gags of cinematic reflexivity or to experiments in the development of special effects, is becoming commonplace in big-budget Hollywood cinema. Thanks to the aesthetic influence of other immersive media—video games, Virtual Reality (VR), AR—digital cinema is reorganizing the spatial relationship between the world of the spectators and the diegetic world of films, building "profilmic" spaces that feel more navigable than ever, engaging with the spectator's sense of proprioception with simulated camera movements that send us careening into film space along the Z-axis,[25] and making us feel perhaps closer than ever to the action we see on screen. This occurs not only by referencing the "surface" of the simulated camera lens in a film's story-world or by piercing through it, but especially by vacillating between the two modes, turning the picture plane into a porous site of circulation.

One notable example of this porous picture plane is Steven Spielberg's *Ready Player One* (2018), a film that already thematizes immersion and the mobile boundaries of virtual spaces through its plot about characters who live out their lives in virtual reality. In a scene that takes the protagonists on a side quest to the world of Stanley Kubrick's *The Shining* (1980), characters—and spectators—are immersed in a digital Overlook Hotel recreated with perfect photorealism. After a cinephilic exploration of the hotel's decor, the avatar Aech finds himself in front of the infamous bleeding elevators.

23 Frank, *Frame by Frame*, p. 68.
24 Frank, *Frame by Frame*, pp. 45–49.
25 On this, see Richmond, *Cinema's Bodily Illusions*.

As the doors open, an immense wave of blood overflows into the corridor and sweeps up Aech—and the simulated camera with it. While Aech and the simulated camera are rushed through the halls of the Overlook by this river of blood, blood droplets fly at and drip down the virtual lens on several occasions. It is a paradoxical gesture: on the one hand, the simulated camera seems to play the role of a fantasy character present in the scene (and acknowledged by the character's eye-line) and, thanks to these subjective first-person shots, the spectator is immersed into the action. On the other hand, the blood droplets on the lens also serve as a reminder that there still is a surface that separates us from this space, even if the surface is only a virtual one. It is a curious—but increasingly common—set up, organized to simultaneously provoke a sense of proximity *and* separation, immersion *and* awareness of cinematic artifice, intense sensation *and* reflective commentary on cinematic materiality. The picture plane accomplishes all of these things, and thus serves as a mediator: (1) it immerses us in the scene and configures our point of view; (2) it extends cinematographic space toward and around us; and (3) it nonetheless circumscribes a boundary that separates us from the action, all at once. In other words, it configures three different scales of proximity at the same time, without editing, without reframing the shot, and without changing points of view.

But if the picture plane is visually represented as a diegetic surface here, it dissolves and becomes immaterial just a few seconds later. Carried away by the river of blood, Aech finds himself in front of the door of the notorious room 237, where he meets the old woman from the bathtub who had once seduced Jack Torrance. She starts chasing Aech with an axe and the two are suddenly transported to the Overlook's snowy hedge maze. During the ensuing chase sequence, the porosity of the picture plane is highlighted: there is a constant alternation between moments in which it is represented diegetically—for example with snow flakes that land on the virtual lens—and moments in which the picture plane is pierced by objects that seem to emerge from the frame in negative parallax—for example, the old woman's axe, as she raises it over her shoulder before swinging it down at Aech. The latter is even more remarkable in the 3D version of the film, which greatly exaggerates the depth of field and allows characters and objects to stereoscopically emerge from the screen, all while maintaining the screen's integrity as a surface upon which debris can accumulate. The virtual surface rapidly alternates between being visible and invisible, present and absent, creating multiple contrasting means of accessing the film space.

Intermedialities of Virtual Porosity

The porosity of the picture plane, or, in other words, the increasingly visible ambivalence in contemporary digital cinema between visually representing the picture plane and dissolving it, is not only a legacy of cartoon animation, but also an intermedial gesture borrowed from video games, especially first-person shooters (FPS). In these games, the spectator-player is immersed, via an avatar, in the navigable spaces of the game and moves along the Z-axis into the deep space of the image. But it is also a medium that abounds with examples in which the virtual surface of the picture plane is rendered visible as an in-game surface, especially with the trope of splashes on the "camera" lens.

In games that feature frequent third-person—often cinematic—views of the game world, the paradigm of the picture plane links video games back to a cinematic articulation of spatiality and materiality. In *Witcher 3: Wild Hunt* (CD Projekt Red, 2015), for example, during scenes in which protagonist Geralt explores the game world by boat, droplets of water splash from the river and accumulate on the surface of the virtual camera lens of the third-person shots. Likewise, in *Detroit: Become Human* (Quantic Dream, 2018), also played from a third-person point of view, scenes featuring the playable character Markus as he is reborn in Android Hell reveal the accumulation of moist condensation on the surface of the virtual camera lens. In *Metal Gear Solid 3: Snake Eater* (Konami Computer Entertainment Japan, 2004), the virtual camera lens is often rendered a diegetic, physical surface upon which raindrops can collect and wasps can land. It is noteworthy that these video game examples are not equally dedicated to photorealism, and yet, they are each concerned with the materiality of the virtual camera lens and with its function in spatial organization. They expand the in-game space by directly acknowledging the external gaze of the spectator-player, overcoming the distancing effect of these third-person shots in order to stage a closer proximity between the player and the objects and substances of the game world.

However, some of the most frequent uses of this trope are in FPS games, in which the camera and the player coincide to create a first-person point of view within the spaces and action of the game's story-world. The first-person, subjective nature of these games creates a link between the player's body in the real world and the avatar's in-game body, which is articulated at the precise location of the virtual lens, here occupying the role of the picture plane. In these cases, the virtual lens does not function merely as a window into the diegetic world, but also as a sort of optical device: typically, it is justified in the narrative as being the goggles or helmet visor that an avatar might wear. When blood and debris coagulate on the transparent surface in front of the player-avatar's eyes in these game situations, it is not meant as a meta-material reference to the game's virtual lens, but is instead meant to be read as if the splatter were landing on the avatar's diegetic goggles or helmet visor. The illusion of complete immersion into the game and of the co-embodiment of the avatar's body by the player and the game character thus remain intact. Furthermore, the device enhances the player's immersion by providing sensorial signifiers regarding the effects of the game-play violence on the body of the virtual avatar. As Håvard Vibeto explains in "The Spectacular Design of First-Person Shooters: Remediating Cinematic Spectacle in *Call of Duty: Advanced Warfare* and *Battlefield 4*" (2019), such devices "adapt spectacle to suit and support [the] interactive nature" of game play.[26] Indeed, by invoking an "impact aesthetics," these games work to use visual clues for the player that provide information cues about the state of the avatar that they embody. Vibeto details:

> The blood-soaked first person view in *CODAW* and *BF4* notifies the player that the avatar is nearly dead and must take cover. Whereas blood splatter on the camera is a well-known convention in action films, in these games, blood splatter also functions as a life-meter. In *CODAW*, when shot, the player can hear the avatar's heavy breathing and the heartbeat growing louder, muzzling out all other sounds. The audiovisual assault thus not only intensifies the perception of peril, but also indicates how to escape the situation and what is happening in terms of gameplay.[27]

Therefore, these devices are not merely aesthetic or meta-cinematic, or even spatially immersive, but also narratively immersive, in that they provide sensory signs to the player that reveal information about the state of their avatar in the game world. In this light, the surface that gets spattered with blood is transformed from a virtual camera lens into a user interface; as a picture plane, it serves as a point of articulation between the player's body and the character's body, and between the game world and our own, in order to influence subsequent game play.

26 Vibeto, "The Spectacular Design of First-Person Shooters," p. 16.
27 Vibeto, "The Spectacular Design of First-Person Shooters," p. 24.

A Last Frontier

As cinema becomes increasingly dematerialized and distanced from the photographic processes of image capture, the picture plane stands as a sort of last frontier. Not only does it stage the spatial and material relations discussed in this chapter, but it also represents the enactment of relations between previous modes of cinema and contemporary ones. By directly invoking cinema's material evolution diegetically and interdiegetically, the picture plane operates as an interface between cinematic technologies and the forms that they produce. The porosity of the picture plane, which oscillates between rendering visible and invisible, between presence and absence, negotiates between the familiar codes established by the camera in traditional cinema and the new possibilities for experimentation opened up by the potentials of virtual production, allowing these to co-exist within a single work.

As contemporary films elaborate new means of immersion and new spatialities, the picture plane provides us with a discourse for situating recent developments within the singular continuum of film and media history. In our current age of technological shifts, it is increasingly complicated to speak of "cameras" or "lenses"; discussions of screen phenomena are complicated by shifts from traditional projection screens to the screens of our mobile devices to screenless immersive images; theorizations of profilmic space are complexified by the fact that there is often no actual space *in front of the camera* nor any camera. However, the polyvalent notion of the picture plane, which speaks to the issues and effects of each of these, can be excavated from art history's past to provide us with a framework for addressing the media of today and tomorrow. In the creation and spectatorship of images which negotiate between two-dimensionality and three-dimensionality—from painting to cinema to video games and VR/AR—the picture plane acknowledges the codeterminacy and cooperation between the technologies which transcribe perspective, the aesthetic forms they produce, and the optical and affective perceptions they solicit.

Bibliography

Abbott, Stacey. *Celluloid Vampires: Life after Death in the Modern World.* Austin, TX: University of Texas Press, 2007.

Alberti, Leon Battista. *On Painting.* New York, NY: Cambridge University Press, 2011.

Augello, Nick, Michael Juarez, Kelly Shay, and Arunachalam Somasundaram. "Hair Effects in *Trolls World Tour.*" In *DigiPro '20. The Digital Production Symposium.*

New York, NY: Association for Computing Machinery, August 2020. https://doi.org/10.1145/3403736.3403940.

Bédard, Philippe. "Le cinéma sans caméra ? Repenser les usages de la 'caméra' à l'ère du cinéma virtuel." *Écrans* 1, no. 13 (2020): 207–223. https://doi.org/10.15122/isbn.978-2-406-10932-7.p.0207.

Bédard, Philippe. "Virtual Production and the Transformation of Cameras Mechanical, Virtual, and Actual." *Animation* 17, no. 2 (July 2022): 226–243. https://doi.org/10.1177/17468477221102498.

Bredow, Rob, David Schaub, Daniel Kramer, Matthew Hausman, Danny Dimian, and R. Stirling Duguid. "*Surf's Up*: The Making of an Animated Documentary." In *ACM SIGGRAPH 2007 Courses (SIGGRAPH '07)*, 1–123. New York, NY: Association for Computing Machinery, 2007.

Brown, William. "Digital Cinema's Conquest of Space." In *Supercinema: Film-Philosophy for the Digital Age*, edited by William Brown, 21–50. Oxford-New York, NY: Berghahn Books, 2013.

Conley, Tom. *Cartographic Cinema*. Minneapolis, MN: University of Minnesota Press, 2007.

Crockett, Tobey. "The 'Camera as Camera.' How CGI Changes the World as We Know It." In *Cinephilia in the Age of Digital Reproduction*, edited by Scott Balcerzak and Jason Sperb, 117–139. London: Wallflower Press, 2009.

Deleuze, Gilles. *Bergsonism*. New York, NY: Zone Books, 1991.

DisneyFan18111928. "*A Bug's Life* Outtakes", *Pixar.fandom.com*. Video, 4:56. https://pixar.fandom.com/wiki/Outtakes?file=A_Bug%27s_Life_Outtakes#A_Bug's_Life_(1998).

Fordham, Joe. "*Children of Men*: The Human Project." *Cinefex* 110 (July 2007): 3–44.

Frank, Hannah. *Frame by Frame: A Materialist Aesthetics of Animated Cartoons*. Oakland, CA: University of California Press, 2019.

Gaudin, Antoine. *L'espace cinématographique. Esthétique et dramaturgie*. Malakoff: Armand Collin, 2015.

Gray, Tim. *@iamtimgray*. https://www.instagram.com/iamtimgray/.

Gray, Tim. "Projects." *I Am Tim Gray*. http://iamtimgray.co.uk.

Gunning, Tom. "Re-Newing Old Technologies: Astonishment, Second Nature, and the Uncanny in Technology from the Previous Turn-of-the-Century." In *Rethinking Media Change: The Aesthetics of Transition*, edited by David Thorburn and Henry Jenkins, 39–60. Cambridge, MA: The MIT Press, 2003.

Ivins, William M. Jr. "On the Rationalization of Sight: With an Examination of Three Renaissance Texts on Perspective." *The Metropolitan Museum of Art Papers*, no. 8 (1938): 7–44.

Jones, Mike. "Vanishing Point: Spatial Composition and the Virtual Camera." *Animation* 2, no. 3 (2007): 225–243.

Kuan-Ting, Lu. "Organizing the Fuzziest Concerts for *Trolls World Tour*." In *ACM SIGGRAPH 2020 Talks (SIGGRAPH '20)*. New York, NY: Association for Computing Machinery, 2020. https://doi.org/10.1145/3388767.3407365.
Laner, Barbara. "The Vampiric Film: Intermedial Incorporation in *Nosferatu, A Symphony of Horror* (1922)." *Post Script: Essays in Film and the Humanities* 32, no. 1 (2012): 28–38.
Lindahl Elliot, Neils. *Mediating Nature*. London: Routledge, 2006.
Massumi, Brian. *Parables for the Virtual: Movement, Affect, Sensation*. Durham, NC: Duke University Press, 2002.
"The Minions Announce *Despicable Me 2*." *IMDb*. Video, 00:53. https://www.imdb.com/video/vi3244728601/.
Movieclips. "Penguins of Madagascar (2014). Mutant Penguins Scene (7/10) | Movieclips." May 8, 2019. YouTube video, 3:37. https://www.youtube.com/watch?v=pAwdeWy9yYM.
Richmond, Scott C. *Cinema's Bodily Illusions: Flying, Floating, and Hallucinating*. Minneapolis, MN: University of Minnesota Press, 2016.
Riesman, Abraham. "The Vulture Transcript: Alfonso Cuarón on *Children of Men*." *Vulture*, January 6, 2017. https://www.vulture.com/2017/01/alfonso-cuaron-children-of-men-transcript.html.
Ross, Miriam. "Stereoscopic Visuality: Where Is the Screen, Where Is the Film?" *Convergence. The International Journal of Research into New Media Technologies* 19, no. 4 (July 22, 2013): 406–414. https://doi.org/10.1177/1354856513494178.
Vibeto, Håvard. "The Spectacular Design of First-Person Shooters: Remediating Cinematic Spectacle in *Call of Duty: Advanced Warfare* and *Battlefield 4*." In *Intermedia Games—Games Inter Media: Video Games and Intermediality*, edited by Michael Fuchs and Jeff Thoss, 15–36. New York, NY: Bloomsbury Academic, 2019.
Whissel, Kristen. "Parallax Effects: Epistemology, Affect and Digital 3D Cinema." *Journal of Visual Culture* 15, no. 2 (2016): 233–249.

About the Author

C. E. Harris is a maître de conférences (assistant professor) in visual studies, media studies and English at Université Jean Moulin Lyon 3 (Lyon, France) and a member of the Institut d'Études Transtextuelles et Transculturelles (UR 4186). The author of several articles and book chapters on the history of film forms, particularly with regard to their evolution into the digital age, and co-editor of Cinématérialismes : Nouvelles approches matérialistes de l'audiovisuel (Éditions Mimésis, 2024), Harris is interested in film philosophy, media archeology, aesthetics, and formalism in the digital age.

12. "Let's Cross This Portal Together": Immersion Mechanisms in VR Performances and Their Current Limitations

Ágnes Karolina Bakk

Abstract: This chapter analyzes three theatrical VR productions, namely *Tempest* (2020) by Tender Claws, *Finding Pandora X* (2020) by Double Eye Studio, and *Welcome to Respite!* (2021) by CoAct Productions and Ferryman Collective. After offering an overview of various design suggestions for contemporary theatrical VR productions, the chapter will show that this type of performance, which has the capacity to bring remote audiences together, can offer new possibilities for co-presence and a sense of liveness, even though these productions still face technological barriers. This can be achieved with the help of the particular mechanics of Live Action Role-Playing (LARP) and immersive theatre. The article concludes by pointing out the limitations of virtual immersive theatre compared to analogue immersive theatre performances and offering possible design suggestions to enhance the immersive experience.

Keywords: theatre, experience design, VRChat

Introduction

Since 2020, COVID-19 lockdowns created new challenges but also new and imaginative mediated theatrical possibilities for performance makers. Audiences had the chance to experience theatrical events that involved liveness and co-presence in many forms, especially in video conferencing and other web-based forms. Meanwhile, new and affordable stand-alone Virtual Reality (VR) headsets appeared on the market.

Bédard, Philippe, Alanna Thain, and Carl Therrien, eds. *States of Immersion Across Media: Bodies, Techniques, Practices*. Taylor & Francis Group, 2025
DOI 10.5117/9789048558766_CH12

In this chapter, I will offer an analysis of these new immersive strategies. I will overview various design suggestions for contemporary theatrical VR productions and present three immersive theatrical VR productions as case studies, namely *Tempest* (2020), by Tender Claws, *Finding Pandora X* (2020), by Double Eye Studio, and *Welcome to Respite!* (2021), by CoAct Productions and Ferryman Collective. I will show that this type of performance, which has the capacity to bring remote audiences together, can offer new possibilities for co-presence and the sense of liveness, even though these productions are still fighting the barriers of technology. This can be achieved with the help of the particular mechanics of Live Action Role-Playing (LARP) and immersive theatre. I will conclude by pointing out the limitations of virtual immersive theatre compared to analogue immersive theatre performances.

Immersive Theatre

Despite the term "immersive" being associated most often today with VR and its associated technologies, "immersive theatre" has existed as a unique genre for many years. And while immersive theatre in its traditional form has nothing to do with VR, reviewing its characteristics will prove useful for this study of contemporary VR performances. For one, while it shares many elements with traditional theatre, the structure of immersive theatre focuses more on the audience and on how the theatrical situation can enable them to become participants and have a sense of agency. According to Josephine Machon, "[i]mmersive theatre is discernible as that practice which actually allows you to be in 'the playing area' with the performers, physically interacting with them. [...] The direct participation of the audience member in the work ensures [they] inhabi[t] the immersive world created."[1] Based on this definition of immersive theatre and on definitions offered by others,[2] I have elsewhere offered an overview of the most important specifics of the genre.[3] These specifics are the following:

- a totally surrounding, physical, and explorable environment
- a multisensorial experience;
- a story world enabling unforeseen events to be embedded into the storyline if needed;
- no fixed text for the performers;

1 Machon, *Immersive Theatres*, p. 68.
2 See Biggin, *Immersive Theatre and Audience Experience*.
3 Bakk, "Epiphany through Kinaesthetics," p. 215.

- a sense of agency for the audience members (which is mostly illusory, as they cannot influence the unfolding of the events in a substantive way);
- and an intimacy that arises from the improvisations of the actors as well as from the atmosphere creating effects in the set design.

An important element of such performances is the onboarding procedure, in which the audience members can be acquainted with the internal rules of the performance and step into its "magic circle." The magic circle is a concept borrowed from Johan Huizinga's game theory that is also used in LARPing: in the latter, the magic circle is the protocol according to which the LARP game starts or ends (in a physical and temporal sense) and, within this protocol, there are various rules pre-established by the given game's framework.[4] This magic circle helps the audience members acquire an understanding of the limits and possibilities of this world and of the community composed of performers and other audience members.

Although the origin of immersive performances can be traced back to the mid-twentieth century,[5] the use of the adjective "immersive" in the context of performance began with the rise of the UK-based company, Punchdrunk. The company creates site-specific performances where the audience members can roam freely in a given space and attend scenes as they want. This free-roaming aspect is an important feature in immersive performances as participants are allowed to discover the story world on their own. In Punchdrunk's performances, audience members usually wear a mask in order to be distinguished from the performers. Their most famous performance is *Sleep No More* (2011–present), which is based on William Shakespeare's *Macbeth.* The Danish-Austrian company SIGNA also creates free-roaming site-specific experiences: these are not the adaptation of any existing fictional works, but rather newly built story worlds in which the audience can also roam around freely, initiate spontaneous dialogues with the performers, and even make a moral choice that can influence their experience.

Early Theatrical VR Productions and Their Characteristics

Since the 1990s, various theatre labs and creators have been experimenting with the relationship between VR and theatre. One of the most prominent

4 Huizinga, *Homo Ludens*, p. 10.

5 See Bishop, *Artificial Hells*, p. 94.

pioneers in this field is Brenda Laurel, the author of *Computers as Theatre*. In her VR production *Placeholder* (1992), Laurel puts her ideas about VR in practice by reinventing "the sacred spaces where we collaborate with reality" in order "to transform it and ourselves."[6] In this performance, the experiencer arrives at a place, puts on the head-mounted display (HMD) and sensors with the help of an assistant, and enters a space where they embody various characters (Crow, Snake, Spider, and Fish). Inside the production, they are guided by the Goddess, a live-performed character usually played by Laurel herself. This offers a highly personalized experience with a clear specification of the role of the experiencer: in this performance, the participant can also enter the magic circle, where they encounter a protocol that presents them with procedures that are different from everyday life. Although it lacks many elements one might traditionally associate with theatre (a stage, clear distinction between audience and performers, etc.), I consider this experience as theatrical, as the participant is guided by an orchestrator that embodies another role—the role of the Goddess. In order to unfold the experience, both the performer and the experiencer have to collaborate, being in the roles pre-defined by the designers. As we shall see later in this chapter, the fact that participants and performers are co-present (both physically and virtually) in *Placeholder* has an impact on the theatricality of the experience.

Another VR immersive production that provides a strong sense of theatricality by way of the embodiment it affords its users is *Osmose* (1995), by Char Davies. This production is a single user experience, where the breathing of the "immersant" (as Davies calls the participant) is monitored, the environment changing accordingly. Here, the participant has an experience where, even if they are a single user, they can interact with an environment that offers them an immediate response. It is also important to note that the immersant's experience is monitored by the orchestrator, as they guide the viewers throughout the whole experience and take care that everything happens as it is planned. Although this can be considered theatrical only to a limited extent, for our purposes it serves as a suitable example of a VR experience that requires the live presence of an orchestrator and participant, and that can only be achieved by relying on the latter's active attitude.

After the mid-1990s, the production of artistic VR works started to decrease.[7] This was mainly because the gap between the formal possibilities for digital visualization and the users' expectations was too big. After 2010, the

6 Laurel, *Computers as Theatre*, p. 197.

7 On this topic, see also Philippe Bédard's chapter in this volume.

industry experienced a new *élan*: the long-awaited Oculus Quest 2 headset, which launched in autumn 2020, brought the stand-alone experience at a more affordable price for users, thereby revitalizing the industry's dream of mass-market VR. If we compare the productions from the early 1990s to those that were made in the last two years, mainly during the pandemic, it is clear that beyond the purely technical differences, more recent performances leverage social VR platforms to organize performances which forego the need to be present in a physical setting, allowing users to experience immersive theatre performances remotely. In the performances offered on social VR platforms, audience members are presented with a coherent world in a digital environment that appears to be explorable. Digital avatars and their capacity to interact also seem to be features that the participants want to investigate as much as possible, although they usually realize quite rapidly that their avatar's interaction possibilities are quite narrow. Due to these technical limitations, I will show in the latter half of this chapter that the agency of the audience members is rather illusory.

By taking place over the Internet, these performances also tend to foster feelings of community and togetherness, namely by providing a shared sense of presence and a central role to the audience. This sensation is explored by recent theatrical VR performances in order to find new meanings for theatricality in virtual realms.

Telepresence, Agency, Social Interaction

Thomas Sheridan defines the phenomenon of "telepresence" as "feeling like you are actually 'there' at the remote site of operation."[8] He also defines virtual presence as a state in which one feels present due to multisensory stimuli of the computer-generated environment.[9] As we can see, these definitions relate to both interpretations of the term "immersion," combining an understanding of immersion as, on the one hand, a property of a technological device and, on the other, the participant's perceptual response to the technological device.[10] The limitations of the device and of the platform play a big role in determining the participants' behaviour and their perceptual answers. The possibilities offered by the device and the platform—that is, what the participants are enabled to do—define their

8 Sheridan, "Musing on Telepresence and Virtual Presence," p. 120.
9 Sheridan, "Musing on Telepresence and Virtual Presence," p. 120.
10 On this, see Nilsson, Nordhall, and Serafin, "Immersion Revisited."

"agency." Importantly, differences between the medium of video games, virtual reality, and theatre leads to a new breed of agency that, in the case of contemporary VR performances, may be seen as an illusion.

The question of agency is highly debated, especially in games studies. Marcus Schulzke identifies three approaches to agency. First of all is the determinist approach, which suggests that "players are largely receivers of information from games."[11] Secondly, the voluntarist approach suggests that players have a detached position from which they can make observations as agents. According to this approach, the games are a "realm of self-expression in which players are free to create," and, without this aspect, the game experience would not have any influence on them.[12] Thirdly, Schulzke suggests a more centrist approach that reconciles the first two: one should recognize that players are autonomous and "more capable of judging their game experiences than determinists assume, while also seeing the way autonomous agents are always constrained by their circumstances."[13] This can serve as a reconciling description for the term "agency," but it is also worth it to analyze it from another discipline's perspective.

Theresa Jean Tanenbaum and Karen Tanenbaum, who rely on theatrical models when examining agency, propose to shift the concept of the "reader" (or what they call the "reader/player") from a player-centric model to a performer-centric model.[14] In their understanding, interactions between the player and the designer are governed by some of the same rules that are in play between multiple performers in a piece of improvisational theatre. Their notion of "performer" includes the expectation "that the interactor assume[s] a role, engage[s] in the narrative as a character, and act[s] in what might be described as a narratively salient way."[15] Meanwhile, the interactor, as a player, might behave in a self-gratifying, pleasure-centric manner, "subverting the story in his own quest to satisfy his desire for Agency."[16] This means that interactors are less interested in what they can do *with* the game world and more in what they can do *to* the game world. While interactors usually oscillate between the player and performer

11 Schulzke, "Models of Agency in Game Studies."
12 Schulzke, "Models of Agency in Game Studies."
13 Schulzke, "Models of Agency in Game Studies."
14 Tanenbaum and Tanenbaum, "Improvisation and Performance as Models for Interacting with Stories."
15 Tanenbaum and Tanenbaum, "Improvisation and Performance as Models for Interacting with Stories," p. 251.
16 Tanenbaum and Tanenbaum, "Improvisation and Performance as Models for Interacting with Stories," p. 251.

forms, video game designers aim to create primarily for the player type of interactors. Tanenbaum and Tanenbaum suggest that "[i]f we can conceive of the relationship between the author and the interactor as one of equal participants in an improvisation, then the issue of interactor *agency* becomes one of performer *responsibilities*."[17] They rely on a "typology of performative knowledge" developed by Lesa Lockford and Ronald J. Pelias to describe five different types of knowledge on which live performers who are engaged in improvisational situations can rely: communication, playfulness, sedimentation, sensuality, and vulnerability.[18] This knowledge arises from the body and the mind and is a process that can be considered both epistemic and aesthetic. The authors call this process "bodily poeticizing" because intuitive, affective, and cognitive types of knowledge are intertwined with somatic knowledge. Some of the aforementioned types of knowledge are harder to achieve when playing video games (e.g., sensuality) and they require responsibility from both the system and the interactor. In analogue theatrical improvisational situations, there is a social contract between the actors that encompasses a commitment. A similar contract must also emerge in the case of interactive digital narratives. The interactor should be treated as a performer and it is important to "recognize that performing in any context is a specialized skill that is learned over time, and not an intuitive ability that every interactor may draw upon."[19] As it is not possible to expect the interactors to have performance skills, designers of VR performances must explore techniques for training them in the specifics of the narrative. The schema of the given production should be clearly communicated and the interactors should not only be aware of what interactions are possible, but also of how to navigate the virtual space. Tanenbaum and Tanenbaum also state that "[e]ach participant has a responsibility to accept the dramatic offers given by the other, and each has equal responsibility to the ultimate meaning of the narrative. In order for this performative dialogue to work, the author must design the system to be able to co-perform with the interactor."[20] According to them, designing interactive digital narratives that feature a performer type of agency requires two features: firstly, "design systems

17 Tanenbaum and Tanenbaum, "Improvisation and Performance as Models for Interacting with Stories," p. 261, original emphasis.

18 Lockford and Pelias, "Bodily Poeticizing in Theatrical Improvisation," p. 432.

19 Tanenbaum and Tanenbaum, "Improvisation and Performance as Models for Interacting with Stories," p. 260.

20 The authors do, however, recognize that there were fewer interactive narratives that could adopt such designs in 2008. Tanenbaum and Tanenbaum, "Improvisation and Performance as Models for Interacting with Stories," p. 261.

that can train and guide interactors in how to perform within them," and secondly, "reframing *agency* as a shared property of all participants in an interactive drama."[21]

As Tanenbaum and Tanenbaum show in their theatrical approach to immersion, social interactions play a crucial role. Indeed, the social aspect of productions must not be overlooked when defining telepresence or virtual presence, both of which play a crucial role in the VR performances analyzed in the latter half of this chapter. Matthew Lombard and Theresa Ditton's widely quoted study from 1997 is an exceptional case in which the social element of presence is discussed, among other things.[22] Two of the six elements that they define as key elements for presence in a virtual environment are related to social aspects: the first is about how the user can be a social actor within the medium, which is defined by the user's ability to control or interact with their environment; the second is about the medium itself being a social actor, meaning that the digital environment's provides its users with a feeling of interaction. However, these are not the only aspects necessary for the sense of togetherness or community in a multiplayer platform such as VRChat, nor are they sufficient for strengthening immersion. In a 2016 study, Sebastian Xavier Samur compares stage presence to presence in VR. He states that:

> Open world environments, found in *Second Life* [...] create the promise of an experience that is unique to them, reinforcing the feeling of presence as the narratives the users create are tailored to their choices. Interactivity can heighten an audience's sense of fictional presence as they realize what they are currently experiencing is the result of past choices made.[23]

Despite the obvious differences between *Second Life* and current social VR platforms such as VRChat, Samur's argument as to the role played by interactivity in fostering a user's sense of presence—we could also add immersion—merits further attention.

In his seminal work *The Digital Performance*, Steven Dixon distinguishes four types of interaction: (1) navigation; (2) participation; (3) conversation;

21 Note that this feature is about the systemic level, rather than Janet Murray's "scripting the interactor," which is more about the narrative level. Tanenbaum and Tanenbaum, "Improvisation and Performance as Models for Interacting with Stories," p. 262, original emphasis.

22 See Lombard and Ditton, "At the Heart of It All."

23 Samur, "Comparing Stage Presence and Virtual Reality Presence."

and (4) collaboration, the last of which is particularly important to note. Dixon defines interactive collaboration as follows:

> Interactive collaboration comes about when the interactor becomes a major author or coauthor of the artwork, experience, performance, or narrative. The collaboration may be between a single user and the computer/virtual environment, but more usually occurs when users work together with others to create new work by means of computer technologies or within a virtual environment.[24]

This type of interactive collaboration is an important innovation and is characteristic of the theatrical VR productions that I analyze in this chapter. Combined with new types of interaction taxonomies (that are defined specifically for VR theatrical performances), this type of interactivity can help us find new ways of discussing these performances. Finally, interactions can also heighten the sense of telepresence with the help of a well-thought-out design of immersive virtual environments.

An Overview of Design Suggestions for Contemporary Theatrical VR Productions

Jason Ferguson suggests that "virtual reality has a storytelling problem and theatre will save it."[25] He takes it that this problem is due to VR's first-person delivery mode, which makes it harder for the audiences to perceive the story as such when they are involved in it themselves.[26] Although this cannot be denied, I suggest that theatrical VR performances organized in multiplayer settings can enhance the storytelling effect with the help of interaction and agency. I have already shown how early VR works could be considered theatrical because of their installation-like characteristics and the strong sense of embodiment that their users experienced (even though these traits are not unique to theatre). We could also point to their characteristic liveness, experienced in a digital setting and manifested as the performer and the spectators actively taking up a specific role. It is also noteworthy that these works are strongly temporal, a fact that is emphasized by the fragility of the technological apparatus. Notably, the liveness of these experiences is what

24 Dixon, *The Digital Performance*, p. 595.
25 Ferguson, "Virtual Reality Has a Storytelling Problem and Theatre Will Save It."
26 Ferguson, "Virtual Reality Has a Storytelling Problem and Theatre Will Save It."

makes them more than mere installations; they are performances. More importantly, since the recent technological advancements, theatrical VR performances make it possible to bring audiences into the same space, often allowing them to interact with the environment, with each other, and with the performers. Indeed, Sita Popat states about VR environments that they "allow us to ask questions about embodiment and humanity through the experiences of our individual bodies in a way that has never been possible before."[27]

Recent scholarship on theatrical VR performances offers yet more recommendations for designing better experiences. In 2020, Saumya Gupta et al. emphasized that many productions that are tagged as theatrical VR do not specify the role of the audience.[28] Among their suggestions, the authors highlight the importance of managing the audience's expectation by offering them explicit roles. They also recommend getting the audience familiarized with the system through what Theresa Jean Tanenbaum calls "situated rehearsals,"[29] which are meant for the participants to try out the interactive situation in a low-risk environment. Another important suggestion by Gupta et al. is that creators should "invest time in 'high-yield' interactional and visual details that support the *illusion* of a social reality for the player within the scope of the *specific desired interactions called for by the script*."[30] By this, they mean that it is important to design theatrical VR performances in such a way that the participants can experience a co-presence with the help of built-in interactions that allow them to co-act with other audience members. This would be necessary both on the level of narrative framing and on the level of interactions that rely on the affordances of the digital avatars.

Based on their own experiments, David Gochfeld, Kiira Benzing, Kevin Laibson, and Alex Coulombe voice some doubts about the importance of participants identifying with their avatar in such performances. Analyzing theatrical VR performances from the actor's perspective, they argue that "for performance, embodiment is not as important as the ability of the actors to convey the characters' behaviors and emotional states: the actors don't need to feel that the avatar is their own body in order to be able to

27 Popat, "Missing in Action," p. 359.

28 Gupta et al., "Investigating Roleplaying and Identity Transformation in a Virtual Reality Narrative Experience."

29 Tanenbaum, "Identity Transformation and Agency in Digital Narratives and Story Based Games," p. 228.

30 Gupta et al., "Investigating Roleplaying and Identity Transformation in a Virtual Reality Narrative Experience," p. 9, original emphasis.

perform as if it is."[31] The authors suggest that performing in VR is similar to operating a puppet, as the performers must master the control system of a given platform to successfully project their expression through their avatar.[32] It is also important to note that actors with more realistic avatars tend to feel they are in a more rigid context then those with less realistic avatars, while on the audience's side, more realistic avatars create higher expectations vis-à-vis the performers. This constitutes an interesting design guideline that can lead to future research areas.

Another paper, authored in 2021 by Shuo Yan et al., offers a specific taxonomy for interactions in VR theatrical settings: (1) individual-based interaction includes mainly linguistic interaction and physical contact; (2) scenario-based interaction "enables [the] audience to play directly with virtual stage props under the guidance of the performer"; and (3) narrative-based interaction can occur if there are possibilities to make meaningful choices that influence the story, which is possible when there can be multiple storylines in the production.[33] The authors also stress the importance of identification and role-play for the players. Taking the affordances of the medium of VR into consideration, the authors suggest that the freedom of the audience members to move and to explore the VR space is very important. They also suggest that creators should offer a "rich sensual experience as far as possible to let the audience feel that they are focused [on] by the performer all the time."[34]

Case Studies

Altogether, the three design recommendations reviewed in this previous section serve as an important starting point for analyzing VR theatrical performances. In what follows, I present three such productions and analyze them from the perspective of their onboarding process, the role they give their audiences, and the kinds of interactions they offer. Based on these analyses, I will conclude by suggesting a new definition for these theatrical performances.

1. Tender Claws: The Tempest

The Tempest, by Tender Claws, takes place on a bespoke application, Tender Claws' immersive theatre platform *The Under Presents*. The theatre platform

31 Gochfeld et al., "Avatar Selection for Live Performance in Virtual Reality," p. 4.
32 Gochfeld et al., "Avatar Selection for Live Performance in Virtual Reality," p. 4.
33 Yan et al., "Performing with Me," p. 3.
34 Yan et al., "Performing with Me," p. 5.

itself was launched in 2019 for Oculus Quest and offers a variety of artistic-oriented video games in which players can explore fantasy worlds. The participants cannot communicate verbally: they can only use pantomime and finger snapping. The special feature of the platform is that actors also participate in it (either in a prerecorded or live way), a feature that permits them to enhance the experience of the audience members by various means, such as guiding them.

The Tempest is a live, semi-scripted participatory VR performance, in which a small number of audience members are guided through the linear story of a reimagined version of Shakespeare's play *Macbeth*. The participants can pre-purchase their ticket at a ticket counter in the app. Showing up at the time indicated on their ticket, they find themselves in a theatre lobby, which they can explore freely. As I have mentioned, the audience cannot talk and everyone's avatar looks exactly the same.[35] Participants can only communicate by pantomime, by snapping their fingers, or by taking off their mask. Because of this, body language is very important for the audience members as well as for the guiding actor, whose role is to help the audience go through all the scenes and to offer their narrative background.

Throughout the performance, a sole actor takes on all the speaking roles: they sometimes talk as Ariel, and sometimes as Prospero. Later, the audience is transitioned to Prospero's lair, which is a three-dimensional memory palace, where they can interact with the actor by bringing them various assets that trigger anecdotes. These anecdotes help them figure out why Prospero and his daughter Miranda ended up there. Afterwards, everyone is teleported to the island where Prospero and Miranda are exiled, a scene where the audience members find themselves sitting at a feasting table. Then, a harpy shows up from above, which is also puppeted by the actor. The feasting table's assets are interactive: the audience can throw food at the harpy or show it a long sword. In the last scene, the guiding actor directs a mute LARP scene with the audience members: two of them receive Miranda and Ferdinand's costume and have to get married, while the others become spirits who have to act joyfully. By way of an epilogue, everyone is teleported back to the fireplace: the actor encourages the audience to dance around it, and, after several light and sound effects, the audience is teleported back to the theatre lobby, where each participant can have a solitary offboarding.

In this performance, audience members can manifest their agency through communicative and physical interactions: they can snap their

35 This was the case when I participated in the performance. In other instances, participants might have different avatars based on what they unlocked on the *Under Presents* platform.

fingers with the controllers or move their body freely to communicate, and they can move in the space and interact with various assets by holding them, throwing them, etc. Both of these interaction types are individual-based and scenario-based, to use the nomenclature proposed by Yan et al. None of them are necessary to unfold the narrative of the performance, but both offer the illusion of agency and of active contribution. These are not interactions that nurture collaboration among the audience members, but rather ones that give them a sense of presence and of bodily awareness in the world of the performance.

The audience members are onboarded through a traditional theatrical format: they are spawned into the lobby of a theatre and, from there, they automatically arrive at the site where the guiding actor awaits them. The actor helps the audience to understand that they are active viewers and also tells them each time what actions they should perform (e.g., illuminate the scene with the torch, perform a wedding, etc.).

2. *Double Eye Studio:* Finding Pandora X

Finding Pandora X, created by Double Eye Studio and directed by Kiira Benzing, takes place on the VRChat platform. It offers its audience members—who play the members of an ancient Greek choir—a high level of interactivity: they can verbally interact with each other and are all allowed to leave the scene independently to explore the space of the performance. After a short technical onboarding, a procedure presented by Hermes that helps them understand how to move in VRChat and where they can change their avatars, the audience members find themselves on Mount Olympus. Their goal is to help Zeus and Hera retrieve the box of hope from Pandora. The participants split into two groups: one group follows Hera into a utopian-looking futuristic city and tries to help her find the box by solving easy, gamified tasks (such as figuring out a word), while the other group follows Zeus to the base of Mount Olympus. There, the second group can have drinks with him and smash various objects. In both cases, the participants enter the spaces via a portal that heightens the magical effect of the performance. The performance ends with a party, where the performers are introduced and where the audience members can interact with them (by having "drinks") as an offboarding.

At the beginning of the performance, the audience members are spawned into an "antechamber," where the creators can facilitate and properly organize the onboarding process. There, the participants learn about how VRChat functions: they are taught the movements that can be performed and how to handle the technological aspects of the performance (such as changing their avatars or going through a portal). That way, they can smoothly contribute to the unfolding of the narrative. The audience members can perform the following actions:

- various physical actions (such as jumping and running);
- simple interactions with objects (such as picking up objects, smashing them, or looking for the box);
- joint interactions (such as solving puzzle elements in the futuristic city scene, which can only be done with the help of another audience member);
- and verbal interaction with the performers (dialogue).

Even though the audience members can perform several types of interaction (mainly individual-based and scenario-based ones), all of them only involve local agency. In this performance, the audience members have to rely on collaborative types of interactions as well, especially when solving puzzles. They are made to understand that, in order to find Pandora, they have to do everything that the performers encourage them to. This constitutes a big challenge, especially for the performers, as they have to rely heavily on improvisational skills. They have to engage in one-to-one conversations in order to help and guide the audience members, which also creates a certain type of intimacy. The pretext of the joint mission is what helps the participants to clearly identify their role and to have a sense of temporary audience community, as everyone's aim, which is scripted in the performance, is the same.

3. CoAct Productions and Ferryman Collective: **Welcome to Respite!**

The theatrical VR performance *Welcome to Respite!* premiered in 2021. It was developed by CoAct Productions and Ferryman Collective. This production also takes place on the VRChat platform. It is the first part of a four-chapter production, which deals with representing mental disorders, such as dissociative identity disorder. The audience members, when stepping into VRChat, are greeted in a space filled with childhood memory objects, most of which are oversized compared to the size of their avatar. The actor who greets the audience offers them a long onboarding: the participants can interact with the actor, sing songs together, and receive a technical onboarding. The actor engages in one-to-one conversations with the participants, which creates a sense of trust and intimacy between them that will be needed throughout the performance. The audience members are then separated, as one member will be the protagonist of the performance and the others will receive extra technical guidance. The rest of the participants become invisible. In this invisible form, they can navigate in the space at a very fast pace, causing many audience members to feel nauseous during the experience.

The invisible audience members arrive at the terrace of a house where they meet Alex, the protagonist (who is played by the participant who was separated from the group at the beginning of the performance). Alex is a young boy who finds himself waiting for his parents. His mother, and later his father, arrive home. While there is no major narrative arc, the elements of the background story—e.g., small comments about the father's drinking habits and the son's possible mental health issues—create suspense. There are also two interactive moments when the invisible audience members can manifest their agency: they have to try to save Alex twice from amorphous evil forces by pushing their controller's buttons. These forces are represented as colourful abstract forms in order for the audience not to attach any concrete meaning to these representations. After the performance, the audience members can have an offboarding in another VRChat room, where further information about mental disorders as well as the production is given.

The invisible audience members have very limited interaction possibilities, mainly individual- and scenario-based ones. The participant who plays the role of the protagonist, on the contrary, can interact with the performers. They also decide on the narrative-based interactions, which means that they can interact with the other protagonists while remaining within the borders of the scripted narrative. That way, they can have an illusion of agency, the illusion that it is possible for them to change the narrative path of the performance.

This performance requires a highly concentrated improvisational method on behalf of the performers in order to keep the participants in the narrative's pathway. It also creates a sense of great intimacy for the other audience members, who can enjoy the narrative through a more voyeur-like participation that encourages them to pursue the joint mission of saving the protagonist.

Towards an Immersive Virtual Theatre

Taken as key representatives of contemporary VR theatre, the three performances discussed above each demonstrate how different onboarding procedures, types of interactions, and roles given to audience members can lead to vastly different results. In this final section, I offer a new definition for immersive virtual theatre by highlighting some of the key issues and challenges that remain. To begin with, it is important to discuss the technological possibilities allowed by the platforms used for each of these experiences. While *The Tempest* was developed on a game engine and while it is periodically reachable from the bespoke application *The Under Presents*, the other two productions were developed for the VRChat platform, one of the freely reachable social VR platforms on which users can upload personalized worlds. All the productions presented are reachable on game-heavy platforms, as both *The Under Presents* and VRChat are available on Meta Quest headsets. Notably, these platforms have embedded design features that influence the performances (and the choices that the creators had to make):

Lack of moderation. While in *The Tempest* the participants' microphones are automatically set to mute, the other two performances are set in VRChat. In these performances, the audience members are therefore either asked to mute their microphone (as in *Welcome to Respite!*) or they can keep their microphone on (as in *Finding Pandora X*). The creators of the latter production took the advantage (and the risk) of allowing participants to keep their microphone on in order to allow verbal interactions between them. While this can result in a co-performance where the participants and the performers co-create the production (the importance of which, as we have seen, was also underlined by Tanenbaum and Tanenbaum) and in this way deepen their sense of agency, it can also have repercussions. Audience members, for instance, are sometimes able to challenge the unfolding of the events. The lack of unified moderation systems in these platforms is still waiting to be addressed. It presents new creative challenges for small creative teams, who, for the time being, rely on acting tools such as improvisation in order to unfold the flow of the production, but these occasions are rather rare.[36]

Various ways of onboarding. All three performances use elements from community rituals, such as onboarding procedures found in immersive theatre and in LARP, in order to define the audience members' roles in the performance. In all of the productions observed here, the performers or dedicated orchestrators help the audience members to learn about the possibilities of the spaces: they learn how to change their avatar, how to teleport or perform different movements, or how to mute or unmute their microphones. These interactions are usually individual- and scenario-based, but they can also be used to nurture the collaboration between the participants. It is important to mention here that the sense of co-location and co-presence in the virtual space in these productions is also enhanced by VR's three-dimensional rendering and stereoscopic vision. According to

36 On this, see Bakk, "Honesty and Experience."

Tilo Hartmann, the fact that avatars are being co-located can mean that the production offers an imminent threat but also an opportunity to its participants, which can have an impact on their embodied self.[37] This is why co-located avatars are approached with greater care in VR than, for example, in video games. These performances cannot allow narrative-based interactions as there are no possibilities to change the script of the performance. *Finding Pandora X* is an exception to this rule, as its script bifurcates, but such a production requires human resources (that is, more actors) and a storyline designed to permit such a bifurcation. The performances have a fixed and mainly linear narrative that either follows the drama or the natural cause-reaction chain of the events. The narratives are pre-scripted, which means that the audience behaviour is very limited; this is the reason why the available interactions are individual- and scenario-based. The audience's attention is nonetheless guided away from these limitations as all the performers rely on their improvisational skills. These skills are necessary for the participants to feel the liveness of the event. The interaction types, together with the intimate atmosphere, also enhance the live aspect of these productions.

Conclusion

The virtual immersive theatre produced in VR is a specific type of immersive performance that is very similar to the performances taking place in an analogue setting. For now, the immersive performances in VR cannot be very long (partly due to the uncomfortable setting of the HMDs). Because of this, there is less time to build up the story world; therefore, these performances need to be pre-scripted. In order to offer a sense of liveness in VR, what is also necessary is a more complex acting style, where actors have to be ready for improvisational situations and for guiding the audience when required. While immersive theatre offers the audience an original story, physical possibilities to manifest agency, and unique experiences,[38] in VR immersive performances, these unique experiences are harder to achieve due to technological obstacles, such as limited possibility for multiplayer participation and limitations regarding navigation in space and interaction with objects. These obstacles can be concealed for a while by creating spatial open-world structures and by attention-guiding cues (borrowed from the

37 Hartmann, "Parasocial Interactions and Paracommunication with New Media Characters."
38 On this, see Alston, *Beyond Immersive Theatre*, p. 2.

science of magic), but more curious audience members will most likely find out the limitations of the performance after a while. On the long term, the creators of such experiences could create more unique experiences with the help of the actors, a more layered story world, and interaction possibilities.[39]

Bibliography

Alston, Adam. *Beyond Immersive Theatre: Aesthetics, Politics and Productive Participation*. London: Palgrave Macmillan, 2016.

Bakk, Ágnes Karolina. "Epiphany through Kinaesthetics: Unfolding Storyworlds in Immersive Analog Spaces." *Violence | Perception | Video Games: New Directions in Game Research*, edited by Federico Alvarez Igarzábal, Michael S. Debus and Curtis L. Maughan, 213–224, Bielefeld: transcript Verlag, 2019. https://doi.org/10.1515/9783839450512-018.

Bakk, Ágnes Karolina. "Honesty and Experience: Interview with Deirdre V. Lyons." *Zip-Scene.com*, March 8, 2022. https://zip-scene.com/2022/03/08/honesty-and-experience/.

Biggin, Rose. *Immersive Theatre and Audience Experience: Space, Game and Story in the Work of Punchdrunk*. Basingstoke-: Palgrave Macmillan, 2017.

Bishop, Claire. *Artificial Hells: Participatory Art and the Politics of Spectatorship*. London: Verso, 2012.

Gupta Dixon, Steven. *The Digital Performance*. Boston, MA: MIT Press, 2007.

Ferguson, Jason. "Virtual Reality Has a Storytelling Problem and Theatre Will Save It." *UploadVR*, May 8, 2016. https://uploadvr.com/virtual-reality-storytelling-problem-theater-will-save/.

Gochfeld, David, Kiira Benzing, Kevin Laibson, and Alex Coulombe. "Avatar Selection for Live Performance in Virtual Reality: A Case Study." In *IEEE Games, Entertainment, Media Conference*, 1–5. New Haven, CT: Institute of Electrical and Electronics Engineers, 2019.

Gupta, Saumya, Theresa Jean Tanenbaum, Meena Devii Muralikumar, and Aparajita S. Marathe. "Investigating Roleplaying and Identity Transformation in a Virtual Reality Narrative Experience." In *CHI '20: Proceedings of the 2020 CHI Conference on Human Factors in Computing Systems*, 1–13. Honolulu, HI: Association for Computing Machinery, 2020.

Hartmann, Tilo. "Parasocial Interactions and Paracommunication with New Media Characters." In *Mediated Interpersonal Communication*, edited by Elly A. Konijn,

39 I would like to express my gratitude to Işık Sarıhan for proofreading and discussing this chapter.

Sonja Utz, Martin Tanis, and Susan S. Barnes, 177–199. New York, NY: Routledge, 2008. https://doi.org/10.4324/9780203926864.

Huizinga, Johan. *Homo Ludens: A Study of the Play Element in Culture*. Hull, MA: Beacon, 1955.

Kuhn, Gustav. *Experiencing the Impossible: The Science of Magic*. Cambridge, MA: MIT Press, 2019.

Laurel, Brenda. *Computers as Theatre*. Reading, MA: Addison-Wesley, 1991.

Lockford, Lesa, and Ronald J. Pelias. "Bodily Poeticizing in Theatrical Improvisation: A Typology of Performative Knowledge." *Theatre Topics* 14 (2004): 431–443.

Lombard, Matthew, and Theresa Ditton. "At the Heart of It All: The Concept of Presence." *Journal of Computer-Mediated Communication* 3, no. 2 (September, 1997). https://doi.org/10.1111/j.1083-6101.1997.tb00072.x.

Machon, Josephine. *Immersive Theatres: Intimacy and Immediacy in Contemporary Performance*. Basingstoke: Palgrave, 2013.

Minsky, Marvin. "Telepresence." *Omni Magazine*, June, 1980. https://web.media.mit.edu/~minsky/papers/Telepresence.html.

Murray, Janet H. *Hamlet on the Holodeck: The Future of Narrative in Cyberspace*. New York, NY: Free Press, 1997.

Nilsson, Niels, Rolf Nordhall, and Stefania Serafin. "Immersion Revisited: A Review of Existing Definitions of Immersion and Their Relation to Different Theories of Presence." *Human Technology* 12, no. 2 (November, 2016): 108–134.

Popat, Sita. "Missing in Action: Embodied Experience and Virtual Reality." *Theatre Journal* 68, no. 3 (September, 2016): 357–378.

Roose, Kevin. "This Should Be VR's Moment. Why Is It Still so Niche?" *New York Times*, April 30, 2020. https://www.nytimes.com/2020/04/30/technology/virtual-reality.html.

Salen, Katie, and Eric Zimmerman. *Rules of Play: Game Design Fundamentals*. Cambridge, MA: MIT Press, 2004.

Samur, Sebastian Xavier. "Comparing Stage Presence and Virtual Reality Presence." *Performance and Technology* 6, no. 2 (August, 2016). https://doi.org/10.1590/2237-266058902.

Schulzke, Marcus. "Models of Agency in Game Studies" *Technoculture: An Online Journal of Technology in Society* 2 (2012). https://tcjournal.org/vol2/schulzke/.

Sheridan, Thomas. "Musing on Telepresence and Virtual Presence." *Presence: Teleoperators and Virtual Environments* 4, no. 1 (1992): 120–126.

Shriram, Ketaki, Soon Youn Oh, and Jeremy Bailenson. "Virtual Reality and Prosocial Behavior." In *Social Signal Processing*, edited by Judee K. Burgoon, Nadia Magnenat-Thalmann, Maja Pantinc, and Alessandro Vinciarelli, 304–316. Cambridge, MA: Cambridge University Press, 2017.

Slater, Michael D., Benjamin K. Johnson, Jonathan Cohen, Maria Leonora G. Comello, and David R. Ewoldsen. "Temporarily Expanding the Boundaries of the Self: Motivations for Entering the Story World and Implications for Narrative Effects." *Journal of Communication* 64 (2014): 439–455. https://doi.org/10.1111/jcom.12100.

Tanenbaum, Theresa Jean, and Karen Tanenbaum. "Improvisation and Performance as Models for Interacting with Stories." In *Interactive Storytelling: ICIDS 2008. Lecture Notes in Computer Science* 5334, edited by Ulrike Spierling and Nicolas Szilas. Berlin: Springer, 2008. https://doi.org/10.1007/978-3-540-89454-4_32.

Tanenbaum, Theresa Jean. "Identity Transformation and Agency in Digital Narratives and Story Based Games." Phd diss., Simon Fraser University, 2015.

Vorderer Peter, Christoph Klimmt, and Ute Ritterfeld. "Enjoyment: At the Heart of Media Entertainment." *Communication Theory* 14, no. 4 (November, 2004): 388–408.

Yan, Shuo, Saixi Ge, Jinhui Wang, and Xukun Shen. "Performing with Me: Enhancing Audio-Performer Interaction in an Immersive Virtual Play." In *CHI EA '21: Extended Abstracts of the 2021 CHI Conference on Human Factors in Computing Systems*, edited by Yoshifumi Kitamura, 1–6. New York, NY: Association for Computing Machinery, 2021. https://doi.org/10.1145/3411763.3451686.

About the Author

Ágnes Karolina Bakk, PhD, is lead researcher of the Immersion and Interaction hub at the Innovation Center of Moholy-Nagy University of Art and Design. She is the founder of the Zip-Scene conference, cofounder of Random Error Studio and Vektor VR section, Hungary's first VR-focused event. She teaches immersive storytelling and speculative design.

13. Immersive Materiality and Television Production in *The Dark Crystal: Age of Resistance* (2019)

Dan Leberg and Sabrina Sauer

Abstract: This chapter shifts the focus from the user's consumption of immersive media to the immersive experiences of media-makers during production. Through a case study of *The Dark Crystal: Age of Resistance* (2019), it explores how certain media production practices necessitate creators to become immersed in physical, imaginary, and virtual materials, thereby shaping the immersive experiences for users or spectators. Through interviews with core puppeteers and analysis of production processes, the chapter examines how materiality, agency, and animacy intersect to create immersive narratives. Finally, drawing on the work of Andrew Pickering and Tim Ingold, the chapter underscores the significance of considering media production practices in discussions about immersive media.

Keywords: puppetry, production studies, interviews, agency, animacy

Critical conversations about immersive media often focus on media technologies, from advances in screening technologies to digital innovations like virtual and augmented reality, and the experiences they afford to their participating users. The media technology's end user—whether configured as a cinematic spectator, a player, or otherwise—is generally the one who becomes immersed by participating within the immersive experience that the media-maker has created.[1] In this chapter, we shift focus from the

1 See Beier, "Virtual Reality"; Elsaesser, "Pushing the Contradictions of the Digital"; and Kennedy and Atkinson, "Virtual Humanity."

Bédard, Philippe, Alanna Thain, and Carl Therrien, eds. *States of Immersion Across Media: Bodies, Techniques, Practices*. Taylor & Francis Group, 2025
DOI 10.5117/9789048558766_CH13

user's consumption of and participation within an immersive media text by considering media immersion as a precondition for some forms of media production. This is not to suggest that the traditional user or spectator lacks agency in the activation of the immersive text, or that the embodied experience of that immersion is neatly compartmentalized from the aspects of the text which afford and support it. Our goal, rather, is to analyze how some media production practices require creative media-makers themselves to become immersed in the physical, imaginary, and virtual materials through which they create the immersive media texts which will, in turn, afford other users or spectators' immersive experiences. In so doing, we emphasize the role of materiality as a nexus of imagination and action in which immersive media experiences can occur.

Our case study for this production-oriented model of immersive media is the blend of sophisticated puppetry practices and digital filmmaking techniques used to create *The Dark Crystal: Age of Resistance* (Jim Henson Productions and Netflix, 2019), the prequel series to Jim Henson's cult classic film, *The Dark Crystal* (1982).[2] In *Age of Resistance*, a band of elf-like Gelfling protagonists incite a rebellion against their monstrous Skeksis overlords upon learning that the Skeksis are using dark magic to drain Gelflings of their "life essence," which the Skeksis consume to prolong their own lives. Described by executive producer Lisa Henson as likely "the largest puppet production ever made," *Age of Resistance* assembled a group of twelve core puppeteers and dozens more assistant puppeteers, alongside an enormous team of puppet makers, designers, technicians, and regular filmmaking personnel for an intense eleven-month shooting schedule to produce a full season of ten one-hour episodes.[3] Despite some preliminary discussion of making a completely computer-animated imagery (CGI) version of the series, Henson and Netflix producers decided unanimously to pursue a vastly more expensive, time-consuming, and logistically complex option: *Age of Resistance* would echo the production design of the 1982 film by using only physical puppets in elaborate studio settings and custom-made props. At the same time, *Age of Resistance* was designed to meet the aesthetic and stylistic expectations of modern television. Although the bulk of the narrational burden rests on the puppeteers and the puppets, series director Louis Leterrier and cinematographer Erik Alexander Wilson insisted

2 For clarity's sake, the 2019 Jim Henson Productions and Netflix prequel series will be referred to as *Age of Resistance*. The 1982 film, directed by Jim Henson, will be referred to as *The Dark Crystal*.

3 Henson, quoted in Lobb, *The Crystal Calls*, 10:10–10:20.

on filming the puppets with far more dynamic camera movement than is normal for puppet-based film and television.[4] Some of the principal Gelfling and Skeksis characters also received CGI embellishments during post-production on their eyes, mouths, tongues, and hands to compliment the life brought to the puppets by their puppeteers.

This sense of infusing life into a mediating object resonates with Helen W. Kennedy and Sarah Atkinson's analysis of immersive virtual reality (VR) texts wherein the end user steps into environments designed to emulate the embodied experiences of the media-maker.[5] The National Theatre's VR production of *Home: Aamir* (Rufus Norris, Toby Coffey, and Erfan Saadati, 2016), for example, translates the real-life journey of a Sudanese refugee into an immersive experience wherein the user explores migrant camps and the "Calais Jungle" while listening to the refugee's personal description of life in the camps. Kennedy and Atkinson distinguish between two characteristics of immersive VR in terms of how the production process mediates the media-maker's lived experiences into and through immersive technology. Not only does immersive VR *transport* the user into the VR text's world and the media-maker's subjectivity by recreating the producing subject's visual and auditory experiences, but it also *transfers* some aspect of the media-maker's cognitive and phenomenological subjectivity to the user by aligning them with the media-maker's perspective on that immersive world.[6]

In the case of *Age of Resistance*, this sense of transportation works on two simultaneous levels. First, the on-screen absence of the puppeteer transports the spectator into the narrative world. One *Age of Resistance* puppeteer who we interviewed[7] insisted that part of the "magic" of puppetry is the moment where the spectator catches themselves believing that the "performing objects" on the screen are indeed real. The puppeteers imbued their fantastical puppets with life by navigating the mechanical intricacies, expressive potentialities, and logistical constraints of the puppet (size, weight, physical comfortability, etc.) to make the puppet live as a dramatic character rather than as an inert statue. Then, the puppeteers in turn found themselves transported into the diegetic world of Thra not only through their close contact with the puppets, but also through the supporting technological

4 Dave Chapman, quoted in Lobb, *The Crystal Calls*, 1:04:37–1:04:57.

5 Kennedy and Atkinson, "Virtual Humanity."

6 Kennedy and Atkinson, "Virtual Humanity."

7 As part of the research process, the authors held unofficial semi-structured one-hour interviews with some of *Age of Resistance*'s principal puppeteers. Although each interviewed puppeteer was generous with their commentary, they each asked to have all comments anonymized when referenced in this chapter.

infrastructures on set that are crucial to creating compelling audiovisual puppet performances. As the series' puppeteers explained to us, the full production's assemblage of puppets, sets, lights, costumes, etc. collectively transport the puppeteers into the fantastical narrative.

The transference of subjectivities between the puppeteer and the puppet is an iterative exchange that aligns the intentions of the former with the physical affordances of the latter. Whether performing the immensely intricate Gelflings or wearing the larger-than-life Skeksis puppets, the core puppeteers were physically and imaginatively immersed within their puppet characters and the accompanying logistical devices through which they adapted their performances to fit within Leterrier and Wilson's Steadicam cinematography. For example, the puppeteers wore earpieces through which they listened to key production crew members and other puppeteers while wearing full-body Skeksis puppets. More importantly, the puppeteers had personal video monitors that let them see what the cameras were filming, which was essential for making the puppets' staging, choreography, and eyeline align with the shot's composition. Rather than creating immersive media to express their own experiences of personal trauma, such as in *Home: Aamir*, *Age of Resistance*'s puppeteers were co-creative forces in the fictional world that their performances inhabited, surrounded by the life-sized fantastical *mise en scene* and in performative dialogue with their technically complex puppet instruments.

The decision to make *Age of Resistance* with only puppets and the experiences of the core puppeteers while working on such a complex, intense, and prolonged shooting process underlines the importance of materiality in an analysis of production-centred media immersion. In particular, as the puppeteers imbue their puppets with life, they also immerse themselves in the puppet's materiality. We argue that the overarching sense of material fidelity to the original *Dark Crystal* starts to locate the immersive production practices as what Science and Technology Studies scholar Andrew Pickering's calls the "Dance of Agency," also referred to as "the mangle of practice," which addresses the emergent "tuning" or negotiation between intentionalized human actions and inanimate material objects' predispositions towards certain uses.[8] This mangle takes the shape of a "dialectic of resistance and accommodation" wherein human actors adjust their planned actions and the material objects that they interact with in response to the material agency of a technological artifact or object.[9] This dance is emergent, but

8 Pickering, *The Mangle of Practice*.

9 Pickering, *The Mangle of Practice*, pp. 22–23.

does weave itself into patterns that we may "grasp and understand."[10] In this case, *Age of Resistance*'s puppeteers found themselves "dancing" with the mechanical construction and storytelling requirements of the puppets: even if the series' puppets were designed and constructed explicitly to reify the screenplays' characters, the immense technical difficulties that the core puppeteers encountered when trying to get the intricate Gelfling puppets to perform with the intended expressiveness reveals a complicated give-and-take, or resistance and accommodation between the puppeteers' intentions and the puppets' complexity as performing instruments.

At the same time, sociologist Tim Ingold's concept of "Dance of Animacy" helpfully expands the immersive pull of *Age of Resistance*'s puppets for their puppeteers by contextualizing the human-object relationship within its surrounding circumstances.[11] While filming *Age of Resistance*, the puppeteers were not just relating to their puppets, but doing so within the contexts of the life-sized constructed set pieces, the hustle and bustle of a big-budget film set, and the personal technological aids like earpieces and visual monitors. It is within this entire socio-material assemblage that the puppeteers co-created a narrative world while immersing themselves within it, imagining with, through, and often inside the material objects that they animated.[12] Ultimately, we argue that puppetry's ability to solicit fascination with and through animated materiality warrants a reconsideration of media production practices' place in the greater conversation about immersive media.

We approach *Age of Resistance*'s immersive materiality through a combination of critical frameworks and methodological practices. We establish a critical intersection between puppet studies and (digital) materiality, through the lenses of Pickering's Dance of Agency and Ingold's Dance of Animacy. We also draw on personal interviews that we conducted with some of *Age of Resistance*'s core puppeteers to help contextualize immersion during production from the perspective of the media-makers themselves. These core puppeteers each performed a combination of leading, secondary, and ensemble roles throughout the series and spoke with us with great enthusiasm and candour on the condition that their comments would be anonymized throughout this chapter.[13] The interviews are combined with

10 Pickering, *The Mangle of Practice*, p. 24.

11 Ingold, *Making*.

12 On this "assemblage," see Tribble and Dixon, "Distributed Cognition," p. 271.

13 Although each interviewee is a masterful and accomplished professional puppeteer who considers their work on *Age of Resistance* to be a career highlight, the social and professional worlds of all performing arts are precarious and temperamental. The puppeteers' requests for

further commentary from *Age of Resistance*'s accompanying "making-of" documentary, *The Crystal Calls* (Randall Lobb, 2019), to better account for how the on-set production conditions helped to shape the puppeteer's immersive experiences.

Dances of Agency and Animacy

Tim Ingold refers to the Dance of Animacy as an alternative to Pickering's conceptualization of human-material interaction in terms of a Dance of Agency. In an effort to catch how things that seem to "just happen" actually unfold, Pickering focuses on how specific characteristics of a material seem to accommodate or resist particular actions.[14] His use of the concept of agency, as the capacity to act that is discovered when studying how worlds become constructed in a certain way,[15] aligns with Bruno Latour's ideas about constructivist, material semiotic actor networks, which see meaning-making and acting in the world as entwined and ongoing processes of negotiation.[16] Technological artifacts can be seen as scripted objects, with human actors either moving in correspondence with technological scripts or engaging in de-inscription to produce alternative uses of technologies by means of "anti-programs."[17] Whereas Latour famously sees human and non-human—including material—actors as "symmetrical" in processes of enactment, Pickering argues that what separates human from non-human actors is intentionality. In dancing with material agency, humans are forced to "tune" their goals and intentions. Materials may have certain characteristics that resist or accommodate particular ways of engagement, but materials do not "possess" intentionality.

Ingold moves away from Pickering's conceptualization of a Dance of Agency, wherein, through the dance, meaning is stabilized through what the latter refers to as a "mangle of practice," to instead propose a "Dance of Animacy."[18] Using the examples of the experience of flying a kite or of

anonymity are to ensure that nothing they say could be misconstrued by any past, current, or prospective employer as reason to refuse to work with the puppeteer again in the future.

14 Pickering, *The Mangle of Practice*, p. 24.

15 See Cooren, Taylor, and van Every, *Communicating as Organizing*, p. 11, quoted in Orlikowski, "Sociomaterial Practices," p. 1438.

16 See Law, "Actor Network Theory and Material Semiotics."

17 See Akrich, "The De-Scription of Technical Objects"; and Akrich and Latour, "A Summary of a Convenient Vocabulary for the Semiotics of Human and Nonhuman Assemblies."

18 Ingold, *Making*.

creating pottery using a pottery wheel and clay, he notes how both experiences do not solely rely on a human actor and non-human materials, but also, in the first case, on wind and, in the second case, on a choreography of sorts to produce a correspondence in action between the potter, the pottery wheel, and the clay in particular movement.[19] Both examples highlight the choreography of human, non-human, and contextual elements in terms of correspondence, moving alongside one another instead of interacting with one another. The kite "needs to correspond with the air": "the mindful or attentive bodily movements of the practitioner, on the one hand, and the flows and resistances of the material, on the other, respond to one another in counterpoint. [...] [B]odily kinaesthesia interweaves contra-punctually with the flux of materials within an encompassing, morphogenetic field of forces."[20]

Materials become transducers of kinetic energy and intention: they mediate and simultaneously inform and transform action.[21] In a sense, then, Ingold's ideas connect to the ideas of Lucy Suchman who, in anthropological studies of human-machine interaction, looks at interaction in terms of plans that take shape in situated actions.[22] Ingold locates this situated action in time and space by anchoring it explicitly to a temporal unfolding or rhythmic reverberation between and with human and material actors. Almost like Karen Barad's posthuman views on intra-action as a way to draw materiality into human action, Ingold weaves together the material and human in a particularly situated dance.[23]

Dancing with Puppets

Age of Resistance is a vivid example of the Dances of Agency and Animacy because of how puppetry and puppet-based storytelling rely on the relationships between puppet, puppeteers, and their surrounding environments. Modern Puppet Studies treat puppet performances as an egalitarian Dance of Agency: an iterative exchange between puppet and puppeteers without any inherent superiority of one over the other.[24]

19 See Ingold, "The Textility of Making."
20 Ingold, *Making*, p. 101.
21 See Ingold, *Making*, p. 102.
22 Suchman, *Human-Machine Reconfigurations.*
23 Barad, "Posthumanist Performativity."
24 See Parker-Starbuck, "Animal Ontologies and Media Representations," p. 392; and Posner, Orenstein, and Bell, "Introduction," p. 6.

John Bell expands upon Frank Proschan's foundational definition of puppets as "performing objects" by emphasizing the deliberate creative encounter between the human body and the material world. Proschan defines puppets as "material images of humans, animals, or spirits that are created, displayed, or manipulated in narrative or dramatic performance," wherein the power dynamic seems to rest with the puppeteer as the creator, displayer, or manipulator of the puppet.[25] Bell embellishes upon Proschan by highlighting the material commonalities and interactions between puppet and puppeteer:

> The material world resists our human intrusion, until we ourselves ultimately rejoin the material world when our lives cease. We can shape and transform the material world, and we can even imagine that world to be inhabited by spirits, but in the end the different elements (wood, leather, glass, stone, sand, water, bone, plastic, and light) resist our attempts to dominate them. [...] Performance with objects is not simply the realization of a dramatic script, nor the performance of language, although these might be parts of it; and neither is it entirely explainable as acting, dance, or any other performance of the body. *It is humans coming to terms with the material world, a momentary alliance or bargain between humans and the stuff of, or literally stuff in performance.*[26]

Puppets, then, remind us of the flesh-and-bone materiality of our own bodies. The life-force that animates our bodies also immerses our bodies in their performances in the lived world. Puppets, like Henson's foam-and-latex creations, transform from inanimate objects into characters like Gelflings and Skeksis through their "momentary alliances" with their human puppeteers: as the puppeteers immerse themselves in their puppets, the puppeteers' animating life-force imbues the puppets with the capacity to perform.

Dassia N. Posner, Claudia Oreinstein, and John Bell reframe puppetry as a "material performance" in which "inanimate matter contains agency not simply to mimic or mirror, but also to shape and create."[27] Although we certainly agree with them that puppeteers imbue puppets with an animating life-force, we wish to reframe the agency that they assign to the puppet as the correspondence between puppet and puppeteer.[28] For Ingold,

25 Proschan, "The Semiotic Study of Puppets, Masks, and Performing Objects," p. 5.

26 Bell, *American Puppet Modernism*, p. 4, our emphasis.

27 Posner, Orenstein, and Bell, "Introduction," p. 5.

28 Posner, Orenstein, and Bell, "Introduction," p. 5.

agency is not something that one possesses as much as what is "enacted."[29] Instead, we view the correspondence between puppet and puppeteer as a Dance of Animacy, wherein the puppeteers negotiate with the puppets' material affordances and creative potentialities to create the performance. For Basil Jones, one of the principal creators and puppeteers for the theatrical production of *War Horse* (Nick Stafford, Basil Jones, and Adrian Kohler, 2007), the animating life force that puppeteers bring to their puppets is what distinguishes puppetry from traditional acting: "The actor is a living person and therefore automatically possesses life. [...] His or her *livingness* is obvious and doesn't need to be 'performed.' [...] The primary work of the puppet is the *performance* of life, whilst for the actor this fundamental battle is already won."[30] Whereas a human film, television, or stage actor does not normally struggle to convince the audience that they are alive, the puppet only comes alive as a material performer while "dancing" with the puppeteer.

For the puppeteer's part, the life-imbuing Dance of Animacy is the point of immersion in the puppet. The puppeteer is physically immersed within the puppet, whether with a single hand, a whole arm, or—if that puppet is a gargantuan Skeksis—their entire body. Whereas Kennedy and Atkinson's immersive transportation primarily considers the reconstruction of another's experiences through audiovisual VR technology for the user's simulation, the puppeteer's media-producing immersion begins in the tactile experience of the puppet's materiality. This tactile encounter with the puppet transports some of the puppeteer's sensorial experience into the puppet character's narrative world by focusing the puppeteer's experience of the narrative world through the puppet instrument. This tactile transportation is closely intertwined with the visual guidance from the puppet's accompanying monitor(s), through which the puppeteer watches the puppet's performance as the director's camera sees it and thereby checks the efficacy of the Dance of Agency in real time.[31]

The puppeteer's transference of animating life energy to the puppet engages the puppeteer's imagination and broader sensorium by aligning the Dance of Animacy within the needs of the screenplay and of the narrative. In puppetry, the transference and the creation of character subjectivity go

29 Ingold, *Making*, p. 97.

30 Jones, "Puppetry and Authorship," p. 254, quoted in Searls, "Unholy Alliances and Harmonious Hybrids," p. 299.

31 The importance of the puppeteer's monitor will be unpacked in more detail later in this chapter.

hand in hand: by articulating the scripted character's subjectivity through the puppet, the puppet's character emerges within and is articulated by the Dance of Agency between the puppeteer and the performing material. Whereas Kennedy and Atkinson locate immersive transference of subjectivity from the media-maker to the end user through the VR text's soundscape, the puppeteer is living and performing the narrative's soundscape as part of transferring it. *Age of Resistance*'s puppeteers often spoke their puppets' dialogue with full emotion while filming to further connect the imagination with the puppet's performance, even if they knew that a professional voice actor would eventually overdub the puppeteer and become the voice of the character that the audience would hear. The material immersion in the puppet required a full-bodied performance that included giving a voice to the puppet so as to fully commit to the performance.

In this regard, the puppet's material composition and design are key factors for its immersive appeal. The puppet's weight, balance, flexibility, and fragility are all part of the material agency with which the puppeteer must learn to "dance." This places considerable importance on the construction process since the puppet-makers "[build] into the puppet certain ways of moving designed to work with the ergonomics of the human body. In turn, the puppeteer(s) or manipulators must flesh these out, literally."[32] Although generally constructed from lighter-weight materials than *The Dark Crystal*'s puppets, *Age of Resistance*'s puppets were considered just as difficult to perform due to their mechanical intricacy and—according to the puppeteers—still-considerable weight.

Age of Resistance's Skeksis were built to be considerably larger than their *Dark Crystal* ancestors and many required three puppeteers per character to operate. The necks and heads of *Age of Resistance*'s Skeksis, for example, are heavily decorated foam and latex sleeves that the primary puppeteer would operate with one arm stretched up to operate the head, and one arm to gesture with one of its arms. The primary and first assistant puppeteers would then wear the rest of the Skeksis's body with a harness-like contraption and collectively control the character's feet and arm movements from inside the larger-than-life puppet. The eyes and eyebrows were operated externally through remote control by a second assistant puppeteer. Although none of the individual fiberglass, foam, latex, layered fabric costumes, and—for some scenes—metal body armour were particularly heavy on their own, their cumulative weight made considerable demands on the puppeteers' bodies, especially considering the puppets' lack of internal ventilation: "[Y]ou had

32 Tribble and Dixon, "Distributed Cognition," p. 271.

to wear black when performing a Skeksis because you would be dripping in sweat by the end of the day, keenly aware of the assistant's head in your armpit the whole time." Furthermore, what made performing a Skeksis extra challenging was the lack of peripheral vision: while being fully immersed in the creature, the performers had access to three video monitors: two showing what the cameras were recording, and one showing the view from above. Dubbed the "God camera," this third image gave performers a sense of the spatial distribution of the puppets and of the set pieces in a scene. What made these performances particularly satisfying though was the grotesque appearance of the puppets, which gave the puppeteers a clear sense of the puppets' performative energy. This echoes Colette Searls's assertion that some of the most successful puppet and motion-capture characters in film and television take advantage of being non-human.[33] Puppetry has such a natural performative synergy with characters that are fantastically monstrous that the non-humanity of the material performer can benefit the puppet's performance, whereas a CGI-embellished human character or puppet whose appearance more closely resembles a human can start to look artificial, dated, or otherwise unconvincing.[34] In the case of *Age of Resistance*, Searls's spectatorial point also rings true from a media production perspective: many of the core puppeteers relished their performances with the grotesque Skeksis because, as one puppeteer described, the "joyous freedom of horribleness" let the material agency of the puppets' appearance do much of the work. The same cannot be said of all of Thra's inhabitants.

One of the most technically ambitious decisions made for *Age of Resistance* was to centre the narrative on Gelfling characters, which the core puppeteers unanimously described as the most difficult puppets they had ever performed. Jim Henson had not been fully satisfied with the wooden Gelflings of *The Dark Crystal*, finding them rigid, awkward, and unexpressive despite performing the film's lead Gelfling character, Jen, himself.[35] *Age of Resistance*'s Gelflings were constructed with complex mechanical skulls to give the puppeteers considerable control over the puppets' eyes and facial expressions. This made the Gelflings not only quite top-heavy but raised the puppet's overall weight to upwards of 12 kilograms, which the puppeteer would have to perform with one arm lifted above their head. As one core puppeteer jokingly remarked: "They [the Gelflings] were hideous [to perform with]. Initially, they all had metal bars in their heads. It was one

33 Searls, "Unholy Alliances and Harmonious Hybrids," p. 294.

34 Searls, "Unholy Alliances and Harmonious Hybrids," p. 303.

35 Henson, quoted in Lobb, *The Crystal Calls*, 21:58-22:20.

of the most horribly difficult things just to make them look real without wobbling. [...] They were the least successful puppets in the original film, so let's make loads of them, and make [*Age of Resistance*'s] story about them." The challenge of not only achieving a high level of precise control over the facial minutiae required for performing the emotional details of the principal Gelfling characters, despite the uneven weight distribution and cumbersome weight, was considerable. Core puppeteer Neil Sterenberg stated that his secret was to find a way of making Rian, his principal Gelfling character, walk smoothly so that the puppet would appear to possess the plausible mass for a creature of Rian's size.[36] Whereas many of Henson's Muppet characters can bounce when they walk without disrupting the verisimilar illusion, Sterenberg's meticulous control of Rian's material and verisimilar tactility demonstrates a deep immersion within the puppet's material agency. Fortunately, the core puppeteers like Sterenberg had further tools available to them on set to help complete the transfer of life energy into their puppets.

Dancing with Screens

Beyond the tactile connection to the puppet as an immersive instrument, *Age of Resistance*'s puppeteers were also visually transported into their characters' fantastical world through the personal digital monitors that accompanied each puppet. On the one hand, television and film productions with puppets afford a great deal of creative freedom in that puppet performances can be shot and then edited together to create performed moments that would be near impossible to perform live as such. For example, the carriage escape sequence from episode five, "She Knows All The Secrets," was shot in hundreds of incremental short takes on a green screen set, heavily embellished upon by CGI animators, and then edited together in such a way that it created the illusion that the Gelfling characters climbed upon and derailed the rapidly moving vehicle.[37]

On the other hand, during our interviews with the core puppeteers, they said that the general focal point of the puppeteers' performance was their personal monitors, which showed them what the cameras saw, rather than the puppet itself. Since none of *Age of Resistance*'s puppets allowed the puppeteers to see directly through the puppets' eyes, these personal

36 Sterenberg, quoted in Lobb, *The Crystal Calls*, 32:17–32:47.

37 See Vanity Fair, "Puppetry in The Dark Crystal," 4:17–4:59.

monitors were the only way for the puppeteers to ensure that their puppets were looking in the right direction, moving their arms naturalistically with and towards whatever props the puppet might carry, and generally gage their efficacy within the Dance of Animacy in getting the puppet to perform the character as intended. Put another way, once the masterful puppeteers had sufficient tactile understanding of the material puppets' mechanical constraints and affordances, the puppeteers' immersion in the material puppet was deep enough that they could "leave their hands alone" and focus on the performances that appeared on screen. As one puppeteer described: "When you perform a puppet, you suddenly become very unaware of yourself. Because you are so engrossed in the monitor, you go into it, so that you stop consciously thinking: 'Oh, I need to move. I need to adjust my hand that way a little bit.' It is almost like you go into the image: the monitor is like the feedback loop, and when you are watching the monitor, you are in Thra." The monitors then become a VR-like immersive screen for the performing puppeteer. By watching their Dance of Agency with the puppet play out within the "feedback loop" of the monitor, the puppeteer's immersion in the narrative world becomes a double Dance of Agency: they are spectator and performer simultaneously and dissolve into their performance as if they "aren't there." As one puppeteer put it: "You forget about what [you are] doing. You just see it on the screen and go, you know. The body knows."

Since the material world of puppets is often organized as a narrative entertainment, the contextualizing forces of narratives, performance venues, audiences, and other production-based logistics expand puppeteering's Dance of Agency into a collective Dance of Animacy. In puppet-based television productions, the puppets do not simply start moving when the director calls: "Action!" Rather, the puppeteers' navigation of the camera, *mise en scene*, and other on-set material presences, all the while remaining hidden from the camera and immersed in the performing puppet, requires a complicated thinking-together across production personnel to make sure that all the performing materials are indeed performing together.

Dancing on Set with Cameras

The filming process for *Age of Resistance* represents a compelling Dance of Animacy because of the ramifications of Leterrier's Steadicam cinematography for the puppeteers' capacities to immerse themselves within the puppets. Despite unanimous praise among the puppeteers for the end result of Leterrier's highly mobile cinematography, the cast and crew's Dance of Animacy revealed the boundaries and expectations of puppet immersion in film and television productions.

Many of *Age of Resistance*'s puppeteers were accustomed to working under film and television conventions that dated back to the heyday of Jim Henson's *The Muppet Show* (1976–1981) and of the original *Dark Crystal*. Henson's trust in and respect for his fellow puppeteers resulted in a filming style designed to accommodate the puppeteers' immersions and foreground the puppet's performances, even at the risk of formal rigidity. The single-camera cinematography in *The Dark Crystal*, for example, was generally quite static: Henson sought to support his puppeteers' immersions in the "feedback loop" of their monitors by keeping the camera relatively still, which meant that the puppeteers never had to worry about eyelines or other staging concerns changing within a single shot. According to puppeteers who knew Henson personally, he also assigned *The Dark Crystal*'s most challenging puppetry tasks to himself, whether the camera was moving or not. *The Dark Crystal*'s puppeteers interpreted this as a community-building act of a compassionate professional puppeteer who "knew the limitations [of his puppets] and how to work them out [himself]." Henson also typically shot in long takes to foreground the puppet performances and it was not uncommon for the puppeteers to improvise and joke around with their puppets between takes in the hopes of keeping the creative energy alive on set. Although this sense of playful community-building around the primacy of the puppets carried over into *Age of Resistance*, the Gelfling and Skeksis puppeteers had to adjust their on-set habits to the stylistic demands of modern big-budget television.

Leterrier's highly mobile and multiple camera cinematography in *Age of Resistance* meant that the image in the puppeteers' monitors was continually moving. Although the innovative and dynamic framing was highly evocative in and of itself, it created considerable logistical challenges for the puppeteers within this greater Dance of Animacy. As previously mentioned, the puppeteers inside a Skeksis were entirely reliant on their three internal monitors to guide their movements and performances: the first two monitors showed the contrasting camera angles of Leterrier and Wilson's Steadicams and the third was the "God camera" overhead shot of the entire set. The puppeteers had to simultaneously adjust their material performances within Leterrier and Wilson's monitors while drawing on the "God camera" monitor to give a sense of spatial distribution of other puppets and props as well as a fixed perspective on where the safest pathways across the set were for the immersed puppeteers to walk. Once the monitors began to move, the puppeteers' reference point for eyelines, arm movements, and other gestures was immediately thrown awry.

The Gelfling puppeteers, for instance, faced problems when not being entirely sure which camera they were playing to while performing a scene with more than one puppet. As one puppeteer explained:

> It was very hard to watch two monitors because [...] you'd have to hit a very specific mark on shot for that to look right. And you would get that right in one monitor and then you'd see the other one and go: that doesn't work. So you'd really force the editing because you'd kind of go: "Well, it doesn't look right on that one." So it would definitely have to be on that shot. But for the other performer doing their character, they might not look right in the other shot. So two of [the shots] might become unusable.

Apart from the need to stay "in frame," puppeteers and those filming alike had to negotiate the physical restrictions posed by shooting on a three-tiered set. As puppeteers had to remain invisible, they had to be "under the floor." Depending on the kind of creature that was being performed, the puppeteer had to work from a specific level to ensure that each puppet was performed at the right relative height. This could also lead to multi-character scenes where puppeteers had to negotiate where to place limbs under the floor in a strategic manner to ensure that puppets could perform the required movement, such as clearing a pathway to leave a scene.

Despite the intense logistical challenges posed by the mobile cinematography, however, the puppeteers unanimously approved of Leterrier's own immersion within the narrow pathways amidst the multi-tiered sets: "[Leterrier] was in the trenches with us." By channelling some of Henson's hands-on pragmatism, creative enthusiasm, and refusal to declare any shot too difficult, Leterrier's production process transformed *Age of Resistance*'s performance spaces and became a physical collective choreography of limbs, materials, cameras, and monitors wherein immersion in the on-set materials was simply a fact of life.

Conclusion

The material negotiations within *Age of Resistance*'s Dance of Animacy reveal that the puppeteers' immersion in their material performance was accommodated and resisted by the material characteristics of their puppets and by how these afforded or restricted movement and vision, as well as by the technical opportunities and challenges posed by camera-handling and geographies of the physical film set. Furthermore, the monitors used and worn by puppeteers effectively accommodated the assessment of their own performance while simultaneously acting as their entry point into the diegetic imaginary. These monitors also resisted parts of the production process: while using Steadicams to move around, the puppets were aimed at making the performance even more dynamic, which worked against craft conventions of televisual puppeteering, thereby forcing puppeteers into adjusting or "tuning" into a new kind of craft or skilled performance.

Age of Resistance therefore harkens back to the "crafted wonder" of pre-digital special effects in film and television while proposing a new direction for modern immersive media production that accommodates multiple Dances of Animacy.[38] The series' preoccupation with materiality goes hand in hand with the cult appeal of the original film: whether the fascination with which audiences and puppeteers invest in material performances, or the fantastically macabre narrative in which a character's life essence is an extractable material for other characters to consume, *Age of Resistance* and *The Dark Crystal* present a unique opportunity to reconsider immersive media as a precondition of media production.[39] Should spectators seek out an immersive narrative experience to parallel that of the puppeteers, the crystal is indeed calling.

Bibliography

Akrich, Madeleine. "The De-Scription of Technical Objects." In *Shaping Technology / Building Society: Studies in Sociotechnical Change*, edited by Wiebe E. Bijker and John Law, 205–224. Cambridge, MA: MIT Press, 1992.

Akrich, Madeleine, and Bruno Latour. "A Summary of a Convenient Vocabulary for the Semiotics of Human and Nonhuman Assemblies." In *Shaping Technology / Building Society: Studies in Sociotechnical Change*, edited by Wiebe E. Bijker and John Law, 259–264. Cambridge, MA: MIT Press, 1992.

Barad, Karen. "Posthumanist Performativity: Toward an Understanding of How Matter Comes to Matter." *Signs: Journal of Women in Culture and Society* 28, no. 3 (2003): 801–831.

38 See Sergeant, "Crafted Wonder."

39 See Bell, *American Puppet Modernism*, p. 5, quoted in Searls, "Unholy Alliances and Harmonious Hybrids," p. 295.

Beier, Klaus-Peter. "Virtual Reality: A Short Introduction" *The Archives of the University of Michigan Virtual Reality Laboratory at the College of Engineering*, 1999–2000. http://websites.umich.edu/~vrl/intro/index.html.

Bell, John. *American Puppet Modernism*. New York, NY: Palgrave Macmillan, 2008.

Cooren, François, James Taylor, and Elizabeth J. van Every. *Communicating as Organizing: Empirical and Theoretical Explorations in the Dynamic of Text and Conversation*. Mahway, NJ: Lawrence Erlbaum, 2006.

Elsaesser, Thomas. "Pushing the Contradictions of the Digital: 'Virtual Reality' and 'Interactive Narrative' as Oxymorons between Narrative and Gaming." *New Review of Film and Television Studies* 12, no. 3 (2014): 295–311. https://doi.org/10.1080/17400309.2014.927182.

Harris, Peter, and Philip Cason, dirs. *The Muppet Show*. Created by Jim Henson and Jack Burns. Aired 1976–1981, in broadcast syndication. ABC Studios, 2016, DVD.

Henson, Jim, dir. *The Dark Crystal*. 1982; Los Angeles, CA: Columbia Tristar Home Entertainment, 2004. DVD.

Ingold, Tim. *Making: Anthropology, Archaeology, Art, and Architecture*. London: Routledge, 2013.

Ingold, Tim. "The Textility of Making." *Cambridge Journal of Economics* 34, no. 1 (2010): 91–102. https://doi.org/10.1093/cje/bep042.

Interviews with *The Dark Crystal: Age of Resistance*'s puppeteers, February 7–18, 2022.

Jones, Basil. "Puppetry and Authorship." In *Handspring Puppet Company*, edited by Jane Taylor, 253–268. Johannesburg: David Krut Publishing, 2009.

Kennedy, Helen W., and Sarah Atkinson. "Virtual Humanity: Empathy, Embodiment, and Disorientation in Humanitarian VR Experience Design." *Refractory: A Journal of Entertainment Media* 30 (2018). https://research.brighton.ac.uk/files/5197859/Virtual_Humanity_empathy_embodiment_KENNEDY_Accepted30May2018_GREEN_AAM.pdf.

Law, John. "Actor Network Theory and Material Semiotics." In *The New Blackwell Companion to Social Theory 3*, edited by Bryan S. Turner, 141–158. Chichester: Blackwell Publishing, 2009.

Leterrier, Louis, dir. *The Dark Crystal: Age of Resistance*. 2019; Los Angeles, CA: Jim Henson Productions and Netflix, 2019.

Lobb, Randall. *The Crystal Calls: Making "The Dark Crystal: Age of Resistance."* 2019; Los Angeles, CA: Netflix, 2019.

Norris, Rufus, Toby Coffey, and Erfgan Saadati, dirs. *Home: Aamir*. 2016; London: National Theatre, 2016. Live VR Performance.

Orlikowski, Wanda J. "Sociomaterial Practices: Exploring Technology at Work." *Organization Studies* 28, no. 9 (2007): 1435–1448. https://doi.org/10.1177/0170840607081138.

Parker-Starbuck, Jennifer. "Animal Ontologies and Media Representations: Robotics, Puppets, and the Real of *War Horse*." *Theatre Journal* 65, no. 3 (2013): 373–393. https://doi.org/10.1353/tj.2013.0080.

Pickering, Andrew. *The Mangle of Practice: Time, Agency, and Science*. Chicago, IL: University of Chicago Press, 1995.

Posner, Dassia N., Claudia Orenstein, and John Bell. "Introduction." In *The Routledge Companion to Puppetry and Material Performance*, edited by Dassia N. Posner, Claudia Orenstein, and John Bell, 1–12. London: Routledge, 2015.

Proschan, Frank. "The Semiotic Study of Puppets, Masks, and Performing Objects." *Sémiotica* 47, no. 1–4 (1983): 3–46.

Searls, Colette. "Unholy Alliances and Harmonious Hybrids: New Fusions in Puppetry and Animation." In *The Routledge Companion to Puppetry and Material Performance*, edited by Dassia N. Posner, Claudia Orenstein, and John Bell, 294–307. London: Routledge, 2015.

Sergeant, Alexander. "Crafted Wonder: The Puppet's Place Within Popular Special Effects Reception." In *The Crafty Animator*, edited by Caroline Ruddell and Paul Ward, 181–201. Cham: Palgrave Animation, 2019. https://doi.org/10.1007/978-3-030-13943-8_8.

Suchman, Lucy. *Human-Machine Reconfigurations: Plans and Situated Actions*. Cambridge: Cambridge University Press, 2007.

Tribble, Evelyn, and Robin Dixon. "Distributed Cognition: Studying Theatre in the Wild." In *The Routledge Companion to Theatre, Performance, and Cognitive Science*, edited by Rick Kemp and Bruce McConachie, 264–275. London: Routledge, 2019.

Vanity Fair. "Puppetry in *The Dark Crystal: Age of Resistance* Explained by Creators." October 21, 2019. YouTube video, 13:18. https://youtu.be/e-6XUIJDh-Y.

War Horse, developed by Nick Stafford, Basil Jones and Adrian Kohler, National Theatre. London, UK. Premiered 17 October 2007.

About the Authors

Dan Leberg is Assistant Professor of Film and Media Studies at the University of Groningen. Prior to his graduate studies, Leberg worked as a professional classical theatre and film actor for almost 25 years. His research focuses on cognition in screen media production, motion capture performance, and Shakespeare on film and television.

Sabrina Sauer is Assistant Professor of Media Studies at the University of Groningen. Her research focuses on understanding how digital tools and platforms facilitate creative, often serendipitous, processes in audio-visual media production. She has extensive experience in public-private-civic interdisciplinary research, often working on the boundary of media studies, science and technology studies, computer science and information retrieval.

14. "More Mbappé Than Me?": Soccer Video Games, Athletic Bodies, and the Performative Displacement

Daniele Iannucci

Abstract: Transposing performance theory to the realm of professional sports, this chapter examines the trajectory of promotional materials for the *FIFA* video game series (now titled *EA Sports FC*) since 2008. It argues that, in response to the technologies developed in these video games and the streaming industries they are connected to, Electronic Arts (EA) has shifted its marketing practices of the series from *feeling like* the athlete, to paradoxically *being more than* the athlete. Immersive sports video game technologies and their industries have increasingly remapped the athlete's body onto users, asking us to reconsider the incorporation of star athletes into these marketing campaigns as more than playful product endorsements, but as sites to analyze the relationship between image rights, digital avatars, and agency.

Keywords: *FIFA* video game series (EA Sports FC), performance theory, body theory, sports star, digital avatars

Before the signature tagline of the Electronic Arts (EA) Sports franchise appears at its end, the closing sequence of *FIFA 10*'s (Electronic Arts, 2009) release trailer intercuts images of everyday fans with some of soccer's biggest stars of the 2009–2010 season, from Wayne Rooney and Giorgio Chiellini to Karim Benzema and Xavi. As a series of passport-like headshots of players and fans alike are rapidly shown in succession (fig. 14.1), a voiceover triumphantly declares: "Ten million Rooneys, ten billion Benzemas, ten zillion Xavis."[1] Though the trailer showcases some of the most advanced

1 Official EA UK, "EA Sports FIFA 10 – The Full 2 Minute TV Ad," 1:48–1:53.

Bédard, Philippe, Alanna Thain, and Carl Therrien, eds. *States of Immersion Across Media: Bodies, Techniques, Practices*. Taylor & Francis Group, 2025
DOI 10.5117/9789048558766_CH14

Figure 14.1: *FIFA 10* Launch Trailer

Figure 14.2: "More Mbappé than me?" from the FIFA 19 Official Launch Trailer

sports gaming technologies available at the time and its collective star power and colossal set pieces remain visually striking over a decade since it first premiered, its closing moments arrive in a technologically simple presentation, borrowing optical illusions from two of the most familiar pre-cinematic devices. The effect of this closing passport-like sequence functions like a zoetrope, with the flickered movement from these seemingly still headshots combining to create a sense of quasi-seamlessness through their montage, while also producing the effect of a thaumatrope, with the rapid intercutting of faces producing a flurry of blended noses, ears, eyes, and mouths that are momentarily superimposed. With one face being almost indistinguishable from the next as these images morph and meld to distort the rush of profiles, the trailer ends with the voice-over asking a most provocative question: "How big can football get?"[2] Almost a decade later, in the "Champions Rise" launch trailer for *FIFA 19* (Electronic Arts, 2018), EA hints at a possible answer to this question by asking another. After a torrent of gamers have star soccer players' digital avatars projected onto their faces, a young man is shown playing the *FIFA 19* video game from his bedroom, taking control of the French international soccer star, Kylian Mbappé. Scoring an in-game goal in spectacular fashion, the young gamer celebrates, crossing his arms with a wry smirk while imitating Mbappé's signature celebratory gesture. The man looks directly into the camera—with the frame modelled after the Twitch streaming service—proclaiming:

2 Official EA UK, "EA Sports FIFA 10 – The Full 2 Minute TV Ad," 1:54.

"My Mbappé is more Mbappé than Mbappé," before Mbappé appears in the flesh, points at himself, and bemusedly asks, "Plus Mbappé que moi ?" [More Mbappé than me?][3]

Over the last decade, the marketing strategies for sports video games—particularly EA Sports' *FIFA* franchise (now titled *EA Sports FC*)—have increasingly focused on a conflation of the celebrity athlete's body, their in-game digital avatars, and the gamers who use them. Unlike previous generations of soccer players, current stars like Kylian Mbappé do not function solely as cover athletes or simply make cameo appearances in the promotional materials for these games, but have grown into their roles as professional soccer players by being enmeshed into the immersive, interactive technologies that these games and their networks have produced.[4] The attempt to entrench players in the world of celebrity sports athletes and marketplaces has shifted EA Games' strategy from *feeling like* the player to—as the Mbappé example presented above paradoxically illustrates—*being more than* the player. This observation is not just a reading of trailers for two games, but a reflection on the markets and technologies these games have actively produced and have slowly implemented worldwide as the popularity of eSports grows. By endorsing and developing a monetized playground for these digitally hybridized bodies—bodies that, in many ways, have been marketed to displace the natural physicality of athletic performability—these leagues and these games have developed and marketed their immersive industries to reimagine, repurpose, and reintegrate the star's athletic body into the marketplace of self-sustaining, user-created streaming communities that alienate the link between flesh and performance.

By situating athletes and the athletic body within an extended version of Vivian Sobchack's theory of the "four bodies"[5] and then merging that theoretical framework with Barry Smart's conception of the "sports star"—understood as an athlete who gains collective recognition for their exceptional and authentic performances and prowess[6]—this chapter

3 EA SPORTS FIFA, "FIFA 19 | Champions Rise | Official Launch Trailer," 0:41–0:49. Mbappé's original French is subtitled in English within the video.

4 Since 2017, EA has increasingly enlisted dozens of young soccer stars (including Mbappé, Mason Mount, and João Félix) to serve as FIFA Ambassadors. Their purpose, according to EA's FIFA Ultimate Team (FUT) website, is to "promote their game" (FIFA Ultimate Team, "FIFA 21 Ambassadors"). On this same webpage, a quote from Mbappé on his role as a FIFA Ambassador reads, "Being on the cover of FIFA is a dream come true [...] I've been playing this game since I was a kid, and I am honored to represent a whole new generation of footballers [...]." (FIFA Ultimate Team, "FIFA 21 Ambassadors").

5 Sobchack, "Being on the Screen."

6 Smart, *The Sports Star*, p. 3.

will examine the trajectory of the *FIFA* video game franchise and its marketing campaigns from *FIFA 09* (Electronic Arts, 2008) to *FIFA 19*. I argue that during this period, bodies, performances, and the agency of sports stars have been increasingly supplanted in favour of a perversely ventriloquized digital reproduction of the athletic body that exists in an improbable new locale—one that hovers between the flesh of athletic performance and the screen that hosts its digital reproduction. Streamers and users, through the continual development of these games and their technologies, have not just repurposed these digital bodies, but have been primed to engage in an interactive, *performative displacement*, whereby athletically performative bodies are reconstituted by, and remapped onto, these same streamers and users. These games ask us to interrogate the expanding economies that gaming and streaming communities have, and continued to, produce. With the growing inclusion of soccer stars, users, and streamers within the ecosystem of the *FIFA* franchise and its promotional materials, how might we consider the implications of the question "More Mbappé than me?" beyond the realm of playful athlete endorsement, and instead, into the exploration of performing bodies, digital avatars, and streaming communities as part of a growing economy in the technologies of immersion?

Defining Bodies: Performance and the Sports Star

In "Being on the Screen: A Phenomenology of Cinematic Flesh, or the Actor's Four Bodies" (2012), Vivian Sobchack outlines the four bodies of her phenomenological model of onscreen performance to better understand the star-celebrity persona.[7] She cites the "prepersonal," "personal," "impersonated," and "personified" bodies as producing a Maurice Merleau-Ponty-inspired bond between the flesh—that is, the physical body of the performer—and the idea—that which appears in and is the product of performances on screen and in public appearances.[8] This link produces "transparent and habituated bodily modes of 'being in the world' with 'being on the screen'."[9] While Sobchack's template uses screen actors as the objects of its analysis, a transposition of her model and of its four bodies into a reading of sports star and their performances will follow below. The sports star can inhabit

7 Sobchack, "Being on the Screen."
8 Sobchack, "Being on the Screen."
9 Sobchack, "Being on the Screen," p. 430.

a wide variety of on- and off-field personalities and roles that are visible within the sporting spectacle as it unfolds in the stadium, or is reinforced through the presentation of their image in broadcasts, press conferences, commercials, or talk-show appearances.

Sobchack's first body, the "prepersonal," "exists and acts by means of its particular physical and biological structure and the general capacities for movement, action, gesture, and voice this structure both enables and constrains."[10] To an athletically performative body—or frankly, to any body—this serves as a most important factor, given the emphasis sport places on the physical body itself: its composition, penchant for action and movement, and how those actions are evaluated for gains that are both qualitative (visual feats of athleticism) and quantitative (measured statistically). Sobchack's second body, the personal, acts as an awareness of those physical capacities, with the comprehension of the prepersonal body being enlisted into movement or gesticulation. She summarizes it succinctly as "a style of being."[11] The process of recruiting those capacities into a personal body and self is not always cognizantly recognized but is instead materialized through the rhythms of daily life. By being, therefore, one is. It is the third body, the impersonated one, whose role is to purposefully "select and mobilize the capacities, qualities, and affects of both their *prepersonal* and *personal* bodies" and to use those "specific and selectively chosen characteristics adapted by the actor for public display."[12] As an example, Sobchack uses Arnold Schwarzenegger who, in playing the roles of Commando, Terminator, or Governor of California ("The Governator"), casts his physical and emotive characteristics into the muscular, no-nonsense, machine-like persona that we, as spectators, recognize through "a certain visible continuity or sameness that grounds these figural differences."[13] While performances on film or in the public sphere require that performers use their impersonated body—coalescing traits of the prepersonal and personal ones—to produce a series of expressions in acting or public-facing roles, the athlete is no different. For example, in soccer, a striker may be asked to fill a particular role on the field, with certain attributes being required of that player's body. As their particular play style has been developed through the proclivities allowed by their prepersonal and personal bodies, a striker's size, pace, dribbling, or finishing are all traits that can be used

10 Sobchack, "Being on the Screen," p. 431.
11 Sobchack, "Being on the Screen," p. 433.
12 Sobchack, "Being on the Screen," p. 436, original emphasis.
13 Sobchack, "Being on the Screen," p. 437.

across any position on the field, with certain attributes being more utilized than others. That use of the impersonated body often extends off the field as well in interviews or correspondences. The uniqueness of the athletic body produces a performance that does not require the same medium-reliant presence as a film actor does to transform the body of the athlete into a particular role. Thus, in the realm of professional athletes, Kylian Mbappé's ability to use his on-field speed, ball control, and finishing abilities in spectacular fashion showcases the employment of the impersonated body to enlist the prepersonal and personal layers underneath. This allows him to perform as a striker within the space of the stadium, the narrative of the match, and the presentation of that match in media. These traits can be further augmented by his use of wry and sometimes smug gestures, like the aforementioned smirking or arm-crossing, which create an on- and off-field character who exudes a recognizably subtle, cheeky satisfaction in his achievements.[14]

The final body, for which these previous three provide a foundation, is the personified body. It is described by Sobchack as an abstraction, as "an act of metaphoric substitution [that] stands in for an idea [...][,] a reified amalgam of screen images[.] [...] Off-screen, the film star is usually exalted not as one of his or her characters but as the metaphorized and reified body of their overall cinematic accumulations."[15] In this instance, the personified body becomes a superimposition of the first three into a fourth and final public persona. The on- and off-screen bodies merge and become larger than life, to such an extent that a persona is projected onto the body of the star, regardless of whether they are on public display or playing a role. This can overshadow their identity. The very question, "More Mbappé than me?", presumes from its viewer the knowledge of Mbappé's personified body through the collection of his appearances in games, highlight reels, interviews, and promotional materials, all in a marketable package of a mischievously playful style. Though this will be discussed further below, the *FIFA 19* trailer includes a collection of these personified bodies by presenting soccer superstars like Neymar. Neymar's penchant for flashy, on-field displays of skill and trickery are not the real subject of his appearance in

14 A 2020 *Guardian* article by Adam White and Eric Devin recounts Mbappé's well known demeanor as "affable" and "cheeky," but also comments on his emerging "bolshness," possibly as a result of his friendships with teammates who have a "sense of entitlement" and give off the impression of being "very spoiled." According to White and Devin, this results in a more "conceited" version of Mbappé. See White and Devin, "Is Kylian Mbappé Spoiled or Just Spoiling for a Move from PSG?"

15 Sobchack, "Being on the Screen," pp. 438–439.

the trailer; his similarly extravagant off-field lifestyle is.[16] He is shown at a mock red-carpet sequence where he is bombarded by fans and camera flashes before he finally drives off in a sports car with flames violently discharging from its tailpipes.

Before applying Sobchack's theory to these trailers, I want to briefly situate these bodies within Barry Smart's conception of the "sports star." In *The Sports Star* (2005), Smart distinguishes between "well-knownness" and "greatness" as criteria for a celebrity's acquisition of fame and heroism. The sports star uses greatness—defined as the actions, deeds, or outstanding performances that are well regarded in a cultural collective[17]—to achieve well-knownness, which refers to their celebrity status. In the realm of the on-screen performance, the actor's personified body demands a certain level of well-knownness that is dependent upon the amalgamation of screen images to form a persona. However, that persona may not necessarily be built upon greatness. For example, Schwarzenegger is rarely celebrated for his greatness as an actor, but his appearance in, and the mobilization of his many bodies into, numerous roles as an action-hero in films, series, and product endorsements have led to his indisputable well-knownness. And yet, Schwarzenegger's persona is also reliant upon his participation and the greatness he achieved as a professional bodybuilder, winning the Mr. Universe Championship three times. Regardless, the sports star uses greatness and "licenses nostalgic narratives recalling heroes of the past and outstanding performances," and should not be confused with a celebrity, "in whom greatness is a noticeably absent quality."[18] As EA's marketing practices have grown, we observe a development that uses the athlete's body as it is translated into the space of the *FIFA* video games. This translation creates an additional body to Sobchack's theory, one that pounces on the levitation between the performance of the flesh and the idea it produces in order to manufacture something Smart would describe as "marketable human models—modern 'heroes' mass-produced to satisfy the market."[19]

16 For recent examples of Neymar's reputation in popular journalism, see Garcia, "PSG's Neymar"; Payne, "Neymar Doesn't Care If You Think He Parties Too Much"; and Wilson, "Neymar Injury Cements Season of His Unfulfilled Lofty Expectations at PSG." Neymar is also listed as one of the detrimental factors for teammate Mbappé's attitude in White and Devin, "Is Kylian Mbappé Spoiled or Just Spoiling for a Move from PSG?"

17 Smart, *The Sports Star*, p. 7.

18 Smart, *The Sports Star*, p. 6 and 8.

19 Smart, *The Sports Star*, p. 8.

Marketing Immersion and Sobchack's Bodies: Performative Displacement in the *FIFA* Series

In response to the question EA asked over a decade ago ("How big can football get?"), the answer appears to be "as big as EA's marketing team will allow," provided they supply users with the technologies to promote such growth. Based on the popularity of FIFA Ultimate Team (FUT[20]), which first launched in 2009, EA's focus has been on the ballooning profits of their games' downloadable content (DLC) and paid loot boxes targeted to FUT gamers, rather than on the mechanical qualities of the games themselves.[21] With FUT hosting one million players in 2009, that figure quickly rose to 3.9 million in 2011 before exploding to 11.2 million in 2013 (current numbers hover around 9.1 million players).[22] EA's marketing strategies have quickly shifted to acknowledge this popularity, with *FIFA* titles released since *FIFA 09* slowly incorporating streamer personas and the FUT mechanics into their respective trailers. Below, I break these trailers down into four categories based on their chronology and engagement with Sobchack's four bodies. These categories are the *body present* (*FIFA 07–09*), *body blended* (*FIFA 10–12*), *body analyzed* (*FIFA 13–16*), and *body re-personified* (*FIFA 17–19*).

Prior to the release of FUT, annual launch trailers in the *body present* phase focused less on blending the line of performing bodies caught between screen image, game image, and reality, and instead revelled in the joy of simple appearances and endorsements by stars like Wayne Rooney, Ronaldinho, or Sergio Ramos. These stars' greatness and well-knownness is marked by their cameos in *FIFA 07* and *FIFA 08*'s trailers and relies on the mere recognition of their personified bodies. EA's trailer for *FIFA 09* showcased the first in a long

20 FUT is an online, in-game system used to purchase and sell in-game versions of real players as well as to build personalized teams for competitive play. Players can be purchased through loot boxes, bought and sold from other players using in-game currencies, bartering, or real monetary funds. Success in FUT is rewarded with invitations to sponsored competitions by EA and monetary prizes. It has since been implemented across EA Sports' other popular franchises, including *Madden*, *UFC*, and *NHL*.

21 This criticism of EA's methods has been well-documented in current journalism, academic, and clinical studies. For a brief entry point into this discussion, see King and Delfabbro, "Predatory Monetization Schemes in Video Games (e.g. 'Loot Boxes') and Internet Gaming Disorder."

22 These figures were obtained by Polygon Magazine in a retrospective piece with FUT creator, Matt Prior, and EA Creative Director of FUT, Adam Shaikh, as they reflected on FUT, five years after its creation. See Sarkar, "EA Looks Back on Five Years of the FIFA Ultimate Team Juggernaut." While sources to confirm the number of current FUT players are scarce, the 9.1 million figure comes from a statement released in 2021 by EA Sports Group General Manager, Cam Weber, on EA Games' website. See Weber, "EA SPORTS Comments on the Future of Football."

line of promos organized around grandiose set pieces of soccer stars actually playing the video game. *09*'s trailer features Wayne Rooney and Gonzalo Higuaín squaring off against one another in an online *FIFA* contest, with Rooney playing from a public watch party in Manchester, and Higuaín from a watch party in Madrid. However, star, gamer, and game are distinct and set realities within the space of the trailer, with clear delineations marked between what occurs in the video game projected on screens and what occurs in the real space surrounding those screens. By this I mean that there is no confusion that onlookers are reacting to a single game, as sightlines and clear continuity editing orient viewers to the coherent action unfolding onscreen and observed by crowds in Madrid and Manchester respectively. These trailers present the sports star and their four bodies on-screen to be adored and celebrated for simply appearing in them.

FIFA 10 initiates the *body blended* phase with its inclusive tagline, "Let's FIFA 10," which proposes a collective participation among its members, regardless of their role as gamer, sports star, or sports star avatar. Unlike the *FIFA 09* trailer, the *FIFA 10* trailer contains such a collection of stars, gamers, and crowds concurrently observing and participating in a plethora of real and virtual in-game matches that the incomprehensibility of the concurrent events and bodies becomes the trailer's defining trait. A goal scored by Xavi on a pitch in Barcelona is instantly met with images of soccer stars and gamers celebrating across distant countries in Europe and Africa.[23] These celebrations are shown performed by bodies in reality, and avatar bodies in the video game. Even the trailer's narration complicates its visual comprehension, and gestures to the game's seemingly infinite networks of synchronicity and simultaneity of play through lines like "ten-hundred-thousand heartbeats skipping, ten-million-billion silent prayers," and "ten-hundred-million whistles blown."[24] Their numeration is nonsensical, but the description is nonetheless overwhelming, with the constant cutting between actions—actions occurring across a seemingly infinite collection of locales (stadiums, living rooms, town squares, and backyards)—creating a sense of collectivity that can only be read through narrative simultaneity. Through a series of *mismatches*-on-action, the apparent seamlessness between these locations and their unfolding drama is only comprehensible by reading the network's narrative of games as a whole. The sequence that closes *FIFA 10*'s trailer, recounted at the opening of this chapter, describes faces melding together through zoetrope and thaumatrope-like effects, with

23 Official EA UK, "EA Sports FIFA 10 – The Full 2 Minute TV Ad," 1:36.
24 Official EA UK, "EA Sports FIFA 10 – The Full 2 Minute TV Ad," 0:06, 0:16, 0:38.

gamers, soccer stars, games, and places momentarily becoming one. This trailer is the first to initiate a marketing strategy that attempts to blur the line between flesh and idea, through the very technologies that the FUT system puts forth.

Launch trailers for both *FIFA 11* and *12* continued to build upon these blurred bodies and boundaries between the stars who endorse the game and the growing network of gamers who play them. In the TV spot for *FIFA 11,* a narrator declares: "No matter how many we number, we play eleven strong," alluding to an almost endless number of faceless gamers playing the game.[25] Faceless players appear throughout the trailer by way of bodies whose entire heads have been replaced with soccer balls inscribed with an "11," and promote that year's tagline, "We are 11." All bodies, regardless of their owner's status as a soccer star, gamer, or faceless participant, are shown as digital projections, occasionally glitching as they appear or disappear in an all-too-obviously digital arena where the action unfolds. In the trailer's final moments, Wayne Rooney appears with the trailer's collection of sports stars, gamers, and faceless bodies behind him, raises his hand, presses a button on his gamepad, and causes everyone on screen to collapse into one final, stylized, digital flash.[26] The digital collapse signals a visual conflation of all the bodies into one, while the cut to black gestures to the demise of traditional lines that separate gamer and sports star, a tradition that will soon be abandoned entirely.

The end of *FIFA 12*'s trailer lays those traditional lines of comprehension between bodies to rest with a mock funeral procession held in an empty stadium. Prior to this moment, *12*'s trailer exhibits a series of real and *FIFA*-generated soccer matches that are accompanied by a young man's never-ending conditional narration: "If I can't play a match, I'll tell someone how to play a match. If I can't tell someone how to play a match, I'll be at a match. If I can't be at a match, I'll watch a match."[27] This endless stream of conditionals is complemented by footage of fans and soccer stars like Gerard Piqué and Karim Benzema commenting on or taking part in the action occurring in both real and *FIFA*-generated matches. Though there are two versions of this trailer, both include a larger collection of soccer stars turned pallbearers, who accompany the narrating gamer's body and casket through a stadium at the trailer's end. In one version, the procession is cut short by the stadium's floodlights being turned off and the young man transported

25 EA Italia, "EA Sports FIFA 11 – Spot TV Extended," 0:24–0:30.

26 EA Italia, "EA Sports FIFA 11 – Spot TV Extended," 1:53–1:55.

27 Electronic Arts (EA) Singapore, "FIFA 12 TV Ad | Love Football, Play Football," 0:00–0:36.

back to his couch. In the alternate version, the camera cuts to the interior of the casket, where the young narrator has a monitor fixed to the roof of his coffin and a controller in hand. Both conclude with the young man turning to the camera and casually uttering, "Guess I'll be having a match then," before gameplay footage appears, and the EA Sports logo presents itself.[28] The death in both trailers symbolizes a rebirth for the franchise and a new trajectory that continues to complicate the relationship between bodies while attempting to harness the traits of the athletes' physical bodies.

FIFA 13's trailer intensifies this sense of belonging.[29] As the first title in the body *analyzed* phase, it is characterized by a dissection of the athletic body through both an aesthetic exploration of the soccer star's body and an economical exploration of the body and its physical abilities. With its tagline, "Join the Club," *FIFA 13*'s trailer shows a young man's entrance into an exclusive soccer clubhouse, populated by other gamers who are buying and selling real and digital players at markets, describing tactics, and interacting with real-life professional soccer players and coaches. *FIFA 13* matches are being streamed live in the background as these gamers, many of whom are just preteens and teenagers, are interviewed and praised as objects of stardom rather than the actual soccer stars present. Most notably, a scene involving Arsenal star, Alex Oxlade-Chamberlain, shows the young player submitting himself to a FUT auction, with an overall player rating of 82, and a rising value of 315,000 credits visible for all to see. The trailer thus suggests that by appropriating these heroic sport-star bodies, gamers can in fact achieve a level of well-knownness —and perhaps even of heroism or greatness—of their own.[30] The push to heighten these quantifiable abilities while exploring the athletic body gestures to an appropriation of those bodies by gamers in the trailers for *FIFA 14* and *15*, with their respective taglines: "It Just Got Real" and "Feel the Game."[31] These trailers exhibit this conflation through the corporeal superimposition of the athlete's body onto the gamers. However, it is the athlete's digital body—in this case, the body of Lionel Messi—that is projected into the gamer's space of reality, which results in a brutal and sometimes grotesque transformation of the body and space. We are subjected to the gamer's eyes growing wider, their teeth being displaced, or their feet turning to cleats which match Lionel Messi's

28 EA SPORTS FIFA, "FIFA 12 TV Ad with Alternate Ending | Bury Me," 0:48–0:58.

29 EA SPORTS FIFA, "FIFA 13 TV Ad – Join the Club."

30 Moreover, Sony's entire marketing campaign across the PlayStation 4's life cycle was "Greatness Awaits."

31 EA SPORTS FIFA, "FIFA 14 TV Commercial – Next-Gen Lionel Messi"; and EA SPORTS FIFA, "FIFA 15 – Official TV Commercial."

own body. Impact from slide tackles that occurs in the game are shown to be felt in the reality of the gamers' space.

FIFA 16 functions like a scientific exploration of the athletic body and of its transference of skill and athleticism to the video game player. The trailer shows cameramen filming two soccer stars (Lionel Messi and Sergio Agüero) fitted with motion capture technology.[32] The images break down their movement into simple polygon figures, quantify what was mechanically captured, and then translate it into a polished version of the FIFA game. In the next scene, gamers take digital control of the two players who have just had their abilities, greatness, and heroism captured and quantified. The entire sequence shows the speed of transference, with five cuts in just three seconds exhibiting the quick temporal and technological progression that these players and their physical bodies undergo in being digitally translated.

The trailers for *FIFA 17*, *18*, and *19* signal a final shift towards an increasingly cessative obligation to the athlete's physical and real body in the *re-personified* phase of these trailers.[33] The tagline for the *FIFA 17* trailer, "Leave Your Mark," is accompanied by a young gamer stepping out of his living room and into a sports awards ceremony where he is accoladed with praise, prize, and stardom for his in-game efforts—efforts which are projected onto a screen above the stage where the ceremony takes place. Perversely, the very players he uses in the game are attending the awards ceremony, cheering *him* on. There is a full circle movement during which Sobchack's four bodies, amalgamated into the personified body ("the reified amalgam of screen images"[34]), become *re-personified*; they are repurposed by the gamer into an image that *somehow is, yet never truly was*. It is not that the gamer's talent or heroism is entirely absent in the realm of gaming, but that this advertised heroism uses the digital shell of a repurposed, displaced athletic performance. If each of Sobchack's bodies is a matryoshka doll, one inside the other, then the re-personified body is a fifth doll in which the digital spectres of the original possessor's four bodies are contained. In "The Displacement Project" (2016), Brian Price explores the relationship between projection and being. He explains that to project is both an act of display and a process of institutionalization or enlistment of objects into a new social order. He writes that "[i]f I am seeing something that I have never before

32 EA SPORTS FIFA, "FIFA 16 – Play Beautiful – Official TV Commercial."

33 EA SPORTS FIFA, "FIFA 17 – Make Your Mark – Official TV Commercial"; EA SPORTS FIFA, "FIFA 18 El Tornado – More Than a Game – Official Trailer"; and EA SPORTS FIFA, "FIFA 19 | Champions Rise | Official Launch Trailer."

34 Sobchack, "Being on the Screen," p. 438.

been present to in its material form as an object in a particular place, and I describe this experience as an instance of *seeing again*, I do so because the image is an object and also nothing material at the very same time."[35] The marketing trailers for these games affirm that same strategy in their creation, in that *seeing again* leads to the visual process of *making again*, whereby greatness and athleticism are, before our very eyes, supplanted through a reproduction of the heroic athletic body underneath, and subsequently enlisted into a new social order.

In the *FIFA 18* trailer, gamers use an avatar version of superstar Cristiano Ronaldo to create and perform an in-game trick dubbed "El Tornado."[36] Though Ronaldo's bombastic persona makes him an example of a star player capable of such a trick, the real Ronaldo is shown unable to perform it. The trick is completed in the *FIFA 18* game by assumed FUT streamers. It is then recorded and shared on YouTube, Twitch, and Twitter, where it is infinitely memed, hashtagged, and GIF-ed by prominent YouTubers like W2S.[37] Noticing the trick on social media being performed with his own *re-personified* body, Ronaldo ceaselessly trains to complete it. After successfully doing so in a live match, highlights of his El Tornado performed in reality are then re-inscribed into the social media mega-verse through memes, hashtags, and GIFs, while simultaneously being reimagined by other professionals and gamers. The tagline for *FIFA 18*, "More than a Game," along with the El Tornado trailer, markets its network of technology to dictate a new potential reality through its digital setting and avatars. This ventriloquist's act is taken even further in the *FIFA 19* trailer. In addition to the Mbappé example that titles this chapter, streamers are shown throughout the trailer as having soccer stars' digital bodies projected onto their own, to the point where mis-identity becomes the result of the re-personified body.[38] In the *FIFA 19* trailer, a young woman is shown on screen, visibly invested in playing the game. As she uses the gamepad, her face glitches, with a sporadic and momentary projection of professional players imprinting themselves onto her face. During this momentary imprint, the young woman verbally identifies with those very players projected onto her, and is simultaneously controlling; "I am Neymar. I am De Bruyne. I am Dybala," she insists.[39] At

35 Price, "The Displacement Project," p. 236, original emphasis.

36 EA SPORTS FIFA, "FIFA 18 El Tornado – More Than a Game – Official Trailer."

37 With his 16 million subscribers on YouTube alone, W2S has gained notoriety from his videos playing matches on FUT, opening card packs (loot boxes), or emulating soccer stars like Cristiano Ronaldo and Lionel Messi.

38 EA SPORTS FIFA, "FIFA 19 | Champions Rise | Official Launch Trailer."

39 EA SPORTS FIFA, "FIFA 19 | Champions Rise | Official Launch Trailer," 0:18–0:39.

Figure 14.3: *FIFA 19* Launch trailer.

this moment, the real Neymar appears. The two look at each other, sharing a moment of simultaneous (mis)recognition, as the projection of his digital face maps itself onto hers (fig. 14.3). All of these moments gesture towards an overwhelming pattern of perversity that performance scholars like Wendy Donniger might refer to as "self-impersonation."[40] However, where Donniger lists self-impersonation as a "basic human way of negotiating relation, illusion, identity, and authenticity,"[41] these moments simultaneously apply and discard the "self" from that term to negotiate a newly materialized unreality fraught with dis-relation, dis-identity, disillusion, and inauthenticity.

The heavy reliance of these trailers on glitch aesthetics reminds us, through Benjamin Mako Hill's definition of the glitch, that "errors can help reveal the [...] hidden constraints and the power that technology imposes" while also revealing "the presence of this hidden mediation."[42] If, as Hill asserts, the purpose of technology is to hide its own complexity, then a blatant display of these technologies of concealment turns this assumption upside down, particularly in the trailers for *FIFA 17*, *18*, and *19*. By this I mean that the visual glitch, embraced as a normalized aesthetic convention (a trend seen in countless other marketing campaigns for films, commercials, and video games) inverts the error so that an apparently ubiquitous dysfunction masks the truth of the real, functioning, and thriving technology in operation. By normalizing the error, the line between function and disfunction, reality and unreality, human and non-human is placed into flux to the point where (mis)recognition, (dis)relation, and (un)reality are inextricably tied to—and experienced simultaneously in—this new norm. The normalized glitch provides a backdrop to the space where these interactive technologies function in an accessible limbo between the flesh and the product seen on the screen and in the stadium. The technical shortcomings of media

40 Donniger, *The Woman Who Pretended to Be Who She Was*, pp. 10–11.
41 Donniger, *The Woman Who Pretended to Be Who She Was*, p. 10.
42 Hill, "Revealing Errors."

and the gaps they produce, from the earliest flickers of the zoetrope and thaumatrope to the latest flash of the glitch, become the sites through which immersion seams its latest innovations.

It is my hope that this brief survey of entries in the *FIFA/EA Sports FC* series will provide a departure point for the continued analysis of the media that accompany and promote these games, inviting discourse on these objects through an array of critical lenses. How far will these displacing technologies and EA's marketing of them take us? When a commentator in the *FIFA 19* trailer rhetorically asks the viewer, "Can you reach the final in Madrid?" a boy answers, "Madrid? It's right here in L.A.," while another shouts "Tokyo" and a third, "Milan."[43] In October 2021, EA's trailer for *FIFA 22* and its accompanying campaign, "Powered by Football," suggested that its games had created "a second reality, more powerful than the first."[44] With these games no longer being exclusively used to provide leisure-time activities, but instead used as newfound platforms of labour with their own emerging star systems, how might we continue to examine the displacement of bodies as these technologies and their industries grow?[45] Is "More Mbappé than Mbappé" just the start, or is L.A. being more Madrid than Madrid, more Tokyo than Tokyo, or more Milan than Milan a new norm? Is this how big football can get?

Bibliography

Donniger, Wendy. *The Woman Who Pretended to Be Who She Was: Myths of Self-Imitation.* Oxford: Oxford University Press, 2005.

EA Italia. "EA Sports FIFA 11 – Spot TV Extended." September 17, 2010. YouTube video, 2:00. https://www.youtube.com/watch?v=vPwrsAm728E.

EA SPORTS FIFA. "FIFA 12 TV Ad with Alternate Ending | Bury Me." October 21, 2011. YouTube video, 1:00. https://www.youtube.com/watch?v=JAQK9c8sS50.

43 EA SPORTS FIFA, "FIFA 19 | Champions Rise | Official Launch Trailer," 1:28–1:35.

44 PlayStation. "FIFA 22 – Powered by Football – Official Launch Trailer | PS5, PS4."

45 Professional sport governing bodies like the National Basketball Association (NBA) and Major League Soccer (MLS) have partnered with EA to establish the North American 2K League and eMLS League in 2017 and 2018, respectively. This trend has developed in Europe, with the English ePremier League launching in 2019 and the Italian eSerie A league in 2020. While gamers can not only claim prize money by competing in these tournaments, they are also employed by the teams taking part in these leagues. For example, those who take part in the NBA 2K League are employed by professional NBA basketball teams on six-month contracts and can earn a salary between US$35,000 and US$38,000 before additional sponsorship deals. See "NBA 2K League Increases Prize Pool to $1.5 Million for 2021 Season."

EA SPORTS FIFA. "FIFA 13 TV Ad – Join the Club." August 30, 2012. YouTube video, 1:35. https://www.youtube.com/watch?v=7ZZDvBQCfm8.

EA SPORTS FIFA. "FIFA 14 TV Commercial – Next-Gen Lionel Messi." November 15, 2013. YouTube video, 0:57. https://www.youtube.com/watch?v=DYYy8_GCdn4.

EA SPORTS FIFA. "FIFA 15 – Official TV Commercial." September 12, 2014. YouTube video, 2:24. https://www.youtube.com/watch?v=Q21bznzy_HY.

EA SPORTS FIFA. "FIFA 16 – Play Beautiful – Official TV Commercial." September 16, 2015. YouTube video, 1:56. https://www.youtube.com/watch?v=F-OgKyUuFfI.

EA SPORTS FIFA. "FIFA 17 – Make Your Mark – Official TV Commercial." September 22, 2016. YouTube video, 1:26. https://www.youtube.com/watch?v=yRo8DtAfNZ4.

EA SPORTS FIFA, "FIFA 18 El Tornado – More Than a Game – Official Trailer." September 28, 2017. YouTube video, 2:20. https://www.youtube.com/watch?v=pjR51wO7vII.

EA SPORTS FIFA, "FIFA 19 | Champions Rise | Official Launch Trailer." September 18, 2018. YouTube video, 2:06. https://www.youtube.com/watch?v=OumZxTdMq_c&t=31s.

Electronic Arts (EA) Singapore. "FIFA 12 TV Ad | Love Football, Play Football." August 30, 2011. YouTube video, 1:00. https://www.youtube.com/watch?v=mOdzmgg6Ifo.

Electronic Arts Inc. "FIFA 21 Ambassadors." *FIFA Ultimate Team*. https://fifauteam.com/fifa-21-ambassadors/.

Garcia, Adriana. "PSG's Neymar: I Will Never Stop Partying; I'm Not Immature." *ESPN*. February 2, 2021. https://www.espn.com/soccer/paris-saint-germain/story/4304838/psgs-neymar-i-will-never-stop-partying-im-not-immature.

Hill, Benjamin Mako. "Revealing Errors." *M/C Journal* 10, no. 5 (2007). https://doi.org/10.5204/mcj.2703.

King, Daniel L., and Paul H. Delfabbro. "Predatory Monetization Schemes in Video Games (e.g. 'Loot Boxes') and Internet Gaming Disorder." *Addiction* 113, no. 11 (2018): 1967–1969.

"NBA 2K League Increases Prize Pool to $1.5 Million for 2021 Season." *NBA 2K League*, April 27, 2021. https://2kleague.nba.com/news/nba-2k-league-increases-prize-pool-to-1-5-million-for-2021-season/.

Official EA UK. "EA Sports FIFA 10 – The Full 2 Minute TV Ad." November 17, 2009. YouTube video, 2:01. https://www.youtube.com/watch?v=YfdvTU11AkQ.

Payne, Marissa. "Neymar Doesn't Care If You Think He Parties Too Much." *The Washington Post*, July 28, 2016. https://www.washingtonpost.com/news/early-lead/wp/2016/07/28/neymar-doesnt-care-if-you-think-he-parties-too-much/.

PlayStation. "FIFA 22 – Powered by Football – Official Launch Trailer | PS5, PS4." October 3, 2021. YouTube video, 2:30. https://www.youtube.com/watch?v=pA9hutnFuRo.

Price, Brian. "The Displacement Project." *Discourse* 38, no. 2 (Spring, 2016): 235–245.

Sarkar, Samit. "EA Looks Back on Five Years of the FIFA Ultimate Team Juggernaut." *Polygon*, March 19, 2014. https://www.polygon.com/2014/3/19/5525710/fifa-ultimate-team-fifth-anniversary-ea-sports-interview.
Smart, Barry. *The Sports Star: Modern Sport and the Cultural Economy of Sporting Celebrity.* London: SAGE Publications Ltd, 2005.
Sobchack, Vivian. "Being on the Screen: A Phenomenology of Cinematic Flesh, or the Actor's Four Bodies." In *Acting and Performance in Image Culture: Bodies, Screens, and Renderings*, edited by Jorg Sternagel, Deborah Levitt, and Dieter Mersch, 429–446. Bielefeld: Transcript, 2012.
Weber, Cam. "EA SPORTS Comments on the Future of Football." https://www.ea.com/news/fifa-22-and-the-future-of-football.
White, Adam, and Eric Devin. "Is Kylian Mbappé Spoiled or Just Spoiling for a Move from PSG?" *The Guardian.* February 3, 2020. https://www.theguardian.com/football/2020/feb/03/kylian-mbappe-spoiled-move-psg-thomas-tuchel-neymar.
Wilson, Jonathan. "Neymar Injury Cements Season of His Unfulfilled Lofty Expectations at PSG." *Sports Illustrated*, March 1, 2018. https://www.si.com/soccer/2018/03/01/neymar-injury-surgery-psg-real-madrid-brazil-world-cup.

About the Author

Daniele Iannucci is a PhD candidate at the University of Toronto's Cinema Studies Institute. His research considers sports portraiture as a form of interactive documentation and surveillance in world soccer. His other research interests include prank calls, audio verité, and infomercials.

Section 3

The Edge of Sense(s)

15. Walter Benjamin on the Threshold of Immersion

Sonny Walbrou

Abstract: Drawing on Walter Benjamin's philosophy, the chapter interrogates the notion of immersion within the context of cinematic experience. Benjamin's concept of distraction, situated against the backdrop of nineteenth-century visual culture, provides a critical lens through which to examine the immersive tendencies of cinematic representation. By juxtaposing Benjamin's philosophy with Jacques Perconte's experimental film *Après le feu*, this chapter illuminates the complexities of cinematic immersion and its relationship to digital mediation.

Keywords: distraction, nineteenth-century visual culture, Kaiserpanorama, stereoscope, voyage

> In *The Arcades Project*, contemplation must be put on trial. But it should defend itself brilliantly and justify itself.
> —Walter Benjamin

> Perspectives must be fashioned that displace and estrange the world, reveal it to be, with its rifts and crevices, as indigent and distorted as it will appear one day in the messianic light. To gain such perspectives without velleity or violence, entirely from felt contact with objects—this alone is the task of thought.
> —Theodor W. Adorno

Along the tracks of a railroad, a landscape quite similar to those of early cinema's phantom ride shots fleetingly emerges at the edges of the frame. Looking at it from across the windshield of a train, the reflecting sunlight reveals the screen in-between. This scenery was shot on a short section of

Bédard, Philippe, Alanna Thain, and Carl Therrien, eds. *States of Immersion Across Media: Bodies, Techniques, Practices*. Taylor & Francis Group, 2025
DOI 10.5117/9789048558766_CH15

Figure 15.1: *Après le feu* (Jacques Perconte, 2010, © ADAGP).

the railroad connecting Bastia and Ajaccio, in France, amidst a landscape reborn from a blaze that happened a few years before. From the front of the car, the camera frames the railway tracks at the centre and the landscape around. The tracks thus designate for the spectator a vanishing point that absorbs them towards a horizon that the train's movement keeps pushing away. As the train and the film continue, the cinematic space starts to disintegrate: the air becomes solid while the rocks seem to turn to liquid. The train goes on and so does the landscape, but some parts of it remain strangely still: pieces of it seem arrested or blocked. Pixels appear that scrupulously unsettle the surface of the image. Then, flares of a strange nature develop, devouring the landscape, whose reality is no longer obvious. The film seems to turn the reviving landscape into a digital second nature. Until the arrival of the train at the station—which never occurs as such in the film, but is rather indicated through the absolute stillness of indecipherable images—the film crosses the different layers of the image while keeping the railway's perspective in view.

Between the past and the present, between the phantasmagorical trip of early train films[1] and the digital images' breakdown,[2] the experimental film *Après le feu* (Jacques Perconte, 2010) places the viewer in the interval between the image and its medium. By the end of the film, only traces of the film's format remain visible on the screen; the travel leads us rather

1 See Belloï, *Le regard retourné*; Gunning, "Landscape and the Fantasy of Moving Pictures"; Kirby, *Parallel Tracks*; and Musser, "The Travel Genre in 1903–1904."

2 See Denson, *Discorrelated Images*; and Jacobs, *Esthétique du signal*.

downward than forward, or, in others words, the format keeps emerging, thus disturbing the surface of the illusion of immediacy. The film traces a route connecting this illusion to the *substrate* it usually denies. Like the landscape it first depicts, *Après le feu* proceeds unevenly as the deep *strata* find their way towards the surface. Indeed, deeper digital processes seem to disrupt the film's surface.[3] The spectator then sees the film's "digital nature," a nature that the image's deceptive naturalization tends to obscure. What happens then is an inversion of the image and of its traces. A dialectic of the image occurs as Perconte turns the image around on the basis of its essential contradictions: stillness and motion, transparency and opacity, immediacy and computation. Nonetheless, neither of the opposites merely replaces the other, the pixel or the computation never completely reject and erase the initial landscape. Instead, *Après le feu* develops a threshold: it establishes a space "in-between [...] where strange things happen."[4] The film brings together the "here" of the material conditions of the image and the "there" of the landscape it shows in the first place. Because it elaborates on early cinema's phantom ride shots and its viewer's related immersion while directing their attention towards the image's materiality, *Après le feu* corresponds to an aesthetic of the threshold, one I will discuss in this chapter using Walter Benjamin's philosophy.[5] Revealing the image's digital signal is not the end of the journey; it rather (re)places the viewer in between a state of immersion and a state of medium awareness. While the phantom ride shots aesthetic of the film bears early cinema's "ideals of immersion"[6] of the virtual voyage,[7] the gradual disruption of the image unveils its materiality and features the distancing qualities of experimental cinema. This is not, however, experienced as a rupture in *Après le feu*. The viewer does not jump from one state to the other: he or she is rather gradually placed within a space that exists only for its contradictions to be experienced. While this threshold is usually defined as a passage or as a transitional space that leads towards immersion or away from it,[8] it bears consideration in its own right

3 As Bidhan Jacobs writes about Perconte's filmmaking: "Compression is at the core of his work and acts by revealing the signal." Jacobs, *Esthétique du signal*, p. 360.

4 Barberger, *Pensées de passage*, p. 27.

5 I could have mentioned other films by Jacques Perconte, for instance *Uishet* (2012) or *Ettrik: Scottish Borders* (2015). I also could have mentioned other filmmakers whose experimental cinema involves trains and viewers far beyond the boundaries of immersion and disruption, such as in the films *The Georgetown Loop* (Ken Jacobs, 1996), *Phantom Ride Phantom* (Siegfried Fruhauf, 2017), or *Train Again* (Peter Tscherkassky, 2021).

6 See Holmberg, "Ideals of Immersion in Early Cinema."

7 See Ruoff, *Virtual Voyages*.

8 See Murray, *Hamlet on the Holodeck*; and Ryan, "Immersion vs. Interactivity."

as it is opposed neither to immersion nor to disruption. We shall find in Benjamin's work proper thoughts to guide us in that place yet to be outlined.

The connections between Walter Benjamin's philosophy and the concept of immersion are not obvious. Looking only at his work about surrealism, cinema, and Bertolt Brecht, he appears much more as a philosopher of the disruptive or of the distancing effect: his shock aesthetic dispels any effect of illusion or absorption, which he associates with dreams and hallucinations. What he terms "distraction" designates the visual culture emerging alongside cinema, as opposed to a past mode of reception that features meditation and contemplation. Benjamin's approach to immersion thus seems immediately critical. Along with the backdrop of positivism, the misleading immediacy that defines nineteenth-century panoramas, historicism, urbanism, world fairs, and more broadly the pervasive perspective system, compel him to avoid substituting the real for its representations. Benjamin interprets such a state, in which representations are meant to be experienced as life itself, in which the illusion exists only to conceal any hint of work, and in which skilfully produced images feed a broader culture of faint "proximity," as a "dreamworld."[9] This world has its houses, its boulevards, its fairs, its department stores, its machineries, its apostles, and, above all, its visual and spectacular economies.[10] The few immersive apparatuses mentioned by him take part in this overall ideological structure as means of alienation: in that sense, they need to be considered as phantasmagorias. To the philosopher, they are masks, illusions, or veils—as much as they are ideas, concepts, discourses, as much as they are objects, apparatuses, *dispositives*—diverting society from contact with its real conditions of existence: past, present, and future. Marc Berdet underlines the significance of phantasmagoria in Benjamin's philosophy, a concept with which the philosopher institutes "the trial [...] of the production's concealed traces."[11]

9 See the chapter "Dream World of Mass Culture" in Buck-Morss, *The Dialectics of Seeing*, pp. 253–286.

10 Although the "dream" is one of the key concepts of the overall *Arcades Project*; see in particular the section K ("Dream City and Dream House, Dreams of the Future, Anthropological Nihilism, Jung", pp. 388–404) and L ("Dream House, Museum, Spa", pp. 405–415).

11 Berdet, *Le chiffonnier de Paris*, p. 13. We should further develop elsewhere the relationships between Benjamin's approach to phantasmagorias and the concept of "commodity fetishism" in Karl Marx's works. Indeed, immersion considered as a naturalized or obscured form of the artificial features interesting connections with Marx's definition of the "commodity-form." Although this cannot be developed here, the critical stance of Benjamin mentioned in this chapter should be understood in regard to Marx's concept of commodity fetishism, as developed in *Das Kapital* (1867). Indeed, if we look at the immersion phenomena as "sensuous things which are at the same time suprasensible or social," we are likely to insist on its historicity rather than

Considering Benjamin's work on cinema as well as his more scattered comments about the stereoscopic device of the Kaiserpanorama, I will consider the intricate relationships between his philosophy and the notion of immersion. Through the "threshold" concept, I will describe a Benjaminian "threshold aesthetic," characterized by both a proximity to immersion and a critical distance. Just as the *flâneur*, the collector, and the ragpicker are "epistemological characters,"[12] I shall demonstrate that both the cinema and the Kaiserpanorama appear as "epistemological devices." Benjamin studied these apparatuses and others beyond the mere description or the immediate consideration of the devices themselves. Instead, they materialize specific forms of knowledge, as a part of a larger cultural sphere developing new ways of thinking and of engaging with the surrounding world and its phenomenon.

A fragment of "The Work of Art in the Age of Its Mechanical Reproduction" (1935) first requires our attention. In this essay, to underline the shift in the modes of reception of artworks throughout history, Benjamin opposes the attentive perception of painting and the tactile perception of cinema. According to him, while the attentive projection of a spectator towards an artwork characterizes the mode of reception of painting, cinema determines the opposite trajectory: the artwork becomes the projectile that forces its way towards the spectator. He thus clearly distinguishes pictorial meditation and cinematographic distraction, the surgical nature of the latter denying the magical nature of the former. To illustrate this difference between the two modes of reception, he refers to a legend: "Distraction and concentration form an antithesis, which may be formulated as follows. A person who concentrates before a work of art is absorbed by it; he enters into the work, just as, according to legend, a Chinese painter entered his completed painting while beholding it."[13] The spectator who "plunges into" an artwork dives into it in the same way the Chinese painter "disappear[s] in the painted pavilion in the backdrop of the landscape."[14] This legend

to assume an ontological definition based on its immediacy effects. See Marx, *Capital*, p. 165. On the relationships between Benjamin and Marx's concept of "phantasmagoria," see Cohen, "Walter Benjamin's Phantasmagoria"; and Markus, "Walter Benjamin, or The Commodity as Phantasmagoria."

12 Berdet, *Le chiffonnier de Paris*, p. 11.

13 Benjamin, "The Work of Art in the Age of Its Mechanical Reproducibility," p. 119.

14 Benjamin, "L'œuvre d'art à l'époque de sa reproduction mécanisée", p. 215, my translation. The legend he is talking about echoes a short story published by Marguerite Yourcenar in *La Revue de Paris*, in 1936, "Comment Wang-Fô fut sauvé." She states from the start that it is adapted from an old Chinese legend—probably concerning the life of the painter Wu-Daozi. The short story ends with the passage of the painter within his own painting, thus escaping the emperor's

accords the viewer a posture of absorption and of projection. It is a situation in which the viewer's position in real space is entirely replaced by the one he or she now occupies in the image. Here absorbed by perspective and resemblance,[15] the spectatorial experience denies and obscures any kind of confrontational, shocking, and contradictory relationships between actual and represented spaces. In distinction, cinema is "shifting perspectives."[16] These "shifts" define a key element of the cinematographic representation of space and time and outline its mode of reception. For Benjamin, the film's spectator is not fixed or still, nor does their gaze merge entirely with the camera's: they dwell in the interval, as they are aware of the cinematographic apparatus and maintain a dialectical relationship to the image. Benjamin tends to highlight the dislocation at the core of cinema's liberatory aesthetic opposition to the immersion implied by previous modes of reception.[17] He sees in it the liberation to come, which is inherent to modernity, that might overcome the persistent ancient modes of knowledge.

Instead of immersion, and by way of modern techniques including cinema, Benjamin invites us to think in terms of "threshold experiences": "We have grown very poor in threshold experiences," he writes in the 1930s.[18] He draws a clear distinction between threshold and boundary: "A *Schwelle* [threshold] is a zone"[19] and is subjected to swelling or to distension.[20] A zone features a thickness and allows for transformation and variation; a threshold is thus presented as a dynamic space that cannot be reduced to a line nor to a boundary. It is a matter of co-penetration, where contradictions meet and somehow melt each other without any form of absolute substitution of one another. In that gap, the "here" and the "there" interpenetrate, just like the past and the present, or like a dream and an awakening: "Unlike the boundary, the threshold represents a transitional space, a dynamic gap made out of transformations and tensions: anything but the imposition of a line. Its function is not to divide but to put heterogeneous spaces in

death sentence: " [T]he trail vanished in the deserted vastness, and Wang-Fô and his student disappeared on that crystal sea Wang-Fô had just created." Yourcenar, "Comment Wang-Fô fut sauvé," p. 859, my translation.

15 In a short review of a Chinese paintings exposition in 1938, Benjamin insists on multiple occasions on the notion of "resemblance." See Benjamin, "Peintures chinoises à la Bibliothèque nationale."

16 Benjamin, "The Work of Art in the Age of Its Mechanical Reproducibility," p. 114.

17 On this, see Boissière, "Apparence et jeu, valeur cultuelle et valeur d'exposition chez Walter Benjamin."

18 Benjamin, *The Arcades Project*, p. 494.

19 Benjamin, *The Arcades Project*, p. 494.

20 See Weber, *Benjamin's-Abilities*.

contact, to make them meet each other."[21] Then, a mode of perception arises, one that expresses both Benjamin's philosophy and the cinematic experience. In between distance and proximity, the threshold experience refuses spatiotemporal identity as well as the empathetic identification of the spectator.

While the indistinction is concealed so it is never experienced as such in the case of immersion, in the threshold, the indistinction becomes the medium or the *milieu* of a new mode of perception. The correlated dislocation is a theoretical construction that takes a critical stance about the seeming clarity and illusion of immediacy. On the contrary, threshold experiences confront what is never given immediately. One can grasp the significance of the cinematographic spatiotemporal confusion in regards to the "dialogic misunderstanding" that Benjamin identifies in surrealism: "'Misunderstanding' is here another word for the rhythm with which the only true reality forces its way into the conversation. The more effectively a man is able to speak, the more successfully he is misunderstood."[22]

The threshold is explicitly evoked by Benjamin in reference to the Kaiserpanorama, a device invented by August Fuhrmann in 1883 and in use until the eve of World War II. In it, the spectator takes a seat in a circular structure. Each of the wood panels in front of them features a stereoscopic binocular. The overall device contains a mechanism that allows for a series of views to pass from one station to the next. At regular intervals, an image is seen by the spectator and then passes to the next, replaced by the view coming from the previous station. While circulating, the photographs offer different views of remote places to a spectator who remains seated all along. The illusion of depth specific to stereoscopic photography increases the sense of an immobile journey, a spectacular tradition the Kaiserpanorama can be tied to. Through the succession of views, the spectator leaps from one location to the next. As recalled by Benjamin in *Berlin Childhood around 1900* (1937): "One of the great attractions of the travel scenes found in the Imperial Panorama was that it did not matter where you began the cycle. [...] One got used to taking the tour in a half-empty room."[23] He continues:

> When I entered for the first time, the time of the most beautiful *vedute* was long gone. Though the magic whom the children were the last audience was still alive. In front of a transparent view of the city of Aix, it tried to

21 Barberger, *Pensées de passage*, p. 15, my translation.
22 Benjamin, "Dream Kitsch" p. 4.
23 Benjamin, *Berlin Childhood around 1900*, pp. 346–347.

> persuade me I had already played among the olive-green coloured lights that fall through the plane trees of the Cours Mirabeau, at a time that shared nothing with other moments of my life.[24]

While in this memory of the Kaiserpanorama the author seems to submit to the role dictated by the apparatus, he only partially does so. Indeed, he reminds us of details undermining the effectiveness of immersion: he stands both in the actual room and in the virtual place at the same time, somewhere in between here and there. Both the sound of the bell and the tremor of the frame announcing the image's departure are underlined:

> But there was a small, genuinely disturbing effect that seemed to me superior. This was the ringing of a little bell that sounded a few seconds before each picture moved off with a jolt, in order to make way first for an empty space and then for the next image. And every time it rang, the mountains with their humble foothills, the cities with their mirror-bright windows, the railroad stations with their clouds of dirty yellow smoke, the vineyards down to the smallest leaf were suffused with the ache of departure.[25]

The recalled experience, summarized by Benjamin as "genuinely disturbing," remains in a gap between the illusion of a voyage via images and the distraction specific to the device. This position is evident when the philosopher mentions the sound and the shiver surrounding the image that is about to vanish. In *The Arcades Project* (posthumous, 1982), the same bell appears again as a "disturbing power,"[26] but is more closely related to the threshold, as the sound of the doorbell threatens the enclosure of an apartment. Indeed, the sound signals that something is about to happen: the enclosed space is about to open and, even for a short period of time, the boundaries will be denied, or at least they will swell:

> The despotic terror of the hand bell, the terror that reigns throughout the apartment derives its force no less from the magic of the threshold. Some things shrill as they are about to cross a threshold. But it is strange how the ringing becomes melancholy, like a knell, when it heralds departure—as in the Kaiserpanorama, when it starts up with the slight tremor of the receding image and announces another to come.[27]

24 Benjamin, *Enfance berlinoise vers 1900*, pp. 39–41, original emphasis, my translation.
25 Benjamin, *Berlin Childhood around 1900*, p. 347.
26 Barberger, *Pensées de passage*, p. 7, my translation.
27 Benjamin, *The Arcades Project*, p. 88.

What remains ambivalent about the Kaiserpanorama in the act of remembrance particular to *Berlin Childhood around 1900* is then explicitly tied to concentration and contemplation within "The Work of Art in the Age of Its Technological Reproducibility." Thus, the aforementioned device does not anticipate cinema's disruptive effect, but rather forms the very last example of a tradition based on meditation and absorption: "Shortly before the reversal of contemplation in cinema, [...] the image's contemplation principle stands with rigour for the last time,"[28] "as it had been once before in the isolated contemplation of the divine image by the priest in the cella."[29] At the dawn of the twentieth century, the Kaiserpanorama stands in for a nineteenth century that never ends. For Benjamin, cinema is about to wrap up this long-lasting tradition. As with many other spectacular and architectural forms, stereoscopic entertainment is embedded in a tradition dominated by the ideals of perspective. In Benjamin's view, the historicism, the panoramas, the Haussmannian boulevards, the world fairs are all an ideological *trompe-l'œil* obscuring representation's conditions of production in favour of an illusory sense of immediacy through resemblance. Thus, the narcotic function of "the history that showed things 'as they really were'"[30] echoes the perspectival structure of the circular panorama as well as that of stereoscopic photography. As Isabelle Garo reminds us, for Karl Marx and Friedrich Engels, "ideology is first a matter of perspective"[31]; it is "described as a falsified, illusive, and counterscientific representation of the real."[32] In that sense, we can understand why Benjamin emphasizes the nineteenth century's "unquenchable thirst for perspective"[33]; not only does it include pictures, but any project or production—whether urban, architectural, scientific, historical, or spectacular—that works both towards the naturalization of its artifice and a prescribed point of view.

In one of the *Arcades Project*'s early fragments, stereoscopic photography appears as a "stifled" or "suffocated perspective."[34] Quoting French writer and journalist Jules Claretie, Benjamin compares it to the perspective of the Parisian arcades. According to him, it has a twofold meaning. First, it relates to the coercive aspect of the apparatus, whose blinders obscure the edges of

28 Benjamin, "L'œuvre d'art à l'époque de sa reproductibilité technique (première version)" pp. 97–98, my translation.
29 Benjamin, "The Work of Art in the Age of Its Technological Reproducibility," p. 131.
30 Benjamin, *The Arcades Project*, p. 463.
31 Garo, *L'idéologie ou la pensée embarquée*, p. 7, my translation.
32 Garo, *L'idéologie ou la pensée embarquée*, p. 10, my translation.
33 Benjamin, *The Arcades Project*, p. 122.
34 Benjamin, *The Arcades Project*, p. 831.

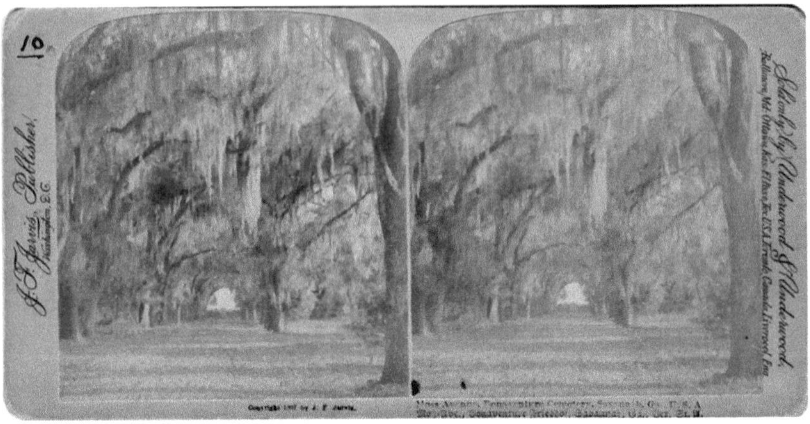

Figure 15.2: Stereo view: Moss Avenue, Bonaventure Cemetery, Savannah, GA, USA (J. F. Jarvis publishers, 1891, author's collection).

the visual field. Second, it also expresses how the gaze is focalized, "stifled," and absorbed towards the depth of the image. All in all, the concealment of the surrounding reality allowed by the blinders gives way to the fictive wonder of the spectator faced with the illusory depth of the photographic images. This aspect has been underlined by Rosalind Krauss: "Stereographic space is perspectival space raised to a higher power. Organized as a kind of tunnel vision, the experience of deep recession is insistent and inescapable. This experience is all the more heightened by the fact the viewer's own ambient space is masked out by the optical instruments he must hold before their eyes."[35] The "ideal isolation" in which the viewer is held reinforces the perspectival space and this "refocusing of attention" allows for better wandering in the depth.[36] Shifting from the foreground to the background, the gaze feels its way through the space; its pauses do not signal a skeptical hesitancy towards the illusions, but are more like the stops along a journey.

This is how writer Oliver Wendel Holmes describes stereoscopic photography and its "stereographic trip."[37] The "stereoscopic gondola"—the author's term for a few views of Venice, which insists on the vehicular nature of the medium—invites the observer to "sail away from one strange scene after another, like disembodied spirits."[38] Nonetheless, the body only seems to disappear within the representation. Indeed, the text is mainly a travelogue

35 Krauss, "Photography's Discursive Spaces," p. 314.
36 Krauss, "Photography's Discursive Spaces," p. 314.
37 Holmes, "Sun-Painting and Sun-Sculpture."
38 Holmes, "Sun-Painting and Sun-Sculpture."

derived from the stereoscopic views. It depicts an itinerary that leads from New York and the East Coast's charms—"Before we set out, we must know something of the sights of our own country"[39]—to the old continent—"We are bound for Europe, and are to leave *viâ* New York immediately."[40] Holmes maintains the use of the present tense to guide the reader through the visited sites, which consists only of photos. Like a tourist guide, he points out a few details and stories to attract the reader's attention while he leads the way: "Here we are at the foot of Charing Cross."; "We are all ready to embark now. Here is the harbour; and there lies the Great Eastern at anchor [...]"; "Let us go into the Abbey for a few moments"; "If you will stroll over to Shottery now with me, we can see the Ann Hathaway cottage from four different points, which will leave nothing outside of it to be seen."[41]

Holmes' *ekphrasis*—which recalls the poetic and wandering description of Denis Diderot[42]—characterizes photography from a present position rather than from a past one. From this derives the illusion of an actual voyage and with it—and with the stereo views especially—"the very presence of all that is most inspiring to the soul."[43] If we compare Holmes's ideal viewer to Benjamin's Chinese painter, the stereoscopic photography features a space in which the viewer can roam.

We can find another example of this ideal viewer strolling within the stereoscopic view. Jacques Babinet, a member of the French Academy of Sciences, writes that "the stereoscopic beholder [...] strolls with their eyes among the fleeing trees like in a forest; and the columns, the flying buttresses, the monument statues [...] let the gaze turn around and penetrate between the protrusions and the central building of the factory."[44] The view and its relief feature a sense of "hidden depth,"[45] as it gives the viewer the illusion that there is always something behind what they can see, something hidden but virtually tangible. Babinet's fantasy of a penetrable view is equivalent to another description of a table-top stereoscope from ten years

39 Holmes, "Sun-Painting and Sun-Sculpture."
40 Holmes, "Sun-Painting and Sun-Sculpture," original emphasis.
41 Holmes, "Sun-Painting and Sun-Sculpture."
42 On this, See Martin, *Décrire le film de cinéma*.
43 Holmes, "Sun-Painting and Sun-Sculpture."
44 Babinet, "Le Stéréoscope et de la vision binoculaire," my translation.
45 Ryan, "Immersion vs. Interactivity," p. 118. Ryan's concept of "hidden depth" refers to how a reader can be absorbed in a narrative up to the point where they assume situations and emotions yet unwritten. We can use the term to describe how a viewer is engaged in a virtually three-dimensional world whose sense of reality depends on a depth that is both complex and faint.

Figure 15.3: Stereo view: Fairyland, Magnolia Cemetery, Charleston SC, USA (Littleton view co, c. 1880, author's collection).

later: the Cosmorama.[46] In the language of commercial promotion, the article promotes its views as explorable spaces. The gaze produces corresponding sensations, affirming that the photographic image's accuracy and details involve the phantasmatic transport of the viewer within the represented places:

> The travel happens without leaving either house or family [...]. The *cosmorama stéréoscopique* [...] brings every country under its mahogany casing, and allows for the *spectacle in an armchair.* You wanted to see Egypt, there it is! Scotland, Norway, Sweden, Greece, all will unfold before your delighted eyes! The accuracy of this panorama cannot be doubted, for nature has reflected itself in this mirror, and its trace has been preserved thanks to science. The prints' perfection [...] is such that one can imagine being transported in the countries that lay before them; we follow the paths, we hike the mountains, we see ourselves on the surface of the lakes, we believe we hear the snow—so perfectly represented it is a *trompe-l'œil*—cracking under our footsteps, we sense the humidity that seeps between the split of a glacier; we were cold a minute ago on the Mont Blanc's plateau or under the vaults of the Grindwald's glaciers or those of

46 The stereoscope's shapes are varied. Aside from the basic one, which the observer holds in his or her hands, there are the most luxurious stereo-classeurs, which make it possible to organize series of images, the viewer alternating between them with a handle. Like in the Kaiserpanorama, the viewer is no longer required to remove his or her eyes from the binocular to see different images in continuity.

Figure 15.4: Stereo view: in the Partnachklamm, Bavarian Alps, Germany (1903, H. C. White Co., author's collection).

> the Furca, and with the rapidity of the mind we transport ourselves to the charming banks of Lake Garda or the Bay of Naples. One minute later a gondola swings before your eyes and there you are in front of the Rialto Bridge or the Bridge of Sighs. The optic's perfection has reached its peak, it is a wonder in which art forgets itself, because it has become nature.[47]

The discourse here employed by an anonymous writer in *Le Journal Amusant* enacts the "stereoscopic tunnel" analyzed by Krauss.[48] Absorbed by the image, the ideal viewer mentioned here has travelled to the other side: they now stand and move beyond the image's surface, unable to doubt any *sense* of the illusion. Somehow, this situation belies the novelty hyped by much more recent virtual reality discourses or by much of new media culture and publicity from the twentieth century.[49] Mechanical objectivity and the extension of perspectival space correspond to a mode of reception that

47 C.D., in *Le Journal Amusant*, p. 8, my translation.

48 Krauss, "Photography's Discursive Spaces," p. 314.

49 For instance, one can easily see how the recent Metaverse promotes and also shapes a particular type of interaction with images and a corresponding type of knowledge based on immersion and identification. As seen in a few promotional contents released in 2021, the best and only way to acquire historical knowledge is to acquire a high degree of immersion. Reviving the past through new technologies is anything but a new phenomenon and we can trace this topos back to the eighteenth and the nineteenth centuries. The same happens with art, as is shown by the fact that some spectators that one of Metaverse's ads features are projected into an "augmented" version of a Henri Rousseau painting. We could point the ideological aspects here implied as knowledge and its medium do not appear as an act of production but is rather experienced as an act consumption.

deletes the boundaries of time and space to replace them with new and unequivocal ones—i.e., without leaving any hints of such an exchange. It is in this sense that we can understand the immobile voyage *topos*[50] that defines an important part of spectacular visual culture since at least the nineteenth century. The viewer is often informed as a beholder by the accurate and trustworthy representation of photography. The identity and the autonomy of the depicted spaces assign him or her to a specific place based on veiled occlusion. Their point of view is determined in a such a way that there are no breaches nor gaps, no possible threshold, nothing else to gaze at but a funnelled illusion whose enclosure prevents any form of collision.

According to Benjamin, the Kaiserpanorama appears to be the last occurrence of this earlier tradition. The likeness of the photographic voyages, that of the world fairs and of the pavilions of colonial exhibitions, and the sealed perspective they have in common with historicism tend to contain the viewer, away from any "genuinely troubling" threshold, to foreclose any spectatorial interrogation of the scene. The threshold is then a place that allows the viewer to maintain one foot in and the other outside the illusion.[51] Extending Benjamin's remarks regarding the phantasmagoria's "bourgeois interior"[52] and how the nineteenth century steadfastly diverted its gaze away from actual socioeconomic concerns, we can speak of an "encased look"[53] to term the tradition here criticized in a different way and to emphasize its visual aspects. Like the "world in a portfolio"[54] that photography shares with panoramas and world fairs—a world thus objectively held close to hand—the "encased look" is contained within the artificial space of a naturalized representation. The mutual containment of both the image-world and the look is well illustrated by the Kaiserpanorama, especially if

50 On this, see Huhtamo, "Dismantling the Fairy Engine."

51 Ryan mentions something similar regarding the case of a medium awareness situation: "The most efficient strategy for promoting an awareness of the mechanism of fictionality is not to block access to the fictional world, but to engage the reader in a game of in-and-out [...]." Ryan, "Immersion vs. Interactivity," p. 126. As it was the case with Murray, this kind of situation helps to better define immersion as an enclosed place rather than to displace it.

52 Berdet writes: "The law of perspective lays potentially in the phantasmagorias of the interior." Berdet, *Le chiffonnier de Paris*, p. 105, my translation.

53 Benjamin insists on the use of cases from the nineteenth century and thinks of the interior as a case itself. In "The Destructive Character" (1931), he uses the term "étui-man" to define the individual accustomed to live in a traditionalist society drove by numerous forms of containment. Benjamin, "The Destructive Character", p. 542.

54 Cormenin, "Égypte, Palestine, Nubie et Syrie." Here, Louis de Cormenin comments on a photographic album of Maxime du Camp. In the photos, taken from 1849 to 1852 in Egypt, Nubia, Palestine, and Syria, he recognizes the traits of an immobile journey.

we read the inventor's description.[55] August Fuhrmann ties stereoscopic wandering to the geographical, technical, and scientific conquests of the modern age. According to him, stereoscopy extends human progress over nature and its implications for the experience of time and space. It replaces the human eye as the telephone does for their ears by abolishing distance. The march of progress assures a technical improvement of the senses and a more immediate access to the truth. By doing so, it recalls the refrain of the "object lesson" so dear to the nineteenth century.[56] The spectacular use of perspective in the Kaiserpanorama betrays the broader outlines of a perspectival theory of knowledge in Fuhrmann, especially when it comes to the future stakes of historical knowledge.

In that case, the doorbell, the signal Benjamin calls for to free the containment of the look, echoes another signal he summons in his written notes for "On the Concept of History" (1940). He states that revolutions determine the "alarm signal" pulled by humanity to stop the locomotive of progress.[57] It is also in this text that Benjamin assumes the role of a historical materialist and confronts the empathetic approach of history he finds in Fustel de Coulanges. He refuses a historiography whose scientific legitimacy is rooted in identification. Instead, the materialist historian must "brush history against the grain,"[58] or, in other words, we could say they must stand on the threshold, in between the past and the present, preferring "arrest"[59] to illusive movement and false continuity and opposing a more cautious and detached spectator[60] to the triumph of the "cultural treasures."[61] It is this caution or detachment I hold to be critical in Benjamin's philosophy with regard to immersion.

In "The Work of Art in the Age of Its Technological Reproducibility," Benjamin states that "Just as the entire mode of existence of human collectives changes over long historical periods, so too does their mode of

55 We can read Fuhrmann's description in Oettermann, *The Panorama*, pp. 229–232.

56 Meredith A. Bak has demonstrated how the stereo views were involved in education and in correlation with an overall epistemological context of the object lesson. See Bak, "Democracy and Discipline."

57 Benjamin, *Gesammelte Schriften*, I, 3, p. 1432; and Löwy, *Fire Alarm*.

58 Benjamin, "On the Concept of History," p. 392.

59 Benjamin, "On the Concept of History," p. 396.

60 The French translation of "On the Concept of History" has *un spectateur réservé*, a cautious or detached spectator. Benjamin, "Sur le concept d'histoire," p. 432. The English translation is a bit different, as the "spectator" disappears from the text and is replaced by the "historical materialist [who] views them with cautious detachment." Benjamin, "On the Concept of History," pp. 391–392.

61 Benjamin, "On the Concept of History," p. 391.

perception."[62] Along with cinema and its visual capacities, he anticipates the rupture of a visual culture long dominated by a contemplative and an immersive mode of reception. Just as the perspective phantasmagorias and the stable point of view are about to burst, so too is cinema about to explode enclosed spaces: "Our bars and city streets, our offices and furnished rooms, our railroad stations and our factories seemed to close relentlessly around us. Then came film and exploded this prison-world with the dynamite of the split second, so that now we can set off calmly on journeys of adventure among its far-flung debris."[63] This explosive capacity of film is obviously opposed to perspectival containment of both the viewer and the represented world. Cinema appears to have the power to create thresholds through a "destructive character," which Benjamin defines as follows:

> The destructive character knows only one word: make room. And only one activity: clearing away. [...] The destructive character tolerates misunderstanding [...]. Some people pass things down to posterity, by making them untouchable and thus conserving them; others pass on situations, by making them practicable and thus liquidating them. The latter are called the destructive. The destructive character has the consciousness of historical man, whose deepest emotion is an insuperable mistrust of the course of things [...]. [H]e sees a way everywhere. Where others encounter walls or mountains, there, too, he sees a way. But because he sees a way everywhere, he has to clear things from it [...]. Because he sees ways everywhere, he always stands at a crossroads. [...] What exists he reduces to rubble—not for the sake of the rubble, but for that of the way leading through it.[64]

Even though cinema is absent here, this short piece by Benjamin stands at the centre of the aesthetical, political, and historiographical stakes of his philosophy. Cinema is neither a medium for reconstruction nor for reproduction; instead, its liberating power lies in its capacity to destroy or to "liquidate" the world. Its purpose is to "make room" in order to answer to a "need for fresh air and open space." In that sense, Benjamin opposes the destructive character to the traditionalist one. The latter preserves, ideally maintaining things untouched and distant, while the former manipulates and liquidates situations, led by the "misunderstanding" Benjamin holds to

62 Benjamin, "The Work of Art in the Age of Its Technological Reproducibility," p. 104.
63 Benjamin, "The Work of Art in the Age of Its Mechanical Reproducibility," p. 117.
64 Benjamin, "The Destructive Character," pp. 541–542.

be an assurance of truth. To destroy or to liquidate are the gestures by which cinema erodes the dominant mode of representation. They create a breach, a crack in which they establish thresholds so that contemplation vanishes. And, if Benjamin writes that the spectator is summoned to "adventurous journeys," from now on these take place among scattered pieces of the world and he or she is not absorbed anymore by any of their illusions, as was the case with the earlier modes of reception. According to the concept of threshold in Benjamin's philosophy, cinema and history share a "disturbing power" that urges the gaze to work. With all its sensitive dimensions, cinema is able to maintain the spectator in between photographic immediacy and its artificial nature. It establishes a play space for the viewer,[65] a zone, a threshold where the contradictions meet: on one side, we find the image, its landscapes, its relief, and its depth; and on the other, the tremors of the frame and the departure signal.

Benjamin's threshold is not a mere rejection of immersion. Critically, with regard to any form of concealment and dissimulation, his philosophy stands against occultation of the point of view and against the substitution of experience as well as its delimitations. Even though it is opposed to a prescribed immersion, we can nonetheless find in Benjamin's work the outline of an immersion, which arises from the viewer. It is an invitation to think of another state of immersion, which does not consist anymore of an act of consumption but rather of an act of *production*. One may recall the hashish experience Benjamin describes, where, in a state of hallucination, he keeps going back and forth between what is actually there and what is not, between what is past and what is present, thus reconfiguring actual space and discovering through such work the connections from which emerges, as in cinema, an aesthetic of "interpenetration."[66] The threshold is thus the medium for interpenetration. An example like the Kaiserpanorama allows us to grasp how a critique of immersion develops from within, in an undistanced attitude that refers to immersion's other, yet to be produced and imagined.

While the threshold of immersion has since been discussed by Murray and Ryan as directly linked to new media experiences, they do not give this threshold enough credit as it continues to be mainly treated as a passage towards a fixed and idealized immersion.[67] In contrast, Benjamin allows

65 The play space, *Spielraum*, is a term used by Benjamin to consider for instance the cinema experience. On this, see Hansen, "Room-for-Play."

66 Benjamin, *The Arcades Project*, p. 858.

67 See Murray, *Hamlet on the Holodeck*; and Ryan, "Immersion vs. Interactivity."

us to redefine it and to give immersion a necessary quality of uncertainty. Although film history might have proven him wrong about film's fundamental capacities of dislocating both the world and our perception as immersion prevails—whether through the illusion of transportation or that of identification and absorption—nonetheless can we devise from it a tactical space[68] to reclaim the heterogeneity of immersion. It could be another way to make sense of the viewer's involvement, and, as such, to devise a definition of immersion that would go beyond current and dominant consumer-oriented applications. As was seen with Perconte in the beginning of this chapter, immersion as a threshold experience gives the viewer the opportunity to stand in between the sides of a multi-layered image; however, the viewer is never merely repelled from it. The experience implies both doubt and absorption, both distance and proximity. The threshold, then, might prove to be an effective concept to study the various ways in which the directly addressed viewer stands *against* representation. Whether through experimental films, fiction, documentary, video games, or interactive films, we should think immersion through the misunderstandings it involves.

Bibliography

Adorno, Theodor W. *Minima Moralia: Reflections from Damaged Life*. London: Verso, 2005.

Babinet, Jacques. "Le Stéréoscope et de la vision binoculaire." *La Revue des deux mondes* 2nd series of the new period, no. 3 (July, 1853): 383–392.

Bak, Meredith A. "Democracy and Discipline: Object Lessons and the Stereoscope in American Education, 1870–1920." *Early Visual Popular Culture* 10, no. 2 (May, 2012): 147–167.

Barberger, Nathalie. *Pensées de passage*. Villeneuve d'Ascq: Presses universitaires du Septentrion, 2011.

Belloï, Livio. *Le regard retourné. Aspects du cinéma des premiers temps*. Québec: Nota Bene-Méridiens Klincksieck, 2001.

Benjamin, Walter. *Gesammelte Schriften*. I, 3, Berlin, Suhrkamp, 1974.

Benjamin, Walter. "L'œuvre d'art à l'époque de sa reproduction mécanisée." In *Écrits français*, edited by Jean-Maurice Monnoyer, 149–250. Paris: Gallimard, 1991.

68 The term "tactical" should be understood here in the sense given by Michel de Certeau. As strategies usually delimitate and rule a place and its uses, tactics have no proper places: "The space of the tactic is the space of the other. Then it must play on and with a terrain imposed on it and organized by the law of a foreign power." Certeau, *The Practice of Everyday Life*, p. 37.

Benjamin, Walter. "Peintures chinoises à la Bibliothèque nationale." In *Écrits français*, edited by Jean-Maurice Monnoyer, 331–340. Paris: Gallimard, 1991.

Benjamin, Walter. "Dream Kitsch." In *Selected Writings*. Vol. 2, Part 1, 1927–1930, edited by Michael W. Jennings, Howard Eiland, and Gary Smith, 3–5. Cambridge, MA: Belknap Press, 1999.

Benjamin, Walter. "The Destructive Character." In *Selected Writings*. Vol. 2, Part 2. 1931–1934, edited by Michael W. Jennings, Howard Eiland, and Gary Smith, 541–542. Cambridge, MA: Belknap Press, 1999.

Benjamin, Walter. *The Arcades Project*. Cambridge, MA: Belknap Press, 1999.

Benjamin, Walter. "L'œuvre d'art à l'époque de sa reproductibilité technique (première version)." In *Œuvres III*, translated by Maurice de Gandillac, Rainer Rochlitz and Pierre Rusch, 67–113. Paris: Gallimard, 2000.

Benjamin, Walter. "L'œuvre d'art à l'époque de sa reproductibilité technique (dernière version)." In *Œuvres III*, translated by Maurice de Gandillac, Rainer Rochlitz and Pierre Rusch, 269–316. Paris: Gallimard, 2000.

Benjamin, Walter. "Berlin Childhood around 1900." In *Selected Writings*. Vol. 3, 1935–1938, edited by Howard Eiland and Michael W. Jennings, 344–413. Cambridge, MA: Belknap Press, 2002.

Benjamin, Walter. *Enfance berlinoise vers 1900*. Paris: L'Herne, 2012.

Benjamin, Walter. "On the Concept of History." In *Selected Writings*. Volume 4. 1938–1940, edited by Howard Eiland and Michael W. Jennings, 389–400. Cambridge, MA: Belknap Press, 2003.

Benjamin, Walter, "Sur le concept d'histoire." In *Œuvres III*, translated by Maurice de Gandillac, Rainer Rochlitz and Pierre Rusch, 427–443. Paris: Gallimard, 2000.

Benjamin, Walter. "The Work of Art in the Age of Its Technological Reproducibility (second version" In *Selected Writings*. Vol. 3, 1935–1938, edited by Howard Eiland and Michael W. Jennings, 101–133. Cambridge, MA: Belknap Press, 2002.

Berdet, Marc. *Le chiffonnier de Paris. Walter Benjamin et les fantasmagories*. Paris: Vrin, 2015.

Boissière, Anne. "Apparence et jeu, valeur cultuelle et valeur d'exposition chez Walter Benjamin." Apparence(s) 1 (2007). https://doi.org/10.4000/apparences.48.

Buck-Morss, Susan. *The Dialectics of Seeing: Watler Benjamin and the Arcades Project*, Cambridge: MIT Press, 1989.

C.D. In *Le Journal Amusant*, n° 327, April 5, 1862, p. 8. https://gallica.bnf.fr/ark:/12148/bpt6k894262b.

Certeau, Michel de. *The Practice of Everyday Life*. London: University of California Press, 1984.

Cohen, Margareth. "Walter Benjamin's Phantasmagoria." *New German Critique*, no. 48 (Autumn 1989): 87–102. https://doi.org/10.2307/488234.

Cormenin, Louis de. "Égypte, Palestine, Nubie et Syrie. Dessins photographiques par Maxime Du Camp." In *La photographie en France. Textes et controverses: une anthologie. 1816–1871*, edited by André Rouillé, 124–125. Paris: Macula, 1989.

Denson, Shane. *Discorrelated Images*. Durham, NC: Duke University Press, 2020.

Garo, Isabelle. *L'idéologie ou la pensée embarquée*. Paris: La Fabrique, 2009.

Gunning, Tom. "Landscape and the Fantasy of Moving Pictures: Early Cinema's Phantom Rides." In *Cinema and Landscape: Film, Nation, and Cultural Geography*, edited by Graeme Harper and Jonathan Rayner, 31–70. Chicago, IL: Intellect, 2010.

Hansen, Miriam Brau. "Room-for-Play: Benjamin's Gamble with Cinema." *October* 109 (Summer, 2004): 3–45. https://www.jstor.org/stable/3397658.

Holmberg, Jan. "Ideals of Immersion in Early Cinema." *Cinémas. Revue d'études cinématographiques. Journal of Film Studies* 14, no. 1 (Fall 2003): 129–147. https://doi.org/10.7202/008961ar.

Holmes, Oliver W. "Sun-Painting and Sun-Sculpture: With a Stereoscopic Trip across the Atlantic." *The Atlantic Monthly* 8, no. 45 (July 1861). https://www.theatlantic.com/magazine/archive/1861/07/sun-painting-and-sun-sculpture-with-a-stereoscopic-trip-across-the-atlantic/628263/.

Huhtamo, Erkki. "Dismantling the Fairy Engine: Media Archaeology as Topos Study." In *Media Archaeology: Approaches, Applications, and Implications*, edited by Erkki Huhtamo and Jussi Parikka, 27–47. Berkeley, CA: University of California Press, 2011.

Jacobs, Bidhan. *Esthétique du signal. Hacker le filmique*. Dijon: Presses du réel, 2022.

Kirby, Lynne. *Parallel Tracks: The Railroad and Silent Cinema*. Durham, NC: Duke University Press, 1997.

Krauss, Rosalind. "Photography's Discursive Spaces: Landscape/View." *Art Journal* 42, no. 4 (Winter 1982): 311–319.

Löwy, Michael. *Fire Alarm: Reading Walter Benjamin's "On the Concept of History."* London: Verso, 2005.

Markus, Gyorgy. "Walter Benjamin, or The Commodity as Phantasmagoria." *New German Critique*, no. 83 (Spring-Summer 2001): 3–42. https://doi.org/10.2307/827788.

Martin, Jessie. *Décrire le film de cinéma. Au départ de l'analyse*. Paris: Presses Sorbonne Nouvelle, 2011.

Marx, Karl. *Capital: A Critique of Political Economy*. Volume 1. London: Penguin Books, 1976.

Murray, Janet H. *Hamlet on the Holodeck: The Future of Narrative in Cyberspace*. New York, NY: Free Press, 1997.

Musser, Charles. "The Travel Genre in 1903–1904: Moving Towards Fictional Narrative." In *Early Cinema: Space, Frame, Narrative*, edited by Thomas Elsaesser, 123–132. London: British Film Institute, 1990.

Oettermann, Stephan. *The Panorama: History of a Mass Medium*. New York, NY: Zone Books, 1997.

Perconte, Jacques. "Après le feu, vidéo initiale." *Vimeo*, September 10, 2009. Video, 6:24. https://vimeo.com/6513389.

Ruoff, Jeffrey. *Virtual Voyages: Cinema and Travel*. Durham, NC: Duke University Press, 2006.

Ryan, Marie-Laure. "Immersion vs. Interactivity: Virtual Reality and Literary Theory." *SubStance* 28, no. 2 (1999): 110–137. https://doi.org/10.2307/3685793.

Weber, Samuel. *Benjamin's-Abilities*. Cambridge, MA: Harvard University Press, 2008.

Yourcenar, Marguerite. "Comment Wang-Fô fut sauvé." *La Revue de Paris* (January–February, 1936): 848–859.

About the Author

Sonny Walbrou is Lecturer in Film Studies at Université de Lille. His work primarily focuses on the representations of the spectator's body in visual culture around 1900. By considering the cinematic apparatus within the broader network of spectacular practices related to attractions, his research lies at the intersection of media archaeology and studies in visual culture. It also involves exploring the historiographical and critical issues of a confrontation between past and present by examining more recent objects of study. From this perspective, the motif of the "immobile journey" determines a significant part of his work, both in the early cinema and in Universal Exhibitions, amusement parks, stereoscopic photography, video games, and virtual reality.

16. Correlations of Immersion and Authenticity in the Experience of Fencing in *Hellish Quart* and *For Honor*[1]

Michał Mochocki and Ignacy Nowicki

Abstract: This chapter presents an in-depth study of correlations between the dimensions of immersion and the dimensions of authenticity in the representations of historical fencing in *For Honor* and *Hellish Quart*. The study is based on M. Mochocki's dual model of immersion/authenticity, which includes six types of immersion defined by S. L. Bowman, and five types of authenticity by N. Wang. It demonstrates how the immersive experience of authenticity (or vice versa: the authenticity of the experienced immersion) is shaped in audiovisual communication, in simulative game mechanics, in the embodied contact with the interface, and in the textual and paratextual communication between the developers and society.

Keywords: heritage, historical games, fencing, martial arts, realism, simulation, authenticity

Introduction

We view immersion as a six-dimensional experience. Each of these dimensions is described in Sarah Lynne Bowman's adaptation of Gordon Calleja's layers of involvement: spatial, narrative, affective, kinaesthetic, shared, and ludic.[2] Bowman notably divides immersion into: environment, narrative,

1 The authors wish to thank Paweł Schreiber, Jakub Majewski, Maciej Talaga, Karol Zakrzewski, Jerzy Miklaszewski, Tomasz Kraśnicki, Kamil Novac, and the developers of *Hellish Quart* for their valuable comments and information. In Michal Mochocki's part, this research was funded by the OPUS20 grant, 2020/39/B/HS2/01999, from the National Science Centre, Poland.

2 Bowman, "Immersion and Shared Imagination in Role-Playing Games"; and Calleja, *In-Game*.

Bédard, Philippe, Alanna Thain, and Carl Therrien, eds. *States of Immersion Across Media: Bodies, Techniques, Practices*. Taylor & Francis Group, 2025
DOI 10.5117/9789048558766_CH16

character, activity, community, and game. Michal Mochocki aligns five of these dimensions of the human experience, described by Bernd Schmitt as verbs: "Sense," "Think," "Feel," "Act," and "Relate" (with no "Play" verb to go with immersion-into-game).[3]

In our exploration of *For Honor* (Ubisoft, 2017) and *Hellish Quart* (Kubold, 2021) (henceforth: *FH* and *HQ*), we rely on Mochocki's dual model of immersion/authenticity.[4] Mochocki correlates five of the six involvements described by Bowman and Calleja with the five dimensions of authenticity observed by Ning Wang and by Mariana Bueno de Andrade Matos and Maria de Lourdes de Azevedo Barbosa.[5] This assumes that immersion in history/heritage-themed games is correlated with the experience of heritage authenticity—and that their five dimensions are too, translatable through Schmitt's language of "experiential" verbs.

Multiple layers of immersion/authenticity may be part of the same experience, overlapping and interconnected. They are not equal, though, in intensity. One or two may dominate, especially when they are prioritized by specific games, genres, and platforms. The configuration of these forms of immersion translates to particular types of experiential authenticity facilitated by in-game interactions with history/heritage content. For instance, *Attentat 1942* (Charles Games, 2017) is designed primarily for narrative immersion aided by environment immersion, which Mochocki compared to museum visits.[6] In this chapter, we will test Mochocki's framework on radically different games, which are rather based on kinaesthetic activity immersion in a historical martial art. Our point of reference will not be museums, but Historical European Martial Arts (HEMA).

HEMA is a worldwide decentralized martial arts practice rooted in the rediscovery of preserved treatises from the fourteenth to the seventeenth centuries and beyond.[7] While HEMA research began as an amateurish bottom-up initiative, an academic approach is also present.[8] Some discuss wrestling or horseback and armoured combat, but the main focus of these contributions is unarmoured weapon fighting, often called "historical

3 Mochocki, *Role-Play as a Heritage Practice*; and Schmitt, *Experiential Marketing*.

4 Mochocki, *Role-Play as a Heritage Practice*; and Mochocki, "Heritage Sites and Video Games."

5 Wang, "Rethinking Authenticity in Tourism Experience"; and Matos and Barbosa, "Authenticity in Tourist Experiences."

6 Mochocki, "Heritage, Authenticity, and The Fiction/Nonfiction Dualism in *Attentat 1942*."

7 See Jaquet, Verelst, and Dawson, *Late Medieval and Early Modern Fight Books*.

8 See Jaquet, Sørensen, and Cognot, "Historical European Martial Art a Crossroad between Academic Research, Martial Heritage Re-Creation and Martial Sport Practices"; and Talaga, "Crowdsourcing w służbie archeologii wiedzy?"

fencing." Comparable to modern (Olympic) fencing or Kendo, HEMA trains fighters for duelling with replicas and simulators of longswords, rapiers, sabers, and more (an example being the Polish HEMA FEDER league). Fencing with many of the same weapons is the core gameplay in *FH* and *HQ*, and therefore it makes sense to look at HEMA for comparison.[9]

This chapter's five sections follow the five types of heritage authenticity identified by Wang and by Matos and Barbosa: objective, constructive, bodily feelings, interpersonal, and self-making. One by one, we explore the correlations of the five authenticities with the six forms of immersion that we have previously listed. Following Mochocki's model, we assume that they *are* correlated—this is not, however, a reality claim but an interpretive lens. We wish to explore how our understanding of history/heritage games can expand if we see heritage authenticity as a close correlate—or component?—of immersion.

Objective Authenticity, Environment Immersion, and Narrative Immersion

Objective authenticity is "an epistemological experience (i.e. cognition) of the authenticity of originals."[10] Its most obvious form is the material substance of buildings and artifacts which survived from the past. It correlates with immersion-into-narrative (narrative information communicates/validates authenticity) and immersion-into-environment (due to the co-presence of the artifact in the same space). In Schmitt's five experiential modules (the verbs Act, Think, Feel, Sense, and Relate) narrative immersion correlates with Think, and environment immersion with Sense.

In a narrower understanding, objective authenticity resides in the physical history of material objects and places. In Live Action Role-Play (LARP), re-enactment, and HEMA, such authenticity is afforded only by direct contact with genuine historical artifacts (e.g., weapons and treatises) and sites. HEMA researchers study objectively authentic weapons to draw conclusions about their usage.[11] They also study old fencing treatises, though usually not original antiquarian books but digital scans, reprints, and the *Wikentauer*

9 *FH* features more peculiar weapons, such as the kama (Japanese sickle), hammer, or various axes. However, longswords, arming swords, and types of sabers are present in the game as well.
10 Wang, "Rethinking Authenticity in Tourism Experience," p. 351.
11 See Miklaszewski, "Starzewski Treatise 'On Fencing' in the Eyes of His Era"; Talaga (personal communication); and Warzecha, "Viking Axe & Shield Combat."

online library.[12] In video games, the objective authenticity of weapons is impossible, as they only offer digital representations of objects.[13]

We propose to expand objective authenticity to mechanical reproductions of original media. If the original essence of media texts (speech, images, films) resides in the form and arrangement of its semiotic signs—and not in the material substance of paper, canvas or tape— then it is possible in games, as shown in the Renaissance paintings of *Four Last Things* (Joe Richardson, 2017) and *The Procession to Calvary* (Joe Richardson and Nephilim Game Studios, Inc., 2020), as well as in the *Morgan Bible* illuminations of the unreleased *Anno Domini 1250* (Flora Games). *FH* does not feature this (or any) objective authenticity. *HQ*, by contrast, does include copies (scans) of analogue artifacts from Public Domain Images on its starting screen and several scans from fight books in its game menu.[14] HEMA practitioners can recognize them as pictures from the works of Fiore dei Liberi (1404), Joachim Meyer (1568), Vincentio Saviolo (1595) or Michał Starzewski (1840–1932). Non-experts can learn this from paratexts or from knowledgeable players.

Constructive Authenticity, Environment Immersion, and Narrative Immersion

Constructive authenticity is experienced analogously to objective authenticity: through "Sensing" (sensory perception of objects, events, or behaviours) and "Thinking" (judging the degree of their authenticity), which correlate with the narrative immersion and environment immersion as described above. But Thinking works differently. Constructive authenticity is not based on a seemingly objective validation of an object's genuine origins. It is subjectively constructed-as-authentic, based on the experiencer's knowledge, associations, and beliefs, however inaccurate they may be. Alternatively, instead of or next to their own judgment, the experiencer may simply follow reliable information that labels an object as a copy or a reconstruction. In

12 On digital scans, see Bratislavský šermiarsky spolok, "Schermus Pictus"; and Hammer, "Tłumaczenie Traktatu Szermierczego Zawartego w Rękopisie Nürnberger Handschrift 3227a z Wyszczególnieniem Trudności Zaistniałych Podczas Przekładu." On reprints, see dei Liberi, *The Flower of Battle*; Forgeng, *The Medieval Art of Swordsmanship*; and Meyer, *The Art of Sword Combat.*

13 See Mochocki, "Heritage Sites and Video Games."

14 The game notably shows *Pojedynek Paska z Jasińskim* (1850-1851), a painting that can be found in the Cyfrowe zbiory Muzeum Narodowego w Warszawie database.

either case, constructive authenticity is "projected onto toured objects" in multiple versions by audiences, practitioners, marketers, etc.[15]

"Toured objects" are not only artifacts and sites but also events and performances. Any dance, ritual, sport, or craft may be a toured "heritage object." So too can HEMA practices, demonstrations, and historical re-enactments. Non-tourism heritage activities include the reading/watching/playing of history-themed media. In these activities, the novel/film/game is equivalent to a "toured object" that represents a story world (which is itself full of objects). Since they all are digital representations, the weapons, bodies, and all their properties and motions (including fencing techniques) in video games can only be constructively authentic, as they lack the tangible materiality required for objective authenticity.

Constructive authenticity may be judged for historical accuracy. It is evidence-based and measurable: the accuracy of any replica, description, or representation of a historical object/place/person/event can be discussed with reference to historiography, archives, archaeology, and museum collections. For audiences who care about source-based historical accuracy, it can be the main criterion of authenticity—a noted fact in historical game studies.[16]

Alternatively, a subjective impression of authenticity may be constructed like heritage: through "virtually anything by which some kind of link, however tenuous or false, may be forged with the past," as is the case, for instance, in:

a) mythologized national narratives a person believes tell the historical Truth;
b) highly inaccurate historical fiction, e.g., TV costume dramas;
c) fragmentary, inaccurate, and misremembered knowledge from school history and/or museum visits.[17]

Due to lack of space, we focus on the criterion of historical accuracy, but this should not deny the importance of the other criteria.

Fencing techniques in HEMA

We follow Maciej Talaga, Jakub Wrzalik, and Krzysztof Janus in their conceptual separation of fighting techniques from fighting practice.[18] Practice is an actual performance of actions: always situated and embodied,

15 See Wang, "Rethinking Authenticity in Tourism Experience," p. 352.
16 See Copplestone, "But That's Not Accurate."
17 Johnson and Sheehy, "Heritage as Business," p. 170.
18 Talaga, Wrzalik, and Janus, "Archaeology of Motion," pp. 166–168.

therefore unique and ephemeral. We link fighting practice to bodily feelings authenticity and activity immersion.[19] Technique, however, "is not an embodied action but an epistemic framework structuring such actions."[20] It is constrained by the physics of motion, which regulates what is possible for a given body to do with a given object. Assuming the continuity of the laws of physics and human biology, we can assume fencing techniques are repeatable and transmissible across places and times. Therefore, they can be accurately (re)constructed.

For HEMA training to be possible, fencing treatises had to be not only translated but also interpreted. Since fight book translation brings numerous challenges, and since technique reconstruction is highly problematic, fight book interpretations fall into the category of constructive authenticity.[21] HEMA researchers do not restrict themselves to translating sources or interpreting particular techniques, they also take an interest in elements of historical re-enactment, including period-accurate replicas of weapons and costumes.[22]

With regard to historical accuracy, Daniel Jaquet, Claus Frederik Sørensen, and Fabrice Cognot distinguish "experiencing practitioners" from "experimenting practitioners."[23] The former mix historical sources with a modern approach and intuition, constructing their techniques by means of creative interpretation. The latter strive to reconstruct historical fencing via rigorous study and experiments. Talaga's work over the years has made a strong case for the "archaeology of motion" as a valid method of experimental research. Textual interpretations of verbal and pictorial depictions in historical fight books are confronted with multiple experiments of fencing with replicas or simulators of weapons—sometimes with objectively authentic originals.[24]

19 On this, see the "Player's Bodily Feelings Authenticity and Activity Immersion" section of this chapter.

20 Talaga, Wrzalik, and Janus, "Archaeology of Motion," p. 167.

21 See Hammer, "Tłumaczenie Traktatu Szermierczego Zawartego w Rękopisie Nürnberger Handschrift 3227a z Wyszczególnieniem Trudności Zaistniałych Podczas Przekładu"; and Talaga, "Traktaty Szermiercze Jako Źródło Wiedzy Praktycznej o Dawnym Fechtunku."

22 See "Tłumaczenie Ustępu Wernera Dotyczącego Różnic w Szermierce Szablą a Pałaszem"; Jaquet and Deluz, "Moving in Late Medieval Harness"; Jaquet et al., "Range of Motion and Energy Cost of Locomotion of the Late Medieval Armoured Fighter"; Sprechfenster Blog, "Project Hausbuch #6"; Sprechfenster Blog, "Children of the Sun"; Sprechfenster Blog, "Stangeschieben"; Talaga, "Narodziny żargonu"; Warzecha, "I.33 Sword & Buckler Sparring with Sharps"; Warzecha, "Lignitzer's First Play of the Buckler"; and Warzecha, "Viking Axe & Shield Combat."

23 Jaquet, Sørensen, and Cognot, "Historical European Martial Art a Crossroad between Academic Research, Martial Heritage Re-Creation, and Martial Sport Practices," pp. 7–9.

24 See Sprechfenster Blog, "Project Hausbuch #4"; and Talaga (personal communication).

If HEMA weapons and clothes are replicas/simulators, their authenticity is constructive (constructed-as-authentic, literally). Actual fencing as a performed practice of historical martial arts may be seen as constructively authentic (or inauthentic) by spectators, and experienced in bodily feelings authenticity by performers.[25]

The alignment of fencing authenticity with immersion types is complex, and the distinction between technique and practice helps to clarify it. The embodied experience of *fencing practice* correlates with activity immersion.[26] *Fencing techniques*, however, are identified as (in)authentic through cognitive processing of information: by measuring the particular representation of techniques (here, in games) against whatever we know about actual fencing. This is constructive authenticity. We align it with immersion-into-narrative, because the process of constructing-as-authentic relies on story world lore (be it archaeology-of-motion experiments, publications on fencing, cloak-and-dagger blockbusters, HEMA classes, or whatever has shaped the person's vision of authentic fencing techniques). We apply the concept of story world to fiction and nonfiction alike (as in Mochocki's work), therefore histor(iograph)y informs story world lore for historical settings.[27]

Fencing techniques in HQ *and* FH

Hellish Quart is named after a particular technique: it refers to a saber cut from right below.[28] The game's constructive authenticity is one of its pillars, since its character animations are based on fencing treatises. For example, the character Isabella's combat stances can be identified as the combat stances named *Tutta Porta di Ferro*, *Posta longa*, and *Posta breve* in Fiore dei Liberi's *Fior di Battaglia* (1404).[29] Her unique move—input hold LT+↑X—on the gamepad resembles *Colpo di Villano*, a technique demonstrated in 2021 by Antoni Olbrychski.[30] Some characters can punish leg attacks by "shifting the[ir] leg," as in Henry Angelo's *The Guards of the Highland Broadsword* (1799).[31] According to a developer of the game, Marie's character is directly inspired by Salvator Fabris's *De lo Schermo, overa Scienza d'arme* (1606).

25 On this, see the sections "Constructed Representation of Character's Bodily Feelings" and "Player's Bodily Feelings Authenticity and Activity Immersion" of this chapter.

26 Mochocki, *Role-Play as a Heritage Practice*, p. 6.

27 Mochocki, *Role-Play as a Heritage Practice*.

28 Starzewski, "O szermierstwie"; *Ze wspomnień o Michale Starzewskim*.

29 Liberi, *Fior di Battaglia* (*MS Ludwig XV 13*), folio 23v–24r.

30 Olbrychski, "Historical Technique #1."

31 See Rowlandson, "The Guards of the Highland Broadsword as Taught at Mr. H. Angelo's Academy."

Kubold's answer to the comment "Looks very Fabris" on a Facebook post dated January 2021 suggests that the team directly referred to old fencing treatises: "Well ... it was supposed to be a practical version of Fabris ([...] but in real life nobody does those figures like on the drawings, they are a bit extreme) [...] So I modified the poses a bit to be more universal."[32] However, the character named Gedeon seems to be based on a present-day interpretation by Sieniawski Fencing.[33] (Either case counts as constructive authenticity).

In *FH*, we can interpret Warden's side combat stances as *pflug* in its version known from Joachim Meyer's *The Art of Sword Combat* (1570).[34] We can also compare Warmonger's "Vicious Impale" to folio 19r and 19v from Talhoffer's *Talhoffer Fetchbuch* (1467); and Peacekeeper's "Riposting Stab" to figure 41 from Capoferro's *Gran simulacro dell'arte e dell'uso della scherma* (1610).[35] However, these may not be intentional references to HEMA, as HEMA-defined constructive authenticity of European martial arts is mostly absent in character animations (whereas Wu Lin faction animations are clearly influenced by Chinese martial arts, as evaluated by a former Shaolin disciple).[36]

Here, again, due to lack of space, we only discussed a specific version of constructive authenticity: the one which is understood as source-validated accuracy, and is therefore available only to knowledgeable audiences. Players with no HEMA knowledge/experience identify constructive authenticity differently, for instance based on film costume dramas.

Fencing and Environment Immersion

Immersion-into-environment is afforded by representations of spatial settings: spaces, structures, objects, and inhabitants, as well as their mutual relations and behaviours. Fencing environments have a well-defined structure and dynamics: the weapons, the duelling bodies (with gear), the floor beneath them, and the obstacles (if any) on the floor. Environmental features include the physical/technical affordances of weapons as actionable objects, as well as properties and behaviour of those objects in motion. Additionally, it often includes spectators (sometimes referees), as well as formal and social rules of duelling that regulate actions and behaviours.

32 Hellish Quart, "Well..."

33 Sieniawski Fencing, "Sztuka Krzyżowa / Cross-Cutting Art (POLISH SWORDSMANSHIP)."

34 Meyer, *The Art of Sword Combat.*

35 Talhoffer, "Talhoffer Fetchbuch," f. 19r and 19v; and Capoferro, "Gran simulacro dell'arte e dell'uso della scherma."

36 See Ranton, "Real Shaolin Disciple Reacts to Marching Fire Expansion."

If the objects, properties, rules, and behaviours (including techniques) are experienced as (constructively) authentic, then the entire environment can be. Immersion into such an environment affords an authentic heritage experience.

Spatial environments in *HQ* are stand-alone battle arenas: grassy meadows, a gallows scaffold, etc. Each arena is autonomous, with no continuity in either topography, weather, or time of day. There is no exploration of the off-screen environment: navigation is restricted to footwork on the small floor/ground space. The environment in indoor arenas is represented by the interior design of buildings and objects; while outdoor arenas—natural features of places and surrounding landscapes, often with an exterior view of the built infrastructure—are inhabited "anthropocentric landscapes."[37] The costumes, weapons, and bodies of the duellist characters are constructed-as-authentic, with digital models of weapons and clothes created from 3D scans of historically accurate replicas provided by private collectors, a film centre, a HEMA weapon-maker, and three re-enactor groups according to the *HQ* credits. They represent standard types (locals and foreign guests) from the Polish-Lithuanian Commonwealth: Polish, Ruthenian, Prussian, Hungarian, French, Spanish, and African; noblemen, ladies, and commoners. One element of the game, however, lowers its believability: a random selection of opponents sometimes puts the same character on both sides of the fight as mirror images. Otherwise, in all aspects, as virtual models of inhabited environments—simultaneously physical, social, and cultural—*HQ* battle arenas allow for an immersive experience of "cultural presence."[38] This also extends to the wooden weapons mode, fought with swords that do not cause bleeding: a practice known from the historical period.[39]

In *FH*, the characters navigate through large locations (a medieval castle for knights), moving from one combat encounter to the next in adjacent rooms and outdoor arenas. Continuous exploration of the complex setting supports immersion-into-environment. Character classes are counterparts of real-world historical warriors: Vikings, knights, and samurai (in the 2017 starter), as well as Wu Lin (from the 2018 expansion). This is a fantasy story world loosely based on historical cultures; however, its particular animations, costume and weapon designs, and architecture look believable enough to be constructively authentic. Experts could debate the accuracy of historical detail, but everyone familiar with globalized popular culture is likely to

37 Vlachos, "Of Ecosystems and Landscapes," p. 191.

38 See Champion, *Critical Gaming*.

39 See Kitowicz, "Opis obyczajów i zwyczajów za panowania Augusta III."

identify the European medieval castle and knights, Vikings, samurai, and Chinese warriors.

In these games, environment immersion (based on sensory perception of the represented space) is coupled with narrative immersion (interpretation of perceived objects in the light of story world information).

Fencing Environments and Narrative Immersion

In some conceptualizations,[40] narrative immersion includes the character.[41] We follow Bowman and Marie-Laure Ryan, who move character-focused immersion to a separate category, thus limiting narrative immersion to involvement with the unfolding story (driven by tension and suspense related to the flow of events), and to cognitive processing of story world information.[42] The corresponding heritage authenticity is of the constructive type: after all, it is in/through narratives that images of the past are constructed-as-authentic (and deconstructed, too).

Both *FH* and *HQ* have a story mode (in *HQ* under development) that provides a narrative context. *HQ* currently offers one sample cutscene (a quarrel in a tavern) and character backstories. In *FH*, the story consists of linear levels featuring combat sequences and cutscenes, and an overarching structure of missions and campaigns. The game's backstory tells the tale of a fantasy apocalypse, from which three (then four) civilizations emerge locked in never-ending warfare. However, the only narrative present in *FH* multiplayer gameplay is a story of progression and outcome of a certain battle on a given battleground. Its connection to the war between the factions is rarely visible, since players usually form mixed-faction teams, and player-characters do not need to match non-playable troops representing factions on the battlefield. Story bits delivered by seasonal events, paratexts and game modes bear little connection to gameplay and hardly count as a well-developed narrative in the sense of an unfolding sequence of events. With our focus on fencing, we will not analyze the narrative as it is almost entirely limited to non-combat scenes.

Beside narrative involvement in actual gameplay, which Calleja calls the "micro phase," there is the "macro phase," before and after gameplay.[43] Macro-phase narrative involvement happens when players read, talk, and

40 Calleja, *In-Game*; and Thon, "Immersion Revisited."

41 Calleja, *In-Game*; Thon, "Immersion Revisited."

42 Bowman, "Immersion and Shared Imagination in Role-Playing Games"; and Ryan, *Narrative as Virtual Reality*.

43 Calleja, *In-Game*.

watch media related to the game story—or story world. *FH* is an original medievalist fantasy creation, so its macro story is self-contained. Its fictionalized portrayal of historical warrior cultures may be enough to be drawn into debates on history, heritage, and media representation of these cultures, but its own narrative does not position itself in real-world history—quite unlike *HQ*, whose digital recreation of costumes, weapons, duels, and wooden sword sparring brings to mind similar endeavours of historical re-enactors. *HQ*'s story mode is yet to come, but its story world is well-defined. This inevitably makes *HQ* linked to general debates on the history and heritage of the Polish-Lithuanian Commonwealth. Engagement with debates on historical story worlds belongs to macro-phase narrative involvement/immersion (and its social aspect to shared community immersion[44]).

Existential Bodily Feelings Authenticity and Activity Immersion

"Bodily feelings" is one of three subtypes of Wang's "existential authenticity," and it can be correlated with kinaesthetic immersion-into-activity.[45] Its experiential verb is "Act." Prolonged involvement in an activity can be immersive—and direct visceral sensations of "doing" validate this experience as authentic.

This authenticity is pursued in HEMA and historical re-enactment. Fear, pain, and physical and mental exhaustion experienced in the combat sport have raw, immediate, unmediated bodily feelings authenticity. In video games such as *HQ* or *FH*, directly experienced bodily feelings are limited to manual operations on game controllers (including vibrations and isomorphic mouse movement), and to embodied psychosomatic reactions to gameplay and story. These reactions emerge in the player's body; yet they are partially conditioned by how the game constructs the representations of the character/avatar's bodily feelings.

Constructed Representation of Character's Bodily Feelings

The player perceives the character's body through interactive digital mediation: visuals, audio, text, and game mechanics. This authenticity—if truly authentic it is—must be classified as constructive: constructed-as-authentic in its depiction of bodily feelings on a computer interface. Remembering Talaga's distinction between fencing technique and fencing practice, we can

44 On this, see the section "Existential Interpersonal Authenticity and Community Immersion" of this chapter.

45 Wang, "Rethinking Authenticity in Tourism Experience."

say that the theoretical model of fencing techniques developed in HEMA can be converted into an interactive model coded into the game system. This model, composed of game mechanics and audiovisual-plus-textual depictions, represents fencing techniques. Actual gameplay, in which the player manually controls the visual-mechanical performance of actions, represents fencing practice—and the corresponding bodily feelings.

The visual-interactional model in *HQ* includes historically accurate techniques from various fencing traditions, with aspects such as distance, tempo, and fatigue. The mechanics are relatively simple: move and attack; the characters will defend themselves automatically if you let them. Complexity emerges in clever use of distance, timing, simulated collisions, and advanced moves, often unique for particular characters. The developers declare that they actually practiced fencing with analogical weapon replicas.[46] The physics of weapon collisions, cut trajectories, and human motion looks "completely realistic" to Matt Easton's expert eye.[47] Patterns for character animation are based on motion capture (mocap) sessions with the historian and accomplished fencer Antoni Olbrychski.[48] This information supports the perception of embodied movement and fatigue as an accurate (constructively authentic) model of bodily feelings.

With no information on *FH* developers' fencing experience, there might have been no conversion of actually experienced bodily feelings to in-game representations. However, *FH* complicated mechanics, explained in the developer's glossary, utilize the full range of the "tactical wheel" of modern fencing theory: simple attacks can be parried, parries can be deceived by feints, feints can be neutralized by counterattacks, counterattacks by countertime, but simple attacks preclude countertime.[49] In *HQ*, the tactical wheel is not directly included. It is better to say it is evocated in its less rigid and physics-based mechanics: certain actions in *HQ* (examples: feints, attacks with unknown endings) are much harder to perform consciously than in *FH* or in real-life fencing.

The pain and shock of wounds in *HQ* are represented by a visual depiction of bleeding and facial-gestural expressions, as well as noticeable impact on stamina, and sometimes on mobility (leg wounds hamper movement). Any wound drains stamina permanently for the duration of the duel, with three degrees of exhaustion reflected in animations. The stamina bar can be

46 See Kubold, "Hellish Quart."
47 IGN, "Sword Expert Reacts to Realistic Sword Fighting Game."
48 See Kubold, "Hellish Quart."
49 See "For Honor"; and Szabo, *Fencing and the Master*, p. 246.

switched on or off. There are multiple hit locations on the body, the head, and the sword-wielding arm being critical. Their hit points are tracked separately, but not displayed to the player. One may receive many blows and still fight back dangerously, and even win a duel, but a single blow may also be an instant kill (a detail that, according to Roderick Hamilton Burgoyne and Donald McBane's writings, is historically accurate).[50] Double (simultaneous) hits also happen (this is historically accurate according to Paulus Hector Mair's *Opus amplissimum de arte athletica* [1540]).[51] Particular attention is given to the final animation of the victorious character: a triumphant pose if they are in perfect health, or clumsy staggering if wounded. Visible wounds and smears of blood on clothes and blades add to the graphic realism. By comparison, in *HQ* practice with wooden swords produces no blood, just a symbolic burst of impact when a hit is scored. Moreover, fencers tap their fingers on the body part, showing precisely where they were hit: a popular practice in HEMA sparrings.

FH has no safe weapon option. Its tutorial is a practice sequence of particular attacks and defences, introduced one at a time, with clear guidelines and set goals. In one stage, the player only practises parrying; in another, only simple attacks. Wounds in *FH* are represented with a far greater symbolism: a health bar and stamina bar, a momentary red or black filter on screen, and splashes of blood that sometimes immediately disappear, and sometimes visible marks of damage on flesh and armour. HEMA fencers have criticized *FH* for its lack of realism in techniques and wounds.[52] Most of the time, we do not see facial expressions or wounds, as characters are clad in armour and helmets. Pain, exhaustion, and fear are expressed by sounds: grunts, screams, or panting. Wounds can be healed by symbolic resting between combat encounters in *FH*. In *HQ*, there is no healing, but each new combat begins at full health.

Player's Bodily Feelings Authenticity and Activity Immersion

As stated in the Polish league tournament ruleset, HEMA tournaments are not simulations.[53] Consequences of cutting and stabbing wounds are impossible to simulate in combat sport, and the fencers obviously do not want to risk it. Nonetheless, accidents do happen, and full-contact combat

50 Burgoyne, *Historical Records of the 93rd Sutherland Highlanders*; and McBane, *The Expert Sword-Man's Companion*.

51 Mair, "Opus amplissimum de arte athletica," "Mixed weapons", pic. 3 and 4

52 See Tvgry, "Prawdziwi wojownicy oceniają sceny walki z gier."

53 "Recent Ruleset (2019 with Minor Changes)."

with steel weapon simulators is extremely "bodily" anyway. Personal experience of sweat, fatigue, fear, pain, domination, and triumph—while holding a weapon of historically accurate parameters—fuels the sense of existential authenticity of such combat. So does proficiency in handling sharp replica weapons, which can be used for cutting targets or very slow no-contact sparring.[54] All this perfectly exemplifies Wang's bodily feelings existential authenticity, Calleja's kinaesthetic involvement, and Bowman's immersion-into-activity: yet another proof of their correlation.

Choosing the right action at the right time is the essence of fencing.[55] Simple attacks, renewals, parry ripostes, stop hits, and evasions are featured in many classic action games, e.g., *Mortal Kombat* (Midway Games, 1992), *Prince of Persia* (Ubisoft Montreal, 2003), and *Demon's Souls* (From Software, 2009).[56] Kinaesthetic control in *HQ* and *FH* is mediated the same way as in those classics: the player toggles buttons on the device, and the on-screen character reacts with assigned moves. Both the player's and character's motions are embodied as muscle movements, but each is radically different: the avatar's full-body motion versus player's fingers and arms. Nonetheless, decision-making in *HQ*, *FH*, and real fencing stays mostly the same: the player has to recognize the opponent's intentions, choose a suitable action, and perform it at the right moment. Such combat decision-making can be very authentic in psycho-tactical bodily feelings, as the character's visual perception immediately correlates with the player's visual perception, and so does the assessment of possible actions, the sense of timing, and the decision-making. *FH* and *HQ* mechanics represent it with different degrees of player control.

FH supports bodily feelings authenticity in a similar way as *Kingdom Come: Deliverance* (Warhorse Studios, 2018), that is by a mimetic interface.[57] It is felt with a mouse and keyboard: the player moves the mouse (and thus their hand) to certain positions, so their character on screen will position their hands and weapon accordingly. These positions serve as starting

54 See CombatCon, "CombatCon 2019 HEMA Cutting Finals"; Warzecha, "Sword & Buckler Fencing with Sharp Swords"; and Warzecha, "I.33 Sword & Buckler Sparring with Sharps."

55 See Czajkowski, *Taktyka i psychologia w szermierce*.

56 Names of actions refer to modern fencing theory. See Szabo, *Fencing and the Master*. Names of particular moves in games may differ. For example, in games, "riposte" is often misnamed "counter-attack." "Action games" refers to games with reaction-based combat in real time. They also refer to the notion of "arcade," according to Fullerton's definition. See Fullerton, *Game Design Workshop*.

57 See Farrow and Iacovides, "Gaming and the Limits of Digital Embodiment."

combat stances and as protective parries.[58] According to modern fencing theory, moving the weapon between such basic positions is crucial for all fencing actions.[59] Here lies a remarkable difference: in *HQ* the correct parry is automatically selected by the character, whereas in *FH* the player must choose the correct parry every single time. In consequence, while performing feints, *HQ* players strive to deceive the game system, whereas *FH* players aim to deceive the other player. These tactics require hiding one's own intentions and anticipating the enemy's, which is essential also in real fencing.[60]

If played with a gamepad, *FH* and *HQ* afford the experience of "feeling the blade": the tactile stimuli of two weapon blades meeting.[61] Due to its minimal muscle movement, the gamepad thumbstick in *FH* is far less mimetic than the mouse. Instead, there are frequent gamepad vibrations: during parries, counter-attacks in opposition, and even preparatory actions like "guardbreak," "bashes," and "traps," comparable to "beats" and "binds" in modern fencing theory. FH players may use these stimuli (along with visuals) to recognize if the strike made contact or just cut through the air. Still, they cannot feel how much pressure the enemy applies when crossing swords, unlike real-life fencers. *HQ* comes closer to this experience: the player can feel not only every parry, but also continuous blade contact. Players can seek these stimuli as signals to perform particular fencing action, as in real fencing.[62]

One last important thing to note about bodily feelings concerns the player's psychosomatic reactions. As Marco Caracciolo insists, a mental simulation of a character's experience is always coupled with physiological responses in the body.[63] As we watch someone experience triumph, failure, confusion, etc., we are triggering an echo of corresponding bodily feelings. *HQ* strongly facilitates this with its naturalistic animations of blood, pain, fear, and death[64] (in line with the current naturalist tendency in

58 See Czajkowski, *Teoria i Metodyka Współczesnej Szermierki*, our translation of "zasłona zabezpieczająca."

59 See Czajkowski, *Taktyka i psychologia w szermierce.*

60 See Czajkowski, *Taktyka i psychologia w szermierce.*

61 See Borysiuk and Waśkiewicz, "Information Processes, Stimulation, and Perceptual Training in Fencing."

62 See Borysiuk and Waśkiewicz, "Information Processes, Stimulation, and Perceptual Training in Fencing."

63 Caracciolo, "Ungrounding Fictional Worlds."

64 See the vivid reactions of Matt Easton, HEMA expert, to *HQ* gameplay videos: IGN, "Sword Expert Reacts to Realistic Sword Fighting Game."

European games[65]). By contrast, in *FH* many visual representations are highly symbolic, not stimulating embodied reactions so directly (though gamepad vibrations to mimic blades penetrating flesh are implemented in both games). Both games may biosemiotically trigger physiological reactions related to triumph, failure, sense of danger, and sports-like competition. The recognition of bodily feelings (and also constructive) authenticity may be especially powerful for practitioners, who can relate the animations and mechanics to their personal experience.[66]

Majkowski, Kozyra, and Prokopek (2023) interpret *HQ* fencing as highly sanitized, without the messy random accidents, mistakes, and imperfections of real bodily performance.[67] We disagree. Their label of "a fencing manual fantasy" could be applied to *FH*, whose mechanics interact with one another in predetermined ways, but not to *HQ*, whose mechanics interact via complex simulated physics, including the position of internal organs, the leverage one weapon applies to another, and various levels of fatigue as ever-changing variables. This results in a highly realistic simulation of embodied practice: with botched attacks, messy scuffles, and unpredictable collisions.

Bodily Feelings Authenticity and Game Immersion

Bowman's immersion-into-game means that the player's attention, effort, and emotion are focused on competition with(in) the game system. It is about winning and scores: optimizing strategies and tactics, collecting collectibles, defeating opponents. Game immersion is the odd one out in the dual model of authenticity/immersion. It does not have a counterpart either in Schmitt's experiential modules or in Wang's authenticities.

Immersion-into-game gives primary importance to game goals and game mechanics at the expense of the narrative layer (story, character, setting). Lack of attention to narrative logic and to psychological realism is considered to be potentially conflicting with narrative, character, and environment immersion. However, in fencing games like *HQ* and *FH*, involvement with the narrative, story world, and character is of marginal importance anyway. The main form of involvement that drives moment-to-moment gameplay decisions is kinaesthetic activity immersion—which does not seem to clash with ludic game immersion. Every game score or achievement is only attainable by fencing actions. Any tactical calculation is about optimizing the choice of fencing actions. Therefore, immersion-into-game in the micro phase

65 See Zagalo, "Naturalist Tendency in European Narrative Games."
66 See Mochocki, *Role-Play as a Heritage Practice*, p. 207.
67 Majkowski, Kozyra, and Prokopek, "Finish – Spare the Shame."

is inseparable from immersion-into-activity. Consequently, the authenticity of the ludic layer of fencing (as a tactical sports game) seems to be the same as in the layer of kinaesthetic activities: bodily feelings authenticity in moment-to-moment action, and constructive authenticity in the analysis of its visual-mechanical representation.[68]

This may be different in the macro phase of fencing tournaments, where calculations happen on the abstract level of numerical results of multiple duels. There, in HEMA or Olympic fencing and in *HQ/FH* tournaments alike, ludic involvement with scores, achievements, and rankings diverges from kinaesthetic involvement in action.

Existential Interpersonal Authenticity and Community Immersion

Interpersonal-existential authenticity emerges on the micro scale of immediate interactions with co-participants and in-game existents, and in the macro phase of *communitas*: the sense of collective identity, memory, values, and connectedness with particular (living and/or historical) communities. This perfectly correlates—via the experiential verb "Relate"—with the micro and macro phases of Calleja's shared involvement and with Bowman's immersion-into-community, whose micro phase takes place during gameplay, while the macro phase covers related pre- and post-game social interaction.

Interpersonal authenticity is the heart of HEMA, as any research group, weapon-making workshop, or fencing club is made by and for the community. Any HEMA topics are gladly discussed by beginners, advanced practitioners, and professionals.[69] Workshops, demonstrations, and lectures follow combat sport events, and, since 2013, the open access journal *Acta Periodica Duellatorum* supports academic research on the matter. Similar sharing of techniques and tactics between players is common in *HQ* and *FH* communities. While *HQ* allows for direct multiplayer gaming on one device and for online multiplayer via intermediary technology, *FH* is developed mostly for multiplayer gaming, and contains different game modes. These facilitate group fighting—which may or may not be implemented in *HQ*

68 On this, see the sections "Player's Bodily Feelings Authenticity and Activity Immersion" and "Constructed Representation of Character's Bodily Feelings" of this chapter.

69 Unlike other combat sports (for example MMA or boxing), competitive historical fencing is not currently a profitable profession, so by "professionals" we mean weapon-makers, researchers, fencing instructors, and judges.

and is absent in HEMA combat sport, unless we count Buhurt practice as such.[70] Official and sponsored *FH* tournaments have already taken place, as well as fandom-driven *HQ* competitions.

Beside the gamer community, the macro phase of community immersion encompasses interactions and connectedness with other communities related to the game—for instance, to its historical content. History/heritage-themed games are thus meaningful to non-game communities whose history or heritage is represented, misrepresented, or ostensibly erased there.[71]

HQ, on this matter, is a case in point. Set in the Polish-Lithuanian Commonwealth, it strongly promotes its cultural heritage. The creative team includes re-enactors and fencing practitioners, role-players, as well as novelist Jacek Komuda. Some are involved in writing and filmmaking about the Commonwealth, with an outspoken dedication to its legacy, including the legendary Polish saber skills. *HQ* players are likely to share a large part of its audience with those of the *Dzikie Pola* (Wydawnictwo MAG, 1997) tabletop role-playing game (RPG), *Born for the Sabre* (Pawel Delag, 2019), a fictionalized documentary, and *The Devil of Łańcut* (2022), a TV series based on Komuda's historical novel (the last one under development). Both film productions rely on the involvement of Polish historical re-enactors, HEMA practitioners, and heritage institutions, as well as on crowdfunding that emphasized two collectively shared sentiments: sympathy for the Noble Commonwealth's heritage, and for historically accurate reconstruction of its fencing art and material culture. *HQ* subscribes to the same values, with some of the same experts and creative specialists as contributors. For players, both gameplay and community participation may constitute meaningful heritage practices contextualized in many others: all in the realm of interpersonal authenticity and immersion-into-community.

Existential Self-Making Authenticity and Character Immersion

The self-making form of existential authenticity emerges in exercising, testing, or searching for one's "true self." This may include various activities, but role-playing is definitely among them, and it therefore partially correlates with Bowman's immersion-into-character. Its experiential verb is "Feel."

70 HEMA does have traditions of combat against multiple enemies, but those are hard to practice in a competitive manner. Though Buhurt practice and other types of knight combat may be linked to historical re-enactment, those practices are usually not considered HEMA.

71 See Mochocki, *Role-Play as a Heritage Practice*, p. 256.

Neither of our games has elaborate plot-driven stories to place their characters within. Nor do they have well-developed characters as personalities which could fully qualify for Ryan's character-centred emotional immersion. The lack of a character-driven story in combat scenes eliminates most venues normally used for identity expression or identity formation in character role-play. Our search for self-making authenticity must turn to the player instead.

For both games, this design is intended, as *HQ* promises "the most realistic fencing experience," and *FH*'s creative director Jason Vandenberghe wanted a game about heroic and empowering combat.[72] Self-making is personal, so our speculation does not exhaust all possibilities. What appears as plausible components of a "true self" in fencing games, real-world HEMA, and re-enactment is self-identification as:

a) a fencer (practitioner) or fencing fan (non-practitioner);
b) an enthusiast or follower of a historical warrior code (chivalry, bushido, etc.);
c) an enthusiast and/or a descendant of a historical culture.

As we saw, a part of self-making is relatedness to a community, actualized in social interactions. We rather talk here about personal identity as a "self-making" internalized in one's mind: existential self-making authenticity experienced via immersion-into-character.

Conclusion

FH and *HQ* feature advanced systems that represent fencing via game mechanics and audiovisuals, supported by haptic feedback and textual descriptions. Contextualized or not in narrative cutscenes, interactive gameplay consists of a sequence of combat encounters. The realism of the fencing experience, declared to be one of the key features of both games, is achieved differently. *HQ* has a complex repertoire of fencing techniques, many unique for one character, but combat mechanics are simplified through automated parrying. It affords "feeling the blade" through gamepad vibrations correlated with on-screen animations. *HQ* simulates tactical situations through physics, with a particular importance accorded to distance and timing. *FH* combines complicated and rigid mechanics structure with mimetic weapon control and a highly embodied "game of deception and

72 Kubold, "*Hellish Quart*"; and Chartier, *Playing Hard*.

anticipation."[73] It is the player who controls all defensive actions in *FH*, with a particular importance given to moving the blade between the right, left, and top basic stances.

The layers of involvement immersion and heritage authenticity of these games are configured as follows:

Environment immersion/involvement is afforded by audiovisual representation of settings (buildings, landscapes, objects, including bodies and weapons), and by the coherent dynamics of the simulated fencing environment (arena, opponents, weapons, and obstacles). Their authenticity as heritage experiences is constructive, with high accuracy of the fencing experience in both games, and with very high historical accuracy of weapons and costumes in *HQ*. (Some images in *HQ* are copies from original fencing treatises: this may count as objective authenticity).

Narrative immersion/involvement reigns in cutscenes and scenarios in the story mode. In combat gameplay, narrative involvement is facilitated primarily by the interpretation of represented objects and events as components of the story world, based on the available story world information. As shown above, the historical authenticity of narrated or depicted story world components can only be constructive (constructed-as-authentic), except for the few objectively authentic reprints found in *HQ*.

Activity immersion/involvement is the core of the fencing experience. It enhances the constructive authenticity (of story world information and audiovisuals) with existential bodily feelings authenticity consisting of:

1. manual operations with a gamepad or mouse;
2. sensations of gamepad vibrations;
3. psychosomatic reactions to what happens on the screen;
4. the complex bodily felt processes of focus, attention shifting and divisibility, as well as split-second decision-making leading to muscle movement.

Ludic game immersion/involvement seems to absolutely merge with immersion-into-activity, as all gameplay actions and achievements are reducible to kinaesthetically controlled fencing manoeuvres. The only exception to this rule is that tournaments draw attention to abstract scores.

Community immersion/involvement goes beyond the shared experience of Player vs. Player gameplay and the relations with gamer communities. *HQ* creators have documented connections with HEMA fencers. *FH* with its

73 This is how English fencer Bob Anderson describes real fencing. See Czajkowski, *Taktyka i psychologia w szermierce*.

fantasy universe leans toward sci-fi and fantasy fandoms related to medieval nostalgia. *HQ* is firmly set in the seventeenth-century Polish-Lithuanian Commonwealth, openly appealing to sympathies for that heritage, which are shared by many non-gamer communities and general populations. Collective sharing fuels the sense of existential-interpersonal authenticity.

Beside character identification in cutscenes, character immersion/involvement in gameplay is predominantly featured by the enactment of fencing. Even if the character is a generic type like *FH*/*HQ*'s fighters, this type may serve as a carrier of values, symbols, beliefs, and legacy the person identifies with. Such imaginary identification may support existential self-making authenticity: the player can project his/her own sense of self on the digital fighter.

A disclaimer is in order: in this analysis we speak about experiential affordances of the games, trying to map ways of facilitating immersion and authenticity. We use a formal analysis (of video games as a medium) and content analysis (of historical themes) to theorize forms of experience potentially facilitated by the games. This does not mean that all players actualize all the affordances in all play sessions, or that all individual experiences follow the same patterns. We make no claims about actual player experiences—this would require empirical studies going beyond what we can tell based on our own experience and players' anecdotal self-reports.

With regard to immersion studies, we have demonstrated that the dual model of immersion/authenticity is applicable even to those history/heritage games that are essentially sports games. Immersion scholars who study historical games may use it to reveal configurations of immersions and authenticities in particular games and genres. Moreover, we framed immersion types as experiential modules of Sense-Act-Think-Feel-Relate. We use these five verbs to align immersions with heritage authenticities—further studies may do the same with game mechanics (which are often discussed as active verbs) and player agency. The language of experiential modules works with any setting, historical or not.

Bibliography

Borysiuk, Zbigniew, and Zbigniew Waśkiewicz. "Information Processes, Stimulation and Perceptual Training in Fencing." *Journal of Human Kinetics* 19 (2008): 63–82. https://doi.org/10.2478/v10078-008-0005-y.

Bowman, Sarah Lynne. "Immersion and Shared Imagination in Role-Playing Games." In *Role-Playing Game Studies: Transmedia Foundations*, edited by José Pablo Zagal and Sebastian Deterding, 379–394. New York, NY: Routledge, 2018.

Bratislavský šermiarsky spolok. "Schermus Pictus." January 28, 2010. YouTube video, 9:47. https://www.youtube.com/watch?v=-82cajZi70s.

Burgoyne, Roderick Hamilton. *Historical Records of the 93rd Sutherland Highlanders*. CITY: Franklin Classics Trade, 1883.

Calleja, Gordon. *In-Game: From Immersion to Incorporation*. Cambridge, MA: MIT Press, 2011.

Capoferro, Ridolfo. "Gran simulacro dell'arte e dell'uso della scherma." *Wiktenauer*. https://wiktenauer.com/wiki/Ridolfo_Capo_Ferro_da_Cagli#The_Use_of_Fencing.

Caracciolo, Marco. "Ungrounding Fictional Worlds: An Enactivist Perspective on the 'Worldlikeness' of Fiction." In *Possible Worlds Theory and Contemporary Narratology*, edited by Alice Bell and Marie-Laure Ryan, 113–131. Lincoln, NE: University of Nebraska Press, 2019.

Champion, Erik. *Critical Gaming: Interactive History and Virtual Heritage*. Farnham: Ashgate, 2015.

Chartier, Jean-Simon. *Playing Hard*. Montreal: MC2 Communication Media, 2018.

CombatCon. "CombatCon 2019 HEMA Cutting Finals." November 5, 2019. YouTube video, 24:00. https://www.youtube.com/watch?v=xHrmQD2Ll28&ab_channel=CombatCon.

Copplestone, Tara Jane. "But That's Not Accurate: The Differing Perceptions of Accuracy in Cultural-Heritage Video Games Between Creators, Consumers and Critics." *Rethinking History* 21, no. 3 (2016): 415–438. https://doi.org/10.1080/13642529.2017.1256615.

Czajkowski, Zbigniew. *Taktyka i psychologia w szermierce*. Katowice: Wydawnictwo AWF, 2007.

Czajkowski, Zbigniew. *Teoria i Metodyka Współczesnej Szermierki*. Katowice: Wydawnictwo AWF, 1968.

Farrow, Robert, and Ioanna Iacovides. "Gaming and the Limits of Digital Embodiment." *Philosophy & Technology* 27, no. 2 (2014): 221–233. https://doi.org/10.1007/s13347-013-0111-1.

Forgeng, Jeffrey L. *The Medieval Art of Swordsmanship: Royal Armouries MS I.33*. Leeds: Royal Armouries Museum, 2018.

"*For Honor*: Terminology Guide." *Ubisoft*, March 11, 2021. https://www.ubisoft.com/es-mx/game/for-honor/news-updates/6XEM2qUiwcokp753uCzgkz/for-honor-terminology-guide.

Fullerton, Tracy. *Game Design Workshop: A Playcentric Approach to Creating Innovative Games*. Boca Raton: CRC Press, 2019.

Hammer, Maciej. "Tłumaczenie Traktatu Szermierczego Zawartego w Rękopisie Nürnberger Handschrift 3227a z Wyszczególnieniem Trudności Zaistniałych Podczas Przekładu." MA diss., Uniwersytet Jagielloński, 2015.

Hellish Quart. "Well... It Was Supposed to Be a Practical Version of Fabris (You Can Call It 'Sloppy,' but in Real Life Nobody Does Those Figures Like on the Drawings, They Are [...]." *Facebook*, January 31, 2021. https://www.facebook.com/hellishquart/photos/p.244277717282069/244277717282069/?type=3.

IGN. "Sword Expert Reacts to Realistic Sword Fighting Game: *Hellish Quart*." February 18, 2021. YouTube video, 6:40. https://www.youtube.com/watch?v=aBaSuw5iXLA.

Jaquet, Daniel, Alice Bonnefoy Mazure, Stéphane Armand, Caecilia Charbonnier, Jean-Luc Ziltener, and Bengt Kayser. "Range of Motion and Energy Cost of Locomotion of the Late Medieval Armoured Fighter: A Proof of Concept of Confronting the Medieval Technical Literature with Modern Movement Analysis." *Historical Methods: A Journal of Quantitative and Interdisciplinary History* 49, no. 3 (2016): 169–186. https://doi.org/10.1080/01615440.2015.1112753.

Jaquet, Daniel, and Vincent Deluz. "Moving in Late Medieval Harness: Exploration of a Lost Embodied Knowledge." *Journal of Embodied Research* 1, no. 1 (2018). https://doi.org/10.16995/jer.7.

Jaquet, Daniel, Claus Frederik Sørensen, and Fabrice Cognot. "Historical European Martial Art: A Crossroad between Academic Research, Martial Heritage Re-Creation and Martial Sport Practices." *Acta Periodica Duellatorum* 3, no. 1 (2015): 5–35.

Jaquet, Daniel, Karin Verelst, and Timothy Dawson. *Late Medieval and Early Modern Fight Books. Transmission and Tradition of Martial Arts in Europe (14th–17th Centuries)*. Leiden: Brill, 2016.

Johnson, T. Bryan, and Thomas P. Sheehy. "Heritage as Business." In *Heritage, Tourism and Society*, edited by David T. Herbert, 170–190. London: Mansell, 1995.

Kitowicz, Jędrzej. "Opis obyczajów i zwyczajów za panowania Augusta III." *Wirtualna Biblioteka Literatury Polskiej*, 1951. https://literat.ug.edu.pl/kitowic/k0020.html.

Kubold. "*Hellish Quart*. Schiavona Attacks Pt. 1. Fencing Game." May 29, 2021. YouTube video, 1:41. https://www.youtube.com/watch?v=PN-TyBvUMvo.

Kubold. "*Hellish Quart*. About." *Hellish Quart*, 2022. https://www.hellishquart.com/#about3-section.

Liberi, Fiore dei. *Fior di Battaglia (MS Ludwig XV 13)*, folio 23v–24r. *Wiktenauer*. https://wiktenauer.com/wiki/Fior_di_Battaglia_(MS_Ludwig_XV_13).

Mair, Paulus Hector. "Opus amplissimum de arte athletica." *Wiktenauer*. https://wiktenauer.com/wiki/Paulus_Hector_Mair.

Majkowski, Tomasz, Magdalena Kozyra and Aleksandra Prokopek. "Finish – Spare the Shame: Realism of *Hellish Quart* and Alt-Samartian Ideology". *Games and Culture*, Online first (2023). https://doi.org/10.1177/15554120231210599.

Matos, Mariana Bueno de Andrade, and Maria de Lourdes de Azevedo Barbosa. "Authenticity in Tourist Experiences: A New Approach Based on Edgar Morin's Complexity Theory." *Revista Brasileira de Pesquisa em Turismo* 12, no. 3 (2018): 154–171. https://doi.org/10.7784/rbtur.v12i3.1457.

McBane, Donald. *The Expert Sword-Man's Companion, or The True Art of Self-Defence*. Glasgow: FFAMHE, 1728. http://www.ffamhe.fr/collectionpalas/expert-swordman-mcbane.pdf.

Meyer, Joachim. *The Art of Sword Combat: A 1568 German Treatise on Swordsmanship*, edited by Jeffrey L. Forgeng. London: Frontline Books, 2016.

Miklaszewski, Jerzy. "Starzewski Treatise 'On Fencing' in the Eyes of His Era." *Academia.edu*, 2021. https://www.academia.edu/45608547/Starzewski_treatise_On_Fencing_in_the_eyes_of_his_era.

Mochocki, Michał. "Heritage Sites and Video Games: Questions of Authenticity and Immersion." *Games and Culture* 16, no. 8 (March, 2021): 951–977. https://doi.org/10.1177/15554120211005369.

Mochocki, Michał. *Role-Play as a Heritage Practice: Historical LARP, Tabletop RPG, and Reenactment*. Abingdon: Routledge, 2021.

Mochocki, Michał. "Heritage, Authenticity, and the Fiction/Nonfiction Dualism in *Attentat 1942*." In *Games and Narrative: Theory and Practice*, edited by Barbaros Bostan, 281–298. Berlin: Springer, 2022.

Olbrychski, Antoni. "Historical Technique #1 Colpo Di Villano." April 15, 2021. YouTube video, 00:20. https://www.youtube.com/watch?v=JUsjo8Hsyl4.

Ranton. "Real Shaolin Disciple Reacts to Marching Fire Expansion." October 19, 2018. YouTube video, 7:36. https://www.youtube.com/watch?v=CCK1h48Pk6g.

"Recent Ruleset (2019 with Minor Changes)." *FEDER, Polish HEMA Federation*, 2021. http://feder.org.pl/en/league-ruleset/.

Rowlandson, Thomas. "The Guards of the Highland Broadsword as Taught at Mr. H. Angelo's Academy." In *Rowlandson the Caricaturist; a Selection from His Works with Anecdotal Descriptions of His Famous Caricatures*, edited by Joseph Grego, 374–375. London: Chatto and Windus, 1880. https://www.britishmuseum.org/collection/object/P_1856-0913-240.

Ryan, Marie-Laure. *Narrative as Virtual Reality: Immersion and Interactivity in Literature and Electronic Media*. Baltimore, MD: Johns Hopkins University Press, 2001.

Schmitt, Bernd. *Experiential Marketing: How to Get Customers to Sense, Feel, Think, Act, and Relate to Your Company and Brands*. New York, NY: Free Press, 1999.

Sieniawski Fencing. "Sztuka Krzyżowa / Cross-Cutting Art (POLISH SWORDSMANSHIP)." June 25, 2020. YouTube video, 1:14. https://www.youtube.com/watch?v=oOBA2FQRqfU.

Sprechfenster Blog. "Children of the Sun: Medieval Strength Conditioning." December 9, 2021. YouTube video, 9:09. https://www.youtube.com/watch?v=mOxWqdydnD4&t=28s.

Sprechfenster Blog. "Project Hausbuch #4. On Hengen, Winden and Other Things." April 19, 2019. YouTube video, 11:13. https://www.youtube.com/watch?v=1of8ihBfCBM.

Sprechfenster Blog. "Project Hausbuch #6. Period Clothing, First Blood." August 6, 2021. YouTube video, 7:15. https://www.youtube.com/watch?v=TM_s59OoA68&t=29s.

Sprechfenster Blog. "Stangeschieben. A Medieval Exercise for Explosive Power." February 15, 2022. YouTube video, 1:57. https://www.youtube.com/watch?v=WPwPMQh3SS8.

Starzewski, Michał. "O szermierstwie." In *Ze wspomnień o Michale Starzewskim*, by Józef Starzewski. Kraków: Drukarnia Narodowa, 1932 [Original manuscript: 1840].

Szabo, Laszlo. *Fencing and the Master.* New York, NY: SKA Swordplay Books, 1998.

Talaga, Maciej, Jakub Wrzalik, and Krzysztof Janus. "Archaeology of Motion: Experiencing the Past through Embodiment." *Historyka Studia Metodologiczne* 51 (2021): 157–181. https://doi.org/10.24425/HSM.2021.138369.

Talaga, Maciej. "Traktaty Szermiercze Jako Źródło Wiedzy Praktycznej o Dawnym Fechtunku." In *Między Nauką a Praktyką. Sesja naukowa towarzysząca wystawie "Szable w dłoń! Szermierka szablą jako polskie dziedzictwo narodowe,"* edited by Małgorzata Taborska, 37–38. Kraków: Wydawnictwo Uniwersytetu Jagiellońskiego, 2018.

Talaga, Maciej. "Crowdsourcing w służbie archeologii wiedzy? Perspektywy integracji pracy naukowców i amatorów na przykładzie rekonstrukcji historycznych sztuki walki." In *Rejestry Kultury*, edited by Ksenia Olkusz, 141–161. Wroclaw: Osrodek Badawczy Facta Ficta, 2019.

Talaga, Maciej. "Narodziny żargonu. Rekonstrukcja znaczenia czasownika schiessen w poźnośredniowiecznym niemieckim technolekcie szermierczym." In *Pojęcie, kategoria, slowo w teorii i praktyce*, edited by Adam Dombrowski, Maria Rudnicka, and Waldemar Zarski, 187–208. CITY: Impuls, 2019.

Talhofer, Hans. "Talhoffer Fechtbuch. / Cod.icon._394a", folio 19r–19v. *Wiktenauer.* https://wiktenauer.com/wiki/Talhoffer_Fechtbuch_(Cod.icon._394a).

Thon, Jan-Noël. "Immersion Revisited: On the Value of a Contested Concept." In *Extending Experiences: Structure, Analysis and Design of Computer Game Player Experience*, edited by Olli Leino, Hanna Wirman, and Amyris Fernandez, 29–43. Rovaniemi: Lapland University Press, 2008.

"Tłumaczenie Ustępu Wernera Dotyczącego Różnic w Szermierce Szablą a Pałaszem." *Silkfencing*, March 6, 2022. https://silkfencing.pl/artykul/tlumaczenie-ustepu-wernera-dotyczacego-roznic-w-szermierce-szabla-a-palaszem.

tvgry. "Prawdziwi wojownicy oceniają sceny walki z gier." March 28, 2021. YouTube video, 19:24. https://www.youtube.com/watch?v=zvKCAYzU3Kg.

Vlachos, George L. "Of Ecosystems and Landscapes: An Essay on Grasping Themes of Environmental History in Video Games." In *Retun to the Interactive Past: The Interplay of Video Games and Histories*, edited by Csilla E. Ariese-Vandemeulebroucke, Krijn H. J. Boom, Bram van den Hout, Angus A. A. Mol, and Aris Politopoulos, 189–200. Leiden: Sidestone Press, 2021.

Wang, Ning. "Rethinking Authenticity in Tourism Experience." *Annals of Tourism Research* 26, no. 2 (1999): 349–370. https://doi.org/10.1016/S0160-7383(98)00103-0.

Warzecha, Roland. "I.33 Sword & Buckler Sparring with Sharps." March 6, 2017. YouTube video, 1:22. https://www.youtube.com/watch?v=ws_pLB1EFd0&ab_channel=RolandWarzecha.

Warzecha, Roland. "Lignitzer's First Play of the Buckler." November 30, 2017. YouTube video, 8:12. https://www.youtube.com/watch?v=tz6aEFdDGi8.

Warzecha, Roland. "Sword & Buckler Fencing with Sharp Swords." January 8, 2016. YouTube video, 6:53. https://www.youtube.com/watch?v=u5dc3sFAJWk&ab_channel=RolandWarzecha.

Warzecha, Roland. "Viking Axe & Shield Combat." November 11, 2018. YouTube video, 19:09. https://www.youtube.com/watch?v=qPs9u3p_k20.

Zagalo, Nelson. "Naturalist Tendency in European Narrative Games." In *Perspectives on the European Videogame*, edited by Víctor Navarro-Remesal and Óliver Pérez-Latorre, 209–26. Amsterdam: Amsterdam University Press, 2022.

About the Authors

Michał Mochocki is an associate professor and director at the Center for Historical Games and Heritage at the Faculty of History, University of Gdansk, Poland. He also works as a game writer and historical consultant at False Prophet video game studio.

Ignacy Nowicki is a game designer at 7LEVELS in Kraków, Poland. He graduated from Game Studies and Design at UKW. He is also a historical fencer and instructor at Krakowska Szkoła Fechtunku and referee of the Polish historical fencing league.

17. New Heritage Ecologies: Embracing Entanglement and Complexity in Virtual Heritage Immersion

Elina Lex

Abstract: As emergent immersive media technologies become increasingly implicated in the field of digital cultural heritage, their use and circulation in these contexts require a growing critical focus. Emerging across discourses of nationalism, colonialism, and armchair travel, immersive media have supported imperial fields of view through heightened and unrestricted vision while valuing spatial and temporal coherence to capture and apprehend authentic representations of historical material. This chapter asks how immersive heritage might challenge and de-centre domineering cultural worldviews and conventions operationalized through immersive media formats while also embracing the radical potentials to change and to make meaningful alternative understandings and experiences for learning. What might it mean to embrace immersive complexity in immersive heritage design practices?

Keywords: embodied museography, polyphony, immersive complexity, ecological thinking, cultural institutions, museum studies

Introduction

Advances in immersive technologies, including the rapid expansion of virtual reality (VR), augmented reality (AR), and extended reality (XR) platforms, have transformed the ways that cultural knowledge and memory can be transmitted, shared, and accessed. The adoption of these technologies by museums, archives, and other memory institutions have increasingly shown the impact of these media technologies in the research, preservation,

Bédard, Philippe, Alanna Thain, and Carl Therrien, eds. *States of Immersion Across Media: Bodies, Techniques, Practices*. Taylor & Francis Group, 2025
DOI 10.5117/9789048558766_CH17

interpretation, and representation of cultural heritage knowledge. The 2009 UNESCO Charter on the Preservation of Digital Heritage articulated this turn, signalling a new generation of technologies that are challenging current archival presentation and historical visioning practices to produce new resources for digital information, historical storytelling, and knowledge creation.[1] From *Beyond 2022* (Trinity College Dublin, 2022), an open-access virtual reality 3D reconstruction of the Irish Record Treasury destroyed by a fire in 1922, *The Library at Night* (Alberto Manguel, Robert Lepage, and Ex Machina, 2016), a 360-degree video experience that transports viewers to libraries across the world, to *Modern Games* (The Mill and *The New York Times*, 2016), a virtual reality experience activating the *New York Times* photo archives of the 2016 Rio Olympics, technologies are beginning to open up affordances for the preservation, interpretation, and access of heritage materials through immersive and interactive techniques. With that said, the use of such formats in these contexts also poses new sets of challenges and risks around their ethical, political, social, and epistemological implications.

Tracing immersive technologies across the wider historical, psychic, and sociopolitical contexts in which they have advanced, immersive technologies have historically emerged alongside discourses of nationalism, colonialism, tourism, and armchair travel. Acts of seeing produced through them have long supported fields of view that have conditioned and fuelled fantasies of dominance, conquest, and a simulated sense of proximity through extended and unrestricted vision.[2] More specifically, the panoramic view of nineteenth-century immersive spectatorship gave rise to an ontology and politics of the viewing experience called the "panoramic observer," which laid the basis for a wide range of immersive viewing formats such as the Cinemascope, IMAX, Omnimax, and the development of more recent 360-degree VR.

The explosion of contemporary immersive media and virtual reality productions in the realms of journalism, documentary, and digital heritage has also generated ongoing critical attention to the political and ethical implications of witnessing in immersive experiences. One of the moral risks involved in this type of witnessing is that it has often offered a new technologically facilitated (Western, settler, white, and middle-class) "tourist gaze" that has centred an unrestricted, objective, and unreflective point of view.[3] Representations produced by immersive technologies also carry within them inherent values, cultural biases, and ideological worldviews.

1 See Cameron and Kenderdine, "Introduction," p. 3.

2 See Huhtamo, *Illusions in Motion*, p. 176.

3 See Miller and Ebrahami Bazaz, "Invited and Implicated."

In practices of virtual heritage, technologies of immersion have often been valued for their spatial and temporal coherence and ability to capture, represent, and apprehend an authentic representation of the world as it was presented. Certain conventions of spatial understandings have also largely been favoured, deriving from Cartesian philosophical traditions, Euclidian spatial understandings, and the reinforcement of mimetic representation, while also privileging the supremacy of vision over other modalities of experience.[4]

As experiential knowledge sites like museums, memory institutions, and heritage sites are increasingly positioned to explore the potential of these new technologies, viewers are being brought new degrees of intimacy, immersion, and relation with the past amplified through new technologies.[5] As such, the cultural, historical, and epistemological dimensions of how immersive technologies have advanced become important as they are increasingly being adopted to communicate and educate about the experiential dimensions of cultural heritage.

This chapter seeks to explore some of the provocative tensions for virtual heritage as a pedagogical tool. On the one hand, practices of immersive virtual heritage have long been grounded in scientific discourses of preservation and in the display of historical material. On the other, advances in technologies have opened new experimental potentials to produce alternative kinds of experiential, immersive, and pedagogical experiences. As approaches to digital storytelling enabled by immersive VR, AR, and XR technologies continue to promise more flexible, participatory, and dialogical modes of experience, they also bring forward the potential for a more complexified vision of virtual heritage to include a wide range of interlocutors, participants, and knowledge systems into them.

Immersive, locative, and networked digital media have therefore been compelling narrative platforms for artists and practitioners in these areas to generate temporally and spatially complex, multi-perspectival, and multi-voiced representations and interventions that can challenge the dominant viewpoints and conventions that have existed through the immersive media lens. While virtual heritage becomes increasingly circulated through immersive technologies, the ever-growing intersection between cultural heritage and immersive technology remains largely unmapped in terms of critical theory and practice. What might it mean to embrace immersive complexity in immersive heritage design practices?

4 See Davies, "Rethinking VR."

5 See Witcomb, "The Materiality of Virtual Technologies," p. 41.

Immersive Complexity

One way of defining immersive complexity in immersive experiences can be the tuning in or the creation of different modes of attention. Different modes of attention can help us grasp at the entanglements of the complex ecosystems that make up the environments around us. Entanglement is a term that has been increasingly used in the environmental humanities as well as in studies of immersion, as it has offered a useful framing for understanding the relationships between humans and environments. Entanglement comes originally from the works of Karen Barad, a particle physicist and feminist philosopher who explores entanglement as the intra-relation between subatomic particles, matter, time, and systems of thought.[6] As entanglement has been increasingly and usefully applied to theories of immersion, it has strengthened the notion that immersion can be an active entanglement with an environment and its actants rather than a separate experience that we visit occasionally.

In a more conventional sense, ideas around immersion have centred on media forms that separate us from the rest of the world, producing a kind of "cocooning" or temporary frame that encloses and wraps us.[7] This is evident when conceptualizing the panorama as well as more contemporary VR formats, where the user is enclosed in time and space within the media's representational centre. But, by collapsing the "out there" and the "in here" frames of immersion, scholars have begun to show how immersive media can bolster a greater sense of entanglement with the environments that already exist around us.[8] Building on immersive sound artist Duncan Speakman's philosophy of immersion, we can ask ourselves the following questions: what if immersion was "something that already exists… Something we live with all the time?"[9] Can immersion be an appropriate frame to describe the way we exist deep inside these complex tangled ecologies? From this perspective, immersive media does not necessarily separate or simulate, but rather is a frame that amplifies, highlights, and reveals. It can create multiple modes of attention that may allow for more critical engagements with the histories, materialities, and environments that we are already immersed and embedded in.[10]

6 Barad, *Meeting the Universe Halfway*.
7 See Speakman, *No Vantage Point*, p. 4.
8 See Scott-Stevenson, "Finding Shimmer."
9 Speakman, *No Vantage Point*, p. 4.
10 See Speakman, *No Vantage Point*, p. 44.

This vision of immersion can be seen operating on multiple levels, scales, and time frames as a series of layered states that a user can move between. According to Duncan Speakman, specialized technology and equipment is not necessary to immersion as it is realized through simple shifts in attention that are powerful enough to generate new forms of knowledge and care for the extant environments that we inhabit.[11] Speakman's immersive AR sound walk *It Must Have Been Dark by Then* (2017) enacts this framework, offering users the opportunity to shift between multi-scalar modes of attention.[12] Bringing stories and sounds collected from remote places and climates around the world, including Latvia's rapidly shrinking population, Louisiana's sinking wetlands, and the desertification in Tunisia, the project layers on these recordings on top of and in response to the listener's own local environment. These temporal and spatial shifts operate in tangible ways for the user: standing in their own urban context, but hearing the acoustic space of a sand dune in Tunisia, a complex kind of layering occurs. This layering invites an ecocritical form of attention that extends beyond the edges of the experiential composition itself, an attentional scale Speakman insists is required to be able to begin comprehending the complexities and time scales of climate change. Combining walking, reading, observing, listening, marking, and locating, the audience shifts between different frames of attention, one that produces a powerful sense of immersion and presence in the immediate local surroundings, yet at the same time producing a constant shift of attention to other global geological time scales. This kind of transitive reading generated through immersive technology allows an engagement with the depth and simultaneity of multiple histories, spaces, identities, meanings, and ecological modes across time and space.[13]

The affordances of creative immersive media demonstrated here—through the multiple shifts possible with temporalities, scales, and meanings—offer ripe avenues for further exploration in the field of virtual heritage. Generating new frameworks of immersion can break down conventional, standardized vantage points that have been ontologically bound to immersive viewing while also generating the potential for new kinds of engagements, embodied knowing, learning through encounter, and practices of care for diverse and complex kinds of heritage and historically salient sites. While these technologies have been increasingly embraced for these new potentials through experimental works like *It Must Have Been*

11 See Speakman, *No Vantage Point*, p. 41.

12 For information on this project see Speakman, *It Must have Been Dark by Then*.

13 See Didur and Fan, "Between Landscape and the Screen," p. 102.

Dark by Then, institutions of digital cultural heritage production have much to benefit from the capacities of immersive technologies to change and radicalize ways of learning and experiencing different times and spaces.

Immersive Virtual Heritage: Histories and Precursors

While the use of immersive media within the field of cultural heritage preservation and presentation is not a new one, to better understand how these practices have emerged and can be complexified, it becomes essential to investigate how the historical, ontological, and epistemological logics of immersive technologies have informed discursive practices of virtual heritage today. Early renditions of virtual heritage can be traced back to the circular panoramas, which were vastly popular spectacles in nineteenth-century Europe and United States. The panoramas, *pan* meaning "all," *orama* meaning "view," were massive hyper realistic hand painted scenes designed to encircle the spectator in order to create the sensation of physical relocation in the centre of an illusionistically rendered space.[14] From the massive scale, hyperrealism, unified field of vision, and the optical illusion of "framelessness" of the representational space, these immersive architectures worked to invoke a sense of presence, of transportation, and of immersion into a different time and place.

Giving rise to an ontology of the viewing experience called the "panoramic observer" that was translated across a range of immersive viewing practices—from the introduction of IMAX technologies to the development of more recent 360-degree VR—the panoramic observer "implies a scopic-ambulatory mastery over the environment, often from an elevated vantage point."[15] From this vantage point, spectators remain relatively immobile but are able to turn their gaze and submit their surroundings to visual scrutiny, a type of discursive perceptual model that has been discussed in relation to alpinism, air travel, observational towers, and the Panopticon.[16] From this vantage point, the visual senses exercise power and mastery over the vast horizon.[17]

The effect of perspectival realism of the panorama also formulated a demand for a totalizing environment that hinted at the fantasy of a complete

14 See Griffiths, *Shivers down Your Spine*, p. 39.
15 Huhtamo, *Illusions in Motion*, p. 337.
16 See Huhtamo, *Illusions in Motion*.
17 See Griffiths, *Shivers down Your Spine*, p. 136.

spectacle and "unmediated" spatial immersion, laying the basis for future large screen formats such as the Cinemascope, IMAX, and Omnimax.[18] The realist lure of immersive media has also privileged certain forms of spatial understandings over others that reflected values of authenticity, and the "auratic" connection to the real. The introduction of photography in the mid-nineteenth century also influenced a new criteria of authenticity for panoramas, in which painting had to defend its role and find a new status.[19] As Walter Benjamin argued, the introduction of mechanical reproduction technologies led to the replacement of the cult value of the "auratic" artistic original by the introduction of media mass production.[20] Intricately painted by hand, panoramas were vast artistic compositions that hoped to preserve the glimmer of the "aura" of the original and that were near-impossible to reproduce.

As a result, historical re-enactments, such as hand-painted scenes of historical sites and events, became a foundational and organizing genres that defined the panoramic experience. This panoramic genre underwrites many approaches to immersive virtual heritage today, from panoramic museum displays to VR simulations. Immersive media have typically been employed for inserting viewers into past environments generated from archaeological and historical data in the virtual interpretation of historical cross-cultural heritage sites. Through immersive panoramas, visitors were given the experience of revisiting and witnessing different times and places, including "authentic" recreations and propagandistic scenes of battles, colonial reign, open frontier landscapes, historical landmarks, and ancient antiquities, which dramatically visualized epic reconstructed representations for citizens to experience back at home. The ability for these audiences "to re-experience an event of enormous nationalist significance, 'to step inside history' was doubtlessly intended to trigger feelings of nationalistic awe and fervour for early-nineteenth-century spectators."[21] Through superior forms of visuality produced via the panorama to the immersive museum exhibit, immersive re-enactments of historical events have produced highly idealized versions of historical events through advanced artistic techniques that constituted viewers as powerful viewing subjects.

The new wave of immersive media brought on by the panorama also reflected a new world that was set in motion: industrial urban environments

18 See Griffiths, *Shivers down Your Spine*, p. 306.
19 See Huhtamo, *Illusions in Motion*, p. 274.
20 Benjamin, "The Work of Art in the Age of Mechanical Reproduction," p. 6.
21 Griffiths, *Shivers down Your Spine*, p. 50.

were evolving and rapidly expanding, new means of communication and travel were proliferating. To reflect these new perceptual experiences, immersive technologies like the panorama served as substitutes for travel, providing audiences a virtual window into the new embodied vantage points generated through rail, ocean, and air travel.[22] Particularly, the popularity of balloon panoramas at the time reflected the investment in views from new horizons and elevated positions as a form of public attraction. Travel and immersive media collapsed in an ontological linkage as categories of *viewer* were often interchangeable with those of *passenger* or *traveller*.[23]

As travelling began to form the content of many of these attractions, across Europe and North America, early panorama technologies were adopted to portray and reflect these nationalistic imaginations through scenes of travel into expanding frontiers and wilderness, experiences of pioneers, and expanding colonial enterprises.[24] Colonial panoramas produced specifically for colonial missions in New Zealand, Australia, Brazil, and India were spectacles specifically designed for emigrating colonial audiences, simulating the embodied experience of accompanying a colonial officer on their trip abroad.[25] As panorama technologies began to dissolve into other forms of spectatorship, such as the Cinéorama, by the end of the nineteenth century, impulses for virtual travel, mobility, and unrestricted tourist gaze remained operationalized in neighbouring forms of spectatorship into the twentieth century, from planetariums, IMAX, and other forms of expanded cinema. The touristic gaze of early travelogue IMAX films also allowed viewers to witness scenes and historical landmarks in emotionally charged, highly aestheticized, and sensory-rich representations.[26]

The pedagogical impact of earlier forms of immersive media becomes essential when we consider its institutional venues of exhibition: world fairs, expositions, and museums, where ideas of industrial progress, colonialism, world domination, and cross-cultural experience provided the cultural, economic, and pedagogical contexts to display these new technologies. As such, many early immersive media works also carry with them the nationalist force that links spectacles of grandness and monumentality with nationalist imaginaries and technological development with progress and colonial domination.[27] While immersive media can be read as the product

22 See Huhtamo, *Illusions in Motion*, p. 192.
23 See Huhtamo, *Illusions in Motion*, p. 112.
24 See Huhtamo, *Illusions in Motion*, p. 176.
25 See Huhtamo, *Illusions in Motion*, p. 194.
26 See Griffiths, *Shivers down Your Spine*, p. 105.
27 See Griffiths, *Shivers down Your Spine*, p. 94.

of many different perceptual habits, historical contexts of exhibition, and politics of visuality, in their epistemological contexts these formats have supported and conditioned a particular viewing practice of dominance, imperialism, and access of knowledge through simulated, extended, and unrestricted vision.

Contemporary Debates on Immersive Virtual Heritage

As immersive architectures of panoramas have collapsed with the evolution of interactive languages, software, and digital interfaces into the twenty-first century using virtual, augmented, and mixed reality technologies, ideas around immersion, interaction, and participation have also evolved. More contemporary understandings of immersion have highlighted the political implications as well as the affordances of immersive subjectivity produced through these experiences. It has now been recognized that having an immersive experience is neither universal nor neutral and is "susceptible to individual subjectivities, proclivities, and biases; it is partisan and political, and thus has the potential to unite or divide."[28] Critiques around the production of 360-degree VR in the realms of journalism, documentary, ethnography, and heritage production have centred on the complexity of its subjects, particularly foregrounding the dangers of cultivating empathy through these immersive vantage points. Such empathetic responses are often connected to the idea of standing, virtually embodied, within the shoes of another. This has been common to 360-degree VR works such as VR documentary film *Clouds over Sidra* (Gabo Arora and Barry Pousman, 2015), a VR documentary co-produced by VRSE and the United Nations depicting a day in the life of a twelve-year-old Syrian girl in the Za'atari refugee camp in Jordan. Criticisms of the film have centred on its amplification of Global North witnessing, in which viewers (often white, settler, and middle-class) peer into the lives of disadvantaged racialized others through an immobile observational vantage point, causing the viewer to do nothing more than to be an onlooker.[29] By interrogating the concept of immersive witnessing as the lens to critique discourses of empathy, some of the moral risks in immersive witnessing is that it often foregrounds the spectacle factor that these technologies can carry, valuing the user's own affective, emotional, or sensationalist response to the technology and reducing an analytic or

28 Yezbick, Yezbick, and Vavarella, "On Immersion."
29 See Rose, "The Immersive Turn," p. 142.

reflective position to properly educate on the situation of the documentary subject.[30]

Cultural heritage VR projects such as *Auschwitz VR* (The Auschwitz Birkenau Foundation, 2019), an immersive walkthrough into the digitally reconstructed grounds of the German Nazi camp Auschwitz II-Birkenau, in 1944, as well as *The Embodying Empathy Project* (Andrew Woolford et al., 2017), immersing users into the experiences of Indigenous students through a virtual reconstruction of the Fort Alexander Indian Residential School, also conflate many of these issues around the educational impact of these technologies, posing new sets of ethical and political challenges for using immersive formats for historical recreation, witnessing, education, and reconciliation.[31] While immersive VR technologies can provide new pathways to access historical knowledge, the question of their accessibility also becomes an essential question. Immersive technologies thus have the power to connect people to and disconnect them from historical knowledge due to various challenges, from access to digital infrastructures, to the required VR headsets and software, as well as through other cultural, linguistic, or economic privileges. Bringing attention to these convergent dynamics and power relationships entrenched in immersive systems is important when drawing in diverse communities into the creation and the consumption of this content.

As immersive media practices begin to collapse with pre-existing heritage practices of collecting, preserving, curating, and exhibiting, this also demands that sites of heritage themselves, particularly national archives, museums, and other heritage sites, continue to be recognized as historical monuments with particular colonial configurations that have perpetuated ongoing extraction and exclusion and where certain concepts, people, and ways of knowing have been privileged over others. With the increased digitization of cultural heritage objects, new ways for accessing and exhibiting knowledge have continued to expand within the immersive media landscape. While these practices are opening up museums' collections in new ways to a range of users, more mainstream immersive historical visioning practices in these contexts have often been utilized for their perceived ability to capture, represent, and apprehend realistic re-creations of historical objects, places, times, people, and realities, as they were presented. These conventions also become firmly rooted when combined with the long-standing discourses

30 See Nash, "Virtual Reality Witness," p. 124.

31 For a discussion of *The Embodying Empathy Project* see Woolford et al., "I Definitely Felt like I Was There.".

of heritage preservation, in which object-centred ontologies centring aura, authenticity, and provenance remain an essential part of museological and heritage preservation discourses. In continuity with the panorama, virtual renditions of museums and heritage sites have largely continued to favour the "reconstructionist approach" to allow viewers to visit a reconstruction of an "authentic" past as an archivable place.[32]

With that said, the interdisciplinarity of immersive media practice has increasingly introduced more experimental and artistic frameworks of embodied museography using virtual, augmented, and mixed reality technologies. These practices have begun to show new kinds of ways that immersion, interaction, participation, and multisensory engagement could be formulated with collections and representational frameworks that incorporate intangible heritage and embodied knowledge systems into museum spaces. This also builds on longer legacies in the field of new museography, which has worked to disrupt ocularcentrist regimes as the dominant framework in which museums have provided knowledge and exerted governmental power through an "exhibitionary complex."[33] Immersive technologies have therefore been embraced for preserving more corporeal and intangible cultural expressions and have provided more radical approaches to traditional heritage interpretation practices that have focused on visual and textual narratives from the contents of glass display cases.[34] Drawing on other sensory modalities from sonic, physical, and olfactory, immersive media can radicalize and transform the passive ocularcentric "knowledge tourist" museum's visitor into an active multisensory participant, in which experiential knowledge spaces can become open to fluid, dynamic, active, and reflective interpretations.

How immersive media can amplify other forms of multi-modal knowledge systems has also been influenced by polyphonic interpretive models in which more dialogical processes can become foregrounded in immersive media experiences.[35] Taken from sound studies, polyphony evokes a kind of choral palimpsest that de-centres and redistributes a singular dominant worldview or system of knowledge so that multiple perspectives can be heard and interpreted simultaneously.[36] Highlighted by Stó:lō scholar Dylan Robinson, polyphony foregrounds the idea that listening is a kind of

32 See Cameron and Robinson, "Digital Knowledgescapes," p. 170.

33 See Bennett, *The Exhibitionary Complex*.

34 See Geismar, *Museum Object Lessons for the Digital Age*, p. 6.

35 See Cameron and Robinson, "Digital Knowledgescapes," p. 169.

36 See Miller and Ebrahimi Bazaz, "Invited and Implicated."

perception that is inclusive of intersectionality and that it is necessary to render a more apt understanding of human engagement with the world—one that is full of and open to productive difference and complexity.[37]

More intersectional production practices of co-creation have therefore increasingly been catching onto the possibilities of immersive technology and onto the polyphonic affordances of these technologies. At its core, co-creation confronts systems of power and singular authority to interrogate fundamental ideas of ownership, meaning making, and knowledge production. Embracing the form of an ecosystem, co-creative models of production embrace process, connection, and interaction to create narrative and representational complexities that offer diverse and polyphonic forms of world-building and narrative structures. These polyphonic structures allow for the inclusion of a kaleidoscope of voices and viewpoints that trade the constraints of a unified narrative for more open, layered, multiple, and complex viewpoints and temporalities.[38] Applied to virtual heritage formats, polyphonic structures can generate multi-authored assemblages of diversity and complexity into historical narratives.[39]

Post-structural and post-colonial understandings have also contributed to a critique of monolithic narratives of historical representations, arguing towards "polyphony as a strategy to turn historical explanations away from causality, linearity, and unity, those elements often linked to hegemonic power that minimizes difference."[40] A polyphonic history can therefore break a single historical event into many fragments from disparate parts to form a mosaic that comes together in a kaleidoscopic mode of inquiry and interpretation. Beyond simply illustrating what happened when, immersive media storytelling can provide an accumulation of representational layers that may tell multiple stories that a user can explore. AR and VR are interesting frameworks from this perspective as they have the advantage of moving beyond the transmission of historical tidbits about objects and environments, instead allowing for a multi-sensory experience of a specific site that can both support and challenge traditional narratives by producing multiple alternative views.[41] Such systems could give users the sensation of being able to navigate beyond the official story of a heritage site, into a web of interconnected complexity, where history is not merely

37 See Robinson, *Hungry Listening.*

38 See Zimmerman and De Michiel, *Open Space New Media Documentary*, p. 57.

39 See Auguiste et al., "Co-Creation in Documentary," p. 63.

40 Zimmerman and De Michiel, *Open Space New Media Documentary*, p. 59

41 See Szabo, "Apprehending the Past," p. 376.

transmitted to the viewer but presented as a collection of meanings that contextualize the landscape. In this way, immersive media can serve as landscape reading aids, as a lens that can amplify alternative views and readings and do justice to complex and often conflicting histories and perspectives around a space.

An example of how immersive media can prioritize polyphonic approaches and articulate immersive complexity is the VR project *Swampscapes* (Elizabeth Miller, Kim Grinfeder, and Juan Carlos Zaldivar, 2018), a virtual reality exploration of Florida's most endangered ecosystem and significant cultural heritage site, the Everglades. Through this contested zone of migration and climate crisis, the project brings attention to the variety of lifeforms that inhabit and hold the fragile site together. Through a VR perspective, the viewer inhabits an interplay of scales and vantage points across human and non-human agencies, from underwater, in a tree, through the viewpoints, voices, and histories of locally informed "knowledge guides."[42] Rejecting tropes of character identification, unrestricted travel, and empathy, *Swampscapes* avoids being implicated in a new form of voyeurism, harkening back to the "tourist" gaze of early panoramic immersive formats. Rather than reinscribe the panoramic eye's promise of more to see through a distant and unrestricted field of view, the project disrupts acts of seeing in which the seamless 360-degree sphere is fragmented across many scales, viewpoints, and temporalities entirely to produce multiple views that reflect on the technology's own limits and reaches of seeing.

The project thus enacts an ecological mode of engagement required to garner a more complex understanding with the entanglements at play. The flexibility of scale represented through the different human and nonhuman perspectives offers the potential for greater communication with the complexity of natural systems that reminds us of our existence within them. The use of scale here also takes on a more ecocritical form. Previous studies on awe in relation to scale in virtual nature environments define the experience of awe as a combination of the vastness created through a grand sense of scale and the urge for mental accommodation or adjustment to such conditions.[43] For example, a grand mountain scene or tall trees in a forest represented in a virtual environment can significantly elicit awe in a user when compared to experiencing a more neutral landscape. Julia Scott-Stevenson argues that the danger of using immersive vantage points to engender awe in this way is that it further risks a distancing, valorization,

42 Miller and Ebrahimi Bazaz, "Invited and Implicated."
43 See Chirico et al., "Designing Awe in Virtual Reality."

and othering of the natural environment from its user.[44] She argues that through the affordances of immersive media, from their ability to generate feelings of presence and new kinds of embodiments, to their ability to both represent and alter scale, these technologies could offer new ways *in* to grasp at the complex environments we find ourselves in.[45] In what she terms an immersive "shimmer," drawn from the northern Australian Yolngu Indigenous term *bir'yun*, which translates as "brilliance" or "shimmer," Scott-Stevenson explores how immersion can rather produce a sense of bewilderment as a revelation to forms of complexity and entanglement.[46] Moving beyond the distancing awe, bewilderment encompasses a position where the user is offered the opportunity to grasp and to communicate with the multiple scales, senses, and materialities that make up the vast and complex system of our naturecultures.[47] In other words, if immersive works can bewilder, they might help us connect with something bigger than ourselves. The affordances of immersive media, such as the play of scale and polyvocality, can therefore offer opportunities for users to grasp abstract information, otherwise distant threats of ecological calamity, and forms of heritage at risk in an operable form.[48]

Alpine Garden MisGuide (Jill Didur, Ian Arawjo, and Amanda Tom, 2015), a locative augmented reality app created in Montreal's Botanical Gardens, is another example that helps to unravel the complex historical interrelationships responsible for the development of this biocultural landscape. Location-based media and augmented reality storytelling are other immersive formats that can create complex relationships with our environments. Ubiquitous GPS and mobile technologies have also enabled new avenues for complexifying heritage sites to create dynamic living archives of spatial culture through the collaborative mapping of a space. The notion of space as a living and layered archive deeply informs this approach of engaging with the complex histories of a space. As an aesthetic space of tourism, contemplation, leisure, and botanical conservation utilized in the service of colonial and capitalist expansion, *Alpine Garden MisGuide* attempts to provide a *mis*guide, or rather a countertour of the garden that seeks to unsettle rather than reinforce the entangled narratives associated with nostalgia and "echo-archic" perspectives that underpin the colonial narratives of the site.[49]

44 Scott-Stevenson, "Finding Shimmer."
45 Scott-Stevenson, "Finding Shimmer."
46 Scott-Stevenson, "Finding Shimmer."
47 Scott-Stevenson, "Finding Shimmer."
48 Scott-Stevenson, "Finding Shimmer."
49 See Didur, "Beyond Anti-Conquest."

Through the soundscapes of various voices and "experts," it achieves locative depth by leveraging layers of time, cultural histories, and knowledge through embodied experience and presence. Scale is also used here, but differently from VR; rather than being presented visually, it is illuminated through audio and also temporally. Breaking more traditional models of heritage interpretation, such as didactic text-based information or an ocularcentric view that promotes an objective or total history of the botanical garden, the *Alpine Garden MisGuide*'s multiple voices and vantage points produced through geo-located sound help to shift the visitors' attention towards the historical contexts and power relations that are not immediately legible when visiting the site but that deeply inform its operation. These layers and simultaneities reveal that spaces, especially heritage sites, have alternative relational meanings that linger underneath.

Like the living ecologies that govern these sites, these projects incorporate an ecological ethic to reveal an assemblage of histories, communities, and uses that make up their polyphonic and complex networked narratives. The move towards a philosophy of ecological thinking for immersive media systems sees virtual environments as we would ecosystems, meaning that they are represented through a complex assemblage of agents, objects, processes, relations, beings, and things that make up our environments, real and virtual.[50] For the future of cultural heritage, Fiona Cameron and Helena Robinson predict immersive environments embodying these complex cultural ecosystems that will bring past environments through highly relational, multi-authored, scalable, and dynamic visualizations and navigation environments.[51] Through the collaborative and participative potentials of immersive platforms, ecosystems, and infrastructures, emerging cultural heritage practices have begun to transform, as they become more deeply implicated within these new ecosystems. As such, museums are being transformed into information transmitters rather than solely centrally localized repositories of knowledge.[52] As archival theorists Wendy M. Duff and Jessica Haskell have highlighted, this wave of digital production practices, including immersive 3D technologies, have "provided new paths into the archives through the assistance of new agents who can alter, manipulate, and play with the records."[53] Virtual and augmented reality as well as 3D production technologies have improved the access to and the

50 See Nash, Hight, and Summerhayes, "Introduction," p. 2.
51 Cameron and Robinson, "Digital Knowledgescapes," p. 184.
52 See Mason, "Cultural Information Standards," p. 232.
53 Duff and Haskell, "New Uses for Old Records," p. 48.

use of heritage materials for many different types of users to use, interpret, and remix. For example, Sarah Younan has explored how 3D scanning and the creative re-activation of museum artifacts at the National Museum of Cardiff have enabled new expressive, personal, and political engagements as well as knowledge authorship with heritage collections.[54]

This illustrates a significant shift in the ways that knowledge institutions tend to be organized around authors, experts, and decision-making hierarchies. Challenging the inherent "expertise" and "authority" of knowledge built into the museum space, immersive media can facilitate engagement with "non-experts" within heritage preservation and interpretation, incorporating multiple reflexive positions into these representations.[55] While cultural institutions have long been entrusted as central repositories of knowledge, digital technologies like immersive media extend them into distributed heritage spaces or rather into *heritage ecologies* where people, heritage professionals, and technologies engage over time and across platforms and multiple locations, real and virtual, in creative, emergent, immersive, and entangled understandings of cultural heritage.

Conclusion

As immersive media expands into new complex ecologies of practice, rather than seeing their "messiness" as inhibiting to experiences of immersion and to historical understandings, how might these aspects become useful and essential parts of the design process itself? How might embracing shifting modes of attention to the various agencies that give meaning to a heritage site become an essential and generative part of an immersive experience? Can these aspects alter how a user engages with history to produce new agendas of care for historically salient spaces and inform their possible futures?

These questions become increasingly important as the broadening domain of digital, interactive, and immersive media begins to influence and to reorient foundational spatial, temporal, and institutional understandings and practices of heritage itself into complex entanglements of technologies, places, times, bodies, and subjectivities. While the long-standing discourses and practices surrounding immersive heritage have centred on the production of spatially and temporally cohesive experiences that seamlessly bring

54 Younan, "Poaching Museum Collections Using Digital 3D Technologies," p. 25.

55 See Rodely, "The Distributed Museum Is Already Here," p. 83.

users into past environments and realities through unrestricted vision and access into these spaces, a new terrain of immersive media has begun to embrace more radical forms of knowledge preservation and creation that challenge these perspectives. As virtual heritage practices advance, they simultaneously must account for the complex political and ethical implications that these technologies carry, as well as for the growing entanglements between cultures, identities, materialities, and technologies that continue to disorient and decentre fixed vantage points and positionalities or stable understandings of place, history, and cultural identity as fixed entities.[56]

As immersive media increasingly intersect in spaces of heritage preservation, display, and education, embracing these complexities into virtual heritage design becomes an important task for the practitioners creating and co-creating immersive experiences about the lives, places, and histories of others. As the highly embodied, experiential, and relational affordances made possible through immersive media formats advance, immersive technologies are arguably the very tools available to provide practical and meaningful ways *in* to navigate and make better sense of these entanglements that make up our new heritage ecologies.

Bibliography

Arawjo, Ian, Jill Didur, and Amanda Tom. *Alpine Garden MisGuide.* Jardins Alpins Autrement. 2015. https://www.alpinemisguide.org/.

Aroroa, Gabo, and Chris Milk. *Clouds over Sidra.* Barry Pousman et al. 2015. https://www.unicefusa.org/stories/clouds-over-sidra-award-winning-virtual-reality-experience/29675.

Auguiste, Reece, Helen De Michiel, Brenda Longfellow, Dorit Naaman, and Patricia R. Zimmermann. "Co-Creation in Documentary: Toward Multiscalar Granular Interventions Beyond Extraction." *Afterimage* 47, no. 1 (2020): 34–71.

Auschwitz VR. Auschwitz Birkenau Foundation. 2019. https://auschwitz-vr.pl/en/.

Barad, Karen. *Meeting the Universe Halfway: Quantum Physics and the Entanglement of Matter and Meaning.* Durham, NC: Duke University Press, 2007.

Benjamin, Walter. "The Work of Art in the Age of Mechanical Reproduction." In *Illuminations*, edited by Hannah Arendt, translated by Harry Zhon, 1–26. New York, NY: Schocken Books, 1969. https://web.mit.edu/allanmc/www/benjamin.pdf.

Bennett, Tony. "The Exhibitionary Complex." In *Museums, Power, Knowledge: Selected Essays.* 23–51. London: Routledge, Taylor & Francis Group, 2017.

56 See Rueb, "Restless," p. 242.

Cameron, Fiona, and Helena Robinson. "Digital Knowledgescapes: Cultural, Theoretical, Practical, and Usage Issues Facing Museum Collection Databases in a Digital Epoch." In *Theorizing Digital Cultural Heritage: A Critical Discourse*, edited by Fiona Cameron and Sarah Kenderdine, 165–193. Cambridge, MA: MIT Press, 2010.

Cameron, Fiona, and Sarah Kenderdine. "Introduction." In *Theorizing Digital Cultural Heritage:. A Critical Discourse*, edited by Fiona Cameron and Sarah Kenderdine, 1–19. Cambridge, MA: MIT Press, 2010.

Chirico, Alice, Francesco Ferrise, Lorenzo Cordella, and Andrea Gaggioli. "Designing Awe in Virtual Reality: An Experimental Study." *Frontiers in Psychology* 8 (January 22, 2018). https://www.frontiersin.org/articles/10.3389/fpsyg.2017.02351.

Davies, Char. "Rethinking VR: Key Concepts and Concerns." In *Hybrid Reality: Art, Technology, and the Human Factor*, edited by Hal Thwaites, 253–262. Montreal: International Society on Virtual Systems and Multimedia, 2003. http://www.immersence.com/publications/char/2003-CD-VSSM.html#.

Didur, Jill, and Lai-Tze Fan. "Between Landscape and the Screen: Locative Media, Transitive Reading, and Environmental Storytelling." *Media Theory* 2, no. 1 (2018): 79–107.

Didur, Jill. "Beyond Anti-Conquest: Unearthing the Botanical Archive with Locative Media." *Intermédialités. Histoire et théorie des arts, des lettres et des techniques / Intermediality. History and Theory of the Arts, Literature, and Technologies*, no. 35 (2020). https://doi.org/10.7202/1076376ar.

Duff, Wendy M., and Jessica Haskell. "New Uses for Old Records: A Rhizomatic Approach to Archival Access." *The American Archivist* 78, no. 1 (2015): 38–58.

Geismar, Haidy. *Museum Object Lessons for the Digital Age*. London: University College London Press, 2018.

Griffiths, Allison. *Shivers down Your Spine: Cinema, Museums, and the Immersive View*. New York, NY: Columbia University Press, 2008.

Grinfelder, Kim, Elizabeth Miller, and Juan Carlos Zaldiva. *Swampscapes*. Kim Grinfelder, Elizabeth Miller, and Juan Carlos Zaldiva. 2018. http://www.swampscapes.org/vr.html.

Huhtamo, Erkki. *Illusions in Motion: Media Archaeology of the Moving Panorama and Related Spectacles*. Cambridge, MA: MIT Press, 2013.

Mason, Ingrid. "Cultural Information Standards: Political Territory and Rich Rewards." In *Theorizing Digital Cultural Heritage: A Critical Discourse*, edited by Fiona Cameron and Sarah Kenderdine, 223–245. Cambridge, MA: MIT Press, 2010.

Miller, Elizabeth, and Aggie Ebrahimi Bazaz. "Invited and Implicated: Toward Critical Reflexivity in Non-Fiction VR." *Digital Radical: The Center on Digital Culture and Society at the University of Pennsylvania's Annenberg School*, June 18, 2020. https://

www.asc.upenn.edu/research/centers/center-on-digital-culture-and-society/the-digital-radical/Invited-and-Implicated.

Nash, Kate, Craig Hight, and Catherine Summerhayes. "Introduction." In *New Documentary Ecologies: Emerging Platforms, Practices, and Discourses*, edited by Kate Nash, Craig Hight, and Catherine Summerhayes, 1–7. New York, NY: Springer, 2014.

Nash, Kate. "Virtual Reality Witness: Exploring the Ethics of Mediated Presence." *Studies in Documentary Film* 12, no. 2 (2018): 119–131.

Robinson, Dylan. *Hungry Listening: Resonant Theory for Indigenous Sound Studies.* Minneapolis, MN: University of Minnesota Press, 2020.

Rodely, Ed. "The Distributed Museum Is Already Here." In *The Routledge International Handbook of New Digital Practices in Galleries, Libraries, Archives, Museums and Heritage Sites*, edited by Hannah Lewi, Wally Smith, Dirk vom Lehn, and Steven Cooke, 81–91. London: Routledge, 2019.

Rose, Mandy. "The Immersive Turn: Hype and Hope in the Emergence of Virtual Reality as a Nonfiction Platform." *Studies in Documentary Film* 12, no. 2 (2018): 132–149.

Rueb, Teri. "Restless: Locative Media as Generative Displacement." In *Mobility and Locative Media: Mobile Communication in Hybrid Spaces*, edited by Adriana De Souza e Silva and Mimi Sheller, 261–278. London: Routledge, 2014.

Scott-Stevenson, Julia. "Finding Shimmer: Immersive Non-Fiction Media and Entanglements in Virtual Nature." *Digital Culture & Education* 12, no. 2 (2020). https://www.digitalcultureandeducation.com/volume-12-2.

Speakman, Duncan. *It Must Have Been Dark by Then*. Ambient Literature. 2017. https://duncanspeakman.net/projects/imhbdbt/.

Speakman, Duncan. *No Vantage Point: Reflections, Provocations, and Experiments in Immersion*. Bristol: South West Creative Technology Network, 2019.

Szabo, Victoria. "Apprehending the Past: Augmented Reality, Archives, and Cultural Memory." In *The Routledge Companion to Media Studies and Digital Humanities*, edited by Jentery Sayers, 367–379. New York, NY: Routledge, 2018.

Woolford, Andrew, Katherine B. Starzyk, Struan Sinclair, Adam Muller, and David B. MacDonald. ""I Definitely Felt like I Was There": Enacting Empathy and Negotiating a Virtual Reality Indian Residential School." In *AlterNative: An International Journal of Indigenous Peoples* 18, no. 3 (2022): 412–23. https://doi.org/10.1177/11771801221117561.

Witcomb, Andrea. "The Materiality of Virtual Technologies: A New Approach to Thinking about the Impact of Multimedia in Museums." In *Theorizing Digital Cultural Heritage: A Critical Discourse*, edited by Fiona Cameron and Sarah Kenderdine, 35–49. Cambridge, MA: MIT Press, 2010.

Yezbick, Julia, Rachel Yezbick, and Emilio Vavarella. "On Immersion. Editorial Essay." *Sensate Journal*, January, 2021. http://sensatejournal.com/on-immersion/.
Younan, Sarah. "Poaching Museum Collections Using Digital 3D Technologies." *Journal of Science and Technology of the Arts* 7 (December, 2015): 25–32.
Zimmermann, Patricia R., and Helen De Michiel. *Open Space New Media Documentary: A Toolkit for Theory and Practice*. London: Routledge, 2018.

About the Author

Elina Lex is an interdisciplinary researcher and media artist with research interests in immersive media technologies, sensory ethnography, and digital archiving practices. Her research-creation work explores the opportunities that multi-sensory immersive media expressions (XR/AR/VR) might bring to improve forms of access, engagement, and impact with archives and cultural memory material. Exploring concepts of the living archive, she often works with various forms of archives, heritage sites, and XR platforms to activate the layered, entangled, and contradictory narratives that inform a historical site. Elina Lex is an active researcher at the Hexagram Network for Research-Creation, the Center for Oral History and Digital Storytelling, and the Post-Image Cluster at the Milieux Institute.

18. The Contribution of Narrative Puzzles to Player Challenge-Based Immersion: A Case Study of *Abzû*

Gabrielle Trépanier-Jobin and Mathilde Savoie

Abstract: In this chapter, the authors suggest that narrative puzzles in video games favour challenge-based immersion, because of the cognitive effort that players must make to interpret clues, draw hypotheses, or solve enigmas. They also argue that narrative puzzles can enhance ludic macro involvement and shared macro involvement when a group of players investigate the story together on forums. To do so, they draw on concepts such as "enigma-driven narratives," "open texts," and "forensic fandom." They also conduct a close reading of the underwater adventure video game *Abzû* (Giant Squid, 2016), in addition to analyzing testimonies provided by four *Abzû* players during semi-structured interviews, as well as discussions between hundreds of players on internet forums.

Keywords: video game, player interviews, forensic engagement, reception, puzzle design

Introduction

In game studies, imaginative immersion and challenge-based immersion are generally perceived as different experiences. Imaginative immersion is described as a feeling of being transported or present in a fictional world, and is generally studied in relation to avatars, non-playable characters, storylines, and diegetic worlds. Challenge-based immersion is, for its part, comprehended as a feeling of being absorbed in an activity, and is typically

Bédard, Philippe, Alanna Thain, and Carl Therrien, eds. *States of Immersion Across Media: Bodies, Techniques, Practices*. Taylor & Francis Group, 2025
DOI 10.5117/9789048558766_CH18

linked to interactivity, goals, rewards, rules, and mechanics.[1] Some game scholars acknowledge that these types of immersion overlap, but they remain vague on how they intersect.

In this chapter, we focus on the overlap between imaginative and challenge-based immersions by highlighting the contribution of what we call "narrative puzzles" to challenge-based immersion. More precisely, we argue that video game storylines are sometimes constructed as disseminated clues that need to be collected and pieced together. These narrative puzzles do not only enhance imaginative immersion, but also favour challenge-based immersion, because of the effort players must make to interpret clues, draw hypotheses, or solve enigmas. We finally hypothesize that narrative puzzles allow what might be named, following Calleja, "ludic macro involvement" and "shared macro involvement" when a group of players investigate the story together on forums.[2]

To develop our arguments, we compare video games' narrative puzzles to what is called in non-interactive media theories "enigma-driven" narratives that raise questions about past or present events, and to "open texts" that favour "narrative ambiguity" and "textual indeterminacy" over closure and clarity.[3] Drawing on the concept of "forensic engagement," we suggest that the model player who interacts with narrative puzzles adopts a forensic attitude like that of the famous detective Sherlock Holmes.[4] We suggest finally that when players share their theories about the game's lore on forums, it generates what Jason Mittell calls "forensic fandom."[5]

To illustrate and support these ideas, we conduct a close reading of *Abzû* (Giant Squid, 2016), an underwater adventure video game whose mysterious story can only be hypothesized by assigning meaning to enigmatic mural

1 See Ermi and Mäyrä, "Fundamental Components of the Gameplay Experience"; and Therrien, "Immersion."

2 See Calleja, *In-Game*.

3 See Mittell, *Complex TV*; Eco, *Lector in Fabula*; Rimmon-Kenan, *Narrative Fiction*; and Iser, "Indeterminacy and the Reader's Response to Prose Fiction."

4 On the notion of "forensic engagement," see Mittell, "Sites of Participation." The expression "model player" is inspired by Umberto Eco's theory of the "model reader," a theoretical figure whose encyclopedia of knowledge and reading competences allow them to perfectly actualize the meaning of a specific text as intended by its author. See Eco, *Lector in Fabula*, p. 101. Similarly, a "model player" is a theoretical figure, implied by a game structure, whose know-how and gaming competences allow them to navigate this game space and interact with its elements in accordance with the designer's expectations. See Genvo, "Caractériser l'expérience du jeu à son ère numérique," pp. 6–7. In the case of a narrative puzzle, the model player is one who cares about the reconstitution of the story and puts the necessary efforts into it. On Sherlock Holmes, see Ginzburg, "Morelli, Freud and Sherlock Holmes."

5 Mittell, "Sites of Participation."

frescos and puzzling events. Our affirmations are also based on the analysis of semi-structured interviews conducted with four *Abzû* players and on the analysis of discussions between hundreds of players who shared their interpretations of the game's background story on internet forums.

Literature Overview

Our argumentation relies, on the one hand, on theories about player immersion, and on the other hand, on theories about narrative puzzles in non-interactive media.

Theories about Player Immersion

In game studies, the term "immersion" is sometimes used to describe the feeling of being transported to another world, which mostly occurs in 3D environments where players control an avatar from a first- or third-person perspective. The term "immersion" is also utilized to describe the feeling of being entirely absorbed in an activity which can also arise in 2D puzzle or strategy games.

Laura Ermi and Frans Mäyrä distinguish three types of player immersion.[6] "Sensory immersion" is a feeling of transportation that unfolds when audiovisual stimuli from the game world surpass those from the physical world, and when players are completely focused on the sensory information of the game. "Imaginative immersion" is a feeling of transportation that happens when players are captivated by the story and identify with their avatar. Finally, "challenge-based immersion" (sometimes called "systemic immersion") is a feeling of absorption that arises when the difficulty of the mental or motor challenges coincides with the players' mental or motor skills. Challenge-based immersion is similar to Mihály Csíkszentmihályi's notion of "flow," which is the mental state of individuals who are fully absorbed in an activity while acting at the limit of their skills.[7] According to Dominic Arsenault and Martin Picard, challenge-based immersion can be divided in two subcategories: "tactical immersion," which implies being absorbed by real-time micro-decisions and hand-eye coordination, and "strategic/cognitive immersion," which entails being absorbed by the elaboration of long-term strategies or by the analysis of the game's ramifications.[8]

6 Ermi and Mäyrä, "Fundamental Components of the Gameplay Experience."
7 Csíkszentmihályi, *Flow.*
8 Arsenault and Picard, "Le jeu vidéo entre dépendance et plaisir immersif."

Ermi and Mäyrä acknowledge that the three types of player immersion mix and overlap in many ways but remain vague on how this happens.[9] Overall, sensory immersion is generally studied in connection with graphic design, sound design, music, and interfaces; imaginative immersion, in relation to stories, non-playable characters, avatars, and fictional worlds; and challenge-based immersion, in conjunction with challenges, feedback, mechanics, rules, and goals, which are, to be fair, the main elements that respectively impact these three types of immersion.[10]

Gordon Calleja is one of the few scholars who adopt a holistic[11] approach to player immersion, even though he prefers to use the term "incorporation" to refer to this intense but fragile psychological state during which the player's consciousness merges with the game world.[12] This expression allows him to refer at once to the assimilation of the game environment into the player's mind and to the embodiment of the player into the game world. Calleja also argues that a simultaneous involvement with many dimensions of a game is necessary to feel incorporated, and that being involved with one dimension implies stopping to pay attention to it. For instance, "kinesthetic involvement" implies the mastery of the game's controls and the perceived fluidity of the avatar's movements.[13] "Spatial involvement" is derived from the feeling of being able to explore the game environment freely, as well as from the mental mapping of the game space.[14] "Ludic involvement" relates to the autotelic[15] pursuit of a central goal, to the comprehension of how secondary goals allow one to reach the main goal, and to the perception of sufficient feedback from the game system.[16] "Affective involvement" concerns the players' emotions towards the game, "from the calming sensation of coming across an aesthetically pleasing scene to the adrenaline rush of an on-line competitive first-person-shooter round to the uncanny effect of an eerie episode in an action-horror game."[17] "Shared involvement" is

9 Ermi and Mäyrä, "Fundamental Components of the Gameplay Experience."

10 On sensory immersion, see Robertson, Czerwinski, and Dantzich, "Immersion in Desktop Virtual Reality." On imaginative immersion, see Murray, *Hamlet on the Holodeck*; and Ryan, *Narrative as Virtual Reality*. On challenge-based immersion, see Brown and Cairns, "A Grounded Investigation of Game Immersion"; and Martey et al., "Measuring Game Engagement."

11 The term holistic refers to the idea that the parts of something are intimately interconnected and explicable only by reference to the whole.

12 Calleja, *In-Game*, p. 167.

13 Calleja, *In-Game*, p. 45.

14 Calleja, *In-Game*, pp. 73–77.

15 An autotelic activity is its own goal, instead of being pursued to reach another goal.

16 Calleja, *In-Game*, pp. 150–155.

17 Calleja, *In-Game*, p. 44.

derived from players' interaction with human-on-non-human agents and fed by the players' need for collaboration, communication, and teamwork.[18] Finally, "narrative involvement" is based on a sustained interest in the scripted narrative and in its integration into the players' "alterbiography," that is to say the emergent narrative that unfolds in the players' mind during gameplay, as they make choices and alter the course of events.[19]

In his chapter on narrative involvement, Calleja refers to David Levine's distinction between two types of scripted game narratives: "push narratives" are delivered to players through cutscenes, text boxes, imposed dialogues with NPCs, quick-time events, etc., and "pull narratives" through disseminated objects that players can collect, or non-imposed dialogues with NPCs.[20] According to Calleja, pull narratives are more engaging than push narratives.

Calleja also draws a distinction between "micro involvement," which takes place in the moment of playing, and "macro involvement," which occurs before or after playing.[21] Macro involvement concerns "aspects that attracted players to the game initially and kept them returning to the game over time."[22] Ludic involvement, for instance, can extend beyond the game when players keep developing strategies. Narrative involvement can occur at the macro level when players are unravelling the game narrative, planning the development of a character, and reading or writing fan fictions about a game's fictional world. Shared involvement can continue after the game when players interact with each other on external platforms to discuss the game's content.

Even though concepts such as "emergent narratives," "experiential narratives," and "alterbiography" show that game narratives and rules interplay, the link between narrative puzzles and challenge-based immersion remains largely unexplored.[23] The same can be said for the relation between narrative puzzles, ludic macro involvement, and shared macro involvement. If challenge-based immersion can be related to mental challenges, as Ermi and Mäyrä suggest, as well as to players' cognitive effort when they analyze

18 Calleja, *In-Game*, p. 43 and p. 104.

19 See Calleja, "Experiential Narratives in Game Environments," p. 5; and Calleja, *In-Game*, pp. 131–133.

20 Calleja, *In-Game*, p. 123.

21 Calleja, *In-Game*, p. 40.

22 Calleja, *In-Game*, p. 39.

23 On emergent narratives, see Salen and Zimmerman, *Rules of Play*; and Jenkins, "Game Design as Narrative Architecture." On experiential narratives, see Pearce, "Towards a Game Theory of Game." On alterbiography, see Calleja, "Experiential Narratives in Game Environments"; and Calleja, *In-Game*.

the ramifications of a game, as Arsenault and Picard indicate, it seems plausible that narrative puzzles can trigger challenge-based immersion during gameplay, as well as ludic and shared involvement after gameplay, when players collaborate online to investigate the story.[24]

Theories about Narrative Puzzles in Non-Interactive Media

Most theories on narrative puzzles do not come from game scholars, but from academics who study non-interactive media such as literature, television, and cinema.

For instance, Mittell distinguishes non-ambiguous "narrative statements," which move the story forward without raising other questions except the usual "what's next?," from "narrative enigmas," which cast doubt upon what led to these events and raise questions such as "what precisely happened, who was involved, why they did what they did, how this came to be, or even whether it actually happened at all?"[25] The scene from the TV series *Lost* (J. Lieber, J. J. Abrams, and Damon Lindelof, 2004–2010) in which Michael shoots his friends Ana Lucia and Libby to release their prisoner (season 2, episode 20), for example, raises questions not only about the ladies' survival, but also about the reason why Michael betrayed his friends and about what happened to him while he was away from the group. Since all these questions are answered in the following three episodes, explains Mittell, this narrative enigma is relatively small compared to the show's "complex web of serialized mysteries and long-lasting ambiguities."[26] The multiplication of central narrative enigmas in *Lost* makes it an "enigma-driven" TV program, unlike a majority of TV shows that mainly raise questions about future events while occasionally providing key insights about the characters' background story.

In most texts, narrative enigmas can be solved; by the end of the reading process, readers reach a final hypothesis and manage to make sense of the entire text. The epilogues of detective novels, for instance, disclose the answers to previously raised questions such as: who is the murderer, who is the thief, or what is the cause of the death?[27] In some cases, receivers can solve narrative enigmas by discovering hidden clues in the text. In other cases, they must wait until more information is provided to piece together the mystery, even though they start processing information from the very beginning.

24 Ermi and Mäyrä, "Fundamental Components of the Gameplay Experience"; and Arsenault and Picard, "Le jeu vidéo entre dépendance et plaisir immersif."
25 Mittell, *Complex TV*, pp. 24–25.
26 Mittell, *Complex TV*, p. 25.
27 See Rimmon-Kenan, *Narrative Fiction*, p. 121.

Narrative video games, such as adventure games and walking simulators, tend to be enigma-driven; piecing the story together is often their main challenge. Unlike most non-interactive media, the interest of video games does not mainly lie in the story, but also in the exploration of the game environment and in the problem-solving. This allows video games to be more enigmatic and to offer fewer answers regarding their story.

Some game narratives are, therefore, closer to what Eco calls "open texts."[28] Unlike closed texts, which direct the reader's interpretation with the help of interpretative clues, open texts, such as James Joyce's *Finnegans Wake* (1939), favour interpretative freedom, "narrative ambiguity," and "textual indeterminacy."[29] They are designed "to prevent the formation of any 'finalized hypothesis' or overall meaning by making various items undermine each other or cancel each other out, without forming neatly opposed possibilities."[30] Sometimes, several finalized hypotheses complement or mutually exclude each other without providing solid grounds for choosing between them. In a nutshell, open texts are constructed by their creators as a playground for various interpretations.

In the case of enigma-driven narratives, as in the case of open texts, the receivers must use a similar method to reconstruct the narrative through the interpretation of collected clues. This method shares semiotic similarities with the inquiry approach of the famous detective character Sherlock Holmes, which consists of making a series of inferences from details that would seem trivial or unimportant to the unaccustomed eye, but "provide the key to a deeper reality, inaccessible by other methods."[31] Mittell calls this method "forensic engagement," which implies to "embrace a detective mentality, seeking out clues, charting patterns and assembling evidence into narrative hypotheses and theories."[32] This forensic attitude is related to a "hyper-attentive" posture of reception and can lead to "forensic fandom" when several receivers investigate a media production together on online forums or wiki sites.[33] Mittell defines forensic fandom as a mode of "engagement encouraging research, collaboration, analysis, and interpretation."[34]

28 Eco, *Lector in Fabula*.

29 Rimmon-Kenan, *Narrative Fiction*, p. 125; and Iser, "Indeterminacy and the Reader's Response to Prose Fiction."

30 Rimmon-Kenan, *Narrative Fiction*, p. 121.

31 Ginzburg, "Morelli, Freud, and Sherlock Holmes," p. 11.

32 Mittell, "Lost in a Great Story," pp. 128–129.

33 See Mittell, "Sites of Participation."

34 Mittell, "Sites of Participation."

Theories on Interpretations

Interpretation is a dynamic process during which a meaning is given to a set of signs by an interpreter. In media theories that are based on constructivist psychology, interpretation is conceived as an active goal-oriented process.[35] As Eco explains, reading a text implies drawing hypotheses about the writer's intention and constantly validating or invalidating these hypotheses according to their coherence with new textual elements.[36] In the case of open texts, the process of drawing hypotheses, testing them, and revising them is, however, more challenging, as semiosis can hardly be stopped to produce final interpretations.[37]

The same can be said for video games with enigmatic or ambiguous stories. Unlike linear books, films, and television shows, video games can, however, ask players to perform actions, such as digging or scrutinizing, to access interpretative clues. Game narratives are "cybertexts" that more systematically involve the readers who must make choices by performing physical actions during the semiotic process.[38] These cybertexts are "ergodic" (*ergos* means "work" and *hodos* means "path" in greek), because they require their readers to make a significant effort to go through them.[39] The fact that players must actively "dig" for clues also means that some pieces of the narrative puzzle might be missing.

Several authors establish parallels between play and interpretation: Katie Salen and Eric Zimmerman state that interpretation constitutes the act of playing; Jacques Henriot indicates that the idea of interpretation introduces the idea of play; Miguel Sicart depicts playing as a hermeneutic process; and Maude Bonenfant asserts that "playing is interpreting"; even if the game's rules are fixed, players are free to interpret them as they wish.[40] Freedom of interpretation is a fundamental aspect of play as it allows players to be creative and to invent new solutions or modes of resolution.

While these authors believe that play results from a free interpretation of game rules, we want to suggest that play can also arise from the free interpretation of game narratives when these are introduced to players as puzzles to solve and as the main challenges of the game. In this type of game, narrativity is not only a hook that piques the players' curiosity and lures

35 See Bordwell, *Making Meaning*.

36 Eco, *Interprétation et surinterprétation*, p. 42.

37 "Semiosis" refers to the process of producing meaning by interpreting signs.

38 See Aarseth, *Cybertext*, p. 1.

39 See Aarseth, *Cybertext*, p. 1.

40 Salen and Zimmerman, *Rules of Play*, p. 374; Henriot, *Sous couleur de jouer*, p. 78; Sicart, *The Ethics of Computer Games*, p. 88; and Bonenfant, *Sens, fonction et appropriation*, p. 225.

them into the game, but also an element that captures the players' attention and forces them to inquire and hypothesize throughout the gameplay.[41]

Drawing on these theories, we define narrative puzzles in video games as push or pull narratives that are enigma-driven or ambiguous, and that are conveyed to players through disseminated clues in cryptic messages, cutscenes, mysterious objects, symbols, etc. These disseminated clues need to be collected, pieced together, decrypted, or interpreted to elaborate hypotheses that can be validated by upcoming information (or not, in the case of open texts). This ergodic interpretative process requires players to adopt a forensic attitude similar to the detective's and to mobilize their mental resources in a similar way than when they face typical video game challenges and puzzles. And when multiple players inquire about a game's lore together, on internet forums, it can lead to what Mittell calls "forensic fandom."

Methodology

To illustrate our ideas, we conducted a case analysis of the adventure game *Abzû* that involves three methodologies: a close reading of the game, interviews with players, and an analysis of players' discussions on forums.

Several criteria were considered before selecting *Abzû*. We were looking for a game with a strongly developed and enigmatic narrative dimension. From a practical standpoint, we also wanted a game that could be completed in a couple of hours, so that the participants who played it over a period of ninety minutes in the laboratory would almost reach the end. We also opted for a game that was sufficiently popular to generate a great deal of discussions between players online. Finally, we were looking for a game that we, as researchers, would enjoy playing several times. As Mäyrä argues, it is essential to play the game that we study to describe the players' experience adequately.[42]

To show that *Abzû* qualifies as a narrative puzzle, we performed a "close reading" of the game, that is to say a meticulous practice of reading that remains near the text. In this type of analysis, the analyst "typically moves through a selection gradually, using highly specific textual evidence to make broader connections and claims."[43] The interpretations that we formulated

41 See Ryan, *Avatars of Story*, p. 286.

42 Mäyrä, *An Introduction to Game Studies*, p. 165.

43 Schur, *An Introduction to Close Reading*, p. 2.

are therefore not representative of the interpretations made by most players. They are also partly informed by the game analyses that we encountered during our research.[44]

To illustrate how narrative puzzles can favour challenge-based immersion, we analyzed the testimonies of four participants, who were asked to play *Abzû* for one hour and a half in a laboratory, before participating in individual semi-structured interviews of forty-five minutes. These interviews were conducted in the context of exploratory research on player immersion that favoured an inductive approach: we asked several open-ended questions to the participants in the hope that new ideas would emerge from the testimonies. Therefore, participants were not asked to comment on the cognitive effort made to piece the narrative puzzle together or on the contribution of this cognitive effort to their challenge-based immersion. These ideas actually came to light when we analyzed their answers to general questions about their overall appreciation of different game elements, such as the avatar, the non-playable characters, the controls, the interface, the game environment, the gameplay and, of course, the narrative.

Finally, to see if *Abzû*'s narrative puzzle can favour what we called, following Calleja, ludic macro involvement and shared macro involvement, we analyzed players' discussions regarding the game's background story on three different forums: Reddit, Steam Community, and GameFAQs. Overall, more than a hundred posts were combed through to find evidence of forensic fandom.

Case Study: *Abzû*'s Narrative Puzzle

Close Reading of Abzû*'s Story*

The underwater adventure game *Abzû* (which means "ocean of wisdom" in Sumerian) is a good example of a narrative puzzle game, since its mysterious story can only be hypothesized by assigning meaning to enigmatic cutscenes and to different clues that are disseminated in the game environment.

At the beginning of the game, the player-character, dressed in a diving suit, wakes up in the middle of the sea. There is no land on the horizon, no information on who they are, on where they are and on what they are aiming for. The player-character has no other choice than to explore the environment in search of answers. They quickly discover a colourful aquatic

44 The analysis of the YouTuber NIKMOE was particularly helpful. See NIKMOE, "The Hidden Message of *Abzû* (*Abzû* Story Breakdown)."

world teeming with life and get the chance to swim with wriggling fish, joyful dolphins, and many other aquatic creatures. Yet, large areas of the ocean are dark and lifeless. The player-character rapidly notices that they have the power to revitalize these areas by entering sacred wells to share the energy contained in their body. During their adventure, the player-character benefits from the help of small robots with whom they share a similar colour palette. At some point, one of these robots is brutally devoured by a great white shark, which seems, at first sight, to be the player-character's main enemy. Yet, the player-character keeps following the shark and helping it across the game.

The only way to know more about the player-character, the great white shark, and the game world is to carefully observe the murals of a flooded temple that the player-character must swim across halfway through the game. Yet, these are not easy to decipher.

In the first temple's entry room, a painting shows individuals—who look similar but not identical to the player-character—as they collect water from the sea with jars to extract its energy before returning it to the ocean. The fact that fish emerge from the area where the water is dropped provides clues on its power to generate life. Another mural shows a sacred well similar to those previously encountered by the player-character. The fact that this well is surrounded by tall golden seaweeds and is crowned by a giant anemone suggests that the marine ecosystem was healthy at that time.

If players look closely at the picture on which the water is collected, they might notice an immense source of energy flowing from the bottom of the ocean, and might see its resemblance to the sacred wells' portals.

In the second room of the same temple, a fresco depicts the player-character as they stand on the top of an inverted pyramidal structure, made of electrical circuits, that rests on the ocean's source of energy. The second picture shows the player-character as they stand beside the pyramid and hold the power of the ocean in their hands, while the individuals depicted in the previous room kneel before them. Hyper-attentive players might notice a representation of the pyramidal structure on the player-character's diving suit. Even though players do not have, at this point, enough information to understand what this pyramid is, they can deduce that the player-character is related to it somehow and perhaps even controls it. They can hypothesize that the pyramid and the player-character are worshiped for their ability to harness the ocean energy. A third painting, however, provides clues regarding the lethal effect of the pyramid on the marine ecosystem: the anemone on top of the sacred well is now colourless and shattered, the fish are swimming away from it, and the surrounding seaweeds lost their vitality.

Figure 18.1: Murals in the entry room of the first temple.

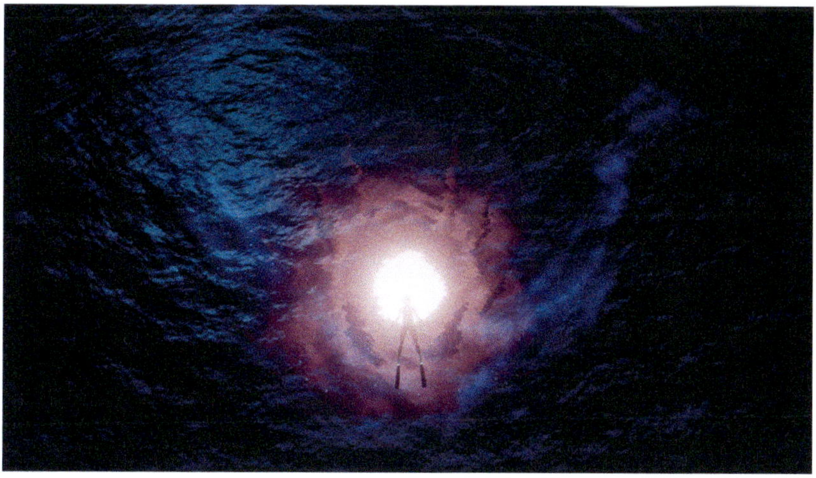

Figure 18.2: The portal of a sacred well.

Its lethal effect becomes obvious when the player-character reaches the actual pyramid that was depicted on the fresco. The contrast between the colourful and lively environment encountered during the first half of the game and the dark and lifeless area where the giant pyramid rests is shocking. This gigantic metal structure looks like an alien, like something that does not belong in the sea. As the player-character ventures into it, they discover a factory where their robot allies are assembled. At this point, hyper-attentive players who are familiar with the topos of a hubristic civilization falling prey to its own technological development—present everywhere in our culture—have enough clues to hypothesize that the individuals depicted on the fresco were the inhabitants of a fallen civilization who used to live in symbiosis with nature before they invented a technology that allowed them to harness the power of the sea more efficiently, but that upset the delicate balance of the ocean and led to the downfall of their civilization. The role of the player-character in this story, however, remains unclear until they light up a hologram that reveals its metal skeleton.

At that moment, players have enough information to infer that the protagonist is an android and, perhaps, the most advanced technological device of the lost civilization. Unlike the pyramidal structures, the player-character, however, operates freely and wishes to restore the sea equilibrium. After exiting the factory, the player-character must avoid hundreds of pyramidal mines that threaten to explode when it gets too close. Players can then hypothesize that these devices are protecting the main pyramid against intruders.

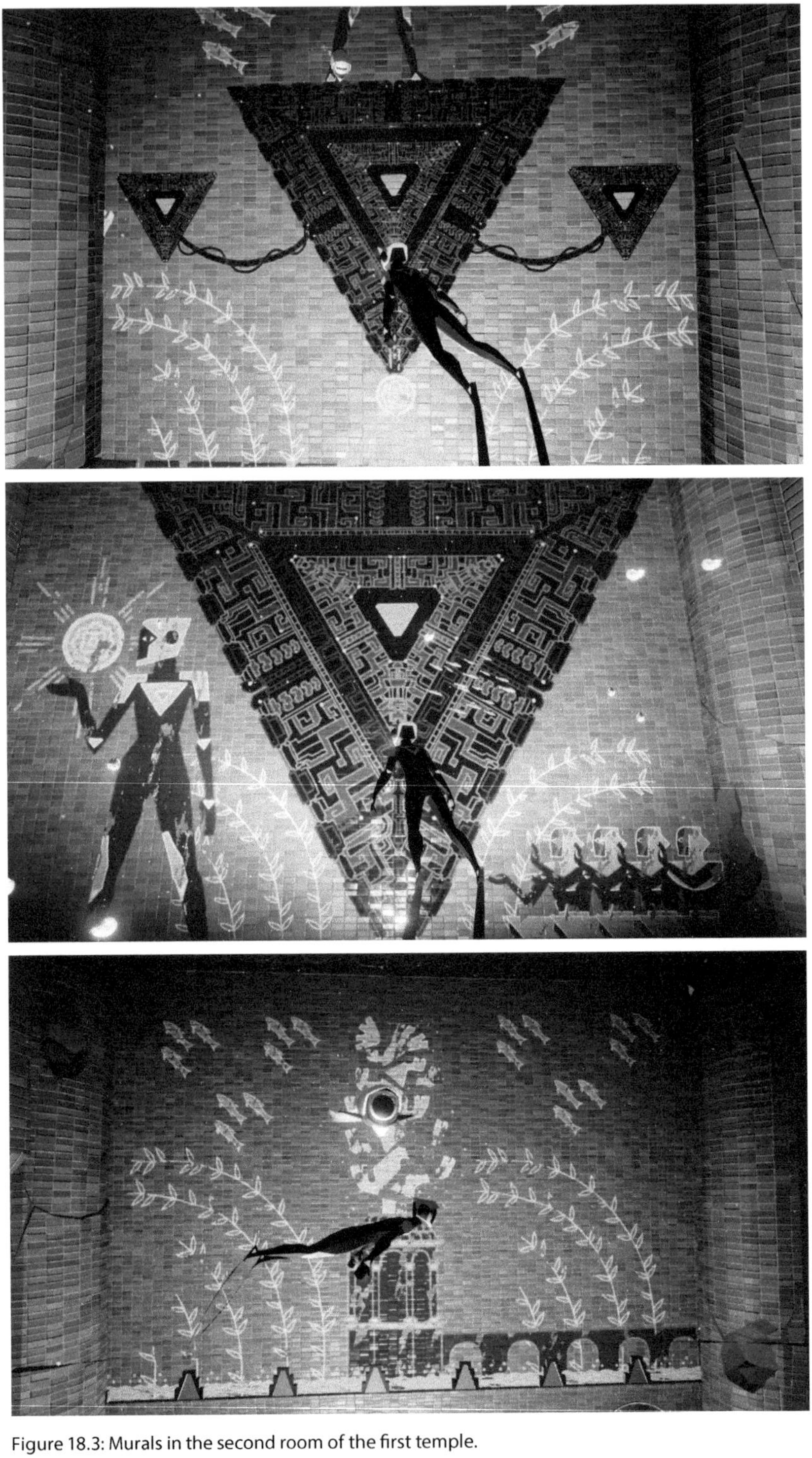

Figure 18.3: Murals in the second room of the first temple.

Figure 18.4: The hologram of the player-character reveals its metal skeleton.

As for the great white shark, it eventually becomes possible to infer that the inhabitants of the vanished civilization had an immense respect for it because of its mastery of the sea's energy. On other temple paintings, we can see these inhabitants following the shark in a god-like fashion and kneeling as it provides them with blue energy.

The great white shark also appears as the noble protector of the sea when it tries to destroy the core of the main pyramid but it is electrocuted along with the player-character and tragically loses consciousness. If players did not realize, at that point, that the main protagonist is an android, this cutscene makes it clear, since its burned diving suit reveals its metal bones and its triangle-shaped heart.

Towards the end of the game, the player-character gathers enough energy from the ocean to gain superpowers. With the help of the great white shark, it destroys pyramidal structures one by one to restore nature and bring back life everywhere in the sea. As the sidekicks make their way back to the main pyramid, they manage to penetrate its red core and miraculously provoke the transformation of this giant metal structure into a green oasis.

Players who pay attention to details and reflect throughout the game can interpret the story of *Abzû* as the cautionary tale of a once-prosperous civilization whose extractive technology led to its downfall. They can also make a parallel with their own world: if humans exploit their natural habitat with no regard for its balance, they will destroy it. They can finally ask themselves the following question: if the ocean energy is the source of all life, could it make the android player-character alive?

Figure 18.5: The hundreds of pyramidal mines that protect the main pyramid.

Figure 18.6: A mural of the great white shark providing energy to the inhabitants.

While solving puzzles and overcoming challenges are most of the time mandatory to progress and to beating the game, the same cannot be said for narrative puzzles such as *Abzû*'s. The players can reach the end of the game without paying attention to its lore. Piecing together *Abzû*'s narrative puzzle is therefore a free undertaking.

Interviews with Four Abzû *Players*

The interviews that we conducted with four *Abzû* players suggest that narrative puzzles such as *Abzû*'s favour challenge-based immersion because of the investigative effort needed to seek out the clues and the cognitive effort required to interpret them, draw hypotheses, and solve the enigma.

Indeed, the rich interpretations of the story formulated by the four research participants right after their gaming session seem to indicate that all of them actively looked out for clues in the game space and reflected during their play session to piece the narrative puzzle together. The four of them came up with a similar interpretation of the general storyline: an evil entity has destroyed the life in the ocean and the player-character must destroy this entity to restore life in the sea. This evil entity was, however, depicted differently from one participant to another: while two participants vaguely described it as "someone or something that killed nature" or as a "malicious triangle," the two other participants provided more detailed definitions, saying it is something "alien" or a "technology" created by an extinct civilization. The same can be said for the avatar, who is sometimes described as a "construction" or a "humanoid robot," and sometimes as "a young person who discovers his own civilization." One of the participants- seemed particularly ambivalent about the nature of the avatar, at first describing it as the only survivor of a lost civilization who tries to fix the mistakes made by their people, and later as a robot with a conscience. For this player, it is unclear if the avatar is the creator of the technology or a technological creation itself, if it is part of the problem or the solution. It should, however, be recalled that the players did not have the chance to complete the game before the interview.

All participants also shared more elaborate hypotheses about *Abzû*'s backstory. Two of them tried to imagine the world beyond the ocean, stating that the entire planet was perhaps destroyed. One player mentioned the "crazy questioning" to which the game gives rise about the combat of nature against technology and perceived that the story is a reflection of our society. Two participants made a connection between the murals on the walls and Egyptian civilization. One player established a link between the red pyramid and the Illuminati, as well as between the pyramid and the Terminator (which, in the eponymous movie, is an intelligent robot that turns against its creators).

The four players agreed on the enigmatic nature of the story. One of them said that we must make our own "definition" of the avatar just like we make our own interpretation of a "poem." Another participant explained that players can have their own "perception" of the story and admitted having trouble "piecing the puzzle together to understand it." One player confided that he was "starting to understand where the avatar comes from and what he is doing, but did not understand yet." They were nonetheless all curious to reach the end and to discover who the avatar really is.

Besides, the four players used expressions that reflected their uncertainty about their interpretations of the story. Three participants identified the drawings on the walls as the main element that allowed them to understand the story. One player mentioned the collection of seashells as a "clue" that helps to understand the value of nature. This is particularly interesting because it shows that, for this player, a gameplay element (collecting seashells) contributed to narrative involvement.

When asked what felt as a reward in the game, three participants described the majestic cutscenes that show the revitalization of destroyed areas after the intervention of the player-character. Considering cutscenes as rewards (an element of gameplay) suggests that this narrative element contributes not only to imaginative immersion, but also to challenge-based immersion.

A recurrent element in players' testimonies led to the idea that narrative puzzles contribute to challenge-based immersion: when asked to describe the main goal of the game during the portion of the interview about gameplay, three players answered that the game's central objective was to understand the story and decrypt the meaning of the murals. Three players mentioned that they would not have wanted more information about the background story, because "it would break the magic" and because the "beauty of the game is to try to understand" the story.

Two participants detailed the thought process behind their interpretations and, while they were doing so, it was obvious that they were still trying to piece the narrative puzzle together. One player mentioned that elaborating hypotheses and revising them is more fun than a text box revealing the story. Another player said that he spent a lot of time looking at the images on the walls to understand their underlying meaning.

When asked which game elements moved them and engaged them the most, the four participants mentioned the enigmatic story. Since all participants highly rated their level of focus and their feeling of being transported in another world during their gaming session, this incites us to believe that *Abzû*'s narrative puzzle had a strong impact on both their imaginative immersion and their challenge-based immersion.

In terms of experience, there is no doubt that fighting games, for example, favour a different kind of challenge-based immersion than narrative puzzles. Arsenault and Picart's (2008) differentiation between "tactical," "strategic," and "cognitive" challenge-based immersion can therefore be useful to describe the immersive experience allowed by narrative puzzles.

Analysis of Abzû *Players' Discussion on Forums*

The analysis of more than one hundred players' testimonies about the story of *Abzû*, posted on *Reddit, Steam Community*, and *GameFAQs*, suggests that its narrative puzzle favoured narrative and ludic macro involvements. Indeed, we observed a great deal of curiosity amongst players regarding *Abzû*'s narrative enigmas and narrative ambiguity. Some players visibly spent a lot of time and effort reflecting on the lore, proposing hypotheses on the story, and solving its mysteries after completing the game. A player claims, for instance, that they thought about the subject "for the last few hours" before explaining their theory in detail on the forum.[45] Several players expressed their desire to play the game again to find more clues: "I'll play it a few more times paying attention to the story..."[46] In this sense, the players' macro engagement is not only narrative, but also ludic.

The analysis of players' discussion of the story of *Abzû* also reveals that narrative puzzles have the potential to generate shared macro involvement, because players investigate the story together online. Indeed, we observed a great spirit of collaboration between players who tried to piece the puzzle together. Some players, for instance, ask other players questions, such as "what exactly is the relationship between human-like race and the machines?" or "have the machines contributed to their [the inhabitants] apparent extinction or simply outlived their creators?"[47] One person explicitly mentions that reading other players' interpretations helped them to better understand the story: "Loved reading these interpretations. I thought the plot was a little too vague the first time around, but I can see the pieces coming together now."[48]

Interestingly, players often share the sources of their theories, which are both in-game materials (like the murals, the cinematic, etc.) and external materials, like mythologies (Sumerian/Babylonian/Greek), pop culture stories (*Wall-E*, Stanton, 2009 and Atlantis, Plato, 360 BC) and similar games (*Journey*, Thatgamecompany, 2012). Some players confront others with different interpretations of *Abzû*'s story: "I had a similar but different conclusion. [...] I was under the impression that the pyramids were constructed by the race. However, I sensed something of an AI/organic struggle, with the original 'creators' dying out for whatever reason [...]."[49]

45 Player 1, *Steam Community*, 2016.

46 Player 2, *Steam Community*, 2016.

47 Player 3, *Steam Community*, 2016.

48 Player 4, *GameFAQs*, 2016.

49 Player 5, *Steam Community*, 2016.

The fragments of interpretation provided by several players eventually served as building blocks to elaborate more advanced theories. In a discussion that took place over a few days on *Steam Community*, for instance, a group of players discussed the potential meaning of the symbols that are visible in the temples and eventually concluded that "the symbols came from Enuma Elish" which "is essentially a Sumerian creation story."[50] Even though players did not necessarily reach a consensus, the ambiguity of *Abzû*'s story opened the door to interpretation games that took place directly on internet forums. In that context, online forums became a playground where players cooperated to solve a mystery while socializing, thereby feeding shared macro involvement for players who took part in the discussion.

Conclusions

Our close reading of *Abzû* suggests that its story can be compared to enigma-driven narratives insofar as it raises questions not only about future events, but also about the past (What happened to this aquatic world? Who or what destroyed it? For what reason?) and about the present (Who is the player-character? What they aiming for?) To some extent, *Abzû*'s narrative can also qualify as an open text that favours narrative ambiguity, because some questions remain unanswered at the end of the game, especially regarding the status of the protagonist (Are they alive and conscious?) With regard to this aspect, the game invites players to draw their own conclusions. Considering all this, and because players must actively seek out clues in *Abzû*, it seems fair to say that this game is an emblematic example of what we call a "narrative puzzle." From this perspective, *Abzû*'s story can be seen as a ludonarrative challenge or a game of interpretation. It also seems appropriate to hypothesize that *Abzû*'s narrative enigmas and ambiguity lead the model player to actively engage in a search for clues and to put significant effort into interpreting them and elaborating theories on the story.

This hypothesis seems to be reinforced by the testimonies of our four research participants who all tried to piece the story together and perceived this activity as part of the game's challenge. All four adopted a hyper-attentive posture towards *Abzû*'s narrative puzzle, which is linked to forensic engagement. Indeed, they noticed many details in the game

50 Player 6, *Steam Community*, 2016.

environment and made connections with elements from other media productions. However, it is obvious that some of them struggled to make sense of the story because they were missing some pieces of the puzzle that they either omitted to collect or did not have the time to collect yet. Regardless of their level of understanding, the extent to which they felt both focused and transported in the game world suggests that narrative puzzles can trigger cognitive challenge-based immersion as much as imaginative immersion.

In light of *Abzû* players' testimonies posted on *Reddit, Steam Community*, and *GameFAQs*, it seems fair to suggest that narrative puzzles can foster not only narrative macro involvement, but also ludic macro involvement, insofar as some players kept reflecting after completing the game and, most interestingly, kept looking for missing puzzle pieces in other players' publications. Players often provided constructive feedback to interpretative efforts, asked questions to better understand it, or simply showed their interest in other players' theories. It therefore seems relevant to hypothesize that narrative puzzles have the potential to enhance shared macro involvement in the form of forensic fandom. Indeed, some groups of players who discussed *Abzû*'s story on *Reddit, Steam Community*, and *GameFAQs* literally played detective together; bringing their own puzzle pieces to the group discussion in order to collectively draw a more complete picture of the story. In that context, forums seem to provide a space where the narrative puzzle-solving process can extend beyond the game's initial framework and where a secondary multiplayer game of interpretation can arise.[51]

More research is needed to describe the immersive experience involved during the resolution of a video game's narrative puzzle. How different does cognitive immersion feel compared to strategic and tactical immersions? How is the immersive experience impacted by the fact that solving a narrative puzzle is an optional challenge?

While concepts such as "challenge-based immersion" and "ludic involvement" are usually reserved for games, our study should encourage researchers in literature, television, or film studies to see if they can also apply them, to some extent, to non-interactive fictions that are based on mystery solving.

51 The notions of ludic macro involvement and of mutiplayer games of interpretation can also be used to describe experiences with mystery-solving novels such as Danielewski's *House of Leaves*.

Bibliography

Arsenault, Dominic, and Martin Picard. "Le jeu vidéo entre dépendance et plaisir immersif: les trois formes d'immersion vidéoludique." *Actes du colloque HomoLudens: Le jeu vidéo: un phénomène social massivement pratiqué* (2008): 1–16.

Aarseth, Espen. *Cybertext: Perspectives on Ergodic Literature.* Baltimore: Johns Hopkins University Press, 1997.

Bonenfant, Maude. "Sens, fonction et appropriation du jeu: l'exemple de World of Warcraft." PhD diss. Université du Québec à Montréal, 2010.

Bordwell, David. *Making Meaning: Inference and Rhetoric in the Interpretation of Cinema.* Cambridge: Harvard University Press, 1989.

Brown, Emily, and Paul Cairns. "A Grounded Investigation of Game Immersion." In *CHI'04 Extended Abstracts on Human Factors in Computing Systems* (2004): 1297–1300.

Calleja, Gordon. *In-Game: From Immersion to Incorporation.* Cambridge: The MIT Press, 2011.

Calleja, Gordon. "Experiential Narratives in Game Environments." In *DiGRA '09 – Proceedings of the 2009 DiGRA International Conference: Breaking New Ground: Innovation in Games, Play, Practice and Theory* (2009).

Csikszentmihalyi, Mihaly. *Flow: The Psychology of Optimal Experience.* New York: Harper Perennial, 1991.

Eco, Umberto. *Interprétation et surinterprétation.* Paris: Presses universitaires de France, 1992.

Eco, Umberto. *Lector in fabula.* Paris: Grasset, 1985 [1979].

Ermi, Laura, and Frans Mäyrä. "Fundamental Components of the Gameplay Experience: Analysing Immersion." *Worlds in Play: International Perspectives on Digital Games Research*, 37, no. 2 (2005): 37–53.

Genvo, Sébastien. "Caractériser l'expérience du jeu à son ère numérique: pour une étude du play design." *Congrès de l'ACFAS: Le jeu vidéo : expérience et pratiques sociales multidimensionnelles* (2008). http://www.expressivegame.com/publications/colloques/caracteriser-lexperience-du-jeu-a-son-ere-numerique-pour-une-etude-du-play-design-s-genvo/.

Ginzburg, Carlo. "Morelli, Freud and Sherlock Holmes: Clues and Scientific Method (translated by A. Davin)." *History Workshop*, no. 9 (1980): 5–36.

Henriot, Jacques. *Sous couleur de jouer. La métaphore ludique.* Paris: José Corti, 1989.

Iser, Wolfgang. "Indeterminacy and the Reader's Response to Prose Fiction." In *Aspects of Narrative*, edited by J. Hillis Miller, 1–45. New York: Columbia University Press, 1971.

Jenkins, Henry. "Game Design as Narrative Architecture." In *First Person: New Media as Story, Performance, and Game*, edited by Noah Wardrip-Fruin and Pat Harrigan, 118–130. Cambridge: The MIT Press, 2004.

Mäyrä, Frans. *An Introduction to Game Studies: Games in Culture*. London: SAGE Publications, 2008.

Martey, Rosa Mikeal, Kate Kenski, James Folkestad, Laurie Feldman, Elana Gordis, Adrienne Shaw, Jennifer Stromer-Galley, Ben Clegg, Hui Zhang, Nissim Kaufman, Ari N. Rabkin, Samira Shaikh, Tomek Strzalkowski. "Measuring Game Engagement: Multiple Methods and Construct Complexity." *Simulation & Gaming*, 45, no. 4–5 (2014), 528–547.

Mittell, Jason. *Complex TV: The Poetics of Contemporary Television Storytelling*. New York: New York University Press, 2015.

Mittell, Jason. "Lost in a Great Story: Evaluation in Narrative Television (and Television Studies)." In *Reading Lost: Perspectives on a Hit Television Show*, edited by Roberta E. Pearson, 119–138. London: I. B. Tauris, 2009a.

Mittell, Jason. "Sites of Participation: Wiki Fandom and the Case of Lostpedia." *Transformative Works and Cultures*, no. 3 (2009b). https://doi.org/10.3983/twc.2009.0118.

Murray, Janet. *Hamlet on the Holodeck: The Future of Narrative in Cyberspace*. Cambridge: The MIT Press, 1997.

Pearce, Celia. "Towards a Game Theory of Game." In *First Person: New Media as Story, Performance, and Game*, edited by Noah Wardrip-Fruin and Pat Harrigan, 143–153. Cambridge: The MIT Press, 2004.

Rimmon-Kenan, Shlomith. *Narrative Fiction: Contemporary Poetics*. London: Routledge, 1989 [1983].

Robertson, George, Mary Czerwinski, and Maarten Dantzich. "Immersion in Desktop Virtual Reality." *Proceedings of the 10th Annual ACM Symposium on User Interface Software and Technology* (1997), 11–19. https://doi.org/10.1145/263407.263409.

Ryan, Marie-Laure. *Avatars of Story*. Minneapolis: University of Minnesota Press, 2006.

Ryan, Marie-Laure. Narrative as Virtual Reality: Immersion and Interactivity in Literature and Electronic Media. Baltimore: The Johns Hopkins University Press, 2001.

Salen, Katie, and Eric Zimmerman. *Rules of Play: Game Design Fundamentals*. Cambridge: The MIT Press, 2003.

Schur, David. *An Introduction to Close Reading*. Harvard University, 1998.

Sicart, Miguel. *The Ethics of Computer Games*. Cambridge: The MIT Press, 2009.

Therrien, Carl. "Immersion." In *The Routledge Companion to Video Game Studies*, edited by Marc J. P. Wolf and Bernard Perron, 451–484. New York: Routledge, 2014.

About the Authors

Gabrielle Trépanier-Jobin is Assistant Professor in Game Studies and Cultural Industry in the School of Media of Université du Québec à Montréal (UQAM). Her main areas of expertise include game studies, media sociology, and gender studies.

Mathilde Savoie is a PhD student in communication (UQAM) and researcher in the Homo Ludens research group on gaming practices and online communication. Her research interests focus mainly on online gaming communities, video game practices and appropriations, as well as video game engagement and immersion.

19. The Opposite of Elsewhere: Home Video as Virtual Reality

Madison Brown

Abstract: This chapter delves into the immersive experience of watching one's own home videos within the familiar confines of home, challenging traditional notions of immersion typically associated with large screens and exotic locales. By connecting the content of home videos with their domestic reception context, the chapter explores how this seemingly mundane activity can offer a profound engagement with memory and self-reflection. Drawing on concepts of virtual reality and spectatorship, the chapter examines the complexities of familial affect and the role of memory in shaping immersive experiences. It argues that while home videos may lack the technological sophistication of mainstream immersive media, they offer a unique form of immersion rooted in personal experience and memory. Through a close analysis of personal home video footage, the chapter demonstrates how the act of viewing one's own past on screen can evoke a sense of virtual presence and introspection.

Keywords: immersion, domesticity, self-reflection, domesticity, amateur media

Moving image media has always had an intimate relationship with elsewhere. From early ethnographic films and amateur travelogues to the television screen's characterization as "your window on the world," moving images have long promised spectators the experience of sight-seeing, exploration, and discovery. Frequently characterized by their large screens, big budgets, and ability to all but physically transport viewers to unique locales, immersive media technologies' capacity to enmesh viewers' sensoria in faraway or imagined places would especially seem to exclude the quotidian experience of watching run-of-the-mill home videos in the domestic space. Nevertheless,

Bédard, Philippe, Alanna Thain, and Carl Therrien, eds. *States of Immersion Across Media: Bodies, Techniques, Practices*. Taylor & Francis Group, 2025
DOI 10.5117/9789048558766_CH19

the unique synonymy of home video's content and its likewise domestic reception context evidences the home mode's unique ability to extend spectatorial engagement beyond the screen. Connecting home video images' there-and-then with the reception context's here-and-now, watching one's own home videos within the home constitutes a counterintuitive but rich instance of spectatorial immersion.

In fact, the experience of viewing one's own home videos within the home sounds eerily like Oliver Grau's pioneering definition of the virtual experience as "immersing oneself in the image space, moving and interacting there in 'real time,' and intervening creatively."[1] Connecting the domestic space *on* the screen with the domestic space *of* the screen, viewing home video within the home uniquely immerses the viewer in the home video image. This experience provides the viewer with a mirror onto themselves as well as a window onto their past, situating them in a virtual *tête-à-tête* with their former selves. These intimacies of memory and repetition—feeling it once, watching it again and again—continually recontextualize home video images within ever more complex formulations of familial affect.

However, home video's ability to elicit three key characteristics of virtual reality—3D immersion, interactivity, and the production of empathy—is not seamless. Structured as much by gaps, lacks, and absences as a virtual experience like walking on the moon, viewing one's own home videos within the home requires the active negotiation of imaginative and affective connections between the past, the present, and the future. For instance, notions of "homes" are constituted as much by actual architectural structures as by personal, cultural, and national ideas about what "home" means; the way we understand ourselves has as much to do with the media technology that we engage with as the visual perspective of those who see us through it; and familial history is structured as much by the likely—and even inevitable—gaps in our own memory as the moments we seem to be able to relive at will. Though these contingencies of home and video might initially appear to reduce immersion, the imaginative labour required of the viewer to fill these gaps actually approximates the intangibility of memory itself.

Compared to IMAX screens, flight simulators, and elaborate museum exhibits, it is not surprising that the quotidian experience of viewing home videos within the home has not been considered under the immersion rubric. Amateur media cultures, analogue media technologies, and domestic media contexts seem the very antithesis of immersive media's appeal to elsewhere, but when taken together, these seemingly unrelated media discourses

1 Grau, *Virtual Art*, p. 3.

provide a new lens on what counts as immersive. In our digital world of increasingly mobile media technologies, viewing home videos recorded on magnetic tape as virtual reality demonstrates that we are not getting "more immersed" as time goes on and as new technologies are developed. In fact, video's capacity to produce a closed-circuit by connecting the camcorder to the television as a monitor introduced users and viewers to a novel visual perspective on oneself, producing the conditions for an analogue experience of virtual reality.

While viewing home videos within the home demonstrates that analogue technologies unrelated to dominant discourses of immersion can offer us immersive experiences, there is a significant caveat: the home videos you're watching must be *yours*. Posing a problem for humanistic scholarship that has yet to embrace the methodological value of personal experience, the immersive potential of familial home video lays dormant in more empirical accounts of amateur media's technological, social, and cultural histories. Rousing the personal and political potential of home mode media scholarship, Michelle Citron's *Home Movies and Other Necessary Fictions* (1999) is an example of how the shared experience—as well as, perhaps counterintuitively, the inimitability—of encountering one's home, oneself, and one's memories on screen is both informed by and informative of larger cultural imaginations about the family.

In this chapter, I analyze home video footage of myself as a child to explore Margaret Morse's evocative suggestion that the virtual realm is nothing but externalized imagination.[2] Drawing from television studies scholarship on domestic reception contexts, theories of embodied spectatorship from 1990s feminist film phenomenology, and critiques of empathy in game studies, I develop a typology for viewing one's own home videos within the home as an immersive experience by tailoring concepts such as scale, frame, and interactive affect to the experience of home video spectatorship. In the process, the spectatorial experience of gaps, lacks, and contingencies illustrates the structuring role that memory plays in immersion and why we never experience other peoples' home videos like our own.

The "home" in home video is more than just a designation of content. Regardless of cultural specificities, all notions of home share two distinct imperatives at their core: shelter and identity.[3] On the one hand, home is practical: it is an architectural structure that functions as an enclosure. Keeping its inhabitants safe from the outside world, home also potentially

2 Morse, *Virtualities*, p. 30.

3 On notions of "home", see Smyth and Croft, *Our House*, p. 13.

keeps its inhabitants forcibly confined within. These divergent functions of the architectural structure are mirrored in home's more affective function. Home is also, on the other hand, "a feeling, an idea, not necessarily located in a fixed place,"[4] serving as a space for the material externalization of identity via use, décor, and location. No matter what form home takes, or whether it exists as a relic of the past rather than any current place of dwelling, both the physical and the psychological elements of home inform the immersive potential of viewing one's own home videos.

While home has long constituted a site of entertainment, demonstrated by Victorian performances of *tableaux vivants* in the drawing room, feminist television scholars first turned their attention to home as a locus for media consumption in the mid-1980s, identifying the ways that televisual spectatorship had been imagined as reifying, repurposing, and even occasionally ruining expectations of feminized domestic labour.[5] Since then, media studies scholarship on the home as a mediated space has grown to include histories of radio consumption, video rental cultures, digital streaming, and domestic computing. In each case, discourses on the home as a mediated space blur public spectacle with private reception context. Viewing one's own home videos within the home amplifies spectatorial intimacy by replacing the public with more private content.

Housed on magnetic tape, inside the cassette, inside their case, inside a bin or a cabinet just like any other household object—forgotten until they're not—home video images are both in and of the home. As if the New York Public Library were filled only with books about the New York Public Library, there is a unique synonymy between home video content and structure, evidencing home video's unique ability to extend spectatorial engagement beyond the screen. When home video images are recorded within the same space as their exhibition context, there's a sense that the diegesis or the world onscreen is being extended into the reception context itself. In this way, watching one's own home videos within the home literalizes Margaret Morse's definition of the virtual as "images that [...] share a temporal and spatial realm and interact with the human beings that are engaged with or in them."[6] The home thus functions as a unique mirroring of the space *on* the screen and the space *of* the screen, evoking one of virtual reality's primary characteristics: 3D immersion.[7]

4 Briganti and Mezei, "Introduction," p. 5.
5 See Spigel, *Make Room for TV*, p. 14.
6 Morse, *Virtualities*, p. 4.
7 On this, see Bolter and Grusin, *Remediation*, p. 162.

It is, as such, not inconsequential that home video's unique brand of spectatorial immersion is televisual. While traditional cinematic reception contexts rely on a darkened room to refocus the viewer's spectatorial attention from the screening environment to the screen itself, including the domestic projection of home movies recorded on small-gauge film, viewers of domestic television sets frequently leave the lights on, move freely around the room, and comment on the images as they are viewed. These more casual viewing conditions have historically rendered the television a centrepiece of domestic space more broadly, figuring the television as a piece of furniture not unlike a couch or a sideboard.[8] However, quite unlike a couch or a sideboard, television also has the potential to reflect home video images of the domestic space back onto itself.

Inspired by television spectatorship's easy integration into the rhythms of everyday life, Tania Modleski's theorization of domestic television reception is predicated on the housewife's experience of distraction: her movement between chores and pleasure, work and play, the responsibilities of home and the escapism of TV land.[9] Splitting her attention between the television and her domestic duties, Modleski's housewife personifies a spectatorial flexibility that can also account for various engagements with home video images. Just as the distraction of the housewife allows her to fold laundry, cook dinner, and catch her stories all at the same time, home video images split the viewer's spectatorial attention between the here-and-now and the there-and-then, fluctuating between the domestic reception context—which distracts them—and the images of domesticity—which they are distracted from. Substantiating Morse's claim that virtual reality requires that users navigate "the co-presence of two or more different, even contradictory, metapsychological effects,"[10] the spectatorial labour of moving between the images from the past and their present display on the television renders both worlds only partially inhabited.

Of course, this experience of distracted spectatorship extends far beyond the restricted subject of the housewife. While Modleski's theory evokes the white, suburban, cis-hetero-woman that soap operas frequently interpellate, Ludmila Lupinacci invokes the very same characteristics to describe the contemporary proliferation of always-on screen media nearly forty years later: a synchronicity of experiences, perpetual disruption, and the constant deferral of expectation.[11] The distinctly domestic figure of Modleski's house-

8 On this, see Spigel, *Make Room for TV*, p. 38.
9 Modleski, "The Rhythms of Reception," p. 43.
10 Morse, *Virtualities*, p. 102.
11 Lupinacci, " 'Absentmindedly Scrolling Through Nothing'," p. 277.

wife thus also represents a more contemporary swath of domestic media consumers, including those that view their own home videos within their home. As such, I will use the figure of the housewife not for its restrictive subjectivity, but as a shortcut to the experience of distracted spectatorship.

Since immersive screens are typically associated with largeness—winding, wrapping, even moving around the viewer—most conceptualizations of immersive media relegate television screens to a historical trajectory of representational aesthetics, counting them alongside paintings, framed photographs, and cinema. Oliver Grau suggests that media with clear borders around the image—"particularly television"—often "leave the observer outside and are thus unsuitable for communicating virtual realities."[12] However, the diegetic extension of viewing home videos within the home actually renders the small-scale television screen a vehicle of perceptual immersion, dissolving the borders between where we watch and what is seen.

Just as 3D screens and experiences are understood to thrust the represented object into the space of the exhibition, video's reliance on the television screen as a monitor reinforces the utility of television studies to understand how viewing one's own home video images within the home is an immersive experience. Whereas celluloid film is projected onto a screen from a distance, video images seem to emerge from within the television itself. Creating depth in contrast to the horizontal and vertical axes typically associated with two-dimensional images, Morse conceptualizes a televisual Z-axis that appears to move "toward or away from the viewer [...] and, by extension, [that moves the image] virtually beyond the glass into viewing space."[13] Not only does this Z-axis provide conceptual depth to home video images, acting as a "powerful virtual zoom" that unites heterogenous worlds of televisual images with their reception context, it also introduces television into a larger legacy of immersive screen technologies.[14]

When screening home video images, the intimate address of the televisual Z-axis follows the directionality of Roland Barthes's punctum, not least for its affective and familial contexts. In *Camera Lucida* (1982), the ur-text of autotheory on familial media, Barthes theorizes the highly personal experience of viewing photography by differentiating what can objectively be identified in an image, what he calls the *studium*, from the infinitely intimate experience of being personally affected by that image, the *punctum*. Like the Z-axis that emerges from the image, the punctum "rises from the

12 Grau, *Virtual Art*, p. 14.
13 Morse, *Virtualities*, p. 16.
14 Grau, *Virtual Art*, p. 81.

scene, shoots out of it like an arrow, and pierces me."[15] Unique to the viewer, to their memories, and to their lived experience, the punctum—like other 3D technologies—follows the directionality of the Z-axis, demonstrating that familial affect is itself immersive. Viewing one's own home video images within the home summons spatial memory and situates it squarely within the private space of the domestic reception.

As is likely the case for many, my family no longer lives in my childhood home, the structure that the majority of our home videos were taken in. This element of memory and imaginative investment is critical, especially at the juncture of home's function as both physical and affective. While the experience of immersion that connects the space on the screen with the space of the screen might seem to constitute a rare and even impossible situation for those without access to that domestic space of home video production, it is not unlike many large-scale immersive experiences that rely on site-specific viewing conditions. As art historian Martha Langford points out, "the real-life domestic experience [of engaging with familial media] is loaded with compensatory pleasures—intimacy, conviviality, emotional investment, perhaps a slice of cake."[16] As such, even if the "real-life domestic experience" is physically unavailable, the compensatory pleasures that Langford describes enhance the role of memory and imagination in the immersive experience of viewing one's own home video within the home.

This interplay between the viewer's affective experience of home and the electronic images that provoke them produces another element characteristic of immersive experience: interactivity. When viewers interact with their own home video images, they are experiencing the recorded familial interactions all over again. Although this element of home video immersion accounts for a more conceptual experience than the one experienced by the typical user of virtual reality, who reaches into the virtual world and physically constitutes it as such, the affective involvement of the viewer is no less interactive. For Morse, the potential for virtual reality to facilitate discourse between individuals is contingent on participants having "the same access to the inside of the same virtual world."[17] This same virtual world, as Morse calls it, is offered by home video images, as well as complicated by them: just as the virtual world of home video is predicated on the liminality that exists between the architectural space of the home and the memorial experience of it, it is the shared element of the immersive experience and viewing

15 Barthes, *Camera Lucida*, p. 26.
16 Langford, *Suspended Conversations*, p. 5.
17 Morse, *Virtualities*, p. 29.

yourself from someone else's perspective that has the potential to develop empathy in the viewer.

This perspective-taking is colloquially understood as "walking in someone else's shoes." But, as Wendy Chun points out, "when you walk in someone else's shoes, you take their shoes."[18] Critiqued as "an empathy machine" for its potential to reinscribe dominant power structures by dictating whose stories get told, by whom, and for whom, virtual reality often minimizes the lived experience of socially and culturally marginalized perspectives, rendering lived experiences of oppression a novel if emotional experience that Lisa Nakamura refers to as "identity tourism."[19] The social and political implications of such virtual perspective-taking shift, however, when we conceive of viewing one's own home videos as virtual reality.

Just as the perspective of virtual reality is rooted in a distribution of looking relations based on who is looking and who is looked at, so too is the home videographer's point-of-view, their spatial relationship to the events they tape, and the corporeal signature of the camcorder's movement, evocative of extant familial relationships. Nearly two decades before critiques of empathy discourse began, Morse pointed out that using virtual reality for entertainment "merely exploits the ability to superimpose the virtual over physical space [...] [and that] the spiritual potential and powerful metapsychological effects of immersion can easily be abused."[20] But, like a child playing dress up in their mother's high heels, viewing one's own home videos has the potential to personalize—and recuperate—the imagined perspective-taking of virtual reality.

If we understand home video as virtual reality, empathy takes on a different valence: this is not empathy with an experience one has never had. Rather, this empathy evokes the corporeal conditions of one's own embodied and emotional memory. I refer to the experience of seeing oneself in home video footage as corporeal specularity. Functioning as both a mirror showing oneself to oneself as well as a window onto another time and space, the television screen that shows the viewer to themselves constitutes an effective tool to enact the "basic human need for reversibility and reciprocity of 'I' and 'you' in discourse" that, according to Morse, fuels the cultural drive for virtual experience.[21] However, the subject and object of looking relations—the discursive shifting between "I" and "you"—is complicated by

18 Chun, "From Cyberspace to No Space."

19 Nakamura, "Feeling Good about Feeling Bad," p. 54.

20 Morse, *Virtualities*, p. 181.

21 Morse, *Virtualities*, p. 10.

watching one's own home videos. One of my family's home videos illustrates the immersive potential of combining virtual reality's perspective-taking with the spectator's own memories.

Once presents had been opened on Christmas Day, 1994, my family went to our Oma's house for an elaborate turkey dinner. From the moment we arrived, I was warned to give Oma's black cocker spaniel, Bailey, some space. After we ate, my brother and my sister began playing with the camcorder. The footage shows them weaving between bodies cleaning up in the kitchen, spying on our father asleep on the couch, and interviewing each other as newscasters. During their videotape escapades, I secretly got down on my belly and stuck my head under the coffee table to talk to Bailey. Overwhelmed by the commotion and not wanting any of it, she snapped at me and sunk her teeth into the tip of my six-year-old nose. I locked myself in the upstairs bathroom to dodge the inevitable "I-told-you-so's." This part of the story went unrecorded, but, when my mother and aunts gathered at the top of the stairs, my camcorder-toting siblings joined them. The subsequent footage shows my mother jimmying the lock, finding me, and checking my forehead for a fever. Everyone laughs in whispers at the sight of me "asleep" in the bathroom, but once my mother gathers up my limp body and walks offscreen, one of my aunts puts her hands over the lens and firmly tells my brother to "turn it off."

As a child playing dead on the bathroom floor, I was the object of my brother's vision—and as a result, of the home video images. Now I am the subject of the spectatorial experience. Like the confluence of the space on the screen and the space of the screen, seeing home video images of yourself on screen produces the immersive experience of shifting between visual positions, of seeing oneself seeing and of feeling oneself feeling. Just as memory itself can adjust to agree with new information about the past, home video not only allows for different interpretations of the past, but actually encourages them, rendering home video spectators interactive participants in the production of meaning.

Of course, seeing oneself on a screen is not unique to home video spectatorship. This experience can be traced back to itinerant filmmakers of the early twentieth century, who travelled from town to town, engaging communities in what Martin Johnson calls "a solipsistic enterprise," in which "to make a movie was also to see one, and those who saw movies often saw themselves in them."[22] Indeed, the origins of cinema themselves would also have offered audiences the opportunity to see themselves in a projected image, such as in the Lumière brothers' *Workers Exiting the Factory* (1895), granted the

22 Johnson, *Main St Movies*, p. 1.

labourers pictured on the screen could afford the price of admission. But, just as Tom Gunning suggests that local films cannot systematically be characterized by formal traits unique to the genre, instead being "dependent on audience recognition to accrue meaning,"[23] the virtuality of one's own home videos is even more reliant on intimate recognition.

While viewing one's own home movies recorded on small-gauge film requires similar recognition, the immersive experience of seeing oneself in a familial home video is augmented by the influence of magnetic videotape's medium specificity on spectatorial experience, but also on what content could be recorded in the first place. When camcorders and magnetic videotape widely replaced home movies in the early 1980s, the addition of sound, extended recording time, and increased camera mobility significantly altered the means by which people represented themselves and their families. Contrary to the prohibitive costs and shorter recording time of home movies on film, home videographers faced less pressure to capture only special occasions. The increased domestic presence of the camcorder resulted in familiarity with its presence and thus in its increased invisibility, more easily inscribing the rhythms of everyday life in the family archive and altering modes of self-representation more broadly.

From my perspective on the bathroom floor, the dog bite situation felt desperate. However, when I rewatch this home video, I see myself from my brother's perspective: wobbly visuals—the corporeal signature of a heavy camcorder held by a twelve-year-old—are accompanied by his laughter at my histrionics. Jay David Bolter and Richard Grusin characterize perspective-taking in virtual reality based on "[a]n insistently visual construction of point-of-view,"[24] and seeing myself through my brother's eyes elicits the very sort of empathy that underlies critiques of virtual reality. This ambiguous empathy, deriving from complex looking relations in an otherwise uncomplicated home video, is a result of navigating perspectival difference. While this new visual perspective does elicit empathy, it shifts almost imperceptibly: at one moment, I empathize with my big brother's experience and visual perspective, but the next, I empathize with my younger self.

This shift in perspective exemplifies Vivian Sobchack's phenomenological concept of the "viewing view."[25] Used to express the viewer's active involvement in the negotiation of spectatorial perception and in direct opposition to theories of passive spectatorship, Sobchack posits that the viewer and

23 Gunning, "Pictures of Crowd Splendour," p. 52.

24 Bolter and Grusin, *Remediation*, p. 246.

25 Sobchack, *The Address of the Eye*, p. 21.

their object of vision constitute each other's existence and purpose through complementary acts of perception and expression.[26] When I watch this home video, I am both viewer and viewed: I express my perception of the camcorder expressing the perception of my brother expressing his perception of my small, sad body on the floor. This reciprocal animation of viewing and viewed in an intimate context takes on a new and amplified applicability. However, just as Bo Ruberg points out that empathy itself "is a messy construct full of ambiguities and slippages,"[27] the navigation of visual perspectives while viewing one's own home videos is rife with ambivalence.

Ruberg notes that the potentially violent and academically overplayed use of "empathy" in game studies actually refers to a complex constellation of emotional and interpersonal skills, including compassion, sorrow, and other queer entanglements.[28] Likewise, the slipperiness of inhabiting different visual perspectives while viewing one's own home videos can produce a whole slew of complicated affects. Immersed in the footage following my dog-bite scenario, I can simultaneously identify embarrassment and scorn, earnestness and disbelief. These affects are indicative of nothing more than a typical big brother-little sister relationship, but immersive affects will always depend on the familial dynamics that structure the home video footage and that continue to shape our intimate engagement with them years later.

Although the affects that derive from viewing one's own home videos within the home are complex, slippery, and ever evolving based on familial dynamics, the relationship between the body viewed on the screen and the body doing the viewing is somewhat more concrete. While memory, as evoked by perspectival shifting and seeing oneself on screen, might constitute a more abstract or conceptual experience, the latter also evokes embodied memory rooted in actual corporeal experiences. Here, the virtual experience we are talking about is not walking in somebody else's shoes: it is the memory of walking in your very own shoes that perhaps gave you blisters, that perhaps are too small now, that perhaps are worn in the video images you are viewing.

The unique experience of seeing yourself seeing extends naturally to the experience of feeling yourself feeling, and this virtuality is amplified by video's medium-specific production of present-ness. In contrast to celluloid film's indexical relationship to the past, electronic media like home video recorded on magnetic tape are associated with the present. Evoking virtual

26 Sobchack, *The Address of the Eye*, p. 21.
27 Ruberg, "Empathy and Its Alternatives," p. 63.
28 Ruberg, "Empathy and Its Alternatives," p. 64.

reality's accent on an "instant present [...] that can be selected, combined, and instantly replayed,"[29] analogue video technologies create the images anew—as if for the first time—every time the videotape's magnetic strip is read. For Sobchack, the dust particles illuminated by film projection remind us of the images palpably passing through space, moving through our homes and theatres, making none of us immune to its corporeal influence.[30] The fact that video images are produced instantaneously and emerge from within the television screen means that they do not exist spatially or move through space in the same way, altering the embodied experience of spectatorship. While Sobchack suggests that electronic media technologies thus disembody the spectator in counterproductive ways, home video's unique electronic temporality that recasts spectatorship as nonlinear, replicable, and simultaneous, is not the pitfall that she laments.[31]

Instead, home video's capacity to "decenter, disperse, and completely extrovert"[32] spectatorial subjectivity opens up a realm of possibilities for immersive experience where former selves and former experiences can coexist within the scene of the screen. The phenomenological animation of viewing and viewed—amplified by the analogue video technologies' production of an instant present—resembles the experience of looking in a mirror. But this corporeal specularity is cracked by the misalignment and perpetual switching between my brother's perspective of me and my own memory of the event. Just as "we see and comprehend and feel films with our entire bodily being, informed by the full history and carnal knowledge of our acculturated sensorium,"[33] I am immersed between spectatorial visuality and remembered corporeal experience.

Like the user of virtual reality is cognizant of the fact that, while proprioceptively convincing, their experience of elsewhere is imaginative instead of actual,[34] viewing my small body lying motionless on the bathroom floor produces the same psychical experience in me as if I were there. Enacting Erkki Huhtamo's description of immersive media as requiring "constant interplay between the user and the machine,"[35] the video images supplement the absent environmental stimuli: my eyes water as I remember the coarse periwinkle bathmat on my cheek, the cold tile on the sides of my stocking

29 Sobchack, *Carnal Thoughts*, p. 155.
30 Sobchack, *Carnal Thoughts*, p. 152.
31 Sobchack, *Carnal Thoughts*, p. 153.
32 Sobchack, *Carnal Thoughts*, p. 158.
33 Sobchack, *Carnal Thoughts*, p. 63.
34 See Velmans, "Physical, Psychological, and Virtual Realities," p. 55.
35 Huhtamo, "Encapsulated Bodies in Motion," p. 165.

feet, the throbbing of my dog-bitten nose. Seeing my younger self feel this experience on screen is perhaps more real to me now than it felt in the moment of my performance, the scar on my viewerly nose a citation of the pain and shame that wrought it.

This capacity of home video images to destabilize notions of onscreen and offscreen recasts Sobchack's notions of perception and expression in new and remarkably virtual ways. Building on her subversion of "onscreen and offscreen as mutually exclusive sites of subject positions,"[36] I watch my own body, mannerisms, and confidence as a child, but from the perspective of an adult, some thirty years hence. Beyond the expression of perception and beyond the perception of expression, I am literally touched and physically moved as I move my hand to touch my long-healed nose. Memory of my now-faded scar corresponds with the home video image on the screen. Not only am I immersed domestically and spectatorially in the video images, but my body itself becomes the locus of immersion.

As a child who grew up in front of the camcorder, I knew all too well that someone would look back at footage of this incident. Indeed, I was—and still am—the one who looks back to this footage most often, a self-fulfilling prophecy of spectatorial expectation. Still, each time I return to my family's videotapes, their images tell a different story. Following Morse's suggestion that "the more abstract and removed information [becomes] from everyday life [...][,] the more virtual the substitute context of subjectivity in a here-and-now,"[37] home video's potential for immersive experience grows with passing time. Immersed in the temporal gap between when the images were recorded and when they were viewed, the virtuality of home video renders the viewer's own body a conduit between the experience of being viewed and of viewing. "The physical and the virtual are complimentary," Huhtamo states, "rather than mutually exclusive realms,"[38] and in this way, home videos need to be watched by people with familial context, or at least with intimate knowledge of where such context is lacking, for the immersion produced by memorial engagement to occur.

But what of home videos for which the viewer has no memory? What of recorded moments for which the viewer was never present? How do these absences inflect the immersive experience of watching one's own home videos within the home? It might seem that this is the end of the road for home video's immersive potential, but this lack only emphasizes

36 Sobchack, *Carnal Thoughts*, p. 66.

37 Morse, *Virtualities*, p. 6.

38 Huhtamo, "Encapsulated Bodies in Motion," p. 177.

the importance of familial context and shared access to the virtual world of memory. The interplay of contextual memory and the home video images—knowing what happened before and knowing what came after, or not knowing and having an intimate investment in that—is imperative to home video's immersive potential. In fact, when home videos picture events that we do not or cannot remember, the work of filling in memorial gaps or accounting for a lack of contextual knowledge may actually involve the viewer more deeply in the immersive experience.

Far from a shortcoming, not remembering home video images is a structural condition of memory. Many theories of mediated memory and memorialization agree that lack and imagination structure memory as much as, if not more than what we can confidently recall: Barthes suggests that "I may know better a photo I remember than a photo I am looking at, as if direct vision oriented its language wrongly";[39] Annette Kuhn notes that "narratives of identity are shaped as much by what is left out of the account—whether forgotten or repressed—as by what is actually told";[40] and Margaret Rose Olin suggests that, "while the details [of familial media] may diminish, the representation viewed distractedly may impart an enhanced sense of presence."[41] Perhaps most useful for understanding how memorial immersion can still occur if the viewer of their own home videos has no memory of the event, Marianne Hirsch's concept of postmemory is "distinguished from memory by generational distance and from history by deep personal connection […] through an imaginative investment."[42] Conceptualized to describe intergenerational trauma, like Morse's notion of sharing access to the same virtual world, the mediation of, in, and through home video images, as well as the familial lore that surrounds them, may well be sufficient proximity to the images for immersion.

But it is impossible to know how much of our memories are produced or inflected by home video images in the first place. Beyond just filling in memorial gaps, home video's spectatorial interaction actively shapes what we think we know. There is no way to know how my memory of the dog-bite incident is modified by my brother's home video footage of its fallout; whether the video has functioned as an aid to memory, or whether the video poses a threat to my memory by reshaping it altogether.[43] Indeed,

39 Barthes, *Camera Lucida*, p. 53.

40 Kuhn, *Family Secrets*, p. 2.

41 Olin, *Touching Photographs*, p. 3.

42 Hirsch, *Family Frames*, p. 22.

43 On this, see van Dijck, *Mediated Memories in the Digital Age*, p. 3.

my confident understanding of how home video corroborates and extends my own memory of the event might simply be evidence of Svetlana Boym's suggestion that "only false memories can be totally recalled."[44] There is no way to know; only to immerse myself in the images, between the cracks, and see what only I can see.

Just as the user of virtual reality's knees knock and palms sweat as they proprioceptively teeter on a precarious cliffside, so too does watching myself play dead on my Oma's bathroom floor evoke an experience more palpable than illusion. Home video's evocation of attributes like 3D immersion, spectatorial empathy, and interactivity demonstrates the virtuality of everyday media technologies, but it also demonstrates the virtuality of taken-for-granted concepts like "home" and "video" themselves. In the decades since analogue video's widespread installation in the domestic sphere, notions of home and of video have both changed drastically, accumulating meanings and representing entirely different things. In addition to home being an architectural structure and an imaginative idea, contemporary notions of home are increasingly mobilized, globalized, and dispersed.[45] Well beyond clunky analogue cassettes, video has become a catch-all for immaterial moving image media, much of it born digital, existing on social media and in "the cloud."[46] As different as they may be, the immersive potentials of these newfangled concepts will remain dormant if we, as scholars, do not opt to get personal.

Family memory can be hard enough to pin down for oneself, let alone theorized for an academic audience. But until we address the domestic, corporeal, and memorial conditions of amateur media cultures, our notions of immersion will be confined to technologies and objects instead of the myriad ways—and myriad places—that we are immersed in media every day. Of course, my own experiences of viewing home videos within the home can only take discussions of immersion so far. Every family will have their own protocols of producing and viewing home videos, and such protocols are also culturally specific, but we need not view such idiosyncrasy as a shortcoming. As Kuhn points out, if our stories and memories are no more interesting than anyone else's, that is precisely the point.[47] Rather than limiting theoretical applicability, distinctions between "my," "your," and "someone else's" home videos only emphasize how spectatorial immersion can be an entirely personalized experience.

44 Boym, *The Future of Nostalgia*, p. 54.
45 On this, see Francisco, "'The Internet Is Magic'."
46 On this, see Berliner, "Whatever Happened to Home Movies?"
47 Kuhn, *Family Secrets*, p. 3.

Bibliography

Barthes, Roland. *Camera Lucida: Reflections on Photography*. New York, NY: Hill and Wang, 1982.

Berliner, Lauren S. "Whatever Happened to Home Movies? Self-Representation from Family Archives to Online Algorithms." *Frames Cinema Journal* 19 (February 18, 2022): 158–184. https://doi.org/10.15664/fcj.v19i0.2381.

Bolter, Jay David, and Richard Grusin. *Remediation: Understanding New Media*. Cambridge, MA: MIT Press, 1998.

Boym, Svetlana. *The Future of Nostalgia*. New York, NY: Basic Books, 2001.

Briganti, Chiara, and Kathy Mezei. "Introduction." In *The Domestic Space Reader*, edited by Chiara Briganti and Kathy Mezei, 3–16. Toronto: University of Toronto Press, 2012.

Chun, Wendy Hui Kong. "From Cyberspace to No Space." Presentation at the Weird Reality Conference, Pittsburgh, PA, October 6–9, 2016.

Citron, Michelle. *Home Movies and Other Necessary Fictions*. Minneapolis, MN: University of Minnesota Press, 1999.

Francisco, Valerie. "'The Internet Is Magic.' Technology, Intimacy, and Transnational Families." *Critical Sociology* 41, no. 1 (2015): 173–190.

Grau, Oliver. *Virtual Art: From Illusion to Immersion*. Cambridge, MA: MIT Press, 2003.

Griffiths, Allison. *Shivers down Your Spine: Cinema, Museums, and the Immersive View*. New York, NY: Columbia University Press, 2008.

Gunning, Tom. "Pictures of Crowd Splendor: The Mitchell and Kenyon Factory Gate Film." In *The Lost World of Mitchell and Kenyon: Edwardian Britain on Film*, edited by Vanessa Toulmin, Simon Popple, and Patrick Russell, 49–58. London: Bloomsbury Publishing, 2004.

Hirsch, Marianne. *Family Frames: Photography, Narrative, and Postmemory*. Cambridge, MA: Harvard University Press, 1997.

Huhtamo, Erkki. "Encapsulated Bodies in Motion: Simulators and the Quest for Total Immersion." In *Critical Issues in Electronic Media*, edited by Simon Penny, 159–186. New York, NY: State University of New York Press, 1995.

Johnson, Martin. *Main St Movies*. Indianapolis, IN: University of Indiana Press, 2016.

Kuhn, Annette. *Family Secrets: Acts of Memory and Imagination*. London: Verso, 1995.

Langford, Martha. *Suspended Conversations: The Afterlife of Memory in Photographic Albums*. Montreal: McGill-Queens University Press, 2001.

Lupinacci, Ludmila. "'Absentmindedly Scrolling through Nothing': Liveness and Compulsory Continuous Connectedness in Social Media." *Media, Culture & Society* 43, no. 2 (2021): 273–90. https://doi.org/10.1177/0163443720939454.

Modleski, Tania. "The Rhythms of Reception: Daytime Television and Women's Work." In *Regarding Television: Critical Approaches. An Anthology*, edited by Ann E. Kaplan, 67–75. Frederick, MD: University Publications of America, 1983.

Morse, Margaret. *Virtualities: Television, Media Art, and Cyberculture*. Bloomington, IN: Indiana University Press, 1998.

Nakamura, Lisa. "Feeling Good about Feeling Bad: Virtuous Virtual Reality and the Automation of Racial Empathy." *Journal of Visual Culture* 9, no. 1 (2020): 47–64.

Olin, Margaret Rose. *Touching Photographs*. Chicago, IL: University of Chicago Press, 2012.

Rogers, Ariel. "Taking the Plunge: The New Immersive Screens." In *Screen Genealogies: From Optical Device to Environmental Medium*, edited by Francesco Casetti, Campe Rüdiger, and Craig Buckley, 135–158. Amsterdam: Amsterdam University Press, 2019.

Ruberg, Bo. "Empathy and Its Alternatives: Deconstructing the Rhetoric of 'Empathy' in Video Games." *Communication, Culture, and Critique* 13, no. 1 (2020): 54–71.

Smyth, Gerry, and Jo Croft. *Our House: The Representations of Domestic Space in Modern Culture*. Amsterdam: Rodopi, 2006.

Sobchack, Vivian. *The Address of the Eye: A Phenomenology of Film Experience*. Princeton, NJ: Princeton University Press, 1992.

Sobchack, Vivian. *Carnal Thoughts: Embodiment and Moving Image Culture*. Berkeley, CA: University of California Press, 2004.

Spigel, Lynn. *Make Room for TV: Television and the Family Ideal in Postwar America*. Chicago, IL: University of Chicago Press, 1992.

van Dijck, José. *Mediated Memories in the Digital Age*. Stanford, CA: Stanford University Press, 2007.

Velmans, Max. "Physical, Psychological, and Virtual Realities." In *The Virtual Embodied: Practice, Presence, Technology*, edited by John Wood, 45-60. New York, NY: Routledge, 1998.

Williams, Raymond. *Television: Technology and Cultural Form*. London: Routledge, 2003.

About the Author

Madison Brown is a writer and curator whose research explores amateur media cultures at the intersection of archives, memory, and everyday life. She received her PhD in Screen Cultures from Northwestern University and is currently the John R. and Barbara Robinson Family Curatorial Fellow in Photography at the Harvard Art Museums.

20. Video Game Immersion as Bodily Atmospheric Attunement

Andrea Andiloro

Abstract: This chapter deploys the concept of atmosphere in its consideration of video game immersion, arguing that a game's atmosphere, a neglected topic until recently within game studies, is critical to the gaming experience. Through a phenomenological approach, the chapter underscores the importance of the player's bodily experiences, and it highlights how these experiences and the game's atmosphere interweave to foster immersion. Using examples from specific video games, such as *Super Mario Bros.*, *Castlevania*, *Dark Souls*, and *Wolfenstein: The New Order*, the author elucidates how design elements influence a game's atmosphere and consequently the player's immersive experience. The study thereby enriches the understanding of immersion in video games, emphasizing the role of atmosphere and its connection to players' embodied experiences.

Keywords: video game, atmosphere, embodiment, affect, phenomenology

Immersion is a controversial and polarizing term within game studies. The word generally refers to feelings of presence, intense absorption, or generally being-there, and it is common to read forum comments or reviews claiming that a video game is "particularly immersive" or that a game mechanic or writing "breaks the immersion." In the literature, the term is often compared, used as synonym for, or contrasted with terms found in VR studies and presence theory, such as "presence" and "telepresence."[1]

Game scholars conceptualize immersion as a complex phenomenon: some understand immersion as a perceptual illusion, others as psychological

1 See Bowman and McMahan, "Virtual Reality"; and Slater, "A Note on Presence Terminology."

Bédard, Philippe, Alanna Thain, and Carl Therrien, eds. *States of Immersion Across Media: Bodies, Techniques, Practices*. Taylor & Francis Group, 2025
DOI 10.5117/9789048558766_CH20

engagement in a task, others still as an absorption in fictional worlds or even as a form of posture adopted by the player, as indicated by Carl Therrien in his review on the topic.[2] This complexity is reflected in the writings of Brenda Laurel, who highlights a visual, aural, and kinaesthetic sensory immersion;[3] in the works on narrative immersion by Janet H. Murray and Marie-Laure Ryan;[4] in the phenomenological reflections of Geoff King and Tanya Krzywinska on how immersion is a matter of immersion in *game worlds*;[5] and in the work on affective immersion of Eugenie Shinkle as well as Ciara Cremin.[6]

Recently, there has been a growing interest in the role of the player body in immersive game experiences. Exemplary, in this regard, is Brendan Keogh's phenomenological critique of immersion as total absorption.[7] We can also include here the historical work of Graeme Kirkpatrick, reflecting upon the subjective *feel* of the immersive state, as it was developing alongside ideas of "deep gameplay" in British game magazines of the 1980s.[8] Magazines started referencing "deep gameplay" as a state made possible by novel control methods and technological developments such as parallax scrolling in home computing. Such developments allowed for a new experience of space and of immersive atmospheres within video games through the pairing of visual and kinaesthetic qualities typical of the arcade with aspects of story and puzzle solving.[9] "Deep gameplay" made it possible to sustain and convey an atmosphere conducive to giving players a sense of being "in" the game.

It is in the spirit of these authors who recentred the discussion surrounding immersion on "experience," "body," "space," and "atmosphere" that I wish to contribute to the debate. I argue that *atmosphere* is a crucial component of the video game experience. While players and reviewers are familiar and talk extensively about atmosphere, for example, in online forums or in game reviews (and have done so for a long time), game studies has only recently started theorizing atmospheres more substantially. Important works in this regard are influenced mainly by the aesthetic theories of Gernot Böhme.[10] Noticeable writings here include those by

2 Therrien, "Immersion," pp. 451–456.

3 Laurel, *Computers as Theatre*, p. 161.

4 Murray, *Hamlet on the Holodeck*; Ryan, *Narrative as Virtual Reality*.

5 King and Krzywinska, *Tomb Raiders and Space Invaders*, p. 118.

6 Shinkle, "Feel It, Don't Think"; and Cremin, *Exploring Video Games with Deleuze and Guattari*, pp. 93–97.

7 Keogh, *A Play of Bodies*, p. 15.

8 Kirkpatrick, *The Formation of Gaming Culture*.

9 Kirkpatrick, *The Formation of Gaming Culture*, p. 62.

10 Böhme, *The Aesthetics of Atmospheres*.

Gregor Herzfeld on video game music;[11] the reflections on the authenticity of historical settings within video games by Felix Zimmermann;[12] and the work by Merlin Seller on the post-human perception of atmospheric video game spaces by part of the coming-together of human and non-human bodies.[13] I too focus on video game atmosphere as an important category for player experience, but I wish to draw a stronger connection between atmosphere and immersion, a connection that is only hinted at by those scholars working within Böhme's paradigm, but is acknowledged much more explicitly by Human-Computer Interaction scholars.[14] Also noticeable in this regard are the reflections by game designers on the link between immersion and atmosphere.[15]

To summarize my position and my goal, I take atmosphere as conducive to immersion, and I explore this connection by addressing how this dynamic plays out bodily and experientially. My approach is therefore largely phenomenological, and my work is also in conversation with authors writing about atmosphere in that tradition beyond Böhme, namely Hermann Schmitz and Martin Heidegger.

Introducing Atmosphere and the Felt-Body

Important insight regarding atmospheres and their relation to immersion is found within the work of humanities scholars who are part of the "atmospheric turn," with Böhme being the best-known theorist.[16] Böhme defines atmospheres as spatial and "in-between" phenomena cutting through the subject-object divide.[17] On the one hand, atmospheres are produced by people and objects present in a space. Böhme suggests thinking of things and people as displaying a "going-forth."[18] These displays are identified as "the ecstasies of things," meaning the colors, smells, shapes, and so on, emanating from things and composing the atmosphere: "atmospheres are

11 Herzfeld, "Atmosphere at Play."

12 Zimmermann, "Historical Digital Games as Experiences"; and Zimmermann, *Virtuelle Wirklichkeiten*.

13 Seller, "Sensing Sims."

14 See Brown and Cairns, "A Grounded Investigation of Game Immersion"; and Ribeiro et al., "Game Atmosphere."

15 See Bentley, "Atmosphere in Games"; Kasavin, "Creating Atmosphere in Games"; and Sabbagh, "Take It All In."

16 Griffero, "Is There Such a Thing as an 'Atmospheric Turn'?", p. 16.

17 Böhme, *The Aesthetics of Atmospheres*, p. 17.

18 Böhme, *The Aesthetics of Atmospheres*, p. 22.

thus conceived not as free floating but on the contrary as something that proceeds from and is created by things, persons, or their constellations."[19] On the other hand, atmospheres always need subjects of perception and are never wholly independent of them. Böhme writes that "atmospheres are quasi-objective, namely they are out there; you can enter an atmosphere and you can be surprisingly caught by an atmosphere. But on the other hand,[...] atmospheres are not beings like things; they are nothing without a subject feeling them."[20]

In between subject and object we find the perceiving felt-body, understood in the phenomenological sense of *Leib*, as opposed to the mere physical body qua *Körper*, as described by Hermann Schmitz.[21] For Schmitz, the *Leib* is a surfaceless space referring to anything requiring more than our traditional five senses to be felt, including embodied emotions (i.e., fear, pain, hunger, itch, disgust, and so on), embodied stirrings (i.e., joy, sorrow, anger, and so on), irreversible embodied directions with or without movement (i.e., glances and breathing), and full-fledged felt movements, "spontaneous or planned, like walking, grasping movement, jumping, dancing, shiver, twitching, and deglutition."[22] For Schmitz, felt-body perceptions constitute a type of atmosphere that may be diffused onto the *Leib* or located on particular islands of it, as in the examples of "headache, stomachache, or tired legs."[23] As such, there is a felt-body atmosphere of laziness, for example: I feel sluggish, unmotivated, my eyelids feel heavy, I emit long sighs and glance idly at my smartphone, every movement feels like a struggle. This is a holistic type of affective bodily atmosphere that is neither purely mental nor purely physical. Perceived atmospheres are felt through our *Leib* present in, and moving through, atmospheric spaces. The way we move through a spatial atmosphere and the kinaesthetic qualities we discern constitute a critical aspect of how an atmosphere subjectively *feels*.[24] Through movement we get the feeling of "being-in-something," a spatial something presenting its own moods. Through movement "we sense what kind of space surrounds us. We sense its *atmosphere*."[25]

Every movement through space presents qualities contributing to that space's perceived atmosphere, and atmospheres require us to attune our

19 Böhme, *The Aesthetics of Atmospheres*, p. 122.
20 Böhme, *The Aesthetics of Atmospheres*, p. 17.
21 Schmitz, "Atmospheric Spaces," pp. 65–66.
22 Schmitz, "Atmospheric Spaces," p. 66.
23 Schmitz, "Atmospheric Spaces," p. 67.
24 Böhme, *The Aesthetics of Atmospheres*, p. 227.
25 Böhme, *The Aesthetics of Atmospheres*, p. 228, original emphasis.

bodies to the specific movements they enable and demand. In relation to video games, we can say that the way players *move* through video game spaces and how that *feels* to them constitutes their atmospheric perception. The idea of "game feel" developed by Steve Swink is relevant here.[26] Game feel refers to how the commands of a video game—the mapping of the buttons, other interface features, and so on—contribute to "a kind of 'virtual sensation,' a blending of the visual, aural, and tactile," contributing to the quality of the feeling of movement in different video game spaces:[27] for example, some video games feel stiff, some feel floaty, and so on. Game feel is achieved through an adequate mix of real-time control, simulated space, and what Swink refers to as adequate levels of polish, such as dust particles, crashing sounds, camera shake, animation, and so on, all in the service of presenting the player with a perceptually convincing experience of movement.[28]

Mario is a typical example of a video game character often considered pleasurable to manoeuvre: the controls in the classic *Super Mario Bros* (Nintendo R&D4, 1985) feel "tight," "precise," and "bouncy," which is unsurprising given the platform-focused gameplay of most *Mario* titles.[29] As a diametrically opposed example to Mario, we can think of *Castlevania* (Konami, 1986), another two-dimensional platformer game published around the same period as *Super Mario Bros.*: the player-character of Simon Belmont, the main character of the game, is stiff; he struts heavily across the levels, and his animations *feel* weighty.[30] Unlike Mario, he cannot change direction mid-air, making *Castlevania* platforming sections particularly tricky.

Game feel understood as such is *felt* but is not quantifiable. We may instead understand it as a perception of the felt-body (*Leib*). Swink states explicitly that game feel does not include emotions or affects, and refers primarily to "a tactile, kinaesthetic sense of manipulating a virtual object."[31] Contrary to this, I argue that we cannot artificially separate affective and kinaesthetic perceptions of the felt-body: we cannot touch or manipulate something or move our body through space without experiencing any sort of affect. It is more useful to think of game feel as a type of felt-body perception that includes kinaesthetic feelings as well as *e-motions*. As Giuliana

26 Swink, *Game Feel.*
27 Swink, *Game Feel*, p. XIII.
28 Swink, *Game Feel*, pp. 2–5.
29 Nintendo R&D4, *Super Mario Bros.*
30 Konami, *Castlevania.*
31 Swink, *Game Feel*, p. XIII.

Bruno writes in regard to the interplay between movement and affect: "motion creates emotion and, reciprocally, emotion creates a movement that becomes communicated."[32] Thinking of kinaesthetic sensations as separated from affects would mean discounting "the felt body with its quite specific dynamics, rhythms of stirrings and corporeal movements, and its ways of being constantly involved in the manifold forms of holistic sensing of situations—rich modes of experience that cannot adequately be narrowed down to perception by means of the sense organs,"[33] or, to put it briefly, it means discounting atmospheric perception.

Atmospheric Attunements

To clarify the link between atmosphere and immersion, I employ the concept of *Stimmung*, or attunement. This is a key term in Martin Heidegger's phenomenology of *Dasein*.[34] Heidegger writes about an ontological *Befindlichkeit*, translated as a "findingness," of which *Stimmung* is the ontical presentation. For Heidegger, finding-oneself-in-a-mood is an existential fact, and the individual moods we may be in (happy, sad, excited, anxious, and so on), are identified as *Stimmung*.

Heidegger explains that *Stimmung* is not a merely psychological, "internal" affair, but neither is it only "out there." Instead, it arises as an atmosphere out of our being-in-the-world, in our engagements with people, objects, and tasks. Heidegger likens *Stimmung* to a musical atmosphere, writing that "an attunement is a [...] melody that does not merely hover over the so-called proper being at hand of man, but that sets the tone for such being, i.e. attunes and determines the manner and way of his being."[35] We may then understand attunements as being-in-atmospheres attending to things that matter to us in the world.

While observing atmospheric attunements, Kathleen Stewart writes that "anything can feel like something you're in. A condition, a pacing, a scene of absorption, a serial immersion in little worlds."[36] Accordingly, attuning to a video game atmosphere means immersing ourselves by attending to something that matters to us, in that moment, in a fully embodied way,

32 Bruno, *Surface*, p. 19.
33 Schmitz, Müllan, and Slaby, "Emotions outside the Box," p. 244.
34 Heidegger, *Being and Time*, p. 172.
35 Heidegger, *The Fundamental Concepts of Metaphysics*, p. 67.
36 Stewart, "Atmospheric Attunements," p. 449.

registering and feeling the qualities, rhythms, forces, relations, and movements of this atmosphere.

James Ash applies *Stimmung* to video games, arguing that immersion in video games is a matter of bodily atmospheric attunement.[37] He explains this by analyzing how players are captivated subjects when playing. The "there" wherein players feel immersed is an intense space designed to generate complex potentials for movement and action in the contingent space between user and game.[38] In the process of play and skill acquisition, players attune to these intense spaces designed for movement and actions, leading to the creation of captivated bodies. More specifically, "attuning oneself to the game involves a self-management of the affective and emotional state of being of the user in an attempt to minimize negative affects such as frustration and vulnerability."[39]

Embodied action and movement are to be understood through affect and cognition, not as opposing categories, as is the case in certain strands of affect theory where affect is associated with pre-conscious, ephemeral, and "unspeakable" bodily capacities and where receptivities inform the ways individuals think and behave.[40] Affects and cognition are instead to be thought as interdependent: according to Ash "processes of cognition can shape affective capacities and affects themselves can work to rewire the relationship between thought and action."[41]

This approach addresses the way skill acquisition in video games is thoroughly affective as well as cognitive through intensive immersions within atmospheric video game arrangements. While we play, affects are rendered material through and through in the ways we flex our fingers, grab our controllers, rock our bodies, shout, and punch the air when we win, swear, and stomp our feet when we lose. In most competitive games, players can self-manage these affects by minimizing the negative ones through processes of becoming-skilful. Designers create intensive spatial environments "designed to generate complex *potentials* for movement and action in the contingent space between user and game," where a possibility for affect is also generated.[42] Attunement, as a concept, helps us "to think through how environments operate to 'matter' affect, and, in turn, how

37 Ash, "Technologies of Captivation."
38 Ash, "Technologies of Captivation," p. 32.
39 Ash, "Technologies of Captivation," p. 28.
40 See Clough, "Introduction"; and Massumi, *Parables for the Virtual*.
41 Ash, "Technologies of Captivation," p. 30.
42 Ash, "Technologies of Captivation," p. 32, original emphasis.

these become central to the type of attunement and bodily capacities that can be developed."[43]

Ash pays particular attention to how the spatial design and the architecture of the intensive spaces of video games create the possibility for particular affects to be "enfleshed" within the body, noticing that these are not just conscious behaviours or acquired bodily skills deployed within a situation. They are rather a feature of our being at any given moment, effectively presenting themselves not only as feelings or emotions, contained "within" a subject, but typically as *atmospheres* with the character of in-betweenness.[44]

We may enter a room and find it has a jovial atmosphere because of the interaction amongst those present, because of its *décor*, because of its lighting and acoustics, and so on. We may then attune to this atmosphere, becoming jovial ourselves. Alternatively, we may be in a foul mood and feel irritated by the joviality of the situation: the world then appears heavier, darker, and more unpleasant to us, because we are attuned to a particular mood (perhaps sadness, depression, or grief) in a way that clashes with the lightness of the room we have just entered, and with which we find ourselves out of tune. Attunements are fundamentally relational, that is, they arise from how we relate to situations, which is always a matter of who we are and how we find ourselves when we arrive at that situation in the first place (what are our past experiences, skills, and socio-cultural background; what is our gender, race, and so on).[45] Likewise, in video games, players always come with a set of pre-existing attunements and bodily skills from their previous life experience, but in playing new video games they also develop new attunements, with their own skilful and bodily intensity informed by the atmosphere of the intensive environments designed by the developers.

Let us look at another example. *Dark Souls* (FromSoftware, 2011) is a third-person action role-playing game, with an emphasis on combat and exploration within a gritty dark fantasy setting. *Dark Souls* is notorious amongst its players for being brutally difficult, because of the near-perfect timing required by its combat system, its treacherous game world filled with traps, and the opaque and unclear instructions on what to do and where to go. Within *Dark Souls*, the level design, the game architecture, the enemy placement, the game mechanics, the controls, and the interface systems together present an intense atmosphere to which players attune. Its intricate game architecture makes it possible for designers to strategically

43 Ash, "Technologies of Captivation," p. 30.

44 Ash, "Technologies of Captivation," p. 34.

45 See Ahmed, *The Cultural Politics of Emotion*.

position enemies in spots from which they can easily surprise players, who are, accordingly, to some degree always on edge. The souls mechanic also functions to maintain high tension. Souls are used as an in-game currency and are obtained by defeating enemies. If players die, they respawn at their latest checkpoints and have one chance to retrieve their souls. If they die twice without retrieving their souls, these souls are irremediably lost. Combat is one of the core gameplay activities. The controls during the combat segments are very deliberate. Players must be precise and strategic on when to attack or defend, as the attack animations cannot be interrupted once initiated: when attacking, one must commit and maintain a high level of alertness. Interface elements contribute to this high-intensity feeling by displaying vitality bars for both players and enemies. Seeing exactly how much damage they are dealing (and taking in return), players may judge whether to take a risk and time an attack (if perhaps they think the enemy is one shot away from being defeated), though they may also leave themselves open to taking damage in that way.

The aforementioned elements are just a few of the ones that coalesce into an environment presenting an atmosphere of a certain intensity to which players attune. During attunement, affects get inscribed into gestures and actions. Through interface elements and level design features such as those mentioned above, but to which we can also add analogue stick sensitivity, animation framerates, sight lines or feedback sound effects, players tacitly attune somatically and analytically to an atmosphere that informs how they move and act within the game. At the same time, "the body of the player [...] is actively shaped by the atmosphere and events to which it is exposed, and it retains aspects of those events across the life course, both intentionally and unintentionally."[46]

Attuning is distinctively tense and unwelcoming for *Dark Souls* novices, for whom the game may feel threatening, irritating, or unwelcoming. This uneasy atmosphere slowly dissipates with skill acquisition. Once successfully attuned to the atmosphere of the intensive environment, negative affects are minimized: players learn where the enemies are placed, memorize the layout of each level, and develop the capacity to discern framerates in a tacitly embodied way. This process of attunement creates "captivated bodies" entering a "state of engaged and focused concentration in order to successfully negotiate the ongoing flow and immediacy of an encounter."[47]

46 Ash, "Technologies of Captivation," p. 36.
47 Ash, "Technologies of Captivation," p. 41.

Beginner players may feel frustration and experience tension and strain while trying to focus their sense of sight and hearing as much as possible, leaning forward towards the screen and squinting their eyes. Without really knowing what to expect, and without having attuned to the game feel, the body of the beginner player is gripped by an intense atmosphere of contingency amplified by the architectural design of the maps themselves, which also convey a sense of dread and anxiety through their gloomy aesthetic. However, during the process of becoming skilful, these negative affects are minimized and countered by more positive ones: satisfaction, sense of achievement, gratification.

Captivation, as described by Ash, wherein "hours become like minutes [...] [and] the recollection of an evening's play is often reduced to memories of one or two particularly intense, contingent, or exciting moments that become foregrounded in a user's mind,"[48] is compatible with the experience of being-there commonly identified as "immersion." This is a type of "intensive processuality" that game studies sometimes also address through the concept of flow,[49] generally experienced and felt as an overall atmospheric tonality rendering possible an immersive absorption in the arrangement.

While they play, players become immersed in an atmospheric arrangement composed of hardware interface, software interface, game architecture, other players, sound effects, graphics, animations, and so on, an arrangement presenting an atmosphere to which players attune. In the case of *Dark Souls*, an intense atmosphere of dread and frustration is generated, resulting in a very tense and unpleasant experience for beginners. By becoming more skilful, they acclimatize to this atmosphere through the acquisition of subtle semi-conscious somato-cognitive attunements and they can minimize the negative affect associated with it. Furthermore, they will carry these bodily attunements in their future, which in practical terms means that they will likely have an easier time when encountering other designs similar to *Dark Souls*', also known as "soulslike" games, compared to players who have never attuned to *Dark Souls*' atmosphere.

Hot and Cool Designs: Saturating and Dissipating the Atmosphere

Video games do not usually present internally homogenous atmospheres. Rather, they often present internal atmospheric variance. At times, the

48 Ash, "Technologies of Captivation," p. 43.

49 See Csikszentmihályi, "Play and Intrinsic Rewards"; and Soderman, *Against Flow*.

atmosphere may be tense, then more relaxed, and then tense again. There is a certain rhythm to a video game's atmosphere. I am referring here to "pacing," which foregrounds the importance of *time* as something to consider when discussing a video game's atmosphere. Steven Conway and Marc Ouellette's work on video game patterns can be usefully applied to this discussion.[50]

Conway and Ouellette build on Marshal McLuhan's taxonomy of "hot" and "cool" media. According to McLuhan, hot media are those extending "one single sense in 'high definition.' High definition is the state of being well filled with data."[51] Furthermore, hot media are low in participation, in that they "do not leave so much to be filled in or completed by the audience."[52] McLuhan's example is that of a photograph conveying lots of visual information to the eye with no need for the viewer to "fill in any gaps." Cool media are the opposite, low in definition and high in participation. McLuhan's example in this case is that of the telephone, conveying little amount of aural information to the ear while requiring the user to fill in the gaps and actively talk back to their interlocutor.[53]

Conway and Ouellette adapt McLuhan's taxonomy to a discussion of game design. First-Person Shooter (FPS) video games are a typical example of hot design, as they are high definition: they provide "strong definitions for what to do (kill), how to do it (shoot) and why one should do so (bad things threaten the avatar's existence), wove in a tight temporal cycle."[54] They also require low participation, since the actions they afford are very limited: shoot, move, repeat. In addition to this, FPSs tend to be linear in both level design and encounters: it is always clear where one needs to go and what one needs to do; players do not need to fill in any gaps.

An example of cool design would be video games belonging to the Walking Simulator (WS) genre. The WS genre effectively cools down the design of the FPS due to the "absence of explicit instruction, evaluative feedback, concise narrative, characters serving as obvious heroes and villains."[55] In WS games, goals are not obvious, motives are ambiguous (thus of low definition) and players need to participate more to fill in the gaps of the narrative (thus requiring a high level of participation).

50 Conway and Ouellette, "Playing It Cool."
51 McLuhan, *Understanding Media*, p. 22.
52 McLuhan, *Understanding Media*, p. 23.
53 McLuhan, *Understanding Media*, pp. 22–23.
54 Conway and Ouellette, "Playing It Cool," pp. 1212–1213.
55 Conway and Ouellette, "Playing It Cool," p. 1213.

Difference in design temperature directly relates to differences in affective experience, with hot designs, on the one hand, tending towards "bursts of emotion, frissons, in response to psychosomatic pressure instigated through high fidelity features such as explicit win/lose conditions, time-sensitive mechanics, and tight cycles of interaction."[56] Cool designs, on the other hand, "tend towards shifts in mood, subtler and longer lasting shifts in one's sense of attunement; anxiety, fascination, pensiveness and boredom."[57]

The hot and cool dialectic is discernible in the way movement feels in the game as afforded through level design, game mechanics, and control scheme, in ways that alter our perception of time and the alternation of embodied rhythms. I suggest that hot designs orient players towards the "now," demanding immediate embodied engagement from them through split-second actions and reactions (jump, shoot, dodge, run, and so on), causing the *Körper* to tense up via sympathetic nerve system activation (muscles contract, heart-beat and breathing accelerate, hair raises on the skin, pupils dilate, and so on) while the *Leib* feels tense, sharp, rigid, and poised. On the other hand, cool design orient players towards an indefinite "past" and "future," allowing them to slow down and take their time (explore, meander, gaze, reflect, and so on), while the *Körper*'s autonomic nervous system activates its parasympathetic circuits (muscles relax, heart-beat and breathing decelerate, and so on) and the *Leib* feels relaxed, placid, even apathetic at times. Pure hot designs tend to be concordant and harmonious considering bodily engagement, offering a distinct rhythm, whereas cool designs are discordant and focus less on motor skills. Game designers may modulate design features such as "interface (hardware and software), narrative and character systems, camera perspective and game dynamics, single and multiplayer (cooperative or competitive) options and scoring systems,"[58] and so on, in order to heat them up or cool them down, and in such a way *saturating* or *dissipating* the bodily atmosphere players attune to.

An example of this would be an FPS alternating intense, hot moments with more relaxed, cooler ones. *Wolfenstein: The New Order* is an example of this.[59] *The New Order* is a reboot of the classic *Wolfenstein 3D*, often credited as being one of the earliest, if not the first FPS game.[60] As such, it largely follows in that design tradition, featuring a first-person camera and

56 Conway and Ouellette, "Playing It Cool," p. 1213.
57 Conway and Ouellette, "Playing It Cool," p. 1213.
58 Conway and Ouellette, "Playing It Cool," p. 1223.
59 MachineGames, *Wolfenstein: The New Order.*
60 id Software, *Doom.*

a gameplay centred around shooting and running across levels, although *The New Order* also presents sections allowing for stealth-based approaches and exploration. Individual missions are linear, goals are always clear, there is no ambiguity as to who the enemy is (the "bad guys" in the game are Nazis, the quintessential villains), and players are limited to a few actions, which are all high-definition features typical of hot designs. The soundtrack reinforces the heat of the design through the industrial metal tunes heard during combat segments. Weapons employ tracer ammunitions, a design choice that results in the screen being filled with the projectile trajectory during particularly intense shootings, which is a good example of a visual interface element contributing to the temperature of the design. The player *Leib* attunes to the tense and sustained bodily atmosphere of the mission typical of many other FPSs, which, as we have seen, is defined by tension and strain of whole bodies. Players don't just "look" or "gaze," they "scan" and "target." They don't just move across the game space, they run frenetically, and take cover. Perhaps players may lean forward or rock sideways while engaged in a tense shootout. In between missions, players come back to their base, where they can interact with non-player character companions, explore the area, and search for collectibles. The design during these segments cools down. Players don't run and hide frenetically, they explore and meander inquisitively; they don't shoot enemies, they talk with allies. The music is subdued, a gentle guitar that is barely perceptible. Players can let go of their bodily tension and instead attune to a more relaxed atmosphere, perhaps one more defined by curiosity and fascination.

Conclusion

Video game atmosphere is a relevant factor in facilitating immersion. Up until now, discussions of video game atmosphere had failed to engage with the wider literature on atmosphere theory, and as such remained poorly theorized. By entering in conversation with other disciplines exploring atmospheres, game studies may encounter concepts and theories useful for a more in-depth understanding of immersion in its bodily felt quality. Atmospheres are neither fully "out there," nor are they a purely mental, subjective phenomenon. Instead, they arise in between our immersive engagements with the world, and result in discernible bodily felt qualities, patterns, and rhythms. Video game atmospheres act very much the same way in that regard. By immersing ourselves within designed possibility spaces, we attune to atmospheres that put a strain on our body, that exert our fingers and eyes, and that make us build up tension in our whole body until we reach a moment of temporary relief, perhaps expressed in a shout or as an air fist. However, not all video game atmospheres are that intense. They are often subdued and ambiguous, and, accordingly, they have less impact on our body, even though they are still affective. Developers may modulate certain of their design features to heat up or cool down the design and thus saturate or dissipate the bodily atmosphere of their game and in such a way achieve a desired atmospheric tempo to which the bodies of the players can attune. This attunement contributes to feelings of immersion. Appreciating players' attunement to the bodily atmosphere of a video game understood as a movement rhythm allows us to further deepen our understanding of the processes involved in immersion and its various nuances.

Bibliography

Ahmed, Sara. *The Cultural Politics of Emotion*. Edinburgh: Edinburgh University Press, 2004.

Ash, James. "Technologies of Captivation: Videogames and the Attunement of Affect." *Body & Society* 19, no. 1 (2013): 27–51. https://doi.org/10.1177/1357034X11411737.

Bentley, Matthew. "Atmosphere in Games." *Game Developer*, June 15, 2013. https://www.gamasutra.com/blogs/MatthewBentley/20130615/194399/Atmosphere_in_Games__Part_1__Atmosphere_introduced.php.

Böhme, Gernot. *The Aesthetics of Atmospheres*. New York, NY: Routledge, 2017.

Bowman, Doug A., and Ryan P. McMahan. "Virtual Reality: How Much Immersion Is Enough?" *Computer* 40, no. 7 (July 2007): 36–43. https://doi.org/10.1109/MC.2007.257.

Brown, Emily, and Paul Cairns. "A Grounded Investigation of Game Immersion." In *CHI EA '04: Extended Abstracts on Human Factors in Computing Systems*, 1297–1300. New York, NY: Association for Computing Machinery, 2004. https://doi.org/10.1145/985921.986048.

Bruno, Giuliana. *Surface: Matters of Aesthetic, Materiality and Media*. Chicago, IL: University of Chicago Press, 2014.

Clough, Patricia T. "Introduction." In *The Affective Turn: Theorizing the Social*, edited by Patricia T. Clough and Jean Halley, 1–33. Durham, NC: Duke University Press, 2007.

Conway, Steven, and Marc Ouellette. "Playing It Cool: Considering McLuhan's Hot and Cool Taxonomy for Game Studies." *Convergence* 26, no. 5–6 (2020): 1211–1225. https://doi.org/10.1177/1354856519880789.

Cremin, Ciara. *Exploring Video Games with Deleuze and Guattari: Towards an Affective Theory of Form.* New York, NY: Routledge Taylor & Francis Group, 2016.

Csikszentmihályi, Mihály. "Play and Intrinsic Rewards." In *Flow and the Foundations of Positive Psychology: The Collected Works of Mihály Csikszentmihályi*, edited by Mihály Csikszentmihályi, 135–153. New York, NY: Springer, 2014.

FromSoftware. *Dark Souls.* Namco Bandai. 2011.

Fuchs, Mathias. *Phantasmal Spaces: Archetypical Venues in Computer Games.* New York, NY: Bloomsbury Publishing, 2019.

Griffero, Tonino. "Is There Such a Thing as an 'Atmospheric Turn'? Instead of an Introduction." In *Atmosphere and Aesthetics: A Plural Perspective*, edited by Tonino Griffero and Marco Tedeschini, 11–62. Cham: Palgrave Macmillan, 2019.

Heidegger, Martin. *Being and Time.* New York, NY: Harper & Row, 1962.

Heidegger, Martin. *The Fundamental Concepts of Metaphysics:. World, Finitude, Solitude.* Bloomington, IN: Indiana University Press, 1995.

Herzfeld, Gregor. "Atmosphere at Play: Aesthetic Consideration on Game Music." In *Music and Game: Perspectives on a Popular Alliance*, edited by Peter Moormann, 147–157. Wiesbaden: Springer Fachmedien Wiesbaden, 2013.

id Software. *Wolfenstein 3D.* Apogee Software. 1992.

Kasavin, Greg. "Creating Atmosphere in Games." Filmed March 8, 2012 at the Game Developers Conference 2012, San Francisco, CA. Video, 1:01:35. https://www.gdcvault.com/play/1015556/Creating-Atmosphere-in.

Keogh, Brendan. *A Play of Bodies: How We Perceive Video Games.* Cambridge, MA: MIT Press, 2018.

King, Geoff, and Tanya Krzywinska. Tomb Raiders *and* Space Invaders*: Video Game Forms and Contexts.* New York, NY: I. B. Tauris, 2006.

Kirkpatrick, Graeme. *The Formation of Gaming Culture.* UK Gaming Magazines, 1981–1995. Berlin: Springer, 2015.

Konami. *Castlevania.* Konami. 1986.

Laurel, Brenda. *Computers as Theatre.* Reading, MA: Addison-Wesley, 1991.

MachineGames. *Wolfenstein: The New Order.* Bethesda Softworks. 2014.

Massumi, Brian. *Parables for the Virtual: Movement, Affect, Sensation.* Durham, NC: Duke University Press, 2002.

McLuhan, Marshall. *Understanding Media: The Extensions of Man.* Cambridge, MA: MIT Press, 1994.

Murray, Janet H. *Hamlet on the Holodeck: The Future of Narrative in Cyberspace.* New York, NY: The Free Press, 2016.

Nintendo R&D4. *Super Mario Bros.* Nintendo. 1985.

Ribeiro, Giovanni, Katja Rogers, Maximilian Altmeyer, Thomas Terkildsen, and Lennart E. Nacke. "Game Atmosphere: Effects of Audiovisual Thematic Cohesion on Player Experience and Psychophysiology." In *CHI Play '20: Proceedings of the*

Annual Symposium on Computer-Huamn Interaction in Play, 107–119. New York, NY: Association for Computing Machinery, 2020.

Ryan, Marie-Laure. *Narrative as Virtual Reality:. Immersion and Interactivity in Literature and Electronic Media*. Baltimore, MD: Johns Hopkins University Press, 2001.

Sabbagh, Michel. "Take It All In: Imbuing Video Games with a Potent Atmosphere." *Game Developer*, June 21, 2018. https://www.gamasutra.com/blogs/MichelSabbagh/20180621/320429/Take_It_All_In_Imbuing_Video_Games_With_a_Potent_Atmosphere.php.

Schmitz, Hermann. "Atmospheric Spaces." In *Atmosphere and Aesthetics: A Plural Perspective*, edited by Tonino Griffero and Marco Tedeschini, 63–76. Cham: Palgrave Macmillan, 2019.

Schmitz, Hermann, Rudolf Owen Müllan, and Jan Slaby. "Emotions outside the Box: The New Phenomenology of Feeling and Corporeality." *Phenomenology and the Cognitive Sciences* 10, no. 2 (June 1, 2011): 241–259. https://doi.org/10.1007/s11097-011-9195-1.

Seller, Merlin. "Sensing Sims: Atmosphere, Aesthetics and the Cyborg Player." In *Materializing Digital Futures: Touch, Movement, Sound, and Vision*, edited by Toija Cinque and Jordan Beth Vincent, 25–44. New York, NY: Bloomsbury Academic, 2022.

Shinkle, Eugenie. "Feel It, Don't Think: The Significance of Affect in the Study of Digital Games." In *Proceedings of the 2005 DiGRA International Conference: Changing Views. Worlds in Play*, 1–7. Vancouver: DiGRA, 2005.

Slater, Mel. "A Note on Presence Terminology." *Presence Connect* 3, no. 3 (2003): 1–5.

Soderman, Braxton. *Against Flow: Video Games and the Flowing Subject*. Cambridge, MA: MIT Press, 2021.

Stewart, Kathleen. "Atmospheric Attunements." *Environment and Planning D: Society and Space* 29, no. 3 (January 1, 2011): 445–453. https://doi.org/10.1068/d9109.

Swink, Steve. *Game Feel: A Game Designer's Guide to Virtual Sensation*. Burlington, MA: Morgan Kaufmann, 2009.

Therrien, Carl. "Immersion." In *The Routledge Companion to Video Game Studies*, edited by Mark J. P. Wolf, 451–484. New York, NY: Routledge Taylor & Francis Group, 2014.

Zimmermann, Felix. "Historical Digital Games as Experiences: How Atmospheres of the Past Satisfy Needs of Authenticity." In *Game | World | Architectonics: Transdisciplinary Approaches on Structures and Mechanics, Levels and Spaces, Aesthetics and Perception*, edited by Marc Bonner, 19–34. Heidelberg: Heidelberg University Publishing, 2021.

Zimmermann, Felix. *Virtuelle Wirklichkeiten. Atmosphärisches Vergangenheitserleben Im Digitalen Spiel*. Marburg: Büchner-Verlag, 2023.

About the Author

Andrea Andiloro is a lecturer in the program of Games and Interactivity at Swinburne University of Technology, Melbourne. He teaches game studies and user-centred design. He researches the phenomenology of video game play, video game ontology, and aesthetics.

Section 4

Interviews

21. Beyond Frontiers: An Interview with Félix Lajeunesse (Co-founder and Chief Creative Officer, Felix & Paul Studios)

Edited by Philippe Bédard

Abstract: To this day, Felix & Paul Studios remain one of the most influential and celebrated players in the field of virtual reality. We spoke with Félix Lajeunesse to document the studio's history and to get his perspective on the medium, the technical developments that have shaped it, and what the future has in store for those working in this sector. The discussion not only offers a practical counterpoint to many of the theories discussed in this volume, but it also touched on the three main topics of *States of Immersion Across Media*: bodies, techniques, practices.

Keywords: virtual reality, immersive media production, 360-degree video, technology, immersion

Felix & Paul Studios is one the best-known names in the virtual reality space. How did you come about working with this medium? What drew you to this format?

Félix Lajeunesse (FL): It started out approximately fifteen years ago. I was working in collaboration with Paul Raphaël (co-founder of Felix & Paul Studios). We were two directors, visual artists, and friends working together doing documentaries, fiction films, commercials, and music videos. We shared a lot of creative interests and we were both fascinated by "experiential cinema"—films that we defined at the time as being more focused on the cinematic experience itself rather than on the plot (the three-act structure storytelling with plot points, turning points, and so on); films in which the narrative appears to be designed as a support for the creation of a highly immersive and sensitive cinematic experience. *2001: A Space Odyssey* (Stanley

---

Bédard, Philippe, Alanna Thain, and Carl Therrien, eds. *States of Immersion Across Media: Bodies, Techniques, Practices*. Taylor & Francis Group, 2025
DOI 10.5117/9789048558766_CH21

Kubrick, 1968), *Platform* (Jia Zhangke, 2000), *Tokyo Story* (Yasujirō Ozu, 1953), and all the films by Apichatpong Weerasethakul are great examples of that specific approach to filmmaking. Those films all share a hypnotic, almost transcendent quality. They make you feel an inherent part of the cinematic experience, instead of making you feel like you are watching something. This experiential approach to cinema certainly inspired the way we envisioned the medium of VR initially.

At the time, we wanted to explore that approach to filmmaking through 3D stereoscopic cinema. Although 3D cinema came and went in waves of hype, and even though much of the general audience's interest in the 3D cinematic experience waned after *Avatar* (James Cameron, 2009), we felt that there was something more profound to unearth and explore with 3D cinema. We felt there was something to learn from an artistic standpoint with stereoscopic imagery, which mainstream American cinema only superficially explored.

In the early days of our exploration, we designed and built 3D cameras ourselves and set out to create cinematic installations that would allow us to reproduce a sense of "pure reality": what happens when you film a character at life-size scale, with a human-to-human distance from the camera, with the right interocular camera calibration, and with the viewer positioned at a specific vantage point from a screen that recreates the scale of real-life? What are the experiential, emotional, and psychological outcomes that can result from synergizing all of those variables into a cinematic installation?

If we look at 3D cinema as it's been utilized by the Hollywood cinema industry over the past few decades, the purpose of using 3D could generally be defined as an attempt to enhance the cinematic spectacle by adding depth to the images and bringing visual assets closer to audiences. 3D has largely been used as an "add-on" to the film experience, rather than as an opportunity to completely rethink the cinematic experience from the ground up. In my opinion, that is why 3D failed. For instance, if you go to the movie theatre to watch a 3D film on the big screen and see a character's face in a close-up, the face will appear to you as a gigantic, oversized human face. It will have 3D volume, which will give it a certain quality of "realism," but that impression will be completely counterbalanced by the fact that a 400-square-foot human face looks nothing like what you would see in real life—not to mention other core aspects of the traditional cinematic language that aren't lifelike, such as cuts and stylized camera motions.

We felt like 3D stereoscopy could be envisioned as a "new medium," much closer to the experience of real life, that could generate a sense of presence for audiences—something that would feel like an evolution from the type of "experiential cinema" we loved so much.

What were your findings with this initial experiment?
FL: Our early experimentations with 3D cinema led us to consider the emotional and psychological output of sitting in front of those immersive experiences. We had guests visiting Paul's house (which was our studio at the time, around 2010–2011) to watch our early 3D experiments and dialogue with us about the work. At the time, we focused on the presentation of a series of immersive 3D shots we had filmed in the backstreets of Shanghai and assembled in a "contemplative," slow-paced edit.

The reaction we got from our guests was very interesting. Most people felt like what they were watching was not cinema, but something of a different nature. The main comments we received was that the installation felt like looking through a window onto another world and that it was so realistic and lifelike that it felt like the characters in the film could "see us" (the audience) as much as we could see them. There was a sense of "reciprocal presence" in the experience—a feeling that you are there, only more so because you're there *with* someone who is a protagonist in the story.

It's through this research and early work that we eventually developed an interest in virtual reality. When VR came around—or came *back* around, I should say—to the forefront of media attention, with Oculus in 2012–2013, we were already deep into our artistic and technological exploration of fully realistic 3D stereoscopic cinematic capture. We wondered: can we deepen our exploration of presence through VR? Can we extrapolate on what we know with 3D cinema to create fully immersive 360-degree 3D cinematic experiences? Can we aspire to tell stories through cinematic VR? Can storytelling synergize with presence, or will it work against it?

With the early hardware coming out of Oculus and after a few experiments on our end, we realized very quickly that there was a path forward. We felt like we had the right technological and creative foundations—and a strong will and creative impulse—to pursue our journey as artists into the fresh new world of virtual reality. Somehow, we felt VR was our destiny.

What project did you create as a result? Did anything change when you released that project to the public?
FL: We did a project at the time titled *Kobe* (Félix Lajeunesse, Paul Raphaël, and Zacharias Kunuk, 2011), which was filmed in Igloolik, an Inuit community. I was working there with producer Stéphane Rituit, co-founder of Felix & Paul Studios, for Igloolik Isuma Productions, the Inuit-owned studio that created *Atanarjuat: The Fast Runner* (Zacharias Kunuk, 2001) and *The Journals of Knud Rasmussen* (Zacharias Kunuk and Norman Cohn, 2007). At

the time, there was increasing convergence between the immersive media work I was doing with Paul on one end, and the cinema work I was doing in the Arctic with Stéphane and Isuma. It is through this convergence that the three of us started working together—a collaboration that led to the creation of Felix & Paul Studios a few years later.

After we launched *Kobe* at the Toronto International Film Festival (TIFF) as a 3D installation, the three of us went on to produce our very first VR experience, titled *Strangers with Patrick Watson* (2014), an uninterrupted, eight-minute VR scene where, as a viewer, you get to be in the close presence of musical artist Patrick Watson as he writes music for his new album in his home studio. This intimate VR experience, considered the first VR film ever made, travelled the world, and suddenly brought our until-then exploratory work into a big spotlight.

With *Space Explorers: The Infinite* (2021), you entered the Location-Based Entertainment (LBE) space. What led you to make that shift?
FL: After *Strangers with Patrick Watson*, we really set the foundations for Felix & Paul Studios as a full-spectrum creative and technology studio and began an intensive exploration of the medium. We created a lot of projects in a matter of just a few years. At the time, we partnered with technology platforms—mainly Oculus/Facebook [now Meta], Google, and Samsung—as business-to-business partnerships for the creation of a slate of premium cinematic VR projects (to help with the at-home adoption of VR glasses). At that time, we were creating work that was designed to be experienced by audiences at home, but unless we went to a film festival to showcase our work or read some of the comments left by users online, we didn't really have a direct connection with an audience at all.

That connection to an audience is something that we started to crave. When you create a VR experience, because it is a presence-based medium, the purpose of the work is ultimately the viewer's experience. The power of the experience is largely defined by the creators' ability to make audiences feel truly, profoundly immersed in the experience. So even though we were creating high-profile VR experiences—we collaborated with President Obama and Michelle Obama (*The People's House*), LeBron James (*Striving for Greatness*), Eminem (*Marshall from Detroit*) and many others—we felt limited in our capacity to interact with an audience and to witness and feel their connection to the work. Something was missing.

A few years later, when we started creating large-scale location-based experiences, mainly *Space Explorers: The Infinite*, we finally found ourselves in the presence of an audience with whom we could interact and from whom we could get feedback. We learned a ton from that. The capacity to feel the audience's response to the work helped us greatly to better understand and improve our craft as VR creators.

How do you differentiate the type of VR projects you are creating from the larger field of immersive media? For instance, I have noticed that quite a few "immersive" galleries have opened which showcase projects that are very different from VR.

FL: For me, this is very simple: any projection show is a variation on the medium of cinema or video. You can add a second, third, or fourth projector to your environment, or you might even point one at the floor and one at the ceiling; in the end what you will end up with is a cinematic experience with multiple screens. That can result in an impressive experience that has an immersive *flavour*; however, the use of 2D screens, 2D projectors, and free-roaming vantage points inherently limits this type of venue's capacity to make you feel truly transported and present in a distinct reality. You always remain "in that room" and the potential for a lifelike digital experience is off the table.

It's the core reason why we gradually moved away from cinema and ventured into the world of VR: to go where cinema could not go and to unlock the emotional potential of real-life experiences through presence-based experiences (instead of via a process of identification to characters that exist beyond the silver screen). So, from my perspective, there are only superficial—and no substantial—similarities between immersive galleries and large-scale VR experiences.

How would you define immersion, then?

FL: True immersion is when you can lose yourself in the experience; when you substitute your reality to the one you're immersed in, in an effortless way. In a good presence-based experience, there is an "experiential tipping point" where you start believing in the experience *enough* to surrender and accept to be emotionally and psychologically transported. It registers in your heart and soul as an experience you might have had in real life. This particularly comes to light when audiences come out of a VR experience (this phenomenon happens systematically with *The Infinite*), describing their experience using solely subjective, first-person language.

After watching a film, we will likely refer to it in a way that is analytical and to a certain extent dissociated from ourselves—"I thought the photography was nice"; "I thought the actors were good"; "the director's signature is great," etc.—as opposed to how we generally refer to our real-life experiences, using the language of direct, first-person experience—"I went to the park and I was listening to the birds, and then I thought about my mother..." When people come out of *The Infinite*, they generally describe their experiences in formulations such as: "I was walking in the Space Station, and then I turned around and I saw this astronaut coming my way and talking directly to me..." Those accounts are far closer to how we discuss real-life experiences compared to how we discuss films we have watched.

There's certainly a level of immersion that you can achieve with literature, theatre, and many other art forms. The difference is that virtual reality aims to *literally* emulate reality in its full sensorial range. Expanding and extrapolating on where this medium will go and hoping it will evolve in the next few years, I think we will see that this transposition between realities—from the physical to the digital world, and from one digital world to another—will become more common and enter the realm of the "normal." There will be a point in time where this artform, this medium, is going to feel like a very natural extension of physical reality itself.

I don't think that any other artform or type of experience can aspire to that.

Another trend we often see is that VR is compared to earlier media such as cinema or video games. Do you think VR will ever become its own medium?
FL: When we started out, we effectively referred to our work and practice as "cinematic" VR. But, now that we've entered the world of large-scale, multiuser, immersive VR experiences, where users find themselves exploring a narrative in which they are participating—and which is also reacting to them—that experience no longer looks or feels like cinema in any way. There are strong cinematic qualities and inspirations to what we do, but it isn't cinema. There are certain elements of gaming and interactivity in our work, but what we produce are not games. I would argue that our work is the result of a fusion between cinema, large-scale multiplayer games, and immersive theatre—and the result is something that doesn't have a satisfying name yet.

The more we explore, define, and refine what that "something" is, the less it will be compared to film and gaming. That conversation is relevant now, but a few years from now, VR will no longer be seen as the cousin of other existing disciplines.

How do other art forms inspire the work you do at Felix & Paul Studios?
FL: I do find that, in order to be a better immersive storyteller, I need to pay attention to reality itself. Of course, there's a lot to learn from other artistic

practices (cinema, music, architecture), but I focus more and more on how things actually happen in real life: on how people behave, on how they react, on their body language. I look at dynamics around a table, at how people position themselves in different spaces. All of that informs the creation of immersive experiences. So, when we get to staging an experience, especially in the context of a large-scale multiuser show, I gravitate towards things I've seen in real life rather than towards points of reference within the world of the arts.

I also generally love art that demands my mental participation or contribution. Art that summons my imagination. If you look at a painting by Jean-Paul Riopelle, the "meaning" of the artwork is for the audience to grasp, to feel, to explore. The audience contributes to the experience of watching and feeling the painting through its own imagination and sensibility. To me, that's when a work of art suddenly feels whole, when it feels fully satisfying.

The immersive market is often associated with the rapid evolution of technologies and the constant shift to new technological paradigms (6DoF, photogrammetry, neural radiance fields, etc.). Why is it that Felix & Paul Studios has remained steadfast in its use of 360-degree video?
FL: It's a question that we've asked ourselves over the years as we have been repeatedly hearing the siren song of new technological possibilities that did not exist until just recently. With that said, when you embark on the creative exploration of a medium, the process requires time and a level of in-depth exploration for any meaningful learning or advancement to be achieved. You experiment and explore, while at the same time you try to rationalize, structure, and operationalize your work.

When we started doing cinematic VR, we realized that we were really just scratching the surface of something that demanded further exploration to be better understood, let alone mastered. Once we had succeeded in shifting from experiential cinema to non-fiction cinematic VR (with *Strangers with Patrick Watson*, then the *Nomads* series), we set out to move towards the world of fiction, to find out how presence could synergize with fictionalized characters and narratives. We then set out to explore the creation of serialized content with *Space Explorers: The ISS Experience.* There was just so much we wanted to explore, validate, and question with the medium of cinematic VR that we made a decision to focus on this discipline first and foremost.

In parallel, every month or so, new possibilities, new tech trends, and new tools would appear on our horizon. We would always take notice, while trying to avoid distraction. Technology doesn't necessarily come with a built-in purpose. Our approach was always to integrate new technologies in our work when it felt like that elevated the work to a new realm or further enabled us to materialize a vision that wouldn't have been possible otherwise. Technology was always perceived as an enabler for the creative.

We operate in a largely tech-driven industry that generally lacks patience and is always on the lookout for the next "big" thing. But we imposed a level of patience and methodology on ourselves in order to stay focused on the incremental acquisition of key learnings. By the time we got to interactive virtual reality with *Space Explorers: The Infinite*, we had been thinking about interactive VR for years while continuing to learn a ton from our cinematic VR work. For our studio, the evolution from cinematic to interactive virtual reality happened when we were ready for it to happen and not when the industry decided that the time had come.

That really defines our relationship with technology. That's why we decided very early on to position ourselves as a creative studio that is tech-enabled, and not as a technology studio led by creative people. This model and approach will continue to define how we integrate the next big leaps in technology that will inevitably happen in the coming years.

About the Author

Philippe Bédard is an independent researcher with a fascination for moving image media and their technologies. As part of his postdoctoral research, he focused on virtual reality, with an emphasis on the notion of empathy. Finally, he has worked extensively in and around the immersive media industry in Montreal, including work with Xn Québec, Festival du Nouveau Cinéma, the National Film Board of Canada's interactive studio, and SODEC.

22. Producing *Intravene*: An Experiment in Immersive Audio[1]

Brenda Longfellow

Abstract: In this chapter, the director recounts the experience of creating *Intravene*, an immersive experience based solely on auditory storytelling around the opioid epidemic. Coming from the perspective of some with experience in film production, this chapter offers a ground-level look at the production of an audio-only immersive experience. In so doing, it raises question about the role that auditory stimuli play in fostering a sense of immersion in audiences. It also serves as a document of the making of *Intravene*, as it was written during—and after—the project's production.

Keywords: immersive media production, opioid epidemic, *Crackdown*, Darkfield

"Episode One: Benzodope"

You put on the headphones and are asked to close your eyes. The signature *Darkfield Radio* bumper fades in: "Connecting… Please wait for a moment… Connecting… connecting… connecting. *Darkfield Radio*: no news, no music, no opinion." There is a loud bang and the sound of water flowing. The dulcet voice of a robotic public address begins:

> Please take a seat and wait for your number to be called. We apologize for the delay, some of our staff are sick. Please be patient. For those of

1 Information on how to experience *Intravene* may be accessed here: *https://www.intravene-show.co*

The *Crackdown* podcast version may be accessed here: https://www.crackdownpod.com/episodes/31-lovedeathbenzodope, https://www.crackdownpod.com/episodes/episode-33-you-will-not-destroy-me

Bédard, Philippe, Alanna Thain, and Carl Therrien, eds. *States of Immersion Across Media: Bodies, Techniques, Practices*. Taylor & Francis Group, 2025
DOI 10.5117/9789048558766_CH22

> you who have not been here before, I can assure you, you are not dead, this is not purgatory ... This room is impossible to imagine, it is a place of forgetting and a place of stolen time, it can be described only as a hole in those you have left behind. Calling overdose number 1,052.
> A woman's voice whispers in your left ear: "Wait here, I'll get you a bundle."

A weave of documentary voices sets up the theme of the episode. Garth Mullins, the Executive Producer of *Crackdown*, recalls when he overdosed on a combination of benzodiazepines (benzos) and opioids and had two days of his life wiped from his memory. Dean Wilson, a longtime activist with the Vancouver Area Network of Drug Users (VANDU), talks about the ever more toxic supply of street drugs and insists we should stop calling them overdose deaths, people are dying of "drug poisoning." Martin Steward, another activist with VANDU adds: "I know more people that have died than are still alive."

The listener is then plunged into the immersive audio environment of a safe injection site in the Downtown Eastside of Vancouver, a rundown neighbourhood on the edge of Chinatown with a large, diverse, and visible population of street drug users. The rain this afternoon is torrential, a dog is barking and folks shout to close the door. Trey Helten, a former drug user and now manager of the Vancouver Overdose Prevention Society, is your guide over the course of the experience. Trey talks us through how the site works, registering people at the door with their "handles," and noting what drugs they are using. "Down" is fentanyl. "Hard" is crack cocaine. His job is to make sure no one dies from an overdose.

From time to time, a woman whispers in your ear. "Do you want to use this table"; "Do you want a test strip?" The voice is eerie and startling, you have a visceral reaction to being interpellated in such an intimate way. You are not outside this experience, passively observing: "you" are inside it. In a culminating moment, you realize you are in the throes of an overdose: the sound fades in and out, you hear the peer workers applying oxygen and jabbing naloxone into your thigh. You fade in and out of consciousness as the tannoy voice intones: "Your number is 1,054."

Those sounds are among the many that make up "Benzodope," the first episode of *Intravene* (2022), a two-part immersive audio experience on the overdose crisis in Vancouver. Since 2016, over 20,000 people have died by accidental overdose in Canada. In the US, over 100,000 people die every year. While our project is set in Vancouver, the epicentre of the overdose crisis in Canada, the crisis is exponentially and globally pervasive. Overdose death is a shadow pandemic, often coloured by sensationalism and misunderstanding, while it racks up preventable deaths that frequently exceed those of COVID-19 in all its variants.

Intravene evolved out of the UK/Canada Immersive Exchange, a joint initiative of the Canada Media Fund and StoryFutures UK that brought twelve creatives together from each side of the Atlantic for a series of intensive workshops with the intention of fostering international co-productions. The process was a bit like speed dating and the *Hunger Games*: you had to locate an international partner, come up with a joint project, and pitch your idea to the assembled. Andrea Salazar from Darkfield—a UK theatre company specializing in immersive binaural theatre—and I hit it off and conspired to pitch the idea of an immersive audio experience on the overdose crisis. I had been working with formerly incarcerated women in the Vancouver region, meeting once a month in a circle to share stories and co-create a public art project. Midway through the project, one of our most beloved members, René Chan, suddenly and shockingly died from ingesting drugs laced with fentanyl. The impact was deeply traumatic for all of us. René was a life force, the manager of a recovery house, a leader in the recovery movement, a wise and wise-cracking woman who had been to the brink and had pulled her life together. Her death haunted all of us and maybe we would never know or understand how this happened but I wanted to do something to honour her life and bring that crisis into focus for myself. I brought a community partner on board, *Crackdown*, a Vancouver-based podcast produced by artists, journalists, and drug user activists who had deep connections and links to the community. Darkfield was excited about the possibility of exploring what the binaural form could do with factual and documentary. Much to our surprise and delight, we were chosen as one of two "audience-ready" projects that would receive funding.

The *Crackdown* team took the lead in core messaging for the project. Their mantra is crystal clear: "Drug users are the experts. We know policy better than policy-makers. We know law better than lawmakers. We know pharmaceuticals better than pharmacists. We know nobody's coming to save us. So we gotta save ourselves."

Crackdown taught us that everything we thought we knew about the drug war, the overdose crisis, and addiction was wrong. The overdose crisis is not about the over-prescription of opioids. It is about how the illicit drug market has become increasingly contaminated, most notably with fentanyl, but more recently with benzodiazepines and benzo-like substances. The crisis, moreover, is directly attributable to prevailing drug policies based on criminalization, over-policing, and moral panics. These policies have not only directly contributed to the growth of organized crime syndicates, they have directly contributed to the toxicity of the illicit drug supply through the "Iron Law of Prohibition," which ensures that as law enforcement becomes more intense, the potency and toxicity of prohibited substances increases. COVID-19 and the disruption of supply lines only exacerbated this tendency.

The United Nations, the World Health Organization, the British Columbia government, Ontario's Big City Mayors, the International Red Cross, the American Public Health Association, and a growing international movement of human rights, medical, and progressive public health organizations have argued that the overdose crisis represents a catastrophic failure of current drug policies. Backed by decades of research and advocacy, they are calling for a radical transformation of the frames we use to understand drug use that move from a criminalizing, stigmatizing, and moralizing approach to one that is holistic in its consideration of the social determinants of health, is informed by a harm-reduction perspective, and that critically moves toward de-criminalization, legalization, regulation, and safe supply.

The challenge for our project was to find common ground between the distinctive cultures and practices of Darkfield and *Crackdown*. Darkfield is known for presenting deeply immersive theatrical experiences in pitch darkness inside shipping containers. In 2020, they had three shows set to launch when the pandemic hit and, like everyone, had to pivot to a new mode of delivery. Their pivot created *Darkfield Radio*, a mobile phone app that allows the listener to experience 360-degree audio experiences inside the comfort of their home. The listener downloads the app on their phone, purchases a ticket, and, at the appointed time, shows up with headphones on, ready to enter the intensity of a fictional world. Darkfield's work is often archly theatrical and plays on the thin line between what is real and what is imagined. *Eternal* (2020), for example, is inspired by Bram Stoker's *Dracula* and has you lie in your bed as characters run into your room and breathe down your neck. The Darkfield constituency consists of theatre aficionados, technology geeks, and audiences curious about new immersive technologies.

Crackdown's constituency is made up of a unique community of drug user activists, harm reduction workers, medical and legal advocates, and podcast listeners who skew liberal left. *Crackdown* has the distinction of being one of Canada's most popular podcasts, with over 15,000 subscribers on Patreon. Its episodes have covered an innovative range of topics from drug policy, policing, the limitations of drug substitution therapy to bogus scientific myths about addiction. They foreground the voices of those with lived experience, woven with investigative reporting and expert testimony, all narrated by the ever-congenial Garth Mullins, the Executive Producer of the podcast and a former heroin user. Funded by an ongoing collaboration with the University of British Columbia's Centre for Substance Research, *Crackdown* has a deep commitment to knowledge mobilization and includes extensive reading lists with each episode. They had never ventured into fiction but were intrigued by the possibilities of binaural audio.

The plan had always been to release a podcast and immersive version of each episode. In the initial stages of ideation, research, and production, we worked under the assumption that while these versions would have slight variations, they would essentially follow the same storyline and incorporate identical content. We had collectively determined that the first episode would concentrate on the impacts of benzos, just the latest in a long line of substances that were adulterating the illicit drug supply. The second episode would be more personal and would be framed as a search for Rosa Mullins, Garth Mullin's great-grandmother, who had been incarcerated in the Riverview Asylum in Vancouver for twenty-six years in the 1950s and 1960s. For Garth, Riverview and current calls to banish and incarcerate drug users were spectres that continued to haunt drug users in the region.

Crackdown began its rigorous research process, providing primers on the issue, story outlines, and the first of a series of interviews with drug user activists. But it slowly dawned on us that, at least for the first episode, the podcast and immersive versions would need to be quite different. This was one of the first lessons in discovering what works for immersive storytelling. The interviews, as rich as they were, took place in a number of different locations and tended, as documentary interviews often do, to be reflective and oriented around the recounting of past events. But compelling immersive content, we came to understand, really only works in the present tense and in a singular spatial surrounding where events occur and implicate you as an auditor. Exposition or third-person narration are fundamentally incompatible with immersivity as they distance the auditor from the immediate experience, which works best when it is haptic, embodied, and visceral—aesthetic modalities that are opposite to the cool, more analytic detachment of exposition.

While it meant jettisoning a lot of rich content, we decided to concentrate the immersive version of "Episode One: Benzodope" in one spatial location, the Vancouver Overdose Prevention Site (OPS), and to weave a smaller selection of the documentary interviews with ambient footage from the location. Our ambient recordings at the OPS provided rich immersivity but we needed a singular identifiable voice to guide listeners. We settled on the voice of Trey Helten, the manager of the site. The challenge of editing and sound mixing was to merge "clean" binaural studio interviews with the ambient recordings into a believable whole and to maintain documentary authenticity while shaping the flow of the story. We had always imagined that the dramatic peak of "Benzodope," the first episode of *Intravene*, would somehow simulate the experience of an overdose where a participant might hover on the edge of death and consciousness, only to be pulled back by the intervention of peer workers at the OPS. *Crackdown* was particularly emphatic in insisting that scripting and acting out an overdose was not going to work. The complex and banal compendium of events that comprise an overdose—the nodding off, the breathing that slows to a rasp, the routine responses of peer workers who have experienced hundreds of overdoses—none of that could be rendered with any degree of believability in fiction and we were all wary of replicating the well-worn sensationalism of mainstream representations. We were left with the delicate representational challenge of trying to avoid voyeurism as well as the ethically problematic artifice that a listener could occupy the actual experience of an overdose victim. In other words, we aspired to create what Kate Nash, in her reflection on the social possibilities of immersive documentary, has termed "proper distance."[2] For Nash, "proper distance" represents a complex representational space between nearness and separation that is required to elicit affective connection and relatability to others without the condescension of occupying the other's place. We were helped in resolving this representational conundrum by the fact that Gord and Trey from the OPS had willingly improvised how they typically respond to an overdose on site and their enactment served as the primary audio for the overdose segment. David Rosenberg, the sound editor and co-artistic director of Darkfield, also worked his magic using sound distortion, repetition, and layering of effects to simulate the experience.

Darkfield typically records with a US$12,000 Neumann binaural sound head in a soundproof studio. We rigged a comparable device using US$1,900 binaural ear pods and a wig head outfitted with rubber ears purchased from an acupuncture training school. We called our rig "Bruce." In order to enhance immersivity and exaggerate the implication of the individual auditor, Darkfield recorded an actor whispering offhand comments directly into the "ear" of "Bruce." Edited into the episode, these are experienced as a startingly intimate interpellation of the auditor, who is oriented in acoustic space as a participant ("are you ok?"; "do you want a test strip?").

As with virtual reality (VR) and other kinds of immersive media, binaural audio affords a very particular kind of experience through the conjuring, as the saying goes, of a space one inhabits rather than a scene one watches. But, unlike VR, our binaural project does not involve complex technological

2 Nash, "Virtual Reality Witness."

mediation: a mobile phone and a set of decent headphones enable anyone, anywhere to access media from the comfort of their home. In addition, unlike a podcast, which demands only a modicum of distracted focus, immersive audio requires and then enables the auditor's full and undivided attention. For twenty-three intense minutes, you close your eyes and are enveloped in an acoustic world that is visceral, jarring, and deeply involving. The brilliance of the Darkfield mobile app is that it maintains the eventfulness of the theatrical experience while relocating it into individual personal spaces. Auditors purchase a ticket and wait for the allotted time, when they are instructed to log on. The show begins and the participant enters a story space that is beyond the world of everyday life, even as they may sit in pajamas at their kitchen table. Throughout the experience, distractions fade and one surrenders to the complex pleasures of being transported on a storytelling journey to elsewhere.

In order to create the impression of 3D sound, the sound editor worked strategically with audio perspective and sound effect cues that played with proximity and distance. As with audition in the natural world, we hear sounds enter our left and right ears in miniscule variations. If a dog barks by our right ear, it will take a microsecond or so for the bark to reach the other ear and the sound will be louder in one ear than in the other. In our episode, a dog barks in the distance and then close to you. Trey moves by you and asks how you are doing, and a female peer worker whispers in your ear while visitors to the site inject drugs in the background. Sounds are clearly recognizable coming from all directions and these binaural signals contain embedded spatial cues in the form of phasing, timing, intensity, and what audio engineers call "spectral coloration"—cues that mimic and enhance natural human localization. As many of our test auditors have told us, they begin to actually visualize what the OPS looks like, although they had never in their life stepped into one and that is no doubt the result of the reality effect of 3D sound.

Intravene was invited to launch as part of the Storyscapes Immersive program at the Tribeca Film Festival in June 2022, one of the world's most prestigious venues for immersive experiences. Known for pushing the boundaries of storytelling with technology, Storyscapes featured twenty-nine extended reality (XR) projects artfully displayed on one floor at Spring Studios, a cavernous studio where audience members could wander through the arcade and try various immersive projects.

For the Tribeca launch, we designed and built an installation in a small soundproof room that suggested an overdose prevention site, with five booths separated by plastic dividers, each booth enhanced with suggestive props such as sharpie boxes where "used" needles could be disposed of. Poster art from activists at VANDU and the OPS plastered the walls. Given the nature of our topic, we knew how important it was to provide some form of modified trigger warning, so members of our team were in constant attendance to provide contextualization and answer questions that came up for audience members. The *Intravene* experience began when listeners put on their headphones, the door closed, the lights dimmed, and the room went suddenly and completely dark. The theatricality of the installation and the strange and disorienting experience of being plunged into darkness and losing physical bearings enhanced and deepened the experience of immersivity. Unlike the VR/AR projects at Spring Studios, *Intravene* was not interactive: audience members were not expected to do anything other than to surrender to the binaural acoustic world which enveloped them. For twenty-three dramatic minutes, they are transfixed in this world where they are subtly acknowledged and addressed, centred and interpellated as a being inside the space where events unfold. And, indeed, audience members consistently and frequently remarked on the emotional intensity of the experience when the lights came back on.

As an exhibition format, however, immersive presents certain kinds of logistical challenges. As is the case in most festivals that endeavour to combine immersive or interactive programming along with the main event—the launch of feature films—the Storyscapes program felt less like the future of cinema or media and more like a compelling but highly idiosyncratic if not marginal phenomena. Attendees at the Spring Studio site tended to be festival goers or guests and public traffic was limited. As for *Intravene*, our small room could only support five listeners at a time. And while attendance was intermittent, by the end of the ten-day festival, we had counted over 350 individual users.

The launch of the project on the Darkfield phone app in June was less successful. Most of those who purchased tickets were from the UK and were, no doubt, followers of Darkfield and immersive theatre in particular. Over a three-week initial run, there were only 300 individual purchases. The relative paucity of the immersive audiences contrasted with the extraordinary reach of the *Crackdown* podcast version of "Benzodope," which had over 10,000 individual downloads during the first two weeks. As a free podcast with a massive subscriber base, *Crackdown* was able to leverage a highly motivated and engaged audience in a manner that, to date, has eluded the immersive version.

Intravene, however, provided an extraordinary opportunity to experiment with new technologies and to test the possibilities of binaural audio as a kinetic medium for the evocation of documentary realities and complex social issues. It succeeded in exposing new audiences to a radical reframing of the overdose crisis and demonstrated the affective potential of binaural audio.

Obviously, spending time in a visceral immersion in an OPS won't tell you what kind of drug policies or actions might alleviate the tsunami of preventable deaths that constitutes the overdose crisis, but it might make you curious enough to explore the issue beyond right-wing calls to incarcerate or institutionalize drug users. Indeed, in our media-saturated world of intense distraction and image spectacle, the pause that intense listening offers might very well provide a provocation to reconsider so very much of what we thought we knew.

Bibliography

Nash, Kate. "Virtual Reality Witness: Exploring the Ethics of Mediated Presence." *Studies in Documentary Film* 12, no. 2 (2018): 119–131.

About the Author

Brenda Longfellow is professor in the department of Cinema & Media Arts at York University. She is a co-editor (with Scott MacKenzie and Tom Waugh) of the anthology *The Perils of Pedagogy: The Works of John Greyson* (2013) and *Gendering the Nation: Canadian Women Filmmakers* (1992). Her documentaries have been screened and broadcast internationally, winning several prestigious awards.

23. A Much-Needed Conversation with Women Working in VR

Edited and moderated by Caroline Klimek

Abstract: This interview explores the underrepresented voices of women in the virtual reality (VR) industry. Through interviews with five women actively working in VR, including artists, community researchers, and industry professionals, the discussion sheds light on their experiences, creative processes, recurring themes, and ongoing challenges. From Linda Zhang's exploration of social dreaming and community co-creation to Michelle Cortese's efforts in making virtual spaces safer for female-identifying individuals, each woman brings a unique perspective to the conversation. The chapter delves into their candid discussions about the trials and tribulations they face in the VR industry, aiming to inspire and empower other women considering entering this evolving field.

Keywords: virtual reality, women in technology, diversity in VR, gender disparity, industry studies

I have been studying virtual reality (VR) for seven years and have noted, time and again, that the voices of women—and particularly racialized women—are missing from conversations about the VR industry and its evolution. When working in virtual reality, women and female-identifying artists and professionals face many barriers, such as income equality, discrimination, and accessibility issues. This chapter started as an honest and open conversation about how women were making VR work. I interviewed five women who are currently working in VR about how they are creating work, the themes they return to, and the challenges they continually face. These women came to this conversation with different perspectives and agendas. Some are artists (such as Nadine Valcin and Cat Bluemke), others are gamers (Paloma Dawkins), community researchers (Linda Zhang), and individuals who work in the industry, at Meta (Michelle Cortese).

Bédard, Philippe, Alanna Thain, and Carl Therrien, eds. *States of Immersion Across Media: Bodies, Techniques, Practices*. Taylor & Francis Group, 2025
DOI 10.5117/9789048558766_CH23

Linda Zhang is an assistant professor in the School of Interior Design at Toronto Metropolitan University (TMU). She creates VR with her community in Toronto's Chinatown and is interested in social dreaming and community co-creation, and how they allow her to think critically about culture, meaning, identity, and memory. I was following her *ChinaTOwn Project* and its many iterations over the last three years. **Michelle Cortese** is currently the Design Operations Lead at Meta and teaches at New York University (NYU). I came across her book chapter "Designing Safe Spaces for Virtual Reality" and needed to hear more about her work on making virtual spaces safer for female-identifying people. **Cat Bluemke** is an artist who works alone and as part of the collectives Tough Guy Mountain and SpekWork Studio. One of the earlier VR makers in Toronto, she explores social and working relations with a focus on precarious, feminized, and domestic labour. I've seen her work at Inter/Access and at Trinity Square Video in Toronto. **Paloma Dawkins** is a virtual reality and video game artist. I was familiar with her *Museum of Symmetry* (2018)—one of the first VR games—produced with the National Film Board of Canada (NFB). Paloma creates worlds that incite creative thinking and wonder, inviting audiences to be inspired by otherworldly scenes and scenarios and hypnotizing sounds and visual details. **Nadine Valcin** is a producer, writer, and director, and has gotten into virtual reality after completing a Master of Fine Arts (MFA) at the Ontario College of Art & Design University (OCAD University). She is working on a number of VR projects, including *Ghost of Remembrance*, which tells the forgotten history of slavery in Canada.

Our conversation was full of honesty and humour as these five women described the trials and tribulations of working in VR. Only by talking openly about the challenges and barriers women face working in VR can we hope to overcome and move beyond them. I hope it inspires any woman who is thinking about entering the VR industry.

This interview has been edited and condensed for clarity.

Caroline Klimek (CK): What was everyone's first introduction to virtual reality?

Paloma Dawkins (PD): My first introduction to VR must have been a headset, a cellphone, and a cardboard box type of thing. Me and my friends were starting to learn Unity and we thought it'd be cool if we made something. And then I splurged for this plastic version of the headset, which had magnifying glasses, and it made it so much more effective. And I just explored it like a 360-degree video and I thought it was so much more compelling than the things that I had seen. I was also going to random game festivals and

I was playing VR there… I had never really played the console stuff until later, when I started developing for HTC VIVE. I thought: "There must be something better." And I found it. I think the NFB was the first place that I was introduced to serious VR systems like the VIVE and Oculus.

Cat Bluemke (CB): My partner [Jonathan Carroll] and I were interns at the Holography Lab, an OCAD University-affiliated studio in the 401 Richmond building in Toronto, but I don't think I had access to a headset until that lab. And then we were able to get our hands on a developer kit from Trinity Square Video through a production residency. And so that's when I started to learn Unity and the first time that we had really put on headsets at all. I remember getting super nauseous. I couldn't work in it as long as Jonathan could. I remember reading at the time that was about gendered bodies reacting differently to those early headsets. But I remember feeling that was really, really true. My introduction was putting our own glitchy work on this old headset that I couldn't really experience for very long because I felt sick.

Linda Zhang (LZ): I was working in Berlin, at Studio Olafur Eliasson at the time, and we were doing some VR-related things in the office. I think one of the first things they did was a 3D scan of the whole office and then [they put] it in VR and then you put on the headset. And then … the chair would be misplaced. So, you'd try to sit in the chair and it wouldn't be there, but the table would be in the right place. It was at this early stage of "we don't know what to do with this yet." I also got a 360-degree camera around the same time. And back then, there was nothing you could really do with those images. Nor did we necessarily understand those images as VR because they were flat 2D images with 360-degree projection mapping. So, for a long time, I would take all these photos and then there was no way to share them because you couldn't put them on YouTube. You couldn't put them into Facebook; all the things that they are now integrated into, didn't exist then. Even a lot of the VR platforms now, where you can put in 360-degree images and stitch together an experience, that didn't exist back then. It was really early. I think it sort of happened around me, even though I didn't necessarily recognize it as such in the beginning.

Michelle Cortese (MC): Technically, my first introduction to VR was hearing about the Virtual Boy in the nineties, the sort of failed Nintendo product that gave kids seizures and had to be taken off the market. My first introduction to it as a person who would come to work in it would happen closer to fifteen years later. About the moment that Facebook bought the Oculus brand, it was something that I was more introduced to—I was in graduate school at the time at NYU and it was just an object that we had around. I actually didn't find it super interesting, to be honest, at that exact moment in time, although I had been studying in a class about Unity development and I was interested in building virtual spaces. But I don't think the idea of building virtual spaces and virtual reality had actually quite aligned and didn't do so for me until when I was working in experiential advertising, when folks started to ask for VR projects… That was the moment that it started to actually connect.

Nadine Valcin (NV): You know people ask me that all the time, and I think one of the most striking experiences for me was the first year that Hot Docs had a VR component in 2015. I remember going to Mirvish Village (which no longer exists) and putting on a headset. Felix & Paul made this trilogy called *Nomads* (2016). It was problematic in a lot of ways, as it's kind of ethnographic. To me, the most striking moment was the one about the Maasai… There are people from a village and they're all in a circle, and you are placed at the centre of the circle. And no matter where you look, the people are looking right at you. And to me, it was the power of feeling surrounded by these people and also the power of the gaze. That has stayed ingrained in my memory.

What is it that compels you to this medium both creatively and as a spectator?

CB: I think initially it seemed really cool. As somebody who grew up using the internet in its Web 1.0 form and was very much attracted to sci-fi aesthetics, it seemed really neat. The more that I've engaged with it, the more dystopian it has appeared. And, certainly, in the context in which I have been working in my overall practice, which is really involved with concepts of work and play and how they present themselves through different technologies both historically and now.

But, really, it was virtual reality I was interested in because it was linked to job training and escapism. [And] also, that idea of not being limited by your body anymore. [And] then also of course all the actual barriers that are present to accessing VR, whether that's getting nauseous on the headset or just the financial barriers, and certainly the barriers in developing anything for yourself.

I think it's definitely a complex medium that I enjoy coming up against the problems or the challenges that it presents, but then also the challenges within the discourse of it where a lot of people still want to see it as: "Oh, if you step into somebody's shoes through VR this will solve the world's problems." I'm still so hyper-aware of my own subjectivity in it. And I think its alienation, perhaps, rather than its integration or its ability to transport you away is really what I like about it.

PD: I just love being in a dream space where anything feels possible. And I love how overwhelming it is. And I feel like it's the only medium where that's possible. I had been developing games for a really long time, just 3D games. But it feels really limited. Where you're in the VR space, you're totally immersed and you're basically at the mercy of whoever has put you in. I love that vulnerability.

LZ: I think there's something really powerful about being able to envision something. Coming from an architectural background, I think there's always something about how we can make an image or how we can see what the design will be like. I can picture a cube in my mind and I can rotate it in any direction, and I can see it. It's been trained over many years as a skillset and it's not something that is really accessible to most people. It's actually a barrier. And I think virtual reality is incredibly powerful for that because anyone can visualize it. And so, a lot of the work that I do is around giving that access to community and on being able to build on the power of envisioning something collectively or dreaming collectively, like social dreaming.

Because in my field, we do use it a lot for envisioning and [it] is usually used by the developer to sell things and displace the community that I'm trying to empower. And that same opportunity that it [offers] to sell things I think can equally be used very powerfully to help the community envision and organize, and build communities.

MC: I do think that modular, more immersive, more body-based computing and immersive interfaces in your field of view are likely to be a large part of the probable next fifteen-plus years of consumer technology. To me, VR was always this way to train up in this new space. And the reason I was interested in training up in this new space is because I felt like I watched, in real time, the harm that Web 2.0 caused, the harm of the last ten years of technology and the way that people jumped in without thinking and designed things irresponsibly. And I got really upset about that and I really wanted to try to position myself at the junction point of a new medium and try to get in, intervene, and help those decisions be more thoroughly and vigorously questioned.

I've been working at Meta for five years. I was hired to specifically work on VR projects back in 2017, when things were quite rudimentary, and I have been with it since the moment that it made the small organization of people that ultimately became what is now Reality Labs, which is the big factory of all this XR stuff. I was incredibly disturbed from the get-go at the lack of consideration of the human factor in the design processes. I felt like design culture there was just basically "do something that causes the most retention, just smack an idea on the table and get it made," and it was just wrong to me. I wanted to fight against that. And because I'm in that social VR space, what hit me immediately is really "oh god, it's incredibly dangerous to put people embodied into these virtual spaces together and just sort of let them do whatever." We have to actually practise extreme care. So, this is the first area of huge consideration that really stems from that sensation, just seeing people suggest things like "Oh, you should just be able to pop into someone's private space however you want to," or "You should be able to leave things for people," or "You should be able to get as close as you want to people." There were certain things happening in VR that made me think: "This is explicitly wrong. Someone needs to explain to you why that is." The first bulk of my research is very much in that space.[1] It's very focused on social and spaces; what they mean, how they work, how we can make them feel safer and how we can make people feel more empowered and autonomous.

NV: I wanted to do some kind of interactive documentary about slavery in Canada. I'd been in contact with a former work colleague of mine, and he asked if I would be interested in doing a VR project. And I went: "Sure, why not?" So, we embarked on this VR thing and we ended up making this company and getting the Canada Media Fund (CMF) experimental stream grant to do a prototype for a project called *Ghost of Remembrance*.[2] The experience of making that prototype was a really steep learning curve. And of course, it took longer than we thought it was going to take. I guess because

1 See Cortese and Zeller, "Designing Safe Spaces for Virtual Reality."

2 The CMF's experimental stream grant is one of the major funds available for creators interested in pursuing virtual reality. The purpose of the fund is to encourage the "development of innovative, interactive content and leading-edge software applications for digital media platforms." See "How to Apply." However, all projects must be interactive. The CMF defines interactivity "as a meaningful participatory experience between the user and a product/technology (including but not limited to immersive technologies that engage and stimulate the user's senses to create perceptually-real sensations), or the user and other users as enabled by the product/technology. Projects that use the internet or mobile platforms to distribute linear content without significant interactive feature(s) are not eligible." "Innovation & Experimentation Program Guidelines 2022–2023."

it was CMF, it was partially interactive and partially a game, which I didn't really feel comfortable with in the end. And so, from doing that, I felt like I needed time to reflect and really understand how VR was different from linear and how I wanted to tell stories with it having some interactivity—but not too much gamification—and explore the tools, understand why and... really understand the technology better. So that's why I went and did the MFA and the project *Our Home and Haunted Land* (2021).

CK: Where are you finding inspiration?
CB: I love VRChat. I think that was such an inspiration and [a] revelatory space for me, as somebody who likes working with other people, because of its collaborative nature.

PD: I was finding a lot of inspiration from other festivals, other creators, and in the game world. These days, I don't have a lot of that. So, it has been really hard for me to find inspiration. But lately, I've been finding a lot of inspiration with my collaborators. I feel like it's really important to have collaborators. Also, I'm deeply inspired by cartooning, comic books, animation, and stuff like that.

MC: My inspiration point is really the fire within me of trying to protect this next generation of technology. That's the fire, that's the burning idealism that gets me to the place where I find inspiration for the research that I've done. The other is very weird, widespread reading. And that reading is all over the place. It's fiction, it's non-fiction, it's textbooks, it's philosophy. It's just anything that is slightly adjacent to virtual worlds or phantom interfaces, things like that. I've been reading *Reality+* by David Chalmers for a couple of months.[3] He's a philosopher who deals largely with virtual worlds. He was the guy who, [twenty years ago, wrote all the] philosophical theory on *The Matrix*.

NV: I think inspiration comes from anywhere. I don't know if you know the story behind the idea for my thesis *Home and Haunted Land*. It came a year before I started my MFA. It coalesced from a random conversation that I had on the streetcar with this elderly man from Guyana. I live in Parkdale, so I was on the Queen streetcar and coming home from work. And he said, "Well, you know, I used to work in the Candy Factory lofts. I used to work in this other shoe factory across from where the Drake Hotel is now." And

3 Chalmers, *Reality+*.

then we got to the Gladstone and he goes, “you know, Gladstone was the largest slaveholder in my country.” And I went, “What?” When I got off the streetcar, I Googled that. The Gladstone Hotel is named after William Ewart Gladstone, who was a British prime minister, but his father was the largest slaveholder in both Jamaica and Guyana. And when slavery was abolished in the British Empire in 1833 and all the slaveholders got compensated for their holdings, he is the one who received the largest payment, meaning he is the one who held the most enslaved people. How do we not know that history?

LZ: I find inspiration from other women of colour. And it is usually folks who are either the organizers or cultural theorists. And I draw a lot from their work and that has ... come before and [is] coming from the Asian community to the Black and Brown community. And so, I think a lot of the ways in which I understand speculating about the future, like speculative design, is largely shaped by the speculative fiction tradition that comes from Octavia Butler. And then the Octavia Butler Fan Club, which includes Adrienne Maree Brown (who also has her own fan club) as well as authors like Sara Ahmed, who are writing about affect theory and what this affective economy means. And I think virtual reality has a really amazing way of bringing some of those intangible things to light in this fully immersive experience. A lot of our work [as researchers] is also a way of trying to understand the world. And for me, then, obviously, it’s not surprising that a lot of my inspiration comes from other women of colour authors who’ve done that… I think most people expect you to answer who are the architects that I really like or who are the VR designers they really like. And for me, it’s almost never that. It’s always badass awesome women of colour.

How are women making VR projects? Who are they working and collaborating with? Where are they finding support financially, emotionally, and creatively?

MC: Sometimes, I think it’s worth dividing this up into where does work happen because I think it’s quite different. There are maybe four factions where VR work happens: tech companies, academia, advertising, and media/art. And those are the four types of people and institutions that are investing in VR. And I think the answer for each one of those factions is quite different. Obviously, tech companies all fund projects in their own interests. And then, [in] academia, you’re working on a tiny stipend and you’re just trying to get [expletive] done, and, hopefully, you have access to grants. In advertising, corporations again, and in arts, you’re looking at grants or self-funding. You have two scenarios: fifty percent of it is corporate-funded, based on corporate interests; the other fifty percent of it is basically self-sourced or grant-based, which obviously pushes the industry in a really specific direction where the tech companies and, to a lesser extent, corporations get to decide what this medium means and is. There's this tiny crossover point: film festivals kind of sit in this weird wedge that is sort of overlaid across all four areas.

PD: I've been finding a lot of support from other female game creators. But, honestly, I don't think that there's as much support for us as for other male creators. I feel a lot of the time I'm creating in a vacuum and I kind of rope my friends into a lot of my projects. I work with a poet a lot of the time, but she doesn't know much about VR at all and she's always kind of like, "This is cool but..." I get a lot of support from her just sort of through the ideas and writing dialogue and getting excited about stuff, but I don't have a lot of friends that I can geek out about the actual VR stuff.

CB: I've found myself surrounded by really supportive people, both the people I collaborate with and then also within these larger communities like Dames Making Games, Gamma Space, and Trinity Square Video. And I think that has been really critical... without being in this really absurdly "right place, right time" sort of situation in Toronto, when I was a student and I had access to certain computers. I think having those opportunities to stumble across the medium, especially in such a hyper-consumerist field (where you always have to buy new hardware and where you constantly need to upgrade your own knowledge of the software), is important... to bring more people into creating. I feel like the ladies I know making VR figure out a way to do it somehow, despite all the barriers.

But definitely, Canadian arts funding for it is huge. Having Canadian arts funding that allows for more projects that are less commercial to exist and be developed is also exceedingly key.

LZ: I should say that I don't even know that many women who are working in VR, so I think that it is not the most accessible. VR is pervasive in architecture and interior design in terms of selling things. It's something that architects use to show their projects to clients or that developers would use to try to pitch a project to future investors. And so, my access to it actually kind of came through my industry, as often happens in a lot of the tech-related things in my industry—tech is commonly just accepted and is sort of based on who paid for this development and what it's being used for. And most of the research happens there. [The] relationship between women in VR and BIPOC folks in VR is different. We can't really ever see that as neutral. They always say that a photograph can't lie or a camera can't lie, but a camera also can't tell the truth; it's up to the person who's using it. But I think especially in tech-heavy industries, which are predominantly male, white, and affluent, there's this assumption that what it's being used for is neutral, but it's a very small group of people who have very similar views who are shaping what it should be used for, who it's for, and who [it] is being used on. And I think that if you're not in that group, it's so obvious that there is a huge gap. There is a huge opportunity that is being missed, being able to do so many other things with this technology.

I'm fortunate to have access to it by being a researcher, and by being part of a university with so much institutional access. But I think if you don't already have access to an institution through the same kind of position that I have, that's incredibly difficult. [A] lot of my community work and research-based work is trying to bridge that kind of access. How do we share this technology that we have with others?

NV: I feel it's really hard because in some ways, you need to be on top of the technology all the time. How do you build something? And so, it's hard because obviously, you have to make a living. It's hard to find people who are interested in working on more artsy projects that aren't games. I wish there were other funds aside from the CMF, which stipulates funded projects must have an interactive component.[4] I always feel like I fall between Arts Council funding and then the CMF and Telefilm Canada, but there seems to be nothing in between. And, instead of being able to get funding to just do your project, it feels like there's so much administration and so much thought that needs to go into how you present it to fit in that box. Why is it that CMF has to have heavy interaction? Why can't you just tell this really amazing story in 3D? Why is that not a valid exercise? Part of the frustration comes from the way the industry and the funding bodies define projects; they don't have a lot of flexibility and they're not really geared towards artists or people who want to explore the medium differently.

What are some themes you find yourself returning to within your VR projects?

CB: Every project centres on that weird alienation that you get when you don the VR headset, that role that you take on as a performer. That's something

4 Nadine Valcin articulated this tension later in our conversation: "The CMF, the experimental stream asks you to have something kind of interactive in a way that implies some kind of gamification," which she views to be at odds with storytelling in virtual reality.

that stuck with me thinking about the vulnerability that you ask people to enter into when they put on that headset. And that also has affected how I want to engage with VR in gallery spaces, because when you ask someone to put on a VR headset, there is intimacy and there is vulnerability because you're asking someone to put something onto their own body and you're asking them to be isolated in this space, in the virtual space. How do we work within this medium of isolation?

VR is fun. And the themes that I find myself drawn to are always very medium-specific because the medium allows you to ask people to be vulnerable, goofy, engage in absurd spaces and situations, and always be hyper-aware of themselves and their own role in whatever you're asking them to engage with, like whether they're the intern or whether they're the passive observer or they're the person trying to get the sheet on the bed.

PD: Definitely genderless, androgynous, sort of complete self-expression beyond all of the constraints of reality. So, I feel like that's kind of what VR lends itself to. It's suddenly [as if] the rules are being defined by whoever has created this space. That's what really gets me interested, because I could totally free your brain from all of the notions of reality that you had depended on. And I feel like especially because my work is two-dimensional, it's effective at doing that. Where I feel like a lot of other VR projects are in 3D so your brain kind of looks at 3D objects and it's like: "Is that real? I'm not sure." And so, I feel like that kind of catches your brain. But my stuff is all—if you check it out, it's very 2D and clearly not real. Or at least that's my thesis, that I can use that to free your brain from the constraints of reality in a way.

NV: I'm really interested in memory. I remember history being so boring at school. But I realize how poorly it's taught. But how much of that is how history explains a lot of our structures and a lot of how we live now and how we tend to think that we live in an ahistorical society. And what happened? You know, "one hundred years before it doesn't matter." Well, yeah, it matters because this is why we're here. So, I think that's a theme that I'm interested in exploring and haunting, too.

LZ: I would say that it's all around re-imagining the present, which then touches on the future and the past because those are two kinds of guiding ways that we think about these things. I try to centre things around memory as a tangible way to kind of explain what it is. Or it's like you don't remember things as like a list, right? And so, I think my interest in VR also is being able to [do] that kind of world-building or collaging elements from the present. But it's also as soon as you put it virtually, that dichotomy between reality and virtual is built into [...] that technology. And so, people are able to suspend their disbelief with VR in a way where, if I make an architectural drawing, I can make a planet a section or something. People assume that's real or it's going to be built or it's real. And so, there's this nice way where we can sort of understand that it's almost like a dream or it's just like other space. And then it allows us to kind of bring in these associative qualities that memory has, and so to understand and reimagine our present together.

What are some of the biggest challenges women are facing in VR? How do you address the many barriers (both financial and physical to name two) and issues of access that working in the VR industry implies?
CB: There was that initial hardware barrier and then the barrier that had to do with—I can't remember the science behind it but something like centres of gravity with various bodies being more prone to motion sickness for women, and if you were wearing those old headsets. I remember this one piece I kind of almost passed out in the headset. I was looking, and then slowly my head fell on the desk really gently because I was losing total ability to perceive space. Also, when I talked to various developers who work in VR, the problems like Black women's hair, for example, or anyone with dreads or larger, big hair has trouble putting on the headset. Currently, my job—Jonathan and I are working on a project for Mackenzie Art Gallery that is attempting to make building to headsets and to mobile VR with Unity software easier. We're making templates for getting that to work both in gallery spaces and for independent curators and organizations. So, access is something that I'm thinking about every day, even if it's more so on the technical side of things.

PD: I feel like a lot of the time it's the women themselves. I see myself saying: "I don't know if I can ask for this much money," and then I end up finding myself in a hole because I can't afford the equipment and resources I need. And then I don't empower myself enough. So, that's something that I find myself falling into. And I'm finding ways to get out of that, but I don't feel a lot of support from people being: "No, girl. Go get your worth." Not a lot of people are doing that. I see guys saying: "You've got to ask for $150 an hour." No one's telling me to do that. Why is that? I'm way more talented than some of these guys and I keep seeing myself thinking: "I don't know if I'm there yet," when I'm [expletive] there. So, I want other girls to be like: "No, you're there. Go."

NV: I think funding is definitely a barrier, and especially then if you're trying to fund a project through arts councils unless you're doing most of the work yourself, I can't hire a developer for a $1,000 a day because it just won't work. It also feels like VR hasn't gone mainstream yet. Five or seven percent of people have a headset. And I also feel like if there were ever a moment in history where VR had the potential to really develop it was during the pandemic. People are at home doing nothing, why wouldn't they get a headset? They're not spending money on going to bars, they could afford a $500 headset. But even when you get on the Oculus Store, if you're not a gamer, you can spend an afternoon or two, but I've gone through all the content that's of interest to me on the Oculus Store… it feels like VR is at this very latent stage and it feels like it's been at this stage for a while. And I think until it really kind of flourishes, it's going to be hard to get financing.

LZ: I think one of the biggest challenges that worries me the most is: what's the long-term maintenance, upkeep, and sustainability of a platform, because they evolve so quickly? And how can I ensure that this is still going to be accessible or still useful in fifteen years? It seems in five years, everything is going to be so outdated or even if trying to upkeep I don't even know if they'll still work. That's one of the challenges: what's the longevity of the projects beyond five years?

MC: Look, we're outnumbered in most of the industries that I mentioned. Those are not places that are largely, particularly fun for women to be in. It's very important that we're there, but I have not had great experiences being a woman in all those places. So, you're coming in with the same [expletive] if you want to come into this where people will immediately reject, discount, and assume that you know nothing. Every room I've walked into in the last ten years of my life has involved me understanding that I have to prove myself immediately or else no one is going to take the rest of what I have to say seriously. I have to come in and rattle off a bunch of tech terms and be that person, and then we cross that hurdle that I'm not an idiot and then we can keep going. That is taxing. The privilege of those who don't have to live with it is that they don't have to live with that emotional tax on them all the time.

And then, you multiply that by the fact that VR is an embodied medium. And that is the thing that has made me align with so many of these things that seem very gendered, that seem to affect disproportionate numbers of women and female-identifying people, is that women's space is so much more regularly violated than so many other human archetypes spaces. It's a constant thing. And we have sent people into an environment in which rules have not been defined and people are being taken advantage of. And on top of all the issues, we have [people] in the tech industry working in this and how exhausting that can be, on the other end of it we're making products for the game industry for the most part right now, because we're still trying to get VR off its feet. We're still trying to get it moving. So, a lot of the early work has happened in the game industry, and will continue to happen in the game industry until VR becomes a much larger consumer product. So, until then, we're also stuck in an industry that is absolutely foul to women and has been for a very long time.

The tax on women in this industry, women who are trying to rise up in the ranks in technology, is one of the many contributing factors that makes it incredibly difficult and emotionally draining for us to work up in the ranks high enough to make the actual [expletive] decisions on these products. I have spent so much time in the company that I'm currently trying to make the decisions that I want to make. And I'm at a point where I have decent access and say, but I'm also still basically a mid-level manager. The road to me having a real say in anything is long and it is paved with women who have fallen off. There are almost no women in leadership positions where I am at. And the ones that you do see, there's not always an applicable pathway in itself, so you're stuck in a situation where it's hard to see a version of reality where you actually get to make the decisions. So, you have to keep begging men to take you seriously, which is why when I walk into a room it's such an important experience and constraint. Because at the end of the day, the barrier is the amount of agency, power, control, the ability that I actually have in the place that I am in.

What advice would you give to other women interested in pursuing VR?
NV: I would say don't be intimidated, it's not as difficult as you think it is. Find good collaborators and just go. There are a lot of workshops at Inter/Access and Dames Making Games where you can start to learn things. I would say if you're going to seriously do VR you need access to a very good computer and see a lot of VR.

CB: I think VR is super fun and it is a very interesting place for women to investigate as artists and think about the technology in general and whose presence is there, whose bodies are there, because it is such an embodied medium. I want to see women making VR work that speaks to those experiences but that also isn't afraid to take things from other spaces, references from other experiences that don't have to focus on the highly technological side. I really would like to see the fandom, the role play, and the world-building that exists within VRChat and queer coding in that space; I want to see that enter more art VR. I think this is a weird platform for us to experience. This asks a lot of your users, your audience, and you as an artist. And yeah, more thoughtfulness. And I think that will come from women and genderqueer folks using VR. I think that's where it'll probably come from.

PD: Definitely just ask your worth. Almost over ask because it's a better thing for your own mental understanding of what you're worth just to ask more than what you think. Because that's a game. Also, don't be afraid of making art because I feel like a lot of the time people are like: "This is such a risk, I've got to make something that people are going to like." But when you make actual art, then you're unique. Taking a risk is really, really important. And again, kind of overdoing it in the creative side I feel like… creates a more interesting world. Being able to create a universe around your idea is also a really cool way to pick out the parts that make your world unique.

LZ: I think it would be to trust your intuition, if something feels like it's wrong or it's off, but everyone else is doing it and they're telling you that that's how it is or how it's supposed to be they're probably wrong and you're probably right. And so, to kind of not silence that part of you that knows that something is off here or that it could be wholly other, it could be completely different than things are now and kind of trust that. And if there is something there, pursue that part of it and see where it takes you because your lived experience is important. Your lived experience tells you something that it might take you a long time to figure out how to say in words but is actually really powerful. And you will figure out what those words are by world-building in VR. You won't figure out the words first, even though people will want you to justify what you're doing. So do not let them silence you.

MC: Definitely do it. Before I say don't back down; if something feels wrong, it probably is. Don't doubt yourself. Feel the things you're feeling. Tell people who need to be told that something is wrong. Basically, ring the alarm. And if you don't feel like you have the power to do so, try. The worst-case scenario: you'll at least know what it looks like to try. That's how we make the next generation of technology better is people give a [expletive] and they'll try to stop the line if something isn't working. Just shut that [expletive] assembly line down.

Acknowledgment: A huge thank you to Cat, Michelle, Paloma, Nadine, and Linda for taking the time to participate in this conversation.

Bibliography

Chalmers, David John. *Reality+: Virtual Worlds and the Problems of Philosophy.* New York, NY: W. W. Norton & Company, 2022.

Cortese, Michelle, and Andrea Zeller. "Designing Safe Spaces for Virtual Reality." In *Ethics in Design and Communication Critical Perspectives*, edited by Andrew DeRosa and Laura Scherling, 49–60. London: Bloomsbury, 2020.

"How to Apply." *Canada Media Fund.* https://cmf-fmc.ca/our-programs/how-to-apply/.

"Innovation & Experimentation Program Guidelines 2022–2023." *Canada Media Fund.* https://cmf-fmc.ca/document/innovation-experimentation-guidelines/.

About the Author

Caroline Klimek holds a PhD in Cinema and Media Arts from York University. Her research interests include film festivals, media industry studies, digital archiving, and emerging technologies.

24. *Espace & Présence*: An Interview with Guillaume Perreault-Roy

Edited by Philippe Bédard

Abstract: Guillaume Perreault-Roy has been working in the fields of virtual and augmented reality for many years, both as an artist and researcher. He is currently creative director (and author) of the VR video game *Broken Edge* (2022). As someone who has worked hands-on with early immersive technologies (including some prototypes which have been forgotten by most recent histories of the medium), Perreault-Roy developed a set of pillars for discussing technologically mediated immersion. Our discussion served to highlight some of the fundamental factors in the budding fields of virtual and augmented reality, but also to take a look at where these technologies might one day go.

Keywords: virtual reality, immersion, technology, research-creation

How did you get into the field of immersive media?
Guillaume Perreault-Roy (GPR): Since 2014, I've been involved in augmented reality (AR) and virtual reality (VR) in a variety of roles. This was a critical period for immersive technologies and their evolution has informed my creative process. While I now work as a creative director in an indie video game company specialized in VR, I've previously worked as a researcher, interactive designer, digital artist, and teacher. I've juggled these roles and the technologies, alternating between AR and VR. This plural approach allowed a direct, hands-on appreciation of these technologies and a point of view informed by various perspectives. The potentials and limitations of immersive media are my main interests and I developed a frame of analysis for both.

Bédard, Philippe, Alanna Thain, and Carl Therrien, eds. *States of Immersion Across Media: Bodies, Techniques, Practices*. Taylor & Francis Group, 2025
DOI 10.5117/9789048558766_CH24

Back in 2014, I was working in a start-up where I initially designed an AR platform based on the first iteration of the Meta glasses.[1] My main interest, then, until 2018, was presence, since the notion of "immersion" was thrown around a lot at the time, often in a very abstract sense. That's when I undertook an initial research project, with the first iteration of my project *Espace & Présence*, and when I came up with the idea of the "three pillars of presence" as an analytical framework. The key point of these pillars was to establish common dependencies of the effect of presence, both in augmented and virtual reality.

The second phase of *Espace & Présence* (2019) was supported by the Société des Arts Technologiques (SAT), a major institution in Montreal that promotes research-creation processes. The way I approached research served my main goal: having access to the means, support, and space to validate the personal convictions I developed towards VR and AR.

How did you come to think of these key notions of immersive media?
GPR: After working with augmented and virtual reality for a year, in 2014–2015, I felt frustrated with the language used in the field. It seemed to me that most publications evoked the immersive quality of technologies or devices without any point of reference. I observed that between a vague description of immersion's potential and of its technical specificities, there was a gap in the appreciation, measure, and evaluation of this very potential.

The "eureka" moment for me was when I took what I had learned from working with AR for a year and applied it to a VR headset. At the time, we were testing out a lot of tracking technologies for AR: volumetric cameras, near-field communication, accelerometers, pattern recognition, infrared cameras, GPS, and visual anchors. We initially thought these were a prerequisite for AR, but not for VR, since most VR experiences at the time assumed a fixed user. But when we tried using a tracking prototype with one of the first mobile VR headsets (the Samsung Gear VR), we found out that the effect of walking in an empty virtual room layered over a real room was spectacular. It was by far the most powerful sense of simulation I had experienced at this point. We had essentially stumbled onto room-scale VR.

It's around that time I started a research project focused on the dependencies of the effect of presence. It's also at that point I conceptualized my first frame of analysis: the "three pillars or presence."

1 This Meta (2012--2019) is not to be confused with the rebranding of Facebook as Meta in the fall of 2021.

What are, in your experience, the key elements of immersion?

GPR: The three pillars of immersive technologies relate to three key aspects of the device: the display, the orientation tracking, and the position tracking. While each of these aspects is important, I believe they truly start to shine when we consider how they impact the user's sense of presence. Comparing the weight of each pillar in the user's effect of presence brings some lucidity. From my point of view, it helps avoiding appreciation pitfalls, as it brings a sense of distance. It allows us to separate content from its technological frame, to compare VR and AR objectively.

For example, the bump in quality between VR headsets from 2015 and 2021 is less impactful than the ability to accurately track the orientation of the user's head. Displays have come a long way since then, but a 2015 headset with orientation tracking would offer a greater sense of immersion than a 2021 headset would without it.

Orientation tracking, the second pillar, is often overlooked because of its accessibility: a wide range of common devices, from phones to video game peripherals, are equipped with this technology. That said, the weight of orientation tracking in the effect of presence was made evident by certain kinds of immersive content, namely 180-degree videos, since seeing the edges of the frame shattered the illusion. The notion of coherency comes to light in this example: meeting the boundaries of the simulation reinforces the user's behaviour to avoid them.

The third and final pillar is position tracking, which refers to the ability to track the user's movement through space rather than just tracking where they are looking. We perceptually construct space while navigating; that is how we understand space. Unfortunately, most contents only use positional tracking as a coherency factor, not as a navigation mechanic. In my research for *Espace & Présence*, I was able to observe how people were affected by the ability to freely navigate in a virtual environment even though the visuals were minimalist and low-resolution.

This third pillar is the most important of the three because it's the hardest to appreciate. As in the example of the 180-degree video, the user will avoid meeting the boundaries of the simulation. Those are typically narrow, because of technological or real-space limitations. This limited range in most of the available contents introduces a bias: it masks the perceived weight of the freedom of movement in the immersive technology experience.

An important thing to remember, however, is that the strength of a single pillar can't make up for the weakness of another. The three pillars define the characteristics of the technological frame separately from the contents it supports. It allows a comparison between AR and VR: in my opinion,

AR (which is publicly available) does not have the immersion potential of VR because the displays are not mature enough. Ironically, the high-end display of Hololens (Microsoft, 2017) (with its narrow field of view) and the ultra-accessible phone-based AR both lack maturity as immersive displays.

Have new technologies affected your thinking around these three pillars?
GPR: Of course. When the Oculus Quest was released, in 2019, its positional tracking made it my main working platform. I wasn't limited by the size of [HTC] Vive's tracking area (4 square meters) or by the narrow field of view of the otherwise amazing [Microsoft] Hololens. I reactivated *Espace & Présence*, using the architecture of the SAT as a framework. My intention was to validate the potential of immersion of an experience based on a preexisting physical space. Again, the experience was minimalist, aesthetically speaking. But presence doesn't rely on realism or on high production values.

The Quest 2 is a good example of this. It might not have the rendering capacity of a personal computer, but the fact that it is tetherless and able to track movement in larger spaces makes it a very powerful device. We see just how the maturity of the three pillars holds more weight than graphical fidelity alone.

There's a very interesting comparative in the history of video games: in the early years of three-dimensional game worlds, in the 1990s, the layouts of controllers were very varied and mostly bad. It took many years for a convention to emerge, allowing us to optimally navigate the virtual worlds. Basically, each hand manages one pillar on all controllers (movement with the left hand, orientation with the right). We are now in the same process in AR and VR, realizing the weight of orientation and position in the immersive experience.

What are the limitations of VR you would deem worthy of a designer's attention?
GPR: The first problem would be navigation, i.e., moving around in the virtual reality space. The second is what I commonly call "one-way physics": the real world can inform the virtual one, but not the other way around. The third problem is isolation: the relation of the user to the real world is obstructed by the virtual one.

Navigation is a very peculiar problem in VR: using navigation conventions derived from video games results in discomfort for most users. Some users can build a tolerance (the common expression is "having VR legs"), but it is a tricky process since resisting motion sickness (staying exposed to the stimuli) can make its effects last longer, possibly hours. Several navigation models have been created for VR with the objective to find a balance between comfort, immersion, and function. There's teleportation and sliding, but the best model is free roaming, where the user naturally walks through a virtual world layered over a dedicated space. The main issue with free roaming is the need for dedicated spaces and for content designed specifically for these spaces.

Meanwhile, the concept of one-way physics describes the idea that the relation between physical and virtual realities is not mutual: the real informs the virtual, but the virtual doesn't inform the real. In other words, it's impossible to simulate solid, interactable objects in VR. One solution to this problem has been to introduce coherent, solid objects in the simulation. For example, a real table can be placed in the position of a virtual table. This strategy is very demanding, however, as every solid object has to be precisely placed in the position of its virtual counterpart and every moveable object has to be tracked. It implies a lot of resources directed to a single aspect of the simulation, which introduces its own set of limitations.

The final limitation is isolation. While isolation helps ensure the user is fully immersed in the VR experience, it also means they are disconnected from reality. For one, this means that users must use a secure, dedicated space, since their awareness of their surroundings is forfeited. Isolation also brings up issues in terms of social conventions, since most social roles—professional, familial, etc.—assume a certain degree of presence and attention which would otherwise be monopolized by the experience. It's interesting to see how some of the criticism that VR has faced in this context has also been directed to other emerging forms of entertainment. From books to television, video games, smartphones, and now VR, there's often been a stigma attached to immersivity, presence, and the inevitable isolation from one's immediate surroundings.

Designing for VR is a very demanding challenge. It requires great lucidity with regard to the layers of technology and contents and on their effect on the user's experience. It is still a pioneer's work, full of exciting discoveries.

How have you experimented with these three pillars of immersion in your own work?

GPR: *Espace & Présence* (phase 1) was my original research project from 2015. The objective was to measure the different components of the effect of presence and to establish dependencies for that effect. The project was a virtual path of five interactive rooms based on the 8m x 5m room scale lab I developed as a prototype. Each room explored different dimensions of navigation in virtual environments. The first room was about navigating the room freely with guided interactions. The second introduced a confronting figure: crossing a narrow bridge. The third shrunk the scale of the space dramatically: walls closed in and users had to find their way out of a maze-like structure by following a line. The fourth room expanded the space, offering a 640-square-metre walkable space. In the fifth room, boundaries shifted the axis of gravity, allowing the user to walk on the walls and ceiling of the virtual room. That final room was the inspiration for my next project, *Le Vertige des Autres* (2016).

Le Vertige des Autres is a minimalist, contemplative path; a scenography presenting the sketches of Jeannette Perreault, a renowned painter in Quebec and my mother. The project was developed as part of a research-creation residency at SAT, which I was offered after some of their representatives had tested *Espace & Présence.* The path in *Vertige* is based on an Escher-esque interactive figure: following a charcoal line, the user makes one step in the void and a switch in gravity seemingly allows them to walk on the wall. The main observations after five years deploying this piece are that: #1) users get vertigo, but never motion sickness; #2) after the first gravity flip, the user loses any sense of their own position and orientation in the real world, but their position in the virtual world stays coherent; #3) the user doesn't notice they are walking in a four-square-metre space: the path, designed to fit the available space, seems more open than its narrow boundaries.

The project was part of the original launch of the VR space at SAT in 2016 and has since been presented in many events, more recently at the Festival International d'Art Vidéo (FIAV) in Casablanca (2021).

A second phase of *Espace & Présence* came about after the launch of the Oculus Quest, when I realized it combined the three pillars of VR: decent display and orientation as well as a robust and wide tracking system. So I launched a second research-creation residency at SAT (2019). This new project allowed users to experience a virtual version of the actual space they were navigating. This project allowed me to validate my theory and communicate it to cultural institutions: the key to exceptional presence is dedicated space. The barriers of technology and content production are falling. Space is the key to presence.

What are your thoughts on the future of immersive media?
GPR: Immersive media are still in their infancy. The bases of a mature vocabulary are actually forming in a collective effort. We can compare this to video games and cinema, and the insight of their own emergence as mass media. Those frames of analysis are pertinent now and in the near future because they support the rapid changes in the field and the evolutions that

are to come. VR and AR will gradually merge as display maturity advances (and the use of actual space in VR is arguably part of this merging process). I predict that, in the near future, from the perspective of digital creators, real space will become a canvas, a range of possibilities.

About the Author

Philippe Bédard is an independent researcher with a fascination for moving image media and their technologies. As part of his postdoctoral research, he focused on virtual reality, with an emphasis on the notion of empathy. Finally, he has worked extensively in and around the immersive media industry in Montreal, including work with Xn Québec, Festival du Nouveau Cinéma, the National Film Board of Canada's interactive studio, and SODEC.

25. The Ultimate Potential of Immersive Media: An Interview with Kent Bye (Host of the *Voices of VR* Podcast)

Edited by Philippe Bédard

Abstract: Kent Bye has been working tirelessly over the last decade documenting the emergence of virtual reality through his *Voices of VR* podcast. Over hundreds of episodes and countless hours of interviews, Bye has been able to offer a glimpse at the continuing development of this nascent medium and the related technologies. Topics discussed range from festival coverage and interviews with artists to debates on ethics, privacy, and philosophy as they relate to immersive media writ large. Notably, *Voices of VR* serves as a snapshot of the most significant projects released in a given year, thereby serving as a living archive of projects which most people might never get to see because of ongoing issues with distribution. *Voices of VR* is a must-listen for anyone invested in this field, making it a priority for us to include Bye's perspective in this volume.

Keywords: extended reality, immersive media, *Voices of VR*, process-relational philosophy

For many avid listeners, myself included, you have become a key figure in collecting oral histories and recording the development of this nascent field of immersive media. How did you find yourself working on this topic? More specifically, what led you to undertake this work of documenting VR and immersive media?

Kent Bye (KB): Back in high school, I had an interest in working on film and media projects, but I was also very apt in STEM (science, technology, engineering, and mathematics). Even though I initially opted to go the engineering route, my career has always straddled this dialectic between art and science.

Bédard, Philippe, Alanna Thain, and Carl Therrien, eds. *States of Immersion Across Media: Bodies, Techniques, Practices*. Taylor & Francis Group, 2025
DOI 10.5117/9789048558766_CH25

For example, I was active in the very early phases of videoblogging (before even YouTube had launched), documenting the media's coverage of the war in Iraq for a documentary called the *Echo Chamber Project*. I ended up talking to a lot of journalists, which allowed me to understand what the limitations of journalism were at the time in terms of having to reduce complex events down into a story as opposed to sharing unfiltered information. That's when I started doing more podcasts. I would go to different conferences and just interview people. I found that I was able to understand what was happening within these very technical communities simply by doing over forty interviews.

When virtual reality (VR) came along, I saw an opportunity to repeat that process: I could talk to the people who were the closest to the medium to understand what they were saying about it. The more I spoke to the artists, the more I started to understand what the medium was about and where it was all going. At the same time, within the first couple of weeks of owning my Oculus DK1, I had done a game jam with other creators. It was through that process that I started trying to get a sense of what the design affordances of this new medium were.

Any experience of VR is necessarily subjective. Can you speak to this tension in your work between discussing the piece as it was made by the artist and the very subjective experience that VR entails?

KB: My preference is always for people to see the piece before they listen to the conversations, but, as you know, that's not always possible. Immersive storytelling is a new genre—a blending of narrative and interactive components—that we're still trying to understand. Because of that, it's important to listen to what the artists are doing. At the same time, I have to assume that most people will never have the chance to see all these projects, because most will never get a wider distribution. As I think about the evolution of new media, I see that it goes through four different phases: 1) a new technology is made available; 2) the artists adopt it and try to push the limits of what it affords; 3) those pieces of art then need to get distributed and into the hands of the audience; 4) lastly, you have the audience watching it. These four phases create this feedback loop that helps the medium as a whole to develop. In terms of immersive storytelling, the distribution phase is what's currently missing.

Part of my process is driven by this lack of distribution. I had to find a way to talk about these pieces while assuming that people have not seen them. They're like one-off performances that I had an opportunity to see. So, I have this dual role of not only trying to have a conversation with the artist about their creative process, but also of documenting my own phenomenological experience to make sense of it, both for myself and for those who might never see the project.

I also have to account for the different versions of a project. Unlike a film, which is essentially a static object that you watch without being able to change how it unfolds (unless you get into interactive films like *Black Mirror: Bandersnatch* [David Slade, 2018]), with interactive media, your agency, your participation, and your situated knowledge are a key part of the decisions and choices that you're making. This means you may have a wholly different experience based on the actions that you took versus another person.

Part of the challenge of reviewing immersive media is that, if there is interaction, agency, or different endings, you may not even know about them unless you talk to the creator. A big part of this is having a sense of suspended judgment. Of course, I have an opinion, but I try to the best of my abilities to suspend my judgment until I have an opportunity to see what the artist's intentions were.

There is also a certain degree to which different genres in different media set up certain expectations. But since we're in this phase where we're forging new genres, we don't always know quite what to expect. How do you evaluate a given project against other experiences if it's the new frontier of a new genre that hasn't existed before?

How do you deal with the challenges of reviewing experiences that are unlike anything we've seen before?

KB: I tend to look at these questions in terms of different qualities of presence, whether it's active presence and agency, mental and social presence, emotional presence, or this sense of embodiment and environmental presence that fosters a feeling of deep immersion within a piece. VR as a medium pulls in different modalities from different media. It's pulling agency from video games, mental and social presence from web design, human computer interaction, and social media generally, but also the more literary use of language. As for emotional presence, VR is borrowing from film, lighting, and music. Finally, with embodied and environmental presence, it's pulling in from the traditions of theatre, dance, architecture, as well as different embodied practices that are trying to modulate your sensory experiences.

These qualities of presence are one dimension, but there's also the context of the piece, the character, the story structure, and so on. I tend to look at these pieces through the lens of quality, context, character, and story. Through all that, I'm also trying to have my own experience.

Since I've seen thousands of VR experiences now, I'm able to get a sense of what's novel or what's new. What are they doing that's different? It's also a question of trade-offs. If they do more of one thing, it may be taking away from something else. So, I try to understand the overall sense of the experiential design. As someone who is watching these experiences, I'm trying to pay attention to all these things. Sometimes, I don't know why they made a decision until I have an opportunity to talk to the creators themselves. That gives me a lot more context for how a piece came about. And since I'm in a dialogue with the creator, I'm able to have them explain their process, but also ask them about the things that were the most striking to me as I watched their experience.

Your work takes the form of long-form podcast interviews. Can you explain why you chose this format, rather than video or written interviews?
KB: I think the way I do podcast is different than anyone else does it. If it were up to me, I would be releasing a daily podcast. It probably would be too much, but I feel that in order to keep up with an emerging media, you almost need to keep to that pace. Just in the year 2019, I went to eighteen different events. That's an event almost every three weeks. There was so much stuff that I felt was important to be present for and capture. But then to digest it, edit it, and put it out makes for around four to five hours for each episode, which ends up being quite a bit of work. I can record things faster than I can get them out!

Part of it, then, is just the post-production process, but the other part is: how do you measure what's relevant or emerging in the moment? What's novel? What's interesting? I'm always trying to keep abreast with what's happening in the moment, but that also means not releasing episodes I have recorded until they become more relevant (when the project receives wider distribution, for example). If it were up to me, I would never have to make that decision. I would just release everything right away. But, at that point, it's too much information to consume. I've tried my best to find a balance between this dual role of real-time journalism and oral history. It's difficult to do both at the same time, both because of time pressures and because of how much attention people have as they listen.

As I move forward, I'd love to explore using virtual reality and other technologies. For instance, could I use Artificial Intelligence (AI) transcripts to help people easily digest all these different things and find other related topics? A lot of the tools I would need don't exist yet. So, I either have to invent them or wait for someone else to create that technology. I'm kind of at the limits of what the technology affords me and where people's attention is.

What would making these interviews available in an immersive format entail? What would immersive media allow you to do that a podcast doesn't?

KB: This is a good opportunity to talk about each of the different mediums and their affordances. With a podcast, there is a lot of spoken language. You're able to go into a lot of abstract concepts and ideas, which fits under the idea of mental and social presence. If we focus on language, videoconferencing is actually very good at offering a certain fidelity of emotional presence, because you're able to see the other person's face and their different body language. Finally, if you look at film, you have all these different ways of building and releasing tension through visual editing, lighting, and music. What's also interesting is the idea of using the medium of film to talk about film. I can also imagine using a video game to discuss the affordances of video games, such as agency.

Now if we get into the spatial medium of VR, you get into both embodiment and architectural aspects. There are elements of world-building or world design that are difficult to translate into other (2D) media, because you actually have to be situated within that context to really understand what's happening within it. So, when I think about using the medium of VR to talk about VR, I'm imagining being able to take a 360-degree video or photo sphere and use it to point things out, almost as if you were on a guided tour. It would be difficult in any other medium than VR itself to call attention to spatial design or to the way a piece uses the modulation of space over time to convey a certain emotion.

Would you say that VR is better for this because it encompasses aspects of all these other media (images, sound, spoken content, text, but also spatial dimensions, etc.)?

KB: I think some people have different elemental centres of gravity. Where one person might favour deeper kinaesthetic or embodied experiences, another might be a visual learner, or they may need a story before they care about a given topic. Some may just need to read a book, while others might need to express their agency. This comes back to the various qualities of presence. Each person has different temperamental preferences.

Part of the challenge of VR as a medium is that it integrates all these different modalities from different design disciplines. When you think about making a film, there are distinct phases of pre-production, production, and post-production. But when you get into interactive media, it's much more iterative. It's more of a cycle where you don't really understand what the interaction is going to be, so you have to experience it yourself or have it tested by users. There needs to be much more iteration and prototyping for these different experiences.

Eric Darnell says that you either passively receive a story (e.g., in a time-based medium, with a storyteller who is trying to build and release tension on a given timeline) or you have interactive media, which throw out the timeline and focus on the user's agency and interactivity. I like to think of this opposition through the metaphors of the yang and the yin, where the former is centred on the individual and their actions (their agency going outwards, making choices, taking action, etc.), while the latter is focused on listening and receiving sensory experiences from the world. The existential challenge of interactive media is "how do you allow the user to have this agency within the context of the story, but also allow them to still communicate a story in a way that feels meaningful, that has the trajectory?"

You often mention in your interviews the "experiential framework" that you have developed to think through immersive media. What disciplines and concepts did you borrow from when building this framework?
KB: It was an evolutionary process that was fuelled by the fact that I have been able to speak to a lot of people across a wide range of disciplines and domains. I've had conversations with mathematicians where I tried to look at their core philosophical frameworks. I've had the chance to talk to quite a number of different neuroscientists about embodied cognition. I've also looked at Chinese philosophy and process-relational philosophy, but I find that there's an elemental centre of gravity throughout these fields that brings me back to Jung and his archetypal psychology (as well as Richard Tarnas's work expanding this approach into cultural analysis).

My framework is really a fusion of all these approaches, at the basis of which is this idea of the phenomenological experience of presence and the different qualities thereof.[1] I'm also influenced by how Jung took inspiration from Galen [of Pergamon]'s temperaments to translate different processes into elements like fire, air, water, and earth, which represent intuition, thinking, emotion, and sensory experience, respectively. All this goes back to the seeds of Western philosophy, with the elements being the basis of all reality (with Empedocles).

There's a book by Grant Maxwell titled *Integration and Difference: Constructing a Mythical Dialectic,* which goes through thirteen different thinkers who have worked on the idea of "difference," from Spinoza, Leibniz, Jung,

1 For a more in-depth exploration of this framework, see Bye, "A Primer on Presence, Immersive Storytelling, & Experiential Design."

and Nietzsche to Bergson, Whitehead, Deleuze, and Isabelle Stengers.[2] What he's trying to trace is that most of the different forms of causation (starting from Aristotle's efficient, material, formal, and final causations) have been ignored in modern philosophy. Maxwell points to Whitehead who said that because of the dominance of final causation and destiny during the medieval period, the modern mind has eliminated these other possibilities. But, when I retrace these different philosophical strands, I see that there's a real emphasis on formal causation (meaning these archetypal potentials, eternal objects; these realms of ideal forms that are realms of possibility and potential). Both Bergson and Deleuze talk about the virtual in terms of potentiality. There's a difference between what's possible and what's actual.

What's interesting with virtual reality is that you can explore these realms of potential and these possibilities, which become actualized in some ways. There's this dialectic between creating realms of possibility and what's been actualized into the actual narrative and story. This dialectic has been around in interactive media for a long time, but as we start to integrate that into the narrative, we can learn a lot from these archetypal potentials from Jung, or even Shelling as well as Nietzsche. Maxwell traces these lineages, very much inspired by Richard Tarnas's *The Passion of the Western Mind* (1991) and *Cosmos and Psyche* (2006).

There are also certain ways that ethics has led me to this. In 2016, at the Silicon Valley VR conference, I started to really dive into these ethical questions and to map out different contextual domains. Context ends up being very crucial as a concept for privacy, because privacy is very contextual as to whether you're sharing information or not. Helen Nissenbaum's theory of "contextual integrity" starts to elaborate some of that. Having a fixed taxonomy of contexts allows me to really understand how a piece like *The Book of Distance* (Randall Okita, 2020), which is taking you through a variety of different contexts and embedding you into those context. As you're embedded into those contexts, you have things like embodied cognition—that is, relational dynamics that have associations which the medium is able to communicate at a deeper level, namely by putting you into the shoes of these different characters.

For any approach that I have, there's a multitude of different types of frameworks. Even all these things I'm saying, you can start to kind of break them down into feminist theory or look at them through the lens of race and decolonization. So, you have all these things together, mashing together what I try to describe as this quality, character, context, and story. Those different lenses have helped me create a memory palace. As I go through

2 Bye, "#1147. Thirteen Philosophers on the Problem of Opposites."

an experience, I'm looking at things, and I have a language for these things, all the while paying attention to each of these things, and seeing how they're all interrelated in some ways. And then, through the process of talking to the creators, I'm able to elaborate some of those different thoughts as well.

As we speak, we're still in the early days of this revival of VR and we're currently seeing a lot of influence from different pre-existing media. What is your perspective on VR's future? Do you think it will ever become its own medium?
KB: I think it's an inevitability that it's going to be the next medium for capturing and sharing experiences and stories. For me, it's not a question as to "if," and more of a question of "when" it is going to be a dominant form of media. "Dominant" in the sense that it's the most immersive and visceral. Of course, if you compare different media, each will have its own affordances, and some people might prefer to read the book, or to watch the movie, or to play the game version of a given story. But there is going to be this next option, this next modality, which is to be the most visceral way of creating memories by being embedded into these story worlds.

So, I don't have any question as to whether it's going to happen. It's more a question of when all the different parts will come together. I should also say that there are going to be other technologies (AI, 5G, etc.) that are going to enable other aspects of this medium. We've seen a lot of stuff with AI art generation (being able to speak and to create an image, for instance). This is getting to the heart of what the holodeck was in *Star Trek*. For me, this also goes back to an idea from Jung and his archetypal psychology of the unfolding of consciousness. We are going to be provided situations that will allow us to express these different archetypal potentials and maybe learn something deep about ourselves.

I often come back to a quote by Robert McKee: "True character is revealed in the choices a human being makes under pressure—the greater the pressure, the deeper the revelation, the truer the choice to the character's essential nature."[3] When we watch a film, we're watching someone else make choices and take action, which reveals their character. What VR does is create different experiences and stories that *we* can go into. *We* are put under pressure, make choices, and take action. As long as we suspend our disbelief and really feel immersed and present, then those choices and actions are going to reflect the actions we would really take if we were actually in this situation in physical reality.

3 McKee, *Story*, p. 101.

The potential there is that we're going to be walking out of these different experiences having learned more about ourselves.

How do you see the role of the body in these immersive experiences?
KB: A common theme that comes up again and again when I speak to neuroscientists is the concept of embodied cognition, meaning that cognition is situated within the body but also in the environmental context of the world around us. If you look at the *Stanford Encyclopedia of Philosophy*, there is a whole entry on the "computational model of the mind," and there's also an entry on the "embodied cognition model of the mind." The computational model tends to think about the mind as this abstraction, as concepts totally dissociated from a body. As you look into the research on embodied cognition, there are ways in which even these abstract concepts are always somehow situated in our embodied experiences; into the world through metaphors; or even just in our understanding when we hear abstract constructs that activate our motor cortex.

This is where the adage "moving is thinking" comes into play. Our body was designed to be this moving machine. In 2002, Margaret Wilson wrote an article in which she looked at six different points of view on embodied cognition: cognition is situated, it is time pressured, we offload our cognitive load into the environment, the environment is a part of the cognitive system, cognition is for action, and, finally, offline cognition is also body-based.[4] These six points go back to the idea that the way we think is always already related to the way we are as a body that is situated into an environment. It's this loop of making choices and taking action.

Think of the way a simulation is different from simply reading a description or instructions. In the simulation, you are embedded in a situational context and you are forced to develop a level of situational awareness. You have to understand all the relational dynamics of that context. And when you are presented with a choice, you have to actually physically take that action with the same type of bodily movements you would be making if you were really in that situation. I've been in simulations where my body reacted unconsciously to the stress of the situation. There's something about the way our bodies are combining all these different modalities together that makes this the most powerful mode of training. To actually go through the situation, it's more predictive to see how people react to those situations, more so than if they were to just read about it or take a multiple-choice test. To actually take the actions in the moment is something that is extremely powerful.

4 Wilson, "Six Views of Embodied Cognition."

In terms of the body in VR, there is also the issue of identity, of being able to embody different characters and different avatar representations. In experiences like VRChat, people can express different genders or an aspect of their identity that might go beyond the limitations of what their physical bodily can express. They are able to dive into a truer essential nature of who they are and how they want to identify and relate to other people. Looking back to things like embodied cognition and situated knowledges from feminist theory—meaning that we are situated into these worlds, both in terms of constructs of power and privilege—are informing the way that we're seeing the world.

There are ways that, as we have these situated simulations of other bodies and other modes of situated knowledges, we can get insight from those other perspectives. There are always going to be limitations from that, because you're not embodied into those experiences over long periods of time. But, that kind of combination of situated knowledge and embodied cognition is really quite compelling in terms of speaking to the power of the medium. It goes back to how we are understanding the world, and it's giving us these embodied metaphors to understand experiences in a much deeper way than any other media before.

How do you see the relation between the so-called "real" world and virtual worlds? For instance, David Chalmers insists on opposing virtual reality to physical reality—rather than just reality—to highlight the fact that VR is also real.

KB: I agree with Chalmers's overall argument that virtual reality is a form of reality, but I have different criteria than he has for describing it, namely in terms of the different qualities of presence: active presence, mental and social presence, emotional presence, embodied presence, and environmental presence.

The main differentiating factor between physical and virtual realities is that sensory physical experience is really difficult to replicate in VR (touch, taste, and smell, but also the haptic feedback of physical reality). That's why I focus on *physical* reality as opposed to *real* reality. Obviously, there will always be differences between the physical and virtual representations of a given phenomenon. But, when you think about these different qualities of presence, the biggest gap at this point, in terms of the differentiation between the virtual and physical, is really in terms of embodied and environmental presence.

Another caveat I would put onto this dialectical framing is that it's easy to jump from discussing physical reality into object-oriented substance metaphysics to saying that all the bases of reality are made up of concrete things. There are other metaphysical approaches—from Spinoza, Whitehead, or William James—that focus on panpsychism or affect and emotions as the basis of reality (or as its underlying basis). Whitehead has these prehensions: he states that these emotional feelings are the underlying fabric of reality. Even in the difference between what's actual and what's potential, if you look at quantum mechanics, there are so many dimensions that go beyond what we can understand with this sort of physical reality.

I keep going back to the process-relational orientation because it gets away from the mindset of object-oriented metaphysics. To understand these realms of potential is a lot of what Deleuze, Bergson, Whitehead, James, Spinoza, and Schelling are doing; this whole line of thinkers is trying to help us escape from the trappings of this substance-metaphysics framing of everything into this more process-relational mode of relation. I reiterate that because there's a way you could put the basis of reality as emotion and affect: Can a *physical* experience give you emotion? Can a *personal* experience give you emotion? It's a way to avoid these binary ways of framing things; of saying that things are not real because they are virtual.

Obviously, there's always going to be limitations between the simulation of reality versus the actual emergent properties of being able to fully participate in the dynamic unfolding of reality as we have, but I feel there are ways in which the medium of VR is able to simulate these deeper philosophical questions. I would just caution us against getting too far into these binary ways of framing things and I'd suggest opting more for a pluralistic and process-relational framing.

You conclude all your interviews by asking your interlocutors what they think the "ultimate potential of VR" is. What would you say is the ultimate potential of immersive media? What have you noticed in the hundreds of answers you've received to this question?

KB: I first asked this question at the Silicon Valley Virtual Reality Conference, in May 2014, and I have been consistently asking it since. In episode 1000 of *Voices of VR*, I put together a three-hour digest of the different answers I received over the years. I would refer folks to that episode for a more objective mapping of the different types of answers and conceptual domains.[5]

Personally, I keep coming back to this opposition between a substance metaphysics perspective—where the world is a static, concrete, fixed object that we are distant from—and a more process-relational view of the world

5 See Bye, "#1000. Voices of VR Retrospective on the Ultimate Potential of Virtual Reality."

as an emergent, evolving, and unfolding phenomenon in the context of which we are included. It's much more about understanding the relational dynamics of who we are and how we relate to the world around us, to ourselves, and to other people.

As I have done this process of working in VR, I have much more skepticism about any singular perspective. I'm trying to embrace plurality and diversity. This pluralism includes situated knowledges, meaning the way everybody's experiences are based on the world that they've lived in. There are certain decisions that you make as an artist about what you are framing, what you are including, what you are not including.

The larger point, though, is that I'm really excited about how you could use these immersive media to imagine a better world and to actually build out what that would look like. More than that, it's about having people experience that world, and then take what is in the realm of the potential and move it into the actual, the physical, the embodied experiences that are manifest in our lives.

We're at this crossroads, as a society, that we need all the help that we can get to be able to have us come into right relationship, not only with the world around us, with all the ecological issues, but also culturally, getting beyond a strict polarization. Jung said: "Hold the tension of the opposites until there's this reconciling third." In Hegelian processes, this is the thesis, antithesis, and then the synthesis. Ken Wilber talks about "transcend and include," so you have to transcend the limitations of your own perspective, but also include the truths of other people's perspectives. So, it's the problem of the opposites that Maxwell talks about in *Integration and Difference*, and which all these different philosophers have talked about over time.

I feel like having this kind of process-relational orientation is something that the medium of VR itself is really built for in terms of creating these experiential worlds, these stories that allow us to learn about ourselves, but also, at the end of the day, hopefully come into stronger relationships to the world around us.

Bibliography

Bye, Kent. "#1000. *Voices of VR* Retrospective on the Ultimate Potential of Virtual Reality." *Voices of VR*, August 28, 2021. https://voicesofvr.com/1000-voices-of-vr-retrospective-on-the-ultimate-potential-of-virtual-reality/.

Bye, Kent. "#1147. Thirteen Philosophers on the Problem of Opposites. Grant Maxwell's *Integration & Difference* Book & Archetypal Approaches to Character."

Voices of VR, October 27, 2022. https://voicesofvr.com/1147-thirteen-philosophers-on-the-problem-of-opposites-grant-maxwells-integration-difference-book-archetypal-approaches-to-character/.

Bye, Kent. "A Primer on Presence, Immersive Storytelling, & Experiential Design." September 16, 2022. YouTube video, 56:51. https://www.youtube.com/watch?v=n_2_qvIymt4.

McKee, Robert. *Story: Substance, Structure, Style, and the Principles of Screenwriting.* New York, NY: Harper Collins, 1997.

Wilson, Margaret. "Six Views of Embodied Cognition." *Psychonomic Bulletin & Review* 9 (2002): 625–636.

About the Author

Philippe Bédard is an independent researcher with a fascination for moving image media and their technologies. As part of his postdoctoral research, he focused on virtual reality, with an emphasis on the notion of empathy. Finally, he has worked extensively in and around the immersive media industry in Montreal, including work with Xn Québec, Festival du Nouveau Cinéma, the National Film Board of Canada's interactive studio, and SODEC.

26. An interview with Skawennati

Edited by Carl Therrien

Abstract: (From the artist's website) Skawennati makes art that addresses history, the future, and change from her perspective as an urban Kanien'kehá:ka (Mohawk) woman and as a cyberpunk avatar. Her early adoption of cyberspace as both a location and a medium for her practice has produced groundbreaking projects such as *CyberPowWow* and *TimeTraveller*™. She is best known for her machinimas—movies made in virtual environments—but also produces still images, textiles, and sculpture.

Keywords: Three Sisters, Indigenous futures, machinima, Second Life, Aboriginal Territories in Cyberspace (AbTeC)

In the *TimeTraveller*™ machinima series, technology appears to be completely transparent: the protagonist wears minimalist goggles and types in mid-air to navigate history, moving seamlessly between an embodied ancestor (such as a Kanien'kehá:ka warrior during the Oka crisis), historical documents or TV broadcasts recreated in the 3D engine. In some of your ideal Indigenous futures, technology has evolved as an emancipatory tool for everyone. Do you think our institutions and the immersive tech sector are inspired by the utopian visions that you and other artists provide? What policies would be needed to move forward in that direction?

Skawennati: I certainly don't believe that the tech world is looking at my thing going: "Oh, yeah, let's do that!" I think we have all been inspired by utopian visions of technology seen in *Star Trek* for instance—you know, the tricorder, the transporter, the replicator, etc. A lot of the technologies we have now, like the iPhone, have been inspired by these visions. I do think art has the potential to influence inventors and corporations, at least I hope it does.

In terms of what policies would be needed to move forward: I'm thinking about a policy of degrowth. I believe that the biggest problem we're facing

Bédard, Philippe, Alanna Thain, and Carl Therrien, eds. *States of Immersion Across Media: Bodies, Techniques, Practices*. Taylor & Francis Group, 2025
DOI 10.5117/9789048558766_CH26

in this world is greed, really. If we can figure out a way to overcome that, I think a lot of amazing stuff is achievable, for a lot of people.

***Star Trek* is a fascinating example to reflect on utopia and growth. The original series engaged with the idea of pushing the frontier and the colonial narrative of Manifest Destiny in a critical manner.**
Skawennati: They were way ahead of the game compared to what was happening in our world, in terms of race for instance. Not only was there a woman on the bridge, but it was a black woman [Nyota Uhura]. Both those things were quite amazing at that time.

One problem with our society is that it is very difficult to imagine a world without capitalism. It's like we're soaking in it, we can't get far enough from it to look at it. I think a lot of people are trying to do that, but it's just so difficult. Most people seem to believe it's just like gravity: it's just there, and we have to work around it. I feel like *Star Trek* was trying to figure out a world beyond expanding into the galaxy to conquer it. It's about exploration, science, and knowledge. It's about encountering other people in the galaxy. The creators didn't always explain the backstory all that much in the beginning, it was left to you to determine how they got to that point. I would like to imagine that they got there by figuring out how to make everyone on the planet happy, fed, safe, and then maybe self-actualized? Maybe there was so much leisure time that people were able to think about expanding into space and figured out how to do it following principles of degrowth, or at least with peace and justice. I could imagine that story.

Star Trek and society have been in dialogue for a couple of generations now. This is a long-term engagement, but of course I think that other movies and books have also opened up people's minds, like *Snow Crash* did for the metaverse, or *The Hitchhiker's Guide to the Galaxy* (one of my favorite books!) and its Babelfish. Talking doors!

At least there is still humour in the future, apparently. The "new" metaverse, after a colonization attempt from Facebook / Meta, is now commonly presented in the trade press as something more "decentralized," as new social platforms for work and play akin to what has already been developed in giant online gaming platforms (such as *Fortnite* or *Grand Theft Auto Online*). Many new products sold as VR technology invite us to upgrade our online experience with state of the art interfaces, in which many corporations invested a lot of R&D money. As someone who developed the potential of *Second Life*, in terms of community building, education, art curation, and production, for a very long time, what are your

thoughts about this incentive to upgrade? As an artist, do you believe XR technology will add something significant to everything that you have been doing, in the metaverse and beyond?

Skawennati: I don't have that much to say about it, really. I don't mind being immersed in VR art for ten minutes or even a half an hour. I just don't want to use VR headsets specifically for my work at this time. I recall the number of hours that I sat in front of a computer to learn *The Palace*, and then learn *Second Life*, and then make machinima on this platform. Working with VR would require a lot of time and I don't feel good about it at this point in my career.

My character in *Traveler Traveller* TM uses VR glasses. I imagined that we would figure something out so that VR goggles wouldn't be potentially hurting our vision. I don't want the machine and its images burning into my eyes. Generally speaking, I never want to trick my eyes. I don't really want to trick my body into thinking it's in an open field when it's in a very closed room. This idea of tricking the body is already present in other parts of our society, in modern medicine for instance. To me it feels like breaking a "contract." Not the social contract, but the natural contract, the most basic contracts we have, between our minds and our bodies. Why would I want to trick my poor body? My God, we need to trust each other!

The notion of trickery is indeed central in these developments. We imagine technologies that could trick even more senses, until we are all connected to the matrix or something like that. Yet there seems to be a lot of hesitation right now about how much closer we want to get to these illusions. If I understand you correctly, in your artistic practice, you are very comfortable working within the frame of a computer screen to create immersive experiences?

Skawennati: Very, very comfortable. I also think that we are experiencing some kind of a crisis in truth right now. We watch things on TV and we don't know what's true anymore. We have our politicians lying to us. We have artificial intelligence, deepfakes, all kinds of things happening for us to not know what the [expletive] is real. I think that when I work and play in *Second Life*, within the frame of my computer, I'm sitting at my desk, I'm sitting in a place. I can look through the computer window, but I can also look into the world around it, reminding myself of what is real, of the universe beyond the simulation. I do believe that some of the relationships that we build and the artwork that I make in a virtual space are real. But I think that we need some line, somewhere, for our bodies and natural processes to function well. We need a balance between distance and closeness. I might not know about eyes well enough to go beyond these remarks, but for me there is something about truthfulness that should remain central.

For nearly two decades, AbTeC has worked hard to make new tech accessible for Indigenous youth. As a creative and social platform, *Second Life* is able to keep up with the times in spite of this constant pressure to push forward. Technological progress always comes at a cost which, beyond the price of the gear itself, can be measured through the divides and exclusions it creates. On top of XR technology, the AI shift is raising similar questions. Do you think that the next-gen metaverse(s) and cutting-edge technology can still be reappropriated as a political tool in this context?
Skawennati: I think so, and that is our intention! We, at AbTeC are always trying to learn and teach the tech, see if other people can learn it, appropriate it, but I think we are also trying to get in earlier and earlier. At one point we started developing the tech ourselves.

Getting young Indigenous people to not be afraid of bringing their Indigenous values to this technical place, and allowing the tech to be affected by that presence, remains an important goal. We have always talked about affecting the look and feel of cyberspace. First we show up, we are there on the web. But then we start building our own web pages and virtual worlds. And now a few years later, Jason [Edward Lewis, AbTeC co-founder and co-director] is doing artificial intelligence, talking about how we can create AI that has Indigenous values. Beyond learning, teaching, and creating the tech, how can we affect how other people are making AI? AI requires great data, but where is the data coming from? Right now it is being created with flawed databases. How do we change that? There are a couple of major things to address in order to fix this situation, and people at AbTeC are not afraid to raise these questions and reflect on solutions.

We got a lot of work to do on this planet! We can become less greedy, and here again I think we need to talk about degrowth. In the world of investment, every quarter a company has to show a profit or flaunt its ability to attract more wealth. Investors are supposed to be getting money quickly. The timelines are so short, it is ridiculous. If we could just calm down! All this expectation of return on your investment is powering the rape of the planet and of our resources. A lot of tech innovation is coming from a negative place, which is greed and fear. Fear of scarcity, of poverty. Fear of the future. There are policies that society needs to adopt so we can start being less stressed out and less greed oriented. These companies fighting for the metaverse are powered by greed as well, a lot of the time, but I don't think it has to be. I prefer the utopian *Star Trek* we just talked about. Once people's needs are satisfied, when we are more appreciative and thankful of what we have, we might just innovate for the sake of curiosity. True curiosity can power technological innovation; it can come from that human impulse. Wouldn't it be wonderful? I think it is completely possible.

There is a wonderful gallery space on AbTeC island in *Second Life*, and many real-life exhibitions have been created / recreated there. This allows artworks to travel far beyond the original spaces and communities it addressed, but of course this "transmediation process" changes the material context and can affect the meaning of a piece. In your practice, you have explored the same concepts or stories over many media. For instance, your machinima *She Falls for Ages* (2017) features a very intricate model of the celestial tree evoked in the Turtle Island creation story. You have also created a celestial tree street sign sculpture exhibited in Tiohtià:ke (Montréal). In your work, how do these different material contexts affect meaning, and the way it can immerse us in the creation story it is taken from?

Skawennati: The material aspect is very important indeed. It started in *Second Life*; I very much wanted to have a celestial tree logo on the title page. I had seen the *Josie and the Pussycats* movie (from Archie comics), it was a very silly movie but I thought the set design was amazing. They were talking about product placement a lot, about how capitalism affects art. And so each hotel room they were in had sponsored rooms, for instance the Target hotel room, in which the carpet was a target, then you had target symbols all over the shower curtain, etc. I thought about how it might be interesting if we had non-capitalist symbols, like the celestial tree logo, all around us. In *She Falls for Ages*, we see it in a lot of different places: on patches on the people, on a bed spread, etc. The sculpture in Tiohtià:ke builds on this graphic design, reimagining the tree from the creation story as a sort of street sign in order to integrate it into the physical context of the city.

The other celestial tree I built was the much more organic form that only exists currently as a 3D virtual object. However, when we were designing that tree, we actually went through many material iterations. To begin, I got some playdoh and I rolled it all up, and then I twisted it, and I made the branches. I showed my 3D modeller what I wanted, and then he made that incredible virtual tree. I wanted it to look like what it says in the story, an amazing organic tree. And you had these wonderful bioluminescent flowers that could be harnessed safely to be used by the people. There are many versions of that story, in one of them the tree had blossoms that light up the world. I have considered transferring it to the actual world, I thought maybe it could be made out of glass, with some electric wiring through the branches in order to light it up.

Your work to bring Indigenous youth to engage with tech is echoed by your ability to dialogue with contemporary pop culture. Perhaps as a direct influence from *Second Life*'s popularity, customizable avatars are everywhere in gaming nowadays. Recently you have transferred numerous skills developed through character customization in *Second Life* into the physical world with the creation of clothing collections. Your practice brought Indigenous avatars all the way up to *Vogue* magazine. This new artistic exploration can also be connected to another popular immersive practice: taking virtual avatars in the actual world through cosplaying. When you create, do you reflect on the potential cultural legitimization coming both from the art world and from active youth cultures?
Skawennati: I think that when I get the spark of the idea, usually, I feel like my instructions have been sent to me. Like, I know I want to make this thing. When you start on the road to actually make anything, you have a lot of time to think about different things, including the reception of a piece. How will this work be perceived? How do I want this work to be perceived?

My first audience is myself, and what I want to see in the world. Throughout the creation process, additional instructions come. I don't think "Oh, I'm going to make this because the kids are going to love it." I think "My God, this is going to be so fun to make," and then sometimes someone will say: "yeah, and the kids are gonna love it!", or "there is a Comic-Con coming up, what if you wore it there?" Right now I'm doing the Three Sisters [a common way to refer to corn, bean, and squash in Iroquois and Cherokee culture] as superheroes. Once I got that idea in my head, more instructions are coming. And so I've made them as avatars, and now I'm working on making costumes in real life [editor note: this work, mixing machinima and fashion, premiered as *They Sustain Us* at Grey Area in San Francisco on March 22, 2024].

Did you know I did a Sky Woman cosplay? My team and I recreated the costume in real life, I dressed up as Sky Woman from *She Falls for Ages*, and we took a picture. I don't love the image in its current state, which explains partly why it has not been seen. It was a really fun project, but my collaborator tried making it look more like the avatar, which was not what I intended. Someone else recently did participate at the second Indigenous Comic-Con wearing a similar costume.

On a global scale, pop culture has been obsessed with dystopic fiction for decades (you know what I mean: disaster movies, zombie series, etc.) Listening to the news at this point in time gives the impression that disaster movies were not so unrealistic, after all. By contrast, the notion of Indigenous futures seems a lot more hopeful. Realistic when it comes to addressing major pitfalls in our civilizations, but a lot more hopeful. What part of our immersive futures are you the most excited or hopeful about?
Skawennati: I have a lot of hope. I really do. I think we're going to do the right thing, I do! The future that I want to see is just and sustainable. People are healthy and happy, they can look how they want, love who they want, dress how they want. People are not fed things that hurt them; we eat such terrible food for ourselves! I would love to see a place where we can be thankful for what we have, and we can try to help one another and try to become all that we can be together.

What will it look like once we are able to be all that we can be? I would love to see us be able to visit other people, visit faraway lands, in a manner that is not bad for the earth. How far are we willing to walk? The technology exists. The technology existed before Western contact to travel the world across the oceans in handmade boats and outrigger canoes.

I do want to see the world that is in *She Falls for Ages*. It's funny, because in that world, the narrator says "people hardly knew the meaning of death," and that is a line from one of the oldest versions of the story that I had found. Mohawk speakers don't say "th" so well, so when the narrator says it, it sounds like "debt," and I love that! I love that it could be both. You could think it is "death," or "debt." They hardly knew the meaning of owing anything!

I wrote a story called "Lest We Remember." Often we say "lest we forget," and we must remember the past, otherwise we're doomed to repeat it. But what if we could forget the past? And what if we could let go of the hates or just let go of certain memories, the memory of capitalism. We could just be a new people, and build it from a blank slate. What would we build? Would we live together in multi-generational families, could that be something that would help us? We would have the wisdom of our elders, we would have old people who maybe are done with being "productive," and could watch children and teach the children, and have that different kind of relationship to a child. Parents are so worried all the time, worrying about their kids being in the best school and getting the best grades. What will it take for us to stop worrying, and to just be?

In the spirit of contradiction, another question is: did we have that in the past? From what I'm reading, I think that actually, Shawnee society seems to have obtained something like that. We lived in these multi-generational houses, we didn't have money, we shared. We shared the food that we grew, we shared the food that we hunted. Of course, there was war, but then the peacemaker came, and there was a very long period of peace, for hundreds of years. I'm starting to have a theory: perhaps that is why we don't want to be assimilated, because we had it so damn good. We liked who we were, and we were free.

In the book *The Dawn of Everything*, there is a great chapter about Shawnee society and their relationship with money. A Huron chief heard about money in France and thought it was so disgusting. He never wanted his people to have money. He said, in essence: you have money, and you have poor people in the streets begging for it. You have money, and you have to turn and bow to your king. We don't bow to anybody. There are chiefs here, but they cannot force us to do anything we don't want to do. It's really interesting, how they thought of freedom and money and the how those two things connect.

So in a way, I'm saying: let's just forget everything that we know now and start blank. But in another way, I'm saying: what if we could remember before this? Wouldn't it be really great if we already lived that way?

About the Author

Carl Therrien is Professor of Cinema and Game Studies at Université de Montréal. He has explored the notion of immersion and affective ecologies in his PhD dissertation and in multiple papers and book chapters over the years. He has published extensively on the current challenges of video game historiography through case studies of specific genres such as visual novels, horror games, and first-person shooters. In *The Media Snatcher* (Platform Studies, MIT Press, 2019), Therrien builds upon previous findings through a comparative study of the TurboGrafx-16 / PC Engine, confronting American and Japanese perspectives of this expansive technological platform. He has co-founded the History of Games international conference series and has contributed to the annual Game History symposium in Montreal since 2014.

Index of Artworks

Index of Persons

Index of Concepts and Technologies